Lecture Notes in Computer Science 10112

Commenced Publication in 1973
Founding and Former Series Editors:
Gerhard Goos, Juris Hartmanis, and Jan van Leeuwen

More information about this series at http://www.springer.com/series/7412

Shang-Hong Lai · Vincent Lepetit
Ko Nishino · Yoichi Sato (Eds.)

Computer Vision – ACCV 2016

13th Asian Conference on Computer Vision
Taipei, Taiwan, November 20–24, 2016
Revised Selected Papers, Part II

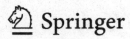
Springer

Editors
Shang-Hong Lai
National Tsing Hua University
Hsinchu
Taiwan

Vincent Lepetit
Graz University of Technology
Graz
Austria

Ko Nishino
Drexel University
Philadelphia, PA
USA

Yoichi Sato
The University of Tokyo
Tokyo
Japan

ISSN 0302-9743   ISSN 1611-3349 (electronic)
Lecture Notes in Computer Science
ISBN 978-3-319-54183-9   ISBN 978-3-319-54184-6 (eBook)
DOI 10.1007/978-3-319-54184-6

Library of Congress Control Number: 2017932642

LNCS Sublibrary: SL6 – Image Processing, Computer Vision, Pattern Recognition, and Graphics

Printed on acid-free paper

This Springer imprint is published by Springer Nature
The registered company is Springer International Publishing AG
The registered company address is: Gewerbestrasse 11, 6330 Cham, Switzerland

Preface

Welcome to the 2016 edition of the Asian Conference on Computer Vision in Taipei. ACCV 2016 received a total number of 590 submissions, of which 479 papers went through a review process after excluding papers rejected without review because of violation of the ACCV submission guidelines or being withdrawn before review. The papers were submitted from diverse regions with 69% from Asia, 19% from Europe, and 12% from North America.

The program chairs assembled a geographically diverse team of 39 area chairs who handled nine to 15 papers each. Area chairs were selected to provide a broad range of expertise, to balance junior and senior members, and to represent a variety of geographical locations. Area chairs recommended reviewers for papers, and each paper received at least three reviews from the 631 reviewers who participated in the process. Paper decisions were finalized at an area chair meeting held in Taipei during August 13–14, 2016. At this meeting, the area chairs worked in threes to reach collective decisions about acceptance, and in panels of nine or 12 to decide on the oral/poster distinction. The total number of papers accepted was 143 (an overall acceptance rate of 24%). Of these, 33 were selected for oral presentations and 110 were selected for poster presentations.

We wish to thank all members of the local arrangements team for helping us run the area chair meeting smoothly. We also wish to extend our immense gratitude to the area chairs and reviewers for their generous participation in the process. The conference would not have been possible without this huge voluntary investment of time and effort. We acknowledge particularly the contribution of 29 reviewers designated as "Outstanding Reviewers" who were nominated by the area chairs and program chairs for having provided a large number of helpful, high-quality reviews. Last but not the least, we would like to show our deepest gratitude to all of the emergency reviewers who kindly responded to our last-minute request and provided thorough reviews for papers with missing reviews. Finally, we wish all the attendees a highly simulating, informative, and enjoyable conference.

January 2017

Shang-Hong Lai
Vincent Lepetit
Ko Nishino
Yoichi Sato

Organization

ACCV 2016 Organizers

Steering Committee

Michael Brown	National University of Singapore, Singapore
Katsu Ikeuchi	University of Tokyo, Japan
In-So Kweon	KAIST, Korea
Tieniu Tan	Chinese Academy of Sciences, China
Yasushi Yagi	Osaka University, Japan

Honorary Chairs

Thomas Huang	University of Illinois at Urbana-Champaign, USA
Wen-Hsiang Tsai	National Chiao Tung University, Taiwan, ROC

General Chairs

Yi-Ping Hung	National Taiwan University, Taiwan, ROC
Ming-Hsuan Yang	University of California at Merced, USA
Hongbin Zha	Peking University, China

Program Chairs

Shang-Hong Lai	National Tsing Hua University, Taiwan, ROC
Vincent Lepetit	TU Graz, Austria
Ko Nishino	Drexel University, USA
Yoichi Sato	University of Tokyo, Japan

Publicity Chairs

Ming-Ming Cheng	Nankai University, China
Jen-Hui Chuang	National Chiao Tung University, Taiwan, ROC
Seon Joo Kim	Yonsei University, Korea

Local Arrangements Chairs

Yung-Yu Chuang	National Taiwan University, Taiwan, ROC
Yen-Yu Lin	Academia Sinica, Taiwan, ROC
Sheng-Wen Shih	National Chi Nan University, Taiwan, ROC
Yu-Chiang Frank Wang	Academia Sinica, Taiwan, ROC

Workshops Chairs

Chu-Song Chen	Academia Sinica, Taiwan, ROC
Jiwen Lu	Tsinghua University, China
Kai-Kuang Ma	Nanyang Technological University, Singapore

Tutorial Chairs

Bernard Ghanem   King Abdullah University of Science and Technology,
                    Saudi Arabia
Fay Huang   National Ilan University, Taiwan, ROC
Yukiko Kenmochi   Université Paris-Est, France

Exhibition and Demo Chairs

Gee-Sern Hsu   National Taiwan University of Science and
                    Technology, Taiwan, ROC
Xue Mei   Toyota Research Institute, USA

Publication Chairs

Chih-Yi Chiu   National Chiayi University, Taiwan, ROC
Jenn-Jier (James) Lien   National Cheng Kung University, Taiwan, ROC
Huei-Yung Lin   National Chung Cheng University, Taiwan, ROC

Industry Chairs

Winston Hsu   National Taiwan University, Taiwan, ROC
Fatih Porikli   Australian National University, Australia
Li Xu   SenseTime Group Limited, Hong Kong, SAR China

Finance Chairs

Yong-Sheng Chen   National Chiao Tung University, Taiwan, ROC
Ming-Sui Lee   National Taiwan University, Taiwan, ROC

Registration Chairs

Kuan-Wen Chen   National Chiao Tung University, Taiwan, ROC
Wen-Huang Cheng   Academia Sinica, Taiwan, ROC
Min Sun   National Tsing Hua University, Taiwan, ROC

Web Chairs

Hwann-Tzong Chen   National Tsing Hua University, Taiwan, ROC
Ju-Chun Ko   National Taipei University of Technology, Taiwan,
                    ROC
Neng-Hao Yu   National Chengchi University, Taiwan, ROC

Area Chairs

Narendra Ahuja   UIUC
Michael Brown   National University of Singapore
Yung-Yu Chuang   National Taiwan University, Taiwan, ROC
Pau-Choo Chung   National Cheng Kung University, Taiwan, ROC
Larry Davis   University of Maryland, USA

Contents – Part II

Deep Learning

People Tracking and Action Recognition

People and Actions

Sparse Code Filtering for Action Pattern Mining

Wei Wang[1]([✉]), Yan Yan[1], Liqiang Nie[2], Luming Zhang[4], Stefan Winkler[3],
and Nicu Sebe[1]

[1] University of Trento, Trento, Italy
wei.wang@unitn.it
[2] National University of Singapore, Singapore, Singapore
[3] Advanced Digital Sciences Center, Singapore, Singapore
[4] Hefei University of Technology, Hefei, China

Abstract. Action recognition has received increasing attention during
the last decade. Various approaches have been proposed to encode the
videos that contain actions, among which self-similarity matrices (SSMs)
have shown very good performance by encoding the dynamics of the
video. However, SSMs become sensitive when there is a very large view
change. In this paper, we tackle the multi-view action recognition problem by proposing a sparse code filtering (SCF) framework which can mine
the action patterns. First, a class-wise sparse coding method is proposed
to make the sparse codes of the between-class data lie close by. Then we
integrate the classifiers and the class-wise sparse coding process into a
collaborative filtering (CF) framework to mine the discriminative sparse
codes and classifiers jointly. The experimental results on several public
multi-view action recognition datasets demonstrate that the presented
SCF framework outperforms other state-of-the-art methods.

1   Introduction

Action recognition has wide applications, such as human-computer interactive
games, search engines, and online video surveillance systems. Videos can be
summarized by labels if the actions can be annotated automatically. Then a
search engine can make better recommendations (*e.g., finding dunks in basketball games*). Usually, the same action observed from different viewpoints has
considerable differences. Therefore, an efficient method to extract robust view-
invariant features is essential for multi-view action recognition. The features can
be roughly grouped into two types, the 2D features [1] and 3D features [2].

Many works employed 3D models to tackle the multi-view action recognition problem. First, the geometric transitions are utilized to obtain projections
across different viewpoints. Then the observations are compared with the projections to find the viewpoint that best matches the observations [3]. However,
how to accurately find body joints to build the 3D model remains an open
problem. Besides, the built model has too many degree-of-freedom parameters,
which must be carefully calibrated. Moreover, the model requires high resolution videos to locate body joints and sometimes may require mocap data [4]. An
alternative solution for multi-view action recognition is to design view-invariant

© Springer International Publishing AG 2017
S.-H. Lai et al. (Eds.): ACCV 2016, Part II, LNCS 10112, pp. 3–18, 2017.
DOI: 10.1007/978-3-319-54184-6_1

2D features. Farhadi *et al.* [5] proposed split-based representations by clustering the similar video frames into splits. The split-based representations can be
transferred among different viewpoints as the change dynamics of the multi-view
videos are the same. Similarly, Junejo *et al.* [6] employed SSMs to encode the
frame-to-frame relative changes. However, the SSMs are robust to view changes
only to a certain extent.

In this paper, to tackle the multi-view problem, we propose a class-wise
sparse coding approach to maintain label consistency. We employ SSM feature
to represent each video. The sparse coding learns a dictionary from SSM representations of the video collections. The dictionary consists of typical action
patterns, and each video is encoded to a code as a linear combination of action
patterns. The label consistency is achieved by penalizing the within class variance of the codes. Thus, the codes of the within-class videos will lie close by,
and accordingly, only the view-invariant action patterns will be learned while

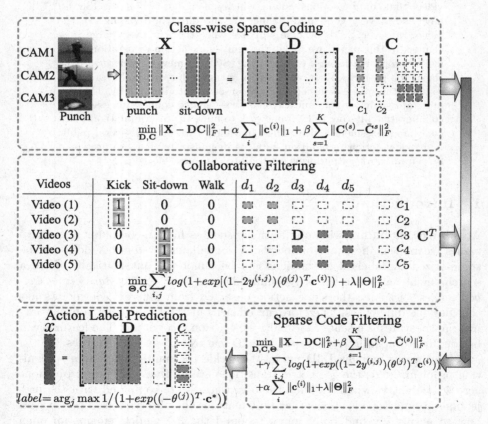

Fig. 1. Overview of sparse code filtering: (top) Class-wise sparse coding. (middle) Collaborative filtering. (bottom right) Sparse code filtering framework. (bottom left) Label
prediction.

the view-dependent information will be suppressed. Then we rely on the codes as video features to do action classification.

To further improve the discriminative power of the codes, we integrate the class-wise sparse coding and classifiers training process into a unified CF framework as shown in Fig. 1. This is because CF can link the dictionary and classifiers together which can optimize them jointly. The dictionary can be adjusted for the classifiers while the classifiers can be adjusted for the dictionary collaboratively. In this way, the learned action patterns in the dictionary can be more discriminative with respect to different actions. Thus, we derive a novel sparse code filtering framework. In the sparse code filtering scheme, each action class is regarded as an user. For the classical collaborative filtering, the entry in the rating matrix (e.g., ranges from 0 to 5) describes how much a user likes the product. In our scheme, however, the entry in the rating matrix, ranging from 0 to 1, represents the probability that a video belongs to an action class. The sparse code filtering framework provides a trade-off between the dictionary reconstruction error and the classification error which derive from the class-wise sparse coding and the logistic classifiers respectively.

To summarize, our work makes the following contributions: (i) We propose a class-wise sparse coding approach to maintain the label consistency by encouraging the sparse codes of the multi-view videos within the same action class to lie close by. (ii) We propose a novel sparse code filtering framework in which the classifiers and dictionary can be optimized collaboratively. Thus, the view-invariant and class-discriminant sparse codes can be learned. (iii) The proposed sparse code filtering framework has a good generalization property and can be applied to other pattern recognition tasks.

2 Related Work

2.1 Action Recognition

Many 3D and 2D based approaches are proposed for action recognition. Through reconstructing 3D human bodies, features can be adapted across different viewpoints through geometric transformation. Weinland *et al.* [7] projected 3D poses into 2D to obtain arbitrary views and employed an exemplar-based HMM to model view transformations. A similar idea is proposed in [8] which employed CRF instead of HMM. Except for designing the 3D models, some works focus on designing view-invariant classifiers, such as linear discriminant analysis [9] and latent multi-task learning [10]. Matikainen *et al.* [11] suggested training models for all the views and then utilizing recommender system to find the suitable model. But the approach in [11] requires huge amount of training samples from different viewpoints. Recently, the recurrent neural network is also applied for the action recognition task [12] as it is good at dealing with signal sequences with various lengths [13,14]. However, these methods can only tackle the single-view action recognition task.

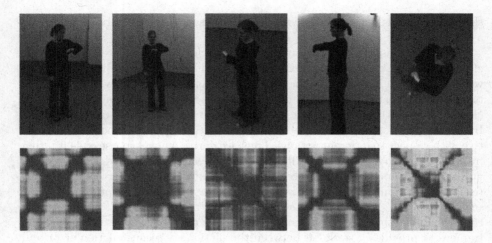

Fig. 2. SSM features extracted from different views.

To achieve view invariance in 2D models, many works try to extract view-invariant features. Farhadi and Tabrizi [5] proposed a split-based representation by clustering video frames into splits. Then videos can be represented by the statistics of the splits, and the split transfer mapping across views can be learned. Based on 2D features, the transfer learning model requires no 3D human reconstructions. Recently, a more robust view-invariant descriptor, self-similarity matrix (SSM) [6] has been proposed. It is relatively stable over the viewpoint changes compared with other features [15]. Similarly to [5], this descriptor encodes the relative changes between pairs of frames, and completely discards the absolute features of each single frame. SSMs can be calculated using different low-level features which have similar properties.

Figure 2 shows the examples from the action videos and their corresponding SSM features. From Fig. 2, we can observe that the SSM features from the 4 side cameras are visually similar, while the feature from camera 5 (on the ceiling) is quite different. Yan et al. [9] revealed that SSMs became less reliable when there was a very large view change. Based on SSMs, Joint Self-Similarity Volume (SSV) was introduced by [16] which utilized Joint Recurrence Plot (JRP) theory to extend SSM. But different from [15], the SSM defined in [16] is the recurrence-plot matrix of the vector representation of each single frame.

2.2 Sparse Coding

Sparse coding, also known as dictionary learning, aims to construct efficient representations of data as a combination of a few typical patterns (dictionary bases). Wang et al. [17,18] used the sparse coding for attribute detection. Raina et al. [19] showed that sparse coding significantly improved classification performance. [20] employed sparse coding for action recognition from depth maps. However, their approach is restricted to the videos which can provide depth information. Qiu *et al.* [21] selected a set of more compact and discriminative bases from the dictionary using Gaussian Process. Guha and Ward [22] investigated different sparse coding strategies, namely, an overall dictionary for all the classes, different dictionaries for each class, and their concatenation. But view changes were not considered. Zheng and Jiang [23] proposed view-specific sparse coding. But sufficient training data from each viewpoint are required. Besides, the label information is discarded. Thus, it can not preserve label consistency. In our work, we add within-class variance into the loss function to preserve the label-consistency. The learned class-wise dictionary can be considered as a more label-smooth feature space compared with the original video feature space.

2.3 Collaborative Filtering

Collaborative filtering is widely used in recommendation systems of commercial websites, such as Amazon and eBay, to recommend products to their consumers. The most attractive characteristic of CF is that it can learn a good set of features automatically [24], which does not require hand-designed features. Taking the movie recommendation system for example, each movie has its own features and each user has its own specific feature preference weights. Given the movie-user rating matrix, CF learn a good set of features for each movie and feature preference weights for each user jointly. During the CF learning process, the features will be adjusted for the feature preference weights for each user, and feature preference weights will also be adjusted for the features iteratively. Inspired by the movie recommendation system, we employ a CF framework to learn class-wise dictionary and classifiers jointly. Thus, the codes and classifiers can be adjusted to better fit each other. The experimental results demonstrate the effectiveness of our framework.

The rest of the paper is organized as follows. We propose the class-wise dictionary learning approach in the Sect. 3. Section 4 presents our sparse code filtering scheme. The experiments are described in Sect. 5. We conclude our paper in Sect. 6.

3   Class-Wise Sparse Coding

The input of the sparse model is the descriptors for n_v videos, where each video is represented by a d-dimension vector $\mathbf{x}_d$. Let $\mathbf{X}_{d \times n_v}$ be the matrix by stacking the all the training video descriptors. In our model, $\mathbf{x}_d$ is the SSM feature. The outputs are the dictionary $\mathbf{D}$ and sparse codes $\mathbf{C}$. The loss of the *classical* sparse coding model, which considers reconstruction error and sparsity, is defined as:

$$\mathbf{L}(\mathbf{X};\mathbf{D},\mathbf{C}) = \|\mathbf{X} - \mathbf{D}\mathbf{C}\|_F^2 + \alpha \sum_{i=1}^{n_v} \|\mathbf{c}^{(i)}\|_1 \tag{1}$$

In Eq. 1, $\mathbf{D}_{d \times n}$ represents the learned dictionary and each column vector in the dictionary represents a typical action pattern, n is the number of typical patterns, $\mathbf{C}_{n \times n_v}$ is the sparse code matrix, whose i-th column, $\mathbf{c}^{(i)}$, is the sparse code of sample i. l_1-norm is a lasso constraint which encourages sparsity, and α balances the reconstruction error and the sparsity penalty.

In order to mine the view-invariant patterns of the SSM feature, we propose a class-wise sparse coding method to encourage the sparse codes of the multi-view within-class videos to lie close by. The closeness is measured by the within-class variance. Given the class labels of the training data, we try to reduce the within-class variance during the learning process. The within-class variance is measured by the Euclidean distance between the videos and their class center. The loss of the class-wise sparse coding model is defined as follows,

$$\mathbf{L}(\mathbf{X};\mathbf{D},\mathbf{C}) + \beta \sum_{s=1}^{K} \|\mathbf{C}^{(s)} - \bar{\mathbf{C}}^s\|_F^2 \qquad (2)$$

The second term in Eq. 2 measures the within-class variance. This term enforces the multi-view within-class videos to have similar sparse codes. K is the number of action classes. $\mathbf{C}^{(s)}$ represents a video collection. s is the class index. Each column vector in $\mathbf{C}^{(s)}$ is the sparse code of the video which belong to action class s. Each column vector in $\bar{\mathbf{C}}^s$ is the mean of all the column vectors in $\mathbf{C}^{(s)}$. $\bar{\mathbf{C}}^s$ has the same size as $\mathbf{C}^{(s)}$. β is the weight of within-class variance penalty.

4   Sparse Code Filtering

4.1   Joint Action Learning

The input to our learning scheme is (1) the learned sparse codes for n_v videos, each represented as a n-dimension vector $\mathbf{c}^{(i)} \in \mathbb{R}^n$, $i = 1, 2, ..., n_v$. (2) the binary action label matrix for all the videos, which is represented as $\mathbf{Y}_{n_v \times n_a}$, n_a is the number of actions. The item $y^{(i,j)}$, is either 1 or 0, which denotes whether or not video i belongs to action class j.

We learn all action classifiers simultaneously in a multi-task learning setting, where each *task* represents one action. The output is the parameter matrix $\mathbf{\Theta}_{n \times n_a}$ whose column vector $\theta^{(j)}$ denotes the parameters of the classifier of action j. In our model, we employ logistic regression classifiers. Given the sparse code matrix and binary action label matrix ($\mathbf{C}_{n \times n_v}, \mathbf{Y}_{n_v \times n_a}$), the loss function is defined as:

$$\mathbf{L}(\mathbf{C}, \mathbf{Y}; \mathbf{\Theta}) = \sum_{i,j} log(1 + exp((1 - 2y^{(i,j)})(\theta^{(j)})^T \mathbf{c}^{(i)})) \qquad (3)$$

Each action classifier has an tuple $\theta^{(j)}$ whose element $\theta_k^{(j)}$ corresponds to the *weight* of the sparse code which is tied to the k-th typical pattern in the dictionary.

4.2   Formulation of Sparse Code Filtering

Usually, the dictionary and classifiers are trained separately. Thus, there is no guarantee that the learned patterns in the dictionary can serve the classification task well. In order to mine the class-discriminative action patterns, we propose a sparse code filtering (SCF) scheme. In our scheme, the prediction function is logistic function whose output denotes the probability that a video belongs to an action. Besides, the parameters are learned by minimizing both the dictionary reconstruction error and classification error. Thus, the dictionary and classifiers are optimized jointly. The learned sparse codes are expected to be view-invariant and class-discriminative. By integrating all the tasks, we can obtain the following loss function:

$$\mathbf{L}(\mathbf{X},\mathbf{Y};\mathbf{D},\mathbf{C},\boldsymbol{\Theta})=\mathbf{L}(\mathbf{X};\mathbf{D},\mathbf{C})+\gamma\mathbf{L}(\mathbf{C},\mathbf{Y};\boldsymbol{\Theta})+\beta\sum_{s=1}^{K}\|\mathbf{C}^{(s)}-\bar{\mathbf{C}}^{(s)}\|_F^2+\lambda\|\boldsymbol{\Theta}\|_F^2 \quad (4)$$

In Eq. 4, γ balances the dictionary reconstruction error and the classification error, the Frobenius norm of $\boldsymbol{\Theta}$ is employed to prevent overfitting. By minimizing the loss function, Eq. 4, a view-invariant and class-discriminative dictionary $\mathbf{D}$, and an action classification parameter matrix $\boldsymbol{\Theta}$ are learned jointly.

Optimization. The input of the SCF framework is video descriptor matrix and binary action label matrix: $[\mathbf{X},\mathbf{Y}]$. The outputs are the dictionary, sparse codes, and parameter matrix for the classifiers: $[\mathbf{D},\mathbf{C},\boldsymbol{\Theta}]$. We propose the following algorithm (Algorithm 1) to solve the framework. When only one variable is left to optimize and the rest are fixed, the problem becomes convex. Thus, we optimize the variables alternatively by fixing the rest.

Initialization in Algorithm 1: we employ k-means clustering to find k centroids as the bases in dictionary $\mathbf{D}_0$. $\boldsymbol{\Theta}_0$ and $\mathbf{C}_0$ are set to $\mathbf{0}$.

The loop in Algorithm 1 consists of three parts:

1. Fix $\mathbf{C}$, $\boldsymbol{\Theta}$, Optimize $\mathbf{D}$. In Eq. (4), only the first term is related to $\mathbf{D}$, and it is a least square problem when the other parameters are fixed. By setting the derivative of Eq. (4) equal to $\mathbf{0}$ with respect to $\mathbf{D}$, we can obtain:

$$(\mathbf{DC}-\mathbf{X})\mathbf{C}^T=0 \Rightarrow \mathbf{D}=\mathbf{X}\mathbf{C}^T(\mathbf{C}\mathbf{C}^T)^{-1} \quad (5)$$

Then we employ the following equation to update $\mathbf{D}$:

$$\mathbf{D}=\mathbf{X}\mathbf{C}^T(\mathbf{C}\mathbf{C}^T+\lambda\mathbf{I})^{-1} \quad (6)$$

λ is a small constant and it guarantees that the matrix $\mathbf{C}\mathbf{C}^T+\lambda\mathbf{I}$ is invertible in case $\mathbf{C}\mathbf{C}^T$ is singular.

Algorithm 1. Solution Structure

1: *Initialization*: $\mathbf{D} \leftarrow \mathbf{D}_0$, $\mathbf{C} \leftarrow \mathbf{C}_0$, $\boldsymbol{\Theta} \leftarrow \boldsymbol{\Theta}_0$
2: **repeat**
3:     *fix* $\mathbf{D}, \boldsymbol{\Theta}$, *update* $\mathbf{C}$:
4:     **for** $\mathbf{C}^{(s)} \in \mathbf{C}$ **do**
5:         *ratio* $\leftarrow 1$
6:         **while** *ratio > threshold* **do**
7:             *run* **FISTA(modified)**
8:             **update** *ratio*
9:         **end while**
10:     **end for**
11:     *fix* $\mathbf{D}, \mathbf{C}$, *update* $\boldsymbol{\Theta}$:
12:     **parallelgradientdescent**
13:     *fix* $\mathbf{C}, \boldsymbol{\Theta}$, *update* $\mathbf{D}$:
14:     **least − squaressolution**
15: **until** converges

---

2. Fix $\mathbf{D}$, $\mathbf{C}$, Optimize $\boldsymbol{\Theta}$. When $\mathbf{D}$ & $\mathbf{C}$ are fixed, we employ the parallel gradient descent method to tackle the problem. Since $\theta^{(j)}$ are independent from each other, we optimize them in parallel. The updating formula is as follows:

$$\theta^{(j)} = \theta^{(j)} - \delta \frac{\partial}{\partial \theta^{(j)}} \mathbf{L}(\mathbf{X}, \mathbf{Y}; \mathbf{D}, \mathbf{C}, \boldsymbol{\Theta}) \tag{7}$$

3. Fix $\mathbf{D}$, $\boldsymbol{\Theta}$, Optimize $\mathbf{C}$. Beck and Teboulle [25] proposed the Fast Iterative Soft-Thresholding Algorithm (FISTA) to solve the classical dictionary learning problem. A soft-threshold step is incorporated into FISTA to guarantee the sparseness of the solution. The complexity for the classical ISTA method is $O(1/k)$, in which k denotes the iteration times. FISTA converges in function values as $O(1/k^2)$, which is much faster. FISTA optimizes $\mathbf{c}^{(i)} \in \mathbf{C}$ independently. However, in our model, $\mathbf{c}^{(i)}$ and $\mathbf{c}^{(j)}$ within the same action class depend on each other. Thus, $\mathbf{c}^{(i)}$, $\mathbf{c}^{(j)}$ must be updated jointly until all of them converge. Thus, we decompose our objective function and modify the original FISTA algorithm to tackle the decomposed sub-objectives.

In Eq. (4), the sparse code matrices with respect to different action classes are independent. Thus, when updating $\mathbf{C} = [\mathbf{C}^{(1)}, ..., \mathbf{C}^{(K)}]$, we decompose the objective function into K sub-objectives, shown as follows:

$$\min_{\mathbf{C}} \sum_{s=1}^{K} \mathbf{L}(\mathbf{C}^{(s)}) = \sum_{s=1}^{K} \min_{\mathbf{C}^{(s)}} \mathbf{L}(\mathbf{C}^{(s)}) \tag{8}$$

Thus, the original objective function is decomposed into K sub-objective functions with respect to each action class. The following shows the details of the deduction of decomposition of Eq. 4. The first two terms in Eq. 4 can be reformulated as follows:

$$\begin{cases} \mathbf{L(X;D,C)} = \sum_{s=1}^{K} \left(\|\mathbf{DC}^{(s)} - \mathbf{X}^{(s)}\|^2 + \alpha\|\mathbf{C}^{(s)}\|_1 \right) \\ \mathbf{L(C,Y;\Theta)} = \sum_{s=1}^{K} \sum_{j=1}^{n_a} \log(1 + \exp(1 - 2y^{(i,j)})(\theta^{(j)})^T \mathbf{C}^{(s)}) \end{cases} \quad (9)$$

Putting the transformed terms from Eq. (9) back into the loss function Eq. (4), we can obtain a new form of the objective function. Because $\mathbf{D}$ and $\mathbf{\Theta}$ are fixed, the term $\lambda\|\mathbf{\Theta}\|_F^2$ becomes a constant. By removing the constant term, we can obtain the loss function as Eq. (8) where

$$\mathbf{L(C}^{(s)}) = \|\mathbf{DC}^{(s)} - \mathbf{X}^{(s)}\|_F^2 + \alpha\|\mathbf{C}^{(s)}\|_1 + \beta\|\mathbf{C}^{(s)} - \bar{\mathbf{C}}^s\|_F^2$$
$$+ \gamma \sum_{j=1}^{n_a} \log(1 + \exp(1 - 2y^{(i,j)})(\theta^{(j)})^T \mathbf{C}^{(s)}) \quad (10)$$

The modified FISTA algorithm is applied to solve the sub-objective functions. The details of the modified FISTA algorithm is as follows:

In the classical dictionary learning model, the sparse codes of training data are independent from each other. Thus, each $\mathbf{c}$ can be optimized independently. However, our new sub-objective needs to optimize a group of training data jointly because these data have dependencies among each other as shown in Eq. (10). For training data $\mathbf{x}^{(i)} \in \mathbf{X}^{(s)}$ in the equation above, its sparse code $\mathbf{c}^{(i)}$ ($\mathbf{c}^{(i)} \in \mathbf{C}^{(s)}$) dependents on other $\mathbf{c}^{(k)}$ ($\mathbf{c}^{(k)} \in \mathbf{C}^{(s)}$). We modify the classical FISTA algorithm to optimize the sub-objectives jointly.

When update $\mathbf{C}^{(s)}$, instead of updating $\mathbf{c}^{(i)}$ independently, all $\mathbf{c}^{(i)} \in \mathbf{C}^{(s)}$ are updated simultaneously using the following form,

$$\mathbf{c}^{(i)} := \mathbf{c}^{(i)} - \delta \frac{\partial L}{\partial \mathbf{c}^{(i)}} \quad (11)$$

This updating procedure of $\mathbf{C}^{(s)}$ will repeat until it converges. Then we apply a soft-threshold step to set the entries in $\mathbf{C}^{(s)}$ whose absolute value is less than the threshold to 0. We repeat the process above until the whole algorithm converges.

Label Prediction. As shown in Fig. 1, in the classical CF framework, when the features of a new movie are given, its ratings by different users can be predicted based on the movie features and the learned feature preference weights. The basic underlying assumption of CF is that users will rate movies which share the similar features with similar scores [26] as we assume that the preferences of the users remain the same. Similarly, each action class can be regarded as one user, and the action videos can be regarded as the movies. The label prediction for a new video $\mathbf{x}$ consists of two steps: sparse coding and probability calculation.

$$\mathbf{c}^* = \arg_{\mathbf{c}} \min \mathbf{L(x, D; c)} \quad (12)$$
$$label = \arg_j \max 1/(1 + exp((-\theta^{(j)})^T \cdot \mathbf{c}^*)) \quad (13)$$

First, given the dictionary $\mathbf{D}$, and video descriptor $\mathbf{x}$ which is the SSM feature, the sparse code $\mathbf{c}^*$ of the new video is calculated by solving the classical sparse coding model as shown in Eq. (12). Then the probability that the new video belongs to action class j can be calculated. The action label is the one which maximizes the probability as shown in Eq. (13).

5   Experiments and Results

5.1   Datasets

We evaluate our framework on three largest public *multi-view* action recognition datasets, as shown in Fig. 3, which are the IXMAS dataset [27], the NIXMAS dataset, and the OIXMAS dataset [28] in which the actions are partially occluded. IXMAS dataset consists of 12 action classes, (*e.g., check watch, cross arms, scratch head, sit down, get up, turn around, walk, wave, punch, kick, point and pick up*). Each action is performed 3 times by 11 actors and is recorded by 5 cameras which observe the actions from 5 different viewpoints. The NIXMAS dataset is recorded with different actors, cameras, and viewpoints, and about 2/3 of the videos have objects which partially occlude the actors. Overall, it contains 1148 sequences.

5.2   Implementation Details

The sparse code filtering is based on SSM descriptors using HOG/HOF features to describe each individual frame. Each video is represented by a 500-dimension vector. Figure 2 shows an example from IXMAS dataset and the corresponding extracted SSM feature. In our experiments, the dictionary size is set to

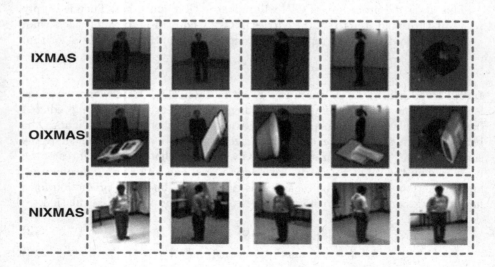

Fig. 3. Multi-view action recognition datasets.

$[600, 700, ..., 1000]$, and all regularization parameters α, β, γ, λ are tuned from $[10^{-3}, 10^{-2}, ..., 10^3]$.

We employ two settings for the experiment, which are *multi-view* setting and *cross-view* setting. For the *multi-view* setting, we have access to the videos from all the viewpoints for training, and use the standard experimental protocol described in [29]: two-thirds and one-third split for training and testing. This experimental protocol is widely used for action recognition. For the *cross-view* setting, one camera view is missing in the training data and we train the model using the data from other four camera views. Then we perform prediction on the missing view.

5.3 Baselines

To evaluate the contribution of the class-wise sparse coding (CWSC), we put the raw features and the codes into two classification scheme: (1) standard radial basis kernel SVM [6] which learns each action classifiers separately, and (2) the multi-task learning approach [9] which learns the action classifiers jointly. The codes and the classifiers are learned separately. We name the two baselines which take the codes as input as (3) CWSC+SVM, and (4) CWSC+MTL, and they are employed as baselines. Then through the comparison between (CWSC+MTL) and our SCF framework, we can observe the extra gain we obtained by training the class-wise dictionary and classifiers jointly. (5) We also choose some other action recognition baselines, such as [9,29,30].

5.4 Results

Multi-view Action Recognition. For the multi-view setting, we use the standard two-thirds and one-third split for training and testing. Table 1 shows the mean action recognition accuracy of all the cameras using different approaches.

We observe that the baselines CWSC+SVM and CWSC+MTL outperform SVM and MTL with raw features respectively. This indicates that the class-wise sparse coding can help encode the view-invariant action patterns which preserve the label consistency. From Table 1, we can also observe that our method has the best performance. This is because our sparse code filtering scheme optimizes the classifiers and dictionary jointly, and it helps learn a class-wise label-discriminative dictionary. Figure 4 shows some qualitative results on IXMAS dataset for our proposed SCF framework and multi-task learning approach for multi-view action recognition.

Cross-View Action Recognition. Tables 2, 3 and 4 show the performances of different approaches on IXMAS, OIXMAS, and NIXMAS dataset.

From Tables 2, 3 and 4, we can observe that our framework achieves better performance compared with other baselines which shows the effectiveness of our learned dictionary. It is also interesting to notice that the fifth camera always has low action recognition accuracy regardless of the classification methods.

Table 1. Multi-view action recognition accuracy of different approaches for 3 datasets.

Methods	IXMAS	OIXASM	NIXMAS
SVM [6]	0.6425	0.4809	0.5680
CWSC+SVM	0.6537	0.5235	0.6026
MTL [9]	0.6883	0.5608	0.6163
CWSC+MTL	0.6889	0.6082	0.6228
Farhadi *et al.* [5]	0.5810	-	-
Huang *et al.* [29]	0.5730	-	-
Liu *et al.* [31]	0.7380	-	-
Reddy *et al.* [32]	0.7260	-	-
Li and Shah [30]	0.8120	-	-
Baumann *et al.* [33]	0.8055	-	-
Ashraf *et al.* [34]	0.8140	-	-
SCF	**0.8594**	**0.7803**	**0.8083**

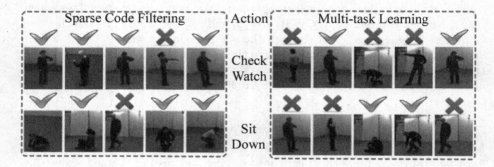

Fig. 4. Qualitative results on IXMAS dataset.

Table 2. Cross-view action recognition performance on the IXMAS dataset

Methods	Missing viewpoints					
	Cam 1	Cam 2	Cam 3	Cam 4	Cam 5	Avg
Junejo *et al.* [6]	0.6663	0.6554	0.6500	0.6243	0.4963	0.6185
CWSC+SVM	0.6880	0.6577	0.6701	0.6187	0.5110	0.6291
Yan *et al.* [9]	0.7554	0.7462	0.7710	0.6973	0.6332	0.7206
CWSC+MTL	0.7559	0.8257	0.8003	0.7759	0.6417	0.7599
SCF	**0.8285**	**0.8322**	**0.8053**	**0.7941**	**0.7384**	**0.7997**

One reasonable explanation is that the fifth camera is placed on the ceiling, and the motion dynamics of different actions observed from this camera are visually similar with each other.

Table 3. Cross-view action recognition performance on the OIXMAS dataset

Methods	Missing viewpoints					
	Cam 1	Cam 2	Cam 3	Cam 4	Cam 5	Avg
Junejo *et al.* [6]	0.5639	0.6250	0.5472	0.4677	0.4423	0.5292
CWSC+SVM	0.5688	0.6477	0.6001	0.5087	0.4511	0.5553
Yan *et al.* [9]	0.5422	0.6540	0.5070	0.5171	0.4730	0.5387
CWSC+MTL	0.5535	0.6826	0.5366	0.5401	0.4867	0.5599
SCF	**0.6080**	**0.6980**	**0.6573**	**0.6957**	**0.5850**	**0.6512**

Table 4. Cross-view action recognition performance on the NIXMAS dataset

Methods	Missing viewpoints					
	Cam 1	Cam 2	Cam 3	Cam 4	Cam 5	Avg
Junejo *et al.* [6]	0.6410	0.6532	0.5912	0.5924	0.5322	0.6020
CWSC+SVM	0.6759	0.6951	0.6226	0.6387	0.5560	0.6377
Yan *et al.* [9]	0.7170	0.6993	0.7542	0.6911	0.6792	0.7082
CWSC+MTL	0.7198	0.7391	0.7559	0.7176	0.6879	0.7240
SCF	**0.8080**	**0.7980**	**0.7573**	**0.7357**	**0.7050**	**0.7608**

5.5 Parameter Tuning

Figure 5 shows the sensitivity study of regularization parameters γ, α, β and λ. In our model, γ balances the dictionary learning loss and the classification loss, α balances the reconstruction error and the sparsity penalty, β provides the trade-off between the dictionary reconstruction loss and intra-class variance penalty, and λ is employed to prevent overfitting of the classifiers. The optimal classification performance can be obtained when dictionary size is set to 800. We observe that the performance changes little (within 0.0015) when we set λ to the different values. So we focus on the other 3 parameter. As shown in Fig. 5(a), when γ is fixed, the mean accuracy varies subtly along the axis of β.

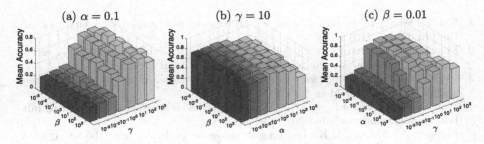

Fig. 5. Sensitivity study of different regularization parameters on IXMAS dataset.

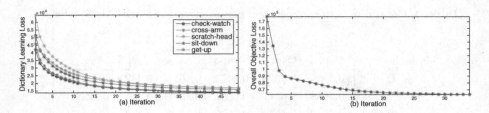

Fig. 6. Convergence of the sparse code filtering algorithm on IXMAS dataset.

However, when β is fixed, the mean accuracy changes dramatically along the axis of β. Thus, γ is more sensitive than β. Similarly, Fig. 5(b) shows that α is more sensitive than β, and Fig. 5(c) shows that γ is more sensitive than α. Thus, we obtain the importance of these parameters $\gamma > \alpha > \beta > \lambda$.

We also analyzes the convergence of our algorithm. Figure 6 plots the convergence curves of the objectives. Figure 6(b) shows that Algorithm 1 converges in 30 iterations. Figure 6(a) plots the convergence curves when updating $\mathbf{C}^{(s)}$ for action classes. It shows that the class-wise dictionary learning converges very fast.

6 Conclusion

In this paper, we propose a novel sparse code filtering framework for multi-view action recognition. First, a class-wise dictionary is learned by encoding label information into the sparse coding process. We integrate class-wise sparse coding and classifier learning into a CF framework. Thus, the classifiers and dictionary are optimized jointly, and they can be adapted for each other. The extensive experimental results illustrate that our proposed method outperforms other important baselines for multi-view action recognition. In the future work, we will take the correlation between the classifiers into consideration. For example, we can suppress the urge of feature sharing between classifiers by adding a l_1 norm penalty to the classifier parameters.

References

1. Cai, Z., Wang, L., Peng, X., Qiao, Y.: Multi-view super vector for action recognition. In: CVPR (2014)
2. Vemulapalli, R., Arrate, F., Chellappa, R.: Human action recognition by representing 3D skeletons as points in a lie group. In: CVPR (2014)
3. Lv, F., Nevatia, R.: Single view human action recognition using key pose matching and viterbi path searching. In: CVPR (2007)
4. Peursum, P., Venkatesh, S., West, G.: Tracking-as-recognition for articulated full-body human motion analysis. In: CVPR (2007)
5. Farhadi, A., Tabrizi, M.K.: Learning to recognize activities from the wrong view point. In: Forsyth, D., Torr, P., Zisserman, A. (eds.) ECCV 2008. LNCS, vol. 5302, pp. 154–166. Springer, Heidelberg (2008). doi:10.1007/978-3-540-88682-2_13

6. Junejo, I.N., Dexter, E., Laptev, I., Perez, P.: View-independent action recognition from temporal self-similarities. TPAMI **33**(1), 172–185 (2011)
7. Weinland, D., Boyer, E., Ronfard, R.: Action recognition from arbitrary views using 3D exemplars. In: ICCV (2007)
8. Natarajan, P., Nevatia, R.: View and scale invariant action recognition using multiview shape-flow models. In: CVPR (2008)
9. Yan, Y., Ricci, E., Subramanian, R., Liu, G., Sebe, N.: Multitask linear discriminant analysis for view invariant action recognition. TIP **23**(12), 5599–5611 (2014)
10. Mahasseni, B., Todorovic, S.: Latent multitask learning for view-invariant action recognition. In: ICCV (2013)
11. Matikainen, P., Sukthankar, R., Hebert, M.: Model recommendation for action recognition. In: CVPR (2012)
12. Du, Y., Wang, W., Wang, L.: Hierarchical recurrent neural network for skeleton based action recognition. In: CVPR (2015)
13. Wang, W., Cui, Z., Yan, Y., Feng, J., Yan, S., Shu, X., Sebe, N.: Recurrent face aging. In: CVPR (2016)
14. Wang, W., Tulyakov, S., Sebe, N.: Recurrent convolutional face alignment. In: ACCV (2016)
15. Junejo, I.N., Dexter, E., Laptev, I., Pérez, P.: Cross-view action recognition from temporal self-similarities. In: Forsyth, D., Torr, P., Zisserman, A. (eds.) ECCV 2008. LNCS, vol. 5303, pp. 293–306. Springer, Heidelberg (2008). doi:10.1007/978-3-540-88688-4_22
16. Sun, C., Junejo, I., Foroosh, H.: Action recognition using rank-1 approximation of joint self-similarity volume. In: ICCV (2011)
17. Wang, W., Yan, Y., Winkler, S., Sebe, N.: Category specific dictionary learning for attribute specific feature selection. TIP **25**(3), 1465–1478 (2016)
18. Wang, W., Yan, Y., Sebe, N.: Attribute guided dictionary learning. In: ICMR (2015)
19. Raina, R., Battle, A., Lee, H., Packer, B., Ng, A.Y.: Self-taught learning: transfer learning from unlabeled data. In: ICML (2007)
20. Luo, J., Wang, W., Qi, H.: Group sparsity and geometry constrained dictionary learning for action recognition from depth maps. In: ICCV (2013)
21. Qiu, Q., Jiang, Z., Chellappa, R.: Sparse dictionary-based representation and recognition of action attributes. In: ICCV (2011)
22. Guha, T., Ward, R.K.: Learning sparse representations for human action recognition. TPAMI **34**(8), 1576–1588 (2012)
23. Zheng, J., Jiang, Z.: Learning view-invariant sparse representations for cross-view action recognition. In: ICCV (2013)
24. Goldberg, D., Nichols, D., Oki, B.M., Terry, D.: Using collaborative filtering to weave an information tapestry. Commun. ACM **35**(12), 61–70 (1992)
25. Beck, A., Teboulle, M.: A fast iterative shrinkage-thresholding algorithm for linear inverse problems. SIAM J. Imaging Sci. **2**(1), 183–202 (2009)
26. Goldberg, K., Roeder, T., Gupta, D., Perkins, C.: Eigentaste: a constant time collaborative filtering algorithm. Inf. Retrieval **4**(2), 133–151 (2001)
27. Weinland, D., Ronfard, R., Boyer, E.: Free viewpoint action recognition using motion history volumes. CVIU **104**(2), 249–257 (2006)
28. Weinland, D., Özuysal, M., Fua, P.: Making action recognition robust to occlusions and viewpoint changes. In: Daniilidis, K., Maragos, P., Paragios, N. (eds.) ECCV 2010. LNCS, vol. 6313, pp. 635–648. Springer, Heidelberg (2010). doi:10.1007/978-3-642-15558-1_46

29. Huang, C.-H., Yeh, Y.-R., Wang, Y.-C.F.: Recognizing actions across cameras by exploring the correlated subspace. In: Fusiello, A., Murino, V., Cucchiara, R. (eds.) ECCV 2012. LNCS, vol. 7583, pp. 342–351. Springer, Heidelberg (2012). doi:10. 1007/978-3-642-33863-2_34
30. Li, R., Zickler, T.: Discriminative virtual views for cross-view action recognition. In: CVPR (2012)
31. Liu, J., Shah, M.: Learning human actions via information maximization. In: CVPR (2008)
32. Reddy, K.K., Liu, J., Shah, M.: Incremental action recognition using feature-tree. In: ICCV (2009)
33. Baumann, F., Ehlers, A., Rosenhahn, B., Liao, J.: Recognizing human actions using novel space-time volume binary patterns. Neurocomputing **173**, 54–63 (2016)
34. Ashraf, N., Sun, C., Foroosh, H.: View invariant action recognition using projective depth. CVIU **123**, 41–52 (2014)

Learning Action Concept Trees
and Semantic Alignment Networks
from Image-Description Data

Jiyang Gao[✉] and Ram Nevatia

University of Southern California, Los Angeles, USA
jiyangga@usc.edu

Abstract. Action classification in still images has been a popular research topic in computer vision. Labelling large scale datasets for action classification requires tremendous manual work, which is hard to scale up. Besides, the action categories in such datasets are pre-defined and vocabularies are fixed. However humans may describe the same action with different phrases, which leads to the difficulty of vocabulary expansion for traditional fully-supervised methods. We observe that large amounts of images with sentence descriptions are readily available on the Internet. The sentence descriptions can be regarded as weak labels for the images, which contain rich information and could be used to learn flexible expressions of action categories. We propose a method to learn an Action Concept Tree (ACT) and an Action Semantic Alignment (ASA) model for classification from image-description data via a two-stage learning process. A new dataset for the task of *learning actions from descriptions* is built. Experimental results show that our method outperforms several baseline methods significantly.

1 Introduction

Action classification in still images has been a popular research topic in computer vision. Traditional fully-supervised learning methods for action classification rely on large amount of fully-labelled data (*i.e.* each image is labelled with one or more action categories) to learn action classifiers. However, labelling image data with action categories requires tremendous manual work, which is time-consuming and hard to scale-up. Another drawback of traditional supervised learning framework is that the action categories are pre-defined and limited, while humans may describe the same action with different phrases, for example, take out the chopping board and fetch out the wooden board. This drawback leads to the difficulty of vocabulary expansion, as CNN [1,2] models or SVM classifiers just assign a label to the test image. Hence, CNN or SVM models would fail to classify the categories that are not in the training set.

We observe that large amounts of images with sentence descriptions are readily available on the Internet, such as videos with captions and social media, such as Flickr and Instagram. Such sentence descriptions can be regarded as weak labels of the images. Sentence descriptions are generated by humans and

© Springer International Publishing AG 2017
S.-H. Lai et al. (Eds.): ACCV 2016, Part II, LNCS 10112, pp. 19–34, 2017.
DOI: 10.1007/978-3-319-54184-6_2

Image-description pairs

| People are playing frisbee | Women are riding a bike | The boy is eating donut |

| Young man about to kick ball | Man are ready to play tennis | A man holding a white surfboard |

Fig. 1. Descriptions, as weak labels to images, contain rich information about actions. Images and corresponding descriptions are from Visual Genome [3]

contain rich information about actions, which could be used to learn an expanding vocabulary of actions. Some example are shown in Fig. 1. Another observation we make is that action concepts are naturally represented as a hierarchy; for example, "play guitar" and "play violin" are subcategories of "play instrument". If such hierarchical structure of action categories is available, classification methods can choose to use detailed knowledge if necessary or generalized knowledge when details are unavailable or irrelevant.

In this paper, we propose a method to tackle the problem of *learning actions from descriptions*: Given a set of image-description data (assuming descriptions containing human action information), learning to recognize human actions. Our method supports hierarchical clustering of action concepts and vocabulary expansion for action classification. Specifically, our method learns an Action Concept Tree (ACT) and an Action Semantic Alignment (ASA) model for classification via a two-stage learning process. ASA model contains a CNN to extract image-level features, an LSTM to extract text embeddings and a multi-layer neural network to align these two modalities. In the first stage, (a) we design a Hierarchical Action Concept Discovery (H-ACD) method to automatically discover action concepts from image-description data and cluster them into a hierarchical structure (*i.e.* ACT); (b) ASA is initialized by the image-description mapping task in stage-1. In the second stage, the target action categories are matched to the nodes in ACT and the associated image data are used to fine-tune ASA for this action classification task to improve the performance. Note that no image data from test domain are used for training.

To facilitate research on this task, we constructed a dataset based on Visual Genome [3], called Visual Genome Action (VGA). Although Visual Genome contains well-annotated region descriptions, we do not use the region information and treat the descriptions as image-level. There are 52931 image-description pairs in the training set and 4689 images of 45 categories in test set. More details of this dataset are given in Sect. 4.1 later.

In summary, our main contributions are:

(1) A Hierarchical Action Concept Discovery (H-ACD) algorithm to automatically discover an Action Concept Tree (ACT) from image-description data and gather samples for each action node in ACT.
(2) An end-to-end CNN-LSTM Action Semantic Alignment (ASA) network which aligns semantic and visual representation to classify actions with expanding vocabulary.
(3) A dataset for the problem of *learning actions from descriptions*, which is built on Visual Genome, containing 52931 image-description pairs for training and 45 action categories for testing.

The paper is organized as follows. Section 2 discusses the related works. In Sect. 3, we will introduce our two-stage framework to learn actions from image-description data. We evaluate our model in Sect. 4 and give our conclusions in Sect. 5.

2   Related Work

Action Classification in Still Images: The use of convolutional neural network (CNN) has brought huge improvement in action classification [4]. [5] fine-tunes the CNN pre-trained on ImageNet and shows improvement over traditional methods. [6] designs a multi-task (person-detection, pose-estimation and action classification) model based on R-CNN. [7] develops an end-to-end deep convolutional neural network that utilizes contextual information of actions. HICO [8] introduces a new benchmark for recognizing human-object interactions, which contains a diverse set of interactions with common object categories, such as "hold banana" and "eat pizza". Ramanathan *et al.* [9] proposes a neural network framework to jointly extract the relationship between actions and uses them for training better action retrieval models. These methods all rely on fully-labelled data.

Weakly Supervised Action Concept Learning: Weakly supervised action concept learning relies on weakly-labelled data, such as video-caption stream data [10,11] and focuses on automatically discovering and learning action concepts. [12] designs a method to automatically discover the main steps for specific tasks, such as "make coffee" and "change tire", from narrated instructional videos. Their method solves two clustering problems, one in text and one in video, applied one after each other and linked by joint constraints to obtain a single coherent sequence of steps in both modalities. Ramanathan *et al.* [13] propose a method to learn action and role recognition models based on natural language descriptions of the training videos. Yu *et al.* [14] discover Verb-Object (VO) pairs from the captions of the instructional videos and use the associated video clips as training samples. The learned classifiers are evaluated in event classification, compared with well defined action categories in HMDB51 [15] and UCF50 [16]. [17] proposes a general concept discovery method from image-sentence corpora and apply the concepts on image-sentence retrieval tasks.

ACD [18] solves a similar problem to ours. It automatically discovers action concepts from image-sentence corpora [19, 20], clusters them and trains classifier for each action concept cluster. However, there are two main drawbacks in this method: (1) no hierarchical clustering: once the action concepts are clustered, the detailed information are lost; (2) no vocabulary expansion: if the target test action categories are missed in the training set, ACD would fail to perform classification.

Language & Vision: Image captioning methods take an input image and generate a text description of the image content. Recently, methods based on convolutional neural networks and recurrent neural networks [21, 22] have shown to be an effective way on this task. VSA [23] is one of the recent successful models. It uses bidirectional recurrent neural networks over sentences, convolutional neural networks over image regions and a structured objective that aligns the two modalities through a multimodal embedding. Besides image captioning, other relevant work includes natural language object retrieval [24] or segmentation [25], which takes an input image and a query description and outputs a corresponding object bounding box or a segmentation mask.

3   Actions from Descriptions

In this section, we introduce the learning framework, which is a two-stage method. In the first stage, our target is to learn a general knowledge base of actions, which contains two parts: a hierarchical structure for action concepts and a general visual-semantic alignment model. In the second stage, the framework learns to classify specific action categories (*i.e.* target categories for test). The classifiers are fine-tuned from the visual-semantic alignment model learned in stage 1. The overall system is shown in Fig. 1.

3.1   Stage 1: Learning General Action Knowledge

As for general action knowledge, we refers to two concepts. The first one is a hierarchical structure of actions, which we call Action Concept Tree (ACT): each node in ACT contains an action concept, such as play frisbee and play basketball, and the related images; the action concepts are extracted from descriptions and the images come from the original image-description dataset. The second one is a general visual-semantic alignment model: the input of the model is an image and a description, and the output is a confidence score of the similarity of the image and the description. The framework of Stage 1 is shown in Fig. 2.

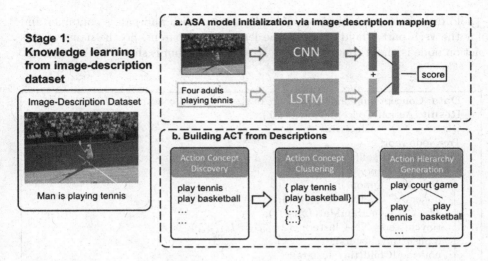

Fig. 2. Stage-1: model initialization via image-description matching and hierarchy action discovery

Hierarchical Action Concept Discovery (H-ACD):. ACD [18] proposed an action concept discovery method working with image-sentence corpora. However, the discovered action clusters are not organized in a hierarchical structure, which may lose the detailed information after clustering. Hence, based on ACD, we propose a Hierarchical Action Concept Discovery (H-ACD) method, which automatically discovers action concepts from image-description data and organizes them in a hierarchical structure using WordNet [26]. The process of action concept discovery and clustering are similar to ACD. First we extract Verb-Object (VO) pairs from sentence descriptions and the visualness of these VO pairs are verified by two fold cross-validation. After visualness verification, we generate a multi-modal representation for each action concept and calculate similarity score for each pair of action concepts.

After computing the similarity, we use the H-ACD algorithm to generate a hierarchical structure for action concepts. Note that nearest neighbor (NN) clustering algorithm is proposed in ACD [18]; we use it as a part of our H-ACD algorithm. We first apply NN-clustering algorithm (we fix the parameter C of NN-clustering as 4.) [18] on all the action concepts to get a list of action clusters. Then, inside each cluster, we continuously apply NN-clustering algorithm to get more smaller clusters; we do this recursively util no new cluster is generated. Each cluster is regarded as a node in the hierarchical structure and the node names are generated following a similar naming strategy of HAN [27] described in the following. For the object part, we find the lowest common hypernym in WordNet. For the verb part, we follow a simple strategy: if the verbs are the same, then the father node keeps the same verb; if the verbs are different, the father node is named as "interact with". For example, for a node containing {hold dish, hold pan}, the least shared parent of dish and pan is container and for the verb part, "hold" itself is the least shared parent. So the name of this action node is "hold container". The H-ACD algorithm is shown in Algorithm 1.

Data: Concept similarity matrix M of size $l \times l$ and concept list L of size l
Result: Action Concept Tree (ACT)
Queue q ← NN-Clustering(M, L);
TreeNode root;
root.addChild(q.all());
while *q not empty* **do**
  $L_{cluster}$ ← q.pop();
  node=ACT.getNode($L_{cluster}$);
  $M_{cluster}$ ← getSimMat($L_{cluster}$);
  tinyclusters=NN-Clustering($M_{cluster}$, $L_{cluster}$);
  q.push(tinyclusters);
  node.addChild(tinyclusters);
end
Generate node names following the naming strategy.

Algorithm 1. Hierarchical Action Concept Discovery (H-ACD) algorithm

---

ASA Model Initialization via Image-Description Mapping: Our final target is to classify action categories. Rather than training classifiers for each category, we want to build a connection between the semantic meaning and visual meaning of actions. Therefore, we formulate the action classification as a visual-semantic alignment problem between the image and action categories. The Action Semantic Alignment (ASA) model contains three parts: a CNN network to extract feature vector of the input image, an LSTM network to extract text embedding and an alignment network to compute the alignment score of the visual and semantic representations. Image-description mapping serves as a parameter initialization method for ASA model, which helps the model to learn a connection between semantic and visual spaces.

The input of ASA is an image I_i and the corresponding sentence description D_i. The image is processed by VGG-16, which outputs a d_{img} dimensional feature v_i. For a text sequence $S = (w1, ..., wT)$ with T words, each word is transformed to a d_{w2v} dimensional vector by the word embedding matrix and then processed by an LSTM module sequentially. The word embedding matrix is trained by skip-gram model on English Wiki data. At the final time step $t = T$, LSTM outputs the final hidden state and we use it as the sentence-level embedding s_i, which is a d_{text} dimensional vector. v_i and s_i are concatenated to vs_{ii} with a length of $d_{vs} = d_{text} + d_{img}$, which is visual-semantic representation of the image and description. Then we train a two-layer alignment network, with a d_{alg} dimensional hidden layer. The alignment network take d_{vs} dimensional input and output a confidence score cs_{ii}, which indicates whether the image and the description is aligned. The alignment network is implemented in a fully convolutional way as two $1 * 1$ convolutional layers (with ReLU function between them).

During the training time, we optimize the model inside each mini-batch. The loss function is as follows.

$$loss_1 = \sum_{i=0}^{N}[\alpha_c\log(1 + \exp(-cs_{i,i})) + \sum_{j=0, j\neq i}^{N} \alpha_w\log(1 + \exp(cs_{i,j}))] \quad (1)$$

where N is the batch size, α_c and α_w are the loss weights for correct and wrong image-description. The loss function encourages the network to output high score of correct image-description pairs and low score of incorrect image-description pairs. In practice, we find that training converges faster using higher loss weights for correct pairs and we use $\alpha_c = 1$ and $\alpha_w = 0.01$.

3.2   Stage 2: Action Classification on Target Categories

Given a set of action categories for classification (without training samples), we adjust the ASA model to the specific action classification task. The first step is to match the given action categories to some existing action nodes in ACT. Then we use the matched action nodes and the associated images to fine-tune our ASA model. The framework of Stage 2 is shown in Fig. 3.

Target Action Categories Matching: We first match the actions via keyword searching. Suppose the target action category is c_i and the action node in ACT is represented by n_j. We extract the verb and object from the target action category c_i and search for them in the discovered action hierarchy to see if there is an exact match. For example, a target action category is "play instrument" and there is a node in action hierarchy named "play instrument", then we match them and use the similarity score (calculated by ASA, see below) between them as a baseline score θ. If there is no exact match via keyword searching, we assign

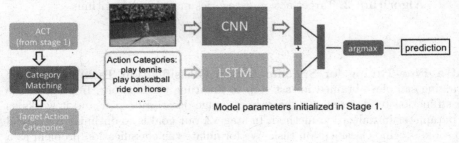

Fig. 3. Stage-2: adjust the model to a specific action classification task.

θ with a constant value. In the second step, we use the ASA model to compute a similarity score between the target action category c_i and all action nodes n_j in ACT. The associated images of n_j are $P_j = \{I_{jk}\}$, which has size m. The similarity score between c_i and n_j is

$$S(c_i, n_j) = \frac{1}{m} \sum_{k=0}^{m} \mathrm{ASA}(I_{jk}, c_i) \tag{2}$$

For a specific action category c_i, we select the action node n_j that has the highest similarity score and if the score is larger than or equal to θ, we match $< c_i, n_j >$. Note that some categories may still not be matched after the second step. After matching, we obtain a list of training samples $\{c_i, P_j\}$. The labels are c_i and the training images are the associated images of the corresponding matched node n_j. We don't assign any training data for the target categories with no matched node in ACT. The matching algorithm is detailed in Algorithm 2 below.

Data: ASA model, ACT and Target action categories $C = \{c_i\}$
Result: Matched pairs $< c_i, n_j >$
for c_i *in* C **do**
    for n_j *in* ACT **do**
        | if $c_i.name=n_j.name$: match $< c_i, n_j >$;
        | break;
    end
    if c_i is matched:
        ExactMatch, $\theta \leftarrow n_j, S(c_i, n_j)$;
        match $< c_i, \text{ExactMatch} >$;
    else:
        ExactMatch, θ=None, InitializationValue;
    MaxScore=0;
    for n_j *in* ACT *and* $n_j \neq ExactMatch$ **do**
        | if $S(c_i, n_j) >$MaxScore:
        | Node, MaxScore $\leftarrow n_j, S(c_i, n_j)$
    end
    if MaxScore $>= \theta$: match $< c_i, \text{Node} >$;
end

Algorithm 2. Target action categories matching algorithm

ASA Fine-Tuning for Specific Action Classification Task: We use the training samples obtained in last step to fine-tune the network. In stage 1, the loss function tends to match the correct image-description pair and it works as a parameter initialization method. In stage 2, our goal is to optimize the model to some specific classification task. We formulate the classification problem as a image-description matching problem. The name of the category is regarded as a text sequence, just like the sentence description. Suppose there are M categories, leading to M corresponding category descriptions $\{CD_j, j = 0, 1, 2..., M - 1\}$. Suppose the label of the input image I_i is t_i, then the loss function of stage 2 is as follows.

$$loss_2 = \sum_{i=0}^{N}[\alpha_c\log(1 + \exp(-cs_{i,t_i})) + \sum_{j=0,j\neq t_i}^{M} \alpha_w\log(1 + \exp(cs_{i,j}))] \quad (3)$$

where $cs_{i,j}$ is the matching score between I_i and CD_j, N is the batch size. We use $\alpha_c = 1$ and $\alpha_w = 0.01$. The loss function encourages the correct image-action pairs to output high positive score and other wrong pairs output low negative score.

Action Category Prediction: At test time, the prediction of an input image I_i is the argmax of the matching scores $cs_{i,j}$ between I_i and CD_j.

$$prediction(I_i) = argmax(cs_{i,j}), \; j = 0, 1, 2...M - 1 \quad (4)$$

4 Evaluation

4.1 Experiments on VGA

Dataset: Visual Genome Action (VGA). There are many image-description datasets, which are suitable for *learning actions from descriptions*. However, none of them contain pre-defined action categories and category annotations for each image. Therefore, we construct a dataset from Visual Genome for this problem, called Visual Genome Action (VGA). We split Visual Genome into two parts: 75% for training and validation and 25% for testing. The training set and test set are carefully checked to ensure that there is no overlap of images between these two sets.

For the training split, since we only focus on human action learning, we filter out the descriptions which don't have verbs or human subjects; for example, "a dog is running on the grass" and "a man with a white shirt" are filtered out. 52931 image-description pairs remain after such filtering. The descriptions in Visual Genome are region based, but we treat them as image-level descriptions. For the test split, we extract Verb-Object (VO) pairs and filter out the ones with very few image samples. After that, we manually filter out the VOs with no visual meaning, such as "do things". Finally, there are 45 categories and 4689 images for testing. The 45 test action categories are listed in Table 1. Some categories overlap; for example, "hold racket" and "play tennis", "hit ball" and "play soccer". We manually checked each image of these categories and added additional labels if necessary. For example, if an image of the category "hold racket" also represents the action of "play tennis", then we also add this image to the category of "play tennis". In other cases, people may be just holding a tennis racket but not playing, then we don't add additional labels to such images.

Table 1. Action categories in VGA test set

boat	brush tooth	color hair	do trick	drink wine
eat fruit	eat pizza	enjoy outdoors	fly kite	hit ball
hold bag	hold banana	hold bat	hold camera	hold controller
hold dog	hold fork	hold kite	hold knife	hold pole
hold racket	hold sandwich	hold umbrella	jump	play baseball
play basketball	play frisbee	play soccer	play tennis	read book
ride elephant	ride horse	ride wave	run	sit
ride skateboard	ski	smile	stand	surf
swim	use phone	walk	watch game	wear necklace

Metric. We tested our model on action classification task on VGA. As for evaluation metric, we report the mean Average Precision (mAP), Recall@1 and Recall@5.

Network Implementation. We implemented ASA network in Tensorflow [28], including CNN network, LSTM network and the multi-layer alignment network. For the CNN part, We use VGG-16 architecture and the parameters are initialized by ImageNet [29] image classification dataset. We use a standard LSTM architecture with 1000-dimensional hidden state. The descriptions input to LSTM have maximum length of 6 for both stage-1 and stage-2. The hidden layer of the visual-semantic alignment network is 500-dimensional. We train a skip-gram [30] model for the word embedding matrix using the English Dump of Wikipedia. The dimension of the word vector is 500. The whole network is trained end-to-end in two stages. We use three Adam optimizers [31] to optimize CNN, LSTM and the visual-semantic alignment network. The learning rates are 0.0001, 0.001 and 0.001 respectively. The model is trained on a Tesla K40 GPU; the batch size is 96. It takes about 1 day to train the whole model for both stage-1 and stage-2.

System Variants. We experimented with variants of our system to test the effectiveness of our method. **ASA (Stage 1):** we only trained the ASA model for stage-1 using the image-description pairs. **ASA (Stage 2):** we only trained the ASA model for stage-2 using the matched action nodes in ACT. **ASA (Stage 1+2, w/o ACT):** ASA model is trained for stage-1 and stage-2, but we only use flat action concepts (*i.e.* only the leaf action nodes in ACT) to match the target action categories. **ASA (Stage 1+2, w/ ACT):** this is our full model; ASA model is trained for stage-1 and stage-2, and full ACT is used to match the target action categories.

Baseline Methods. We introduce the baseline methods we implemented.

ACD [18]+SVM+AdaBoost: In this baseline method, we use ACD [18] to discover a list of action concepts from the training set and train SVM classifiers [32] for each action concept. Then we match the test action categories with the discovered action concepts by keyword searching. Multiple action concepts may matched to the same test categories and each of them can be regarded as weak classifier to the test category. To make use of all the related training data, we further use AdaBoost to build a stronger classifier.

ACD [18]+DeViSE [33]: In this baseline method, we first use ACD to discover a list of action concepts. Instead of training SVM classifiers for each of them, we apply DeViSE [33] methodology. The verb and the object of a action category are transformed to vectors using a word embedding matrix and are concatenated together. The word embedding matrix is trained by wiki dump data and the dimension of the word vector is 500. All the discovered action concepts and the associated images are used to train DeViSE model. At test time, the action categories are transformed to vectors using the same word embedding matrix. The prediction of an input image is the argmax of the matching scores between the image and the test categories.

Visual-semantic alignment [23]: This baseline method is similar to the model in VSA [23]. However, we use regular LSTM instead of BRNN to encode the input description. The image is processed by VGG-16 and 4096 dimensional fc7 vector is extracted as image-level feature. The training data are image-description pairs. The output of the model is a confidence score which indicates whether the image and description are matched. The test image is matched with all action categories and the prediction is the category with the highest score.

Action Concept Tree (ACT). There are totally over 100 action concepts (*i.e.* leaf action nodes) discovered in the training set of VGA. These action concepts are clustered into a 4-layer action concept tree (ACT). Due to the limited space, we can't illustrate the whole ACT. Some nodes in ACT are shown

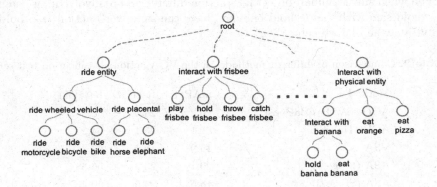

Fig. 4. Example nodes in ACT.

in Fig. 4. Under the node "ride entity", we can see that "ride motorcycle", "ride bicycle" and "ride bike" are clustered together and the automatically generated node name is "ride wheeled vehicle", as "wheeled vehicle" is the lowest common hypernym of "bike", "bicycle" and "motorcycle" in WordNet. Under the node "interact with frisbee", there are four leaf nodes: "catch frisbee", "hold frisbee", "play frisbee" and "throw frisbee". They have common object "frisbee" but different verb actions, so we generate the father node name as "interact with frisbee". The node "interact with physical entity" is illustrated as a poor case of our naming strategy. The child nodes have no common verb and the lowest common hypernym of the objects in WordNet is "physical entity", therefore the father node is named as "interact with physical object", which is a very vague action name. Although, the naming strategy is not ideal in this case, the cluster itself still represents one meaningful action category: "interact with food".

Action Classification Results. The experimental results on VGA are shown in Table 2. From the results, we can see that our 2-stage learning method outperforms several baseline methods. Training models with only stage-1 or stage-2 would lower the performance. Stage-1 only learns general image-description matching knowledge and it does not optimize the model to a specific action classification task; on the other hand, without stage-1, stage-2 optimizes the model from random parameters and it may overfit on such a small dataset of language. Using the hierarchical structure of action concepts (i.e. ASA (Stage1+2, w/ ACT)) brings a 1.7% improvement, compared with the flat structure of action concepts (i.e. ASA (Stage1+2, w/o ACT)). We believe the reason is that ACT and the node matching algorithm together provide a better way to organize and search for the generalized and detailed knowledge of actions. For example, compared with the flat action concept structure, the test category "brush tooth" is matched not only with the node of "brush tooth", but also with the parent node of "hold toothbrush" and "brush tooth" in ACT, which allows ASA to use the additional data provided by "hold toothbrush".

In Fig. 5, some example predictions are shown. We can see that failure could happen when subtle human-object interaction differences are involved; for example, "hold sandwich" and "hold banana" have the same verb action (i.e. hold) and visually similar objects.

Table 2. Comparison of different methods on the VGA action classification test set

Method	mAP(%)	R@1(%)	R@5(%)
ACD+SVM+AdaBoost	20.2	24.5	56.3
ACD+DeViSE	22.1	25.1	54.2
VSA	15.9	18.1	47.3
ASA (Stage 1)	20.1	25.3	56.5
ASA (Stage 2)	18.5	24.6	50.4
ASA (Stage 1+2, w/o ACT)	26.8	29.6	60.4
ASA (Stage 1+2, w/ ACT)	**28.5**	**31.3**	**63.2**

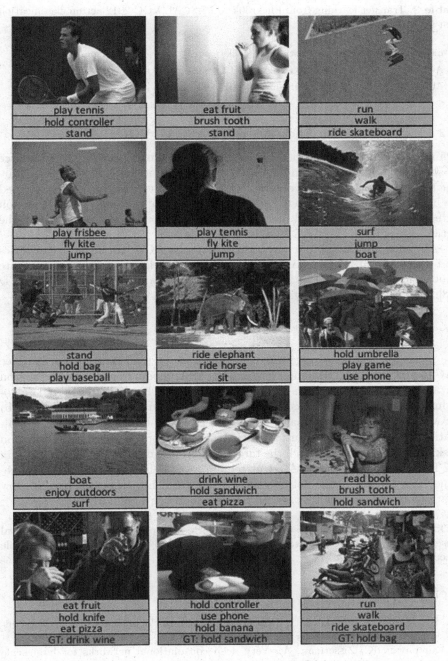

Fig. 5. Prediction examples of the top 3 results on VSA test set. The first four rows are positive examples and green represents the condition when the prediction matches the ground truth. The last row shows some failure cases. (Color figure online)

Table 3. Transfer learning from Flickr30k to PASCAL VOC 2012 action classification test set (AP%).

method	jump	phone	instr.	read	bike	horse	run	photo	comp.	walk	mAP
ACD [18]	62.2	15.4	78.8	**29.6**	84.5	85.9	**60.8**	24.0	69.2	**32.4**	54.3
ASA+ACT	**63.5**	**15.5**	**80.9**	28.9	**86.7**	**92.0**	60.7	**24.1**	**69.3**	30.9	**55.2**

4.2   Experiments on Flickr30k and PASCAL VOC

We use the same experiment setup as ACD [18]: using Flickr30k [20] as source image-description dataset and PASCAL VOC 2012 action classification as target test dataset. Flickr30k contains 30000 images and each image is captioned by 5 sentences. PASCAL VOC 2012 action classification dataset has 10 action categories. We train our full model (ASA + ACT) on Flickr30k and apply the action concepts on PASCAL VOC.

As shown in Table 3, our method outperforms ACD [18] in most categories and by 0.9% in mAP. For example, "ride bike" and "ride horse" are two separate subcategories in our ACT and provide more precise data for training, while ACD [18] may cluster these two with other categories such as "ride skateboard".

5   Conclusion

We presented a two-stage learning framework to learn an Action Concept Tree (ACT) and an Action Semantic Alignment (ASA) model from image-description data. Stage-1 has two steps: (a) ACT is discovered and built by H-ACD algorithm, each node in the tree contains an action name and the relevant images; (b) ASA model is trained by image-description mapping task for parameter initialization. In stage two, we adjust the ASA model to a specific action classification task. The first step is to match the target action categories to the action nodes in ACT discovered in stage-1. After matching, we use the associated data to fine-tune ASA model to this action classification task. Experimental results show that our model outperforms several baseline methods significantly.

Acknowledgement. This research was supported, in part, by the Office of Naval Research under grant N00014-13-1-0493. We would like to thank Chen Sun for valuable discussions.

References

1. Krizhevsky, A., Sutskever, I., Hinton, G.E.: Imagenet classification with deep convolutional neural networks. In: NIPS (2012)
2. Simonyan, K., Zisserman, A.: Very deep convolutional networks for large-scale image recognition. In: ICLR (2015)
3. Krishna, R., Zhu, Y., Groth, O., Johnson, J., Hata, K., Kravitz, J., Chen, S., Kalantidis, Y., Li, L.J., Shamma, D.A., Bernstein, M., Fei-Fei, L.: Visual genome: connecting language and vision using crowdsourced dense image annotations (2016)

4. Guo, G., Lai, A.: A survey on still image based human action recognition. Pattern Recogn. **47**, 3343–3361 (2014)
5. Oquab, M., Bottou, L., Laptev, I., Sivic, J.: Learning and transferring mid-level image representations using convolutional neural networks. In: CVPR (2014)
6. Gkioxari, G., Hariharan, B., Girshick, R., Malik, J.: R-CNNs for pose estimation and action detection. arXiv preprint arxiv:1406.5212 (2014)
7. Gkioxari, G., Girshick, R., Malik, J.: Contextual action recognition with R*CNN. In: ICCV (2015)
8. Chao, Y.W., Wang, Z., He, Y., Wang, J., Deng, J.: Hico: A benchmark for recognizing human-object interactions in images. In: ICCV (2015)
9. Ramanathan, V., Li, C., Deng, J., Han, W., Li, Z., Gu, K., Song, Y., Bengio, S., Rossenberg, C., Fei-Fei, L.: Learning semantic relationships for better action retrieval in images. In: CVPR (2015)
10. Rohrbach, A., Rohrbach, M., Tandon, N., Schiele, B.: A dataset for movie description. In: CVPR (2015)
11. Torabi, A., Pal, C., Larochelle, H., Courville, A.: Using descriptive video services to create a large data source for video annotation research. arXiv preprint arxiv:1503.01070 (2015)
12. Alayrac, J.B., Bojanowski, P., Agrawal, N., Sivic, J., Laptev, I., Lacoste-Julien, S.: Learning from narrated instruction videos (2016)
13. Ramanathan, V., Liang, P., Fei-Fei, L.: Video event understanding using natural language descriptions. In: ICCV (2013)
14. Yu, S.I., Jiang, L., Hauptmann, A.: Instructional videos for unsupervised harvesting and learning of action examples. In: ACM MM (2014)
15. Kuehne, H., Jhuang, H., Garrote, E., Poggio, T., Serre, T.: HMDB: a large video database for human motion recognition. In: ICCV (2011)
16. Reddy, K.K., Shah, M.: Recognizing 50 human action categories of web videos. Mach. Vis. Appl. **24**, 971–981 (2013)
17. Sun, C., Gan, C., Nevatia, R.: Automatic concept discovery from parallel text and visual corpora. In: ICCV (2015)
18. Gao, J., Sun, C., Nevatia, R.: ACD: action concept discovery from image-sentence corpora. In: ICMR (2016)
19. Lin, T.-Y., Maire, M., Belongie, S., Hays, J., Perona, P., Ramanan, D., Dollár, P., Zitnick, C.L.: Microsoft COCO: common objects in context. In: Fleet, D., Pajdla, T., Schiele, B., Tuytelaars, T. (eds.) ECCV 2014. LNCS, vol. 8693, pp. 740–755. Springer, Heidelberg (2014). doi:10.1007/978-3-319-10602-1_48
20. Young, P., Lai, A., Hodosh, M., Hockenmaier, J.: From image descriptions to visual denotations: new similarity metrics for semantic inference over event descriptions. TACL **2**, 67–78 (2014)
21. Donahue, J., Anne Hendricks, L., Guadarrama, S., Rohrbach, M., Venugopalan, S., Saenko, K., Darrell, T.: Long-term recurrent convolutional networks for visual recognition and description. In: CVPR (2015)
22. Vinyals, O., Toshev, A., Bengio, S., Erhan, D.: Show and tell: a neural image caption generator. In: CVPR (2015)
23. Karpathy, A., Fei-Fei, L.: Deep visual-semantic alignments for generating image descriptions. In: CVPR (2015)
24. Hu, R., Xu, H., Rohrbach, M., Feng, J., Saenko, K., Darrell, T.: Natural language object retrieval. In: CVPR (2016)
25. Hu, R., Rohrbach, M., Darrell, T.: Segmentation from natural language expressions. arXiv preprint arxiv:1603.06180 (2016)

26. Miller, G.A.: Wordnet: a lexical database for english. Commun. ACM **38**, 39–41 (1995)
27. Cao, S., Chen, K., Nevatia, R.: Abstraction hierarchy and self annotation update for fine grained activity recognition. In: WACV (2016)
28. Abadi, M., Agarwal, A., Barham, P., Brevdo, E., Chen, Z., Citro, C., Corrado, G.S., Davis, A., Dean, J., Devin, M., Ghemawat, S., Goodfellow, I., Harp, A., Irving, G., Isard, M., Jia, Y., Jozefowicz, R., Kaiser, L., Kudlur, M., Levenberg, J., Mané, D., Monga, R., Moore, S., Murray, D., Olah, C., Schuster, M., Shlens, J., Steiner, B., Sutskever, I., Talwar, K., Tucker, P., Vanhoucke, V., Vasudevan, V., Viégas, F., Vinyals, O., Warden, P., Wattenberg, M., Wicke, M., Yu, Y., Zheng, X.: TensorFlow: large-scale machine learning on heterogeneous systems (2015). https://www.tensorflow.org/
29. Deng, J., Dong, W., Socher, R., Li, L.J., Li, K., Fei-Fei, L.: Imagenet: a large-scale hierarchical image database. In: CVPR (2009)
30. Mikolov, T., Sutskever, I., Chen, K., Corrado, G.S., Dean, J.: Distributed representations of words and phrases and their compositionality. In: NIPS (2013)
31. Kingma, D., Ba, J.: Adam: a method for stochastic optimization. In: ICLR (2015)
32. Fan, R.E., Chang, K.W., Hsieh, C.J., Wang, X.R., Lin, C.J.: LIBLINEAR: a library for large linear classification. JMLR **9**, 1871–1874 (2008)
33. Frome, A., Corrado, G.S., Shlens, J., Bengio, S., Dean, J., Mikolov, T., et al.: Devise: a deep visual-semantic embedding model. In: NIPS (2013)

Pedestrian Color Naming via Convolutional Neural Network

Zhiyi Cheng, Xiaoxiao Li, and Chen Change Loy[✉]

Department of Information Engineering, The Chinese University of Hong Kong,
Hong Kong SAR, China
ccloy@ie.cuhk.edu.hk

Abstract. Color serves as an important cue for many computer vision tasks. Nevertheless, obtaining accurate color description from images is non-trivial due to varying illumination conditions, view angles, and surface reflectance. This is especially true for the challenging problem of pedestrian description in public spaces. We made two contributions in this study: (1) We contribute a large-scale pedestrian color naming dataset with 14,213 hand-labeled images. (2) We address the problem of assigning consistent color name to regions of single object's surface. We propose an end-to-end, pixel-to-pixel convolutional neural network (CNN) for pedestrian color naming. We demonstrate that our Pedestrian Color Naming CNN (PCN-CNN) is superior over existing approaches in providing consistent color names on real-world pedestrian images. In addition, we show the effectiveness of color descriptor extracted from PCN-CNN in complementing existing descriptors for the task of person re-identification. Moreover, we discuss a novel application to retrieve outfit matching and fashion (which could be difficult to be described by keywords) with just a user-provided color sketch.

1 Introduction

Color naming aims at mapping image pixels' RGB values to a pre-defined set of basic color terms[1], *e.g.*, 11 basic color terms defined by Berlin and Kay [5] - black, blue, brown, grey, green, orange, pink, purple, red, white, and yellow. Color names have been widely used as a type of color descriptor for a variety of applications such as image retrieval and image classification [37]. Recent studies [18,25,40] have applied color naming for the task of person re-identification [11,13,17,21,22,42,44] to achieve robust person matching under varying illuminations. Automatic color naming has also been exploited for cloth retrieval and fashion parsing [24].

In this study, we focus on the task of assigning consistent color names to pedestrian images captured from public spaces (see Fig. 1). This task is non-trivial since the observed color of different parts of a pedestrian's body surface

---

[1] A basic color term is defined as being not subsumable to other basic color terms and extensively used in different languages.

© Springer International Publishing AG 2017
S.-H. Lai et al. (Eds.): ACCV 2016, Part II, LNCS 10112, pp. 35–51, 2017.
DOI: 10.1007/978-3-319-54184-6_3

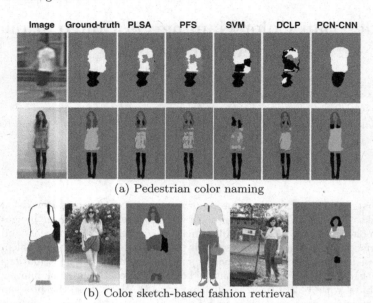

(a) Pedestrian color naming

(b) Color sketch-based fashion retrieval

Fig. 1. (a) State-of-the-art non-learning based color naming method, DCLP [26] and learning-based approaches based on hand-crafted features, including SVM [24] and PLSA [37] fail to extract accurate color names for different regions. In contrast, the proposed Pedestrian Color Naming Convolutional Neural Network (PCN-CNN) generates color labels consistent with the ground truth (PCN-CNN generates color labels over its own predicted foreground region while other methods use ground-truth foreground mask). (b) A meaningful application of our method is the retrieval of outfit matching based on a simple user-provided color sketch (from left to right: sketch, retrieved image, and the corresponding estimated color names map). The application is demonstrated in Sect. 5.3. (Color figure online)

can look totally different under disparate illuminant conditions and view angles. In addition, strong highlights and shadows can make the RGB values of the same surface span from light to dark. Creases and folds in clothing surface can also lead to drastically different predictions of color. Some examples are shown in Fig. 1. Existing methods are not effective for this kind of challenging scenarios. Specifically, some of these approaches are non-learning-based methods [26], they thus cannot effectively capture the uncontrollable variations for specific scenarios. Some other methods rely on hand-crafted features and color histograms, e.g., LAB color space [37], SIFT and HOG features [24], which may have limited expressive power to represent the image content (more details in Sect. 2).

We believe that the key to address the aforementioned problem is a model that is capable of extracting meaningful representation to achieve color constancy [2,6,10], i.e. the capability of inferring the true color distribution intrinsic to the surface. Such a representation needs to be learned from a large-scale training set to ensure robustness for real-world scenes. To this end, we make two main contributions:

- *A large-scale dataset* - Existing color naming datasets either lack of sufficient training samples or do not come with pixel-level annotation (see Sect. 3). To facilitate the learning and evaluation of pedestrian color naming, we introduce a large-scale dataset with careful manual segmentation and region-wise color annotation. The dataset contains 14,213 images in total, which is the largest color naming dataset that we aware to our knowledge. All the images are collected under challenging surveillance scenarios (Market-1501 dataset [43]), with large variations in illumination, highlights, shadow changes, different pedestrian poses and view angles. We show that the dataset is essential for pre-training a color naming deep network for a number of pedestrian-related applications, including person re-identification and cloth retrieval.
- *End-to-end color naming* - We propose a Pedestrian Color Naming Convolutional Neural Network (PCN-CNN) to learn pixel-level color naming. In contrast to existing studies [24,37] that require independent components for feature extraction and color mapping, our CNN-based model is capable of extracting strong features and regressing for color label for each pixel in an end-to-end framework. Conditional random field (CRF) is further adopted to smooth the pixel-wise color predictions. Our network is specially designed to handle images with low resolution, and hence it is well-suited for processing pedestrian images captured from low-resolution surveillance cameras.

Extensive results on the Market-1501 [43] and Colorful-Fashion [24] datasets show the superiority of our approach over existing color naming methods [4, 24,26,37]. We further show the applicability of PCN-CNN in complementing existing visual descriptors for the task of person re-identification (Re-ID). In particular, we demonstrate consistent improvement using the PCN-CNN features in conjunction with different existing Re-ID approaches. In addition, we also highlight an interesting application for outfit matching retrieval. In particular, in the absence of imagery or keyword query, we show that it is possible to retrieve desired fashion images from a gallery through just a simple and convenient 'color sketch'. An example is depicted in Fig. 1(b). Such a color-driven query provides rich region-wise color description and can be used in conjunction with visual attribute-driven query [21,28] for 'zero-shot' retrieval.

2   Related Work

Color Naming. Benavente et al. [4] proposed a pixel-wise color naming model based on lightness and chromaticity distribution, which did not consider cross-pixel relations and intrinsic consistency. Serra et al. [34] and Liu et al. [26] improve the region consistency of color names based on this pixel-wise color naming results. In particular, Serra et al. [34] applied CRF to infer the color intrinsic components from images. They extracted the intrinsic information according to the segmentation results of Ridge Analysis of color Distribution (RAD) [38], and assigned the same color label to pixels connected by a ridge. However, the RAD method only described the RGB histogram distribution and may fail to handle the complicated color distribution. Besides, only with ridge information, the method cannot reliably predict the correct color label from a region if a big portion of the surface's pixels are affected by shadows or highlights. Liu et al. [26] applied the similar CRF model and built a label propagation model where the color labels of pixels in normal region will be propagated to those shadowed and highlighted regions in the same objects' surface. However, their model relied on the detection results of highlights and shadows [16,20,31,36,38] with mainly the intensive and reflectance information, which do not suit for complicated color distribution cases, especially for the challenging pedestrians under real-world settings.

Van de Weijer et al. [37] used LAB histogram features as 'words' and applied them into a Probabilistic Latent Semantic Analysis (PLSA) model to learn for 'topic' color naming. Liu et al. [24] designed a concatenated feature by RGB, LAB color spaces and SIFT, HOG features. Mojsilovic [32] built a multi-level color description model and estimated color naming combined with segmentation. However, this work did not address the issues of shadowed and highlighted regions. All of these hand-crafted features lack robustness to dramatic illumination changes.

Pedestrian Descriptors. Person re-identification [13] aims at recognizing the same individual under different camera views. To tackle the challenging appearance changes by varying viewpoints, illumination and poses, many researchers have proposed different pedestrian descriptors. Gray et al. [15] introduced an ensemble of localized features (ELF) consisting of colors and textures for viewpoint robustness. Layne et al. [21] proposed to use mid-level semantic attributes, fused with low-level features in ELF to obtain improved results. Bazzani et al. [3] exploited three complementary aspects of the human appearance: the overall chromatic content, the spatial arrangement of colors into stable regions, and the presence of recurrent local motifs with high entropy. More recently, a 'mirror representation' [9] is proposed to explicitly model the relation between different view-specific transformations. Chen et al. [7] proposed a Spatially Constrained Similarity function on polynomial feature map and achieved a new state of the art results. Recent studies have explored the illuminant-invariant color distribution descriptors for Re-ID. Kviatkovsky et al. [19] introduced log-chromaticity color space to identify persons under varying scenes. To complement the traditional color information, color naming has been applied to recent studies [18,25,40] and achieved improvements over the state-of-the-art models. Kuo et al. [18] employed the semantic color names learned by [37]. Yang et al. [40] employed the salient color names according to RGB values. However, these applied color naming models did not show region consistency and had limited robustness to dramatic illumination changes.

3   Pedestrian Color Naming Dataset

A well-segmented region-level color naming dataset is essential for both model training and evaluation. A dataset collected from realistic scenes, with diverse illumination, highlights and shadows, and varying view angles, counts heavily to the success of pedestrian color naming.

Table 1. The train/test distribution of 11 basic colors at region level (arranged in alphabetical order) of the Pedestrian Color Naming (PCN) dataset

	Black	Blue	Brown	Grey	Green	Orange	Pink	Purple	Red	White	Yellow
Train	8040	2192	722	3256	1576	318	1138	669	1972	5013	1651
Test	1264	275	113	491	234	37	139	109	249	710	202

There is no public large-scale color naming dataset with pixel-level labels. The Google Color Name [37] and Google-512 datasets [33] contain 1100 and 5632 images, respectively, but both of them are weakly labeled with only image-level color annotations. The Object dataset [26] and Ebay dataset [37] include 350 and 528 images with region-level color annotation. These datasets are far from enough for learning and testing a CNN-based color naming model.

To facilitate the learning of evaluation of pedestrian color naming, we build a new large-scale dataset, named *Pedestrian Color Naming* (PCN) dataset, which contains 14,213 images, each of which hand-labeled with color label for each pixel. The dataset and the annotations can be downloaded at http://mmlab.ie.cuhk.edu.hk/projects/PCN.html.

Image Collection. All images in the PCN dataset are obtained from the Market-1501 dataset [43]. The original Market-1501 dataset consists of pedestrian images of 1,501 identities, captured from a total of six surveillance cameras. Each identity has multiple images with varying scene settings and poses under multiple camera views. These images contain strong highlights and shadows with various illumination conditions and view angles. We carefully select a subset of 14,213 images which have good visibility of the full body and diverse color distribution. We consequently divide the dataset into a training set of 10,913 images, a validation set of 1,500 images and the remaining 1,800 images for testing. Table 1 summarizes the distribution of the different color labels in both the training and test subsets. Note that there may be multiple colors co-exist in the same image. Some colors, namely purple and orange, are relatively lower in numbers since pedestrians tend to wear clothes with more common colors such as black and white.

Super-Pixel-Driven Annotation. Pixel-by-pixel labeling of color labels is a tedious task. We attempted this possibility but found it not scalable. To overcome this problem, we first oversegment each image into 100 super-pixels through the popular SLIC superpixel segmentation method [1]. We found that the super-pixels align well with the object contours most of the time. We then carefully identify the color label for each super-pixel following the 11 color names defined by Berlin and Kay [5], excluding the background, human skin, and hair areas. Note that some super-pixels are originated from the same region (*e.g.* different regions of a pair of jeans). We manually group these super-pixels together to form a single region. Eventually, each coherent region shares the same color label, and the labels for all regions collectively form a label map with the same resolution as of the associated image. Figure 2 depicts some example images and their corresponding pixel-level color label maps.

Fig. 2. Some examples of labeled images in the Pedestrian Color Naming dataset. The images in the first row are the original images and those in the second row are the color label maps where each region is visualized using the corresponding basic colors (black, blue, brown, grey, green, orange, pink, purple, red, white, and yellow). The background region is shown in dark cyan. (Color figure online)

4 Pedestrian Color Naming Convolutional Network

Problem Formulation. Given a pedestrian image $\mathbf{I}$, our goal is to assign each pixel of $\mathbf{I}$ with a specific color name. Specifically, we define a binary latent variable $y_c^i \in \{0, 1\}$, indicating whether an i-th pixel should be named with a color name c, where $\forall c \in \mathcal{C} = \{1, 2, \ldots, 11\}$, representing the 11 basic color names [5].

We approach this problem in a general CRF [12] framework with the unary potentials generated by a deep convolutional network. The energy function of CRF is written as

$$E(\mathbf{y}) = \sum_{\forall i \in \mathcal{V}} U(y_c^i) + \sum_{\forall i,j \in \mathcal{E}} \pi(y_c^i, y_d^j), \tag{1}$$

where $\mathbf{y}$, $\mathcal{V}$, and $\mathcal{E}$, represent a set of latent variables, nodes, and edges in an undirected graph. Here, each node represents a pixel in image $\mathbf{I}$ and each edge captures the relation between pixels. The $U(y_c^i)$ measures the unary cost of assigning a label c to the i-th pixel, and $\pi(y_c^i, y_d^j)$ is the pairwise term that quantifies the penalty of assigning labels c, d to pixels i, j respectively. We define the unary term in Eq. (1) as

$$U(y_c^i) = -\ln p(y_c^i = 1|\mathbf{I}), \tag{2}$$

where $p(y_c^i = 1|\mathbf{I})$ represents the probability of assigning label c to i-th pixel. In this study, we model the probability using PCN-CNN, which will be described next.

For the pairwise term, we let $\pi(y_c^i, y_d^j) = \mu(u, v)D(i, j)$, where $\mu(u, v)$ represents a prior color co-occurrence. Although this prior can be learned from data, to simplify the problem we make a mild assumption that $\mu(u, v) = 1$ for any arbitrary pair of color labels. The $D(i, j)$ measures the distances between pixels,

$$D(i, j) = w_1||f(\mathbf{I}_i) - f(\mathbf{I}_j)||^2 + w_2||(x_i, y_i), (x_i, y_j)||^2, \tag{3}$$

where f is a function that extracts features from the i-th pixel, e.g., RGB values, while (x, y) denote the coordinates of a pixel, and w_1, w_2 are constant weights. The pairwise term encourages pixels that are close and similar to each other to share the same color label.

Network Architecture. Deep convolutional network has shown immense success for various image recognition tasks. Different from existing problems, we need to cope with a few unique challenges. Firstly, we need to deal with the background clutter, which is detrimental to the foreground color prediction. Secondly, our problem requires special care in designing the architecture since pedestrian images are typically low in resolution, e.g. 128 × 64 in the Market-1501 dataset [43]. This challenge is especially crippling since most off-the-shelf deep networks contain pooling layers that could significantly reduce the effective size of the input images. We cannot afford this information loss.

Consequently, we based our solution on the VGG_{16} network [35] but with the following modifications. To handle the background clutter, we additionally consider background as a label and train the network to jointly estimate for both foreground-background segmentation and color naming, resulting in 12-category output. That is, the network output has 11 color names and a background indicator $b \in \{0, 1\}$ to indicate the presence of background at a pixel.

To handle the small input resolution issue, we need to modify the VGG_{16} network. We still initialize the filters in our network with all the learned parameters to make full use of VGG_{16} pre-trained by ImageNet. Nevertheless, for the pixel-to-pixel prediction of low-resolution input, more information should be preserved. Table 2 compares the hyper-parameters of the VGG_{16} network and our network. We use ai and bi to denote the i-th group in Table 2(a) and (b). Our network contains 13 convolutional layers, two max-pooling layers, and the last three layers act as the fully convolutional layers and de-convolutional layers, which generates the final labeling results. As summarized in Table 2, we increase the resolution of convolved data by removing three max-pooling layers from VGG_{16}. As a result, the smallest size of feature map in our model is 32 × 16 (based on input-size of 128 × 64), keeping more information compared with VGG_{16}.

Filters of $b6$ are initialized with the filers of $a7$, where each filter in $a7$ should be convolved with $a5$ on a stride (the stride length is 2). Since the max-pooling layer $a6$ has been removed, the 3 × 3 receptive filed is padded into 5 × 5 with zeros every other parameter in the filter, to keep the resolution identical to one-stride convolution. The following convolutional layers are padded in the similar way. For the fully convolutional layer $b8$, if all the 7×7 parameters are to be applied for initialization, a padded 49×49 receptive filed is needed in the similar way, which needs more padding to the input feature map to keep the output size after up-sampling. Since large zero padding can affect the performance, we down-sample the parameters of receptive-field [8] from 7×7 to 3×3 before applying them for initialization. In this way, the padded 17×17 with zeros from 3×3 is applied in $b8$ as the fully convolutional layer. Finally, the $b10$ layer up-samples the feature maps to 128×64 by bilinear interpolation, and generates the 12-dimensional prediction for each pixel (11 color + background labels).

Table 2. The comparisons between VGG_{16} and our PCN-CNN, as shown in (a) and (b) respectively. The 'fs', '#cha', 'act' and 'size' represent the filter stride size, number of output feature maps, activation function, and size of output feature maps, respectively. And 'conv', 'max', 'dconv', and 'fc' represent the convolution layer, max-pooling layer, deconvolution layer, and fully-connected layer, respectively. The 'relu', 'idn' and 'soft' represent the rectified linear unit, identity and softmax activation functions.

(a) VGG_{16}: $224 \times 224 \times 3$ input image; 1×1000 output labels.

	1	2	3	4	5	6	7	8	9	10	11	12
layer	2×conv	max	2×conv	max	3×conv	max	3×conv	max	3×conv	max	2×fc	fc
fs	3-1	2-2	3-1	2-2	3-1	2-2	3-1	2-2	3-1	2-2	-	-
#cha	64	64	128	128	256	256	512	512	512	512	1	1
act	relu	idn	relu	idn	relu	idn	relu	idn	relu	idn	relu	soft
size	224	112	112	56	56	28	28	14	14	7	4096	1000

(b) Our PCN-CNN: $128 \times 64 \times 3$ input image; $128 \times 64 \times 12$ output label maps.

	1	2	3	4	5	6	7	8	9	10
layer	2×conv	max	2×conv	max	3×conv	3×conv	3×conv	conv	conv	dconv
fs	3-1	2-2	3-1	2-2	3-1	5-1	9-1	17-1	1-1	1-1
#cha	64	64	128	128	256	512	512	4096	4096	12
act	relu	idn	relu	idn	relu	relu	relu	relu	relu	soft
size	128×64	64×32	64×32	32×16	32×16	32×16	32×16	32×16	32×16	128×64

It is worth pointing out that deep convolutional network has been widely used for image segmentation [8,27,29]. Differs from these prior studies, our work is the first attempt to use CNN for color naming. In terms of network architecture, our network shares some similarity to the Deep Parsing Network (DPN) [27]. Unlike DPN that accepts input image of resolution 512×512, we design our network to accommodate for small pedestrian images and remove pooling layers to avoid information loss. We attempted to enlarge pedestrian images to fit DPN's require-ment but the performance of this alternative is inferior to that achieved by our final design.

Training details are given as follows. We start with an initial learning rate of 0.001, and reduce it by a factor of 10 at every 5K iterations. We use a momentum of 0.9, and mini-batches of 12 images.

5   Experiments

In this section, we first evaluate PCN-CNN's performance for color naming. We also examine the effectiveness of color names descriptor extracted from PCN-CNN for the task of person re-identification. Furthermore, we show an interesting application with PCN-CNN, using only simple sketches as probe to retrieve desired outfit matching of fashion images from a real-world image gallery.

5.1   Pedestrian Color Naming

In this experiment, we analyze PCN-CNN's performance for pedestrian color naming.

Datasets. We perform evaluations on the proposed PCN dataset (relabelled Market 1501 dataset [43]) and a cloth dataset, Colorful-Fashion [24], both of which have a test subset of 1,800 and 2,682 images, respectively. The PCN dataset is challenging due to its low image resolution (128×64) and large variations in terms of illumination and pedestrian pose. The Colorful-Fashion dataset contains images with a higher resolution (600×400), but the cloth patterns are more complex and colorful. Images in the Colorful-Fashion dataset comes with region-wise color labels. Note that the dataset also annotates hair pixels with color names, we therefore include the hair region estimation in our evaluation. For the PCN dataset, we label the color names based on the procedure described in Sect. 3.

Evaluation Metrics. To measure the performance of both the pixel-wise and region-wise accuracies, we apply two metrics for model evaluation:
(1) *Pixel Annotation Score* (PNS) - this score [37] measures the percentage of correctly predicted color names at pixel level. We average the PNS for all regions as the final score to measure the consistency of color naming.
(2) *Region Annotation Score* (RNS) - each region's color label is specified by its dominant color names prediction of pixels. We then calculate the averaged accuracy of prediction at the region level.

Results. We compare our PCN-CNN against with state-of-the-art methods, including PLSA [37], PFS [4], SVM-based color classifier [24] and DCLP [26]. Besides, we also adopt CRF to smooth PCN-CNN color names prediction and evaluate the performance. For a better foreground estimation on pedestrian images, the PCN-CNN is first pre-trained on the large-scale pedestrian parsing dataset PPSS [30], which encourages the network to generate binary map composed of pedestrian region and the background. The pre-trained parameters of PCN-CNN are then fine-tuned on the training partition of the PCN dataset and Colorful-Fashion dataset, respectively, for the respective tests on the two datasets. Likewise, all learning based methods, *e.g.* PLSA and SVM, are retrained using the same training partition employed by PCN-CNN to ensure a fair comparison. It is worth pointing out that during the evaluation of PCN-CNN, we employ the foreground masks generated by itself before applying the evaluation metrics. For other baselines (PLSA, PFS, SVM, and DCLP), we use the ground-truth masks for this

Table 3. Performance over PCN and Colorful-Fashion test set. PNS and RNS denote the averaged pixel annotation score and region annotation score respectively. The smoothed color names prediction is denoted by PCN-CNN + CRF.

Method	PCN		Colorful-Fashion	
	PNS	RNS	PNS	RNS
PLSA [37]	63.1	68.4	57.4	71.4
PFS [4]	61.1	68.5	48.6	60.5
SVM [24]	62.8	62.2	43.5	45.4
DCLP [26]	56.8	62.0	47.8	54.8
PCN-CNN	74.1	80.3	70.2	81.8
PCN-CNN + CRF	**74.3**	**80.8**	**71.1**	**81.9**

purpose. Given the more accurate masks compared to PCN-CNN generated ones, these baselines therefore gain additional advantages than PCN-CNN.

Table 3 and Fig. 3 show the performance comparison and confusion matrix (based on RNS), respectively, among different methods. Qualitative results are provided in Fig. 4. As shown in the experimental results, our model achieves superior performance in both PNS and RNS metrics, with outstanding robustness to shadows and highlights, creases and folds. Adding CRF to PCN-CNN further boosts its performance.

5.2   Color Naming for Person Re-identification

Pedestrian color naming provides a powerful feature for person re-identification, even in low-resolution images, due to its robustness to varying illumination and view angles. A robust color naming model with good consistency helps to describe people more accurately by ignoring the minor change in RGB values. In this section, we combine the region-level color names generated by PCN-CNN with several existing visual descriptors for the task of person re-identification, and test the performance on the widely used VIPeR dataset [14].

Feature Representation. Similar to [23], we first partition a pedestrian image into six equal-size horizontal stripes, represented as $H = [h_1, \ldots, h_6]^\mathsf{T}$. For the i-th part h_i, we use a histogram of color names as the feature representation, resulting into a 66-dimensional descriptor for all the parts. The c-th bin of a histogram h_i denotes the probability of all pixels in the corresponding part being assigned to color name c. To minimize the influence of background clutter, we only extract the color distribution of the foreground region. The estimated feature is called pedestrian color naming (PCN) descriptor in the following session. We concatenate the PCN descriptor with several representative visual descriptors for person re-identification. These include one of the most widely used features called ensemble of localized features (ELF) [15,21]; a pure color-based features, named salient color names based color descriptor (SCNCD) [40]; a recent advanced features known as mirror representation [9]. The original ELF, SCNCD and 'mirror representation' descriptors and those concatenated with PCN descriptor are fed

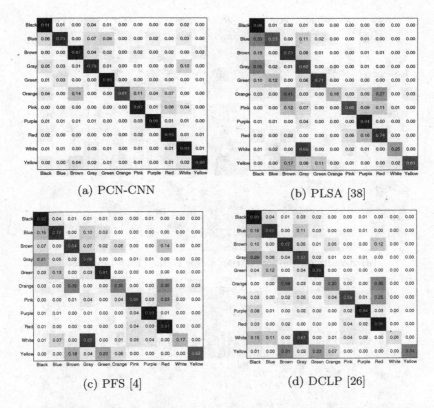

Fig. 3. Confusion matrix of color naming (regional level) on the Pedestrian Color Naming dataset.

into the KMFA metric learning method [39] for matching. Moreover, the PCN feature is also fed into a recent outperforming similarity learning method with spatial constraints (SCSP) [7], fused with other originally used visual cues.

Experiment Settings. The VIPeR dataset contain 632 pedestrian image pairs, with varying illumination conditions and view angles. Each pedestrian has two images per camera view. All the images are normalized to 128 × 48 pixels. We randomly choose half of the image pairs for training and the others for testing. This procedure is repeated for 10 evaluation trials. Averaged performance is measured over the trials by using the typical cumulative matching characteristic (CMC) curve. In particular, we report the rank k matching rate, which refers to the percentage of probe images that are correctly matched with the true positives in the gallery set in the top k rank.

Results. As can be observed from Table 4, the PCN descriptor is capable of improving the performance of a wide range of existing Re-ID visual descriptors, from ensemble of color/texture features (ELF), pure color based features (SCNCD), as well as the more elaborated mirror representation. Moreover, a new state-of-the-art accuracy can be achieved by SCSP learning method, when

(a) Results on the Pedestrian Color Name (PCN) dataset

(b) Results on Colorful-fashion dataset

Fig. 4. Qualitative results on the PCN and Colorful-Fashion datasets. The background is indicated by dark cyan. PCN-CNN generates color labels over its own learned foreground region while other methods use ground-truth foreground mask. (Color figure online)

concatenating with the PCN descriptor. It is interesting to see that PCN descriptors yields large improvement to the SCNCD method, which is also based on color names. The results suggest the robustness of our approach in complementing existing pedestrian descriptors.

Table 4. Comparative results between original person re-identification descriptors vs. descriptors enhanced with PCN descriptor. Results are reported on the VIPeR dataset.

Rank k	1	5	10	20
ELF [15]	23.77	51.17	64.62	78.89
ELF [15] + PCN	**36.36**	**68.92**	**82.69**	**92.63**
SCNCD [40]	21.33	38.86	49.02	59.91
SCNCD [40] + PCN	**28.45**	**52.09**	**64.11**	**75.03**
Mirror-KMFA [9]	42.97	75.82	87.28	94.84
Mirror-KMFA [9] + PCN	**45.03**	**77.56**	**89.05**	**96.04**
SCSP [7]	53.54	82.59	**91.49**	96.65
SCSP [7] + PCN	**54.24**	**82.78**	91.36	**99.08**

5.3   Color Naming for Zero-Shot Cloth Retrieval

One may relatively often has to do with combining different colors of shirts, pants, and shoes together. Or one might want to purchase a particular piece of garment in mind but do not know how to describe its combination of colors and patterns. Instead of elucidating a long textual description of it, one could just draw a sketch! A recent paper [41] has applied this idea for fine-grained shoe retrieval using monochrome sketches. In this section, we show the possibility to 'retrieve with colors'.

Specifically, one simply needs to paint with a few strokes the desired color on a sketch with specific combinations and patterns. The sketch can then serve as a query for cloth/fashion retrieval. This is possible through the following steps: we process a color sketch using PCN-CNN to transform it into a map with 11 color names, and further convert it into a PCN histogram (see Sect. 5.2). We assume all the gallery images have been processed in the same way. We then apply histogram intersection to measure the similarity of features for retrieval.

Experiment Settings. A total of 80 images with rich color and complex patterns are selected from the Colorful-Fashion dataset [24], and we ask volunteers to draw for the corresponding color sketches. The task is to use the sketch as query and correctly retrieve the true image among the 2,682 test images of Colorful-Fashion dataset. The top-k retrieval accuracy is adopted as the metric.

Results. Table 5 shows the cloth retrieval results with different color naming models. Thanks to the robustness of PCN-CNN, our method achieves an impressive top-1 retrieval rate of 42.86%, surpassing other baselines. Some qualitative results are shown in Fig. 5, in which we compare the retrieved results and generated color map of our PCN-CNN and PLSA. With poorer region-level consistency compared to PCN-CNN, which is critical for the cloth retrieval task, PLSA can easily fail to retrieve the ground-truth matching clothes with strong highlights and shadows.

Table 5. Top-k retrieval accuracy on Colorful-Fashion dataset (a subset is selected, see text for details) using color sketch as query.

Rank k	1	5	10	20
PLSA [37]	6.49	12.99	16.99	23.38
PFS [4]	6.48	12.98	18.18	20.78
PCN-CNN	**42.86**	**68.83**	**72.73**	**83.12**

Fig. 5. We show the top-5 retrieval results with sketches as probes, using PCN-CNN and PLSA, respectively. The retrieved images highlighted with red boundary represent the ground-truth matching cloth images. (Color figure online)

6   Conclusion

We have presented an end-to-end, pixel-to-pixel convolutional neural network for pedestrian color naming, named PCN-CNN. To facilitate model training and evaluation, we have introduced a large-scale pedestrian color naming dataset, containing 14,213 images with carefully labeled pixel-level color names. Extensive experiments show that the PCN-CNN is capable of generating consistent color name to clothing surfaces regardless of large variations in clothing material and illumination. The PCN descriptor extracted from the model is not only useful for complementing existing pedestrian descriptors, but also generalizable for sketch-to-image retrieval.

Acknowledgement. We would like to show our gratitude to the authors of [9], for sharing their features and codes of matching procedure for the person re-identification experiments.

References

1. Achanta, R., Shaji, A., Smith, K., Lucchi, A., Fua, P., Susstrunk, S.: SLIC super-pixels compared to state-of-the-art superpixel methods. IEEE Trans. Pattern Anal. Mach. Intell. (PAMI) **34**(11), 2274–2282 (2012)
2. Barron, J.T.: Convolutional color constancy. In: International Conference on Computer Vision (ICCV) (2015)
3. Bazzani, L., Cristani, M., Murino, V.: Symmetry-driven accumulation of local features for human characterization and re-identification. Comput. Vis. Image Underst. **117**(2), 130–144 (2013)
4. Benavente, R., Vanrell, M., Baldrich, R.: Parametric fuzzy sets for automatic color naming. JOSA A **25**(10), 2582–2593 (2008)
5. Berlin, B., Kay, P.: Basic Color Terms: Their Universality and Evolution. University of California Press, Berkeley (1991)
6. Bianco, S., Cusano, C., Schettini, R.: Single and multiple illuminant estimation using convolutional neural networks (2015). arXiv preprint arXiv:1508.00998
7. Chen, D., Yuan, Z., Chen, B., Zheng, N.: Similarity learning with spatial constraints for person re-identification. In: Conference on Computer Vision and Pattern Recognition (CVPR) (2016)
8. Chen, L.C., Papandreou, G., Kokkinos, I., Murphy, K., Yuille, A.L.: Semantic image segmentation with deep convolutional nets and fully connected CRFs (2014). arXiv preprint arXiv:1412.7062
9. Chen, Y.C., Zheng, W.S., Lai, J.: Mirror representation for modeling view-specific transform in person re-identification. In: International Joint Conference on Artificial Intelligence (IJCAI) (2015)
10. Cheng, D., Price, B., Cohen, S., Brown, M.S.: Effective learning-based illuminant estimation using simple features. In: Conference on Computer Vision and Pattern Recognition (CVPR) (2015)
11. Farenzena, M., Bazzani, L., Perina, A., Murino, V., Cristani, M.: Person re-identification by symmetry-driven accumulation of local features. In: Conference on Computer Vision and Pattern Recognition (CVPR) (2010)
12. Freeman, W.T., Pasztor, E.C., Carmichael, O.T.: Learning low-level vision. Int. J. Comput. Vis. (IJCV) **40**(1), 25–47 (2000)
13. Gong, S., Cristani, M., Yan, S., Loy, C.C.: Person Re-Identification. Springer, London (2014)
14. Gray, D., Brennan, S., Tao, H.: Evaluating appearance models for recognition, reacquisition, and tracking. In: International Workshop on Performance Evaluation for Tracking and Surveillance (2007)
15. Gray, D., Tao, H.: Viewpoint invariant pedestrian recognition with an ensemble of localized features. In: Forsyth, D., Torr, P., Zisserman, A. (eds.) ECCV 2008. LNCS, vol. 5302, pp. 262–275. Springer, Heidelberg (2008). doi:10.1007/978-3-540-88682-2_21
16. Guo, R., Dai, Q., Hoiem, D.: Single-image shadow detection and removal using paired regions. In: Conference on Computer Vision and Pattern Recognition (CVPR), pp. 2033–2040 (2011)
17. Hirzer, M., Roth, P.M., Köstinger, M., Bischof, H.: Relaxed pairwise learned metric for person re-identification. In: Fitzgibbon, A., Lazebnik, S., Perona, P., Sato, Y., Schmid, C. (eds.) ECCV 2012. LNCS, vol. 7577, pp. 780–793. Springer, Heidelberg (2012). doi:10.1007/978-3-642-33783-3_56

18. Kuo, C.H., Khamis, S., Shet, V.: Person re-identification using semantic color names and rankboost. In: Winter Conference on Applications of Computer Vision (WACV) (2013)
19. Kviatkovsky, I., Adam, A., Rivlin, E.: Color invariants for person reidentification. IEEE Trans. Pattern Anal. Mach. Intell. **35**(7), 1622–1634 (2013)
20. Lalonde, J.-F., Efros, A.A., Narasimhan, S.G.: Detecting ground shadows in outdoor consumer photographs. In: Daniilidis, K., Maragos, P., Paragios, N. (eds.) ECCV 2010. LNCS, vol. 6312, pp. 322–335. Springer, Heidelberg (2010). doi:10.1007/978-3-642-15552-9_24
21. Layne, R., Hospedales, T.M., Gong, S., Mary, Q.: Person re-identification by attributes. In: British Machine Vision Conference (BMVC) (2012)
22. Liao, S., Hu, Y., Zhu, X., Li, S.Z.: Person re-identification by local maximal occurrence representation and metric learning. In: Conference on Computer Vision and Pattern Recognition (CVPR) (2015)
23. Liu, C., Gong, S., Loy, C.C., Lin, X.: Person re-identification: what features are important? In: European Conference on Computer Vision Workshop (2012)
24. Liu, S., Feng, J., Domokos, C., Xu, H., Huang, J., Hu, Z., Yan, S.: Fashion parsing with weak color-category labels. IEEE Trans. Multimedia **16**(1), 253–265 (2014)
25. Liu, X., Wang, H., Wu, Y., Yang, J., Yang, M.H.: An ensemble color model for human re-identification. In: Winter Conference on Applications of Computer Vision (WACV) (2015)
26. Liu, Y., Yuan, Z., Chen, B., Xue, J., Zheng, N.: Illumination robust color naming via label propagation. In: International Conference on Computer Vision (ICCV) (2015)
27. Liu, Z., Li, X., Luo, P., Loy, C.C., Tang, X.: Semantic image segmentation via deep parsing network. In: International Conference on Computer Vision (ICCV) (2015)
28. Liu, Z., Luo, P., Qiu, S., Wang, X., Tang, X.: DeepFashion: powering robust clothes recognition and retrieval with rich annotations. In: Conference on Computer Vision and Pattern Recognition (CVPR) (2016)
29. Long, J., Shelhamer, E., Darrell, T.: Fully convolutional networks for semantic segmentation. In: Conference on Computer Vision and Pattern Recognition (CVPR) (2015)
30. Luo, P., Wang, X., Tang, X.: Pedestrian parsing via deep decompositional network. In: Proceedings of IEEE International Conference on Computer Vision, pp. 2648–2655 (2013)
31. McHenry, K., Ponce, J., Forsyth, D.: Finding glass. In: IEEE Computer Society Conference on Computer Vision and Pattern Recognition, CVPR 2005, vol. 2, pp. 973–979. IEEE (2005)
32. Mojsilovic, A.: A computational model for color naming and describing color composition of images. IEEE Trans. Image Process. **14**(5), 690–699 (2005)
33. Schauerte, B., Fink, G.A.: Web-based learning of naturalized color models for human-machine interaction. In: International Conference on Digital Image Computing: Techniques and Applications (2010)
34. Serra, M., Penacchio, O., Benavente, R., Vanrell, M.: Names and shades of color for intrinsic image estimation. In: Conference on Computer Vision and Pattern Recognition (CVPR), pp. 278–285 (2012)
35. Simonyan, K., Zisserman, A.: Very deep convolutional networks for large-scale image recognition (2014). arXiv preprint arXiv:1409.1556
36. Tan, R.T., Ikeuchi, K.: Separating reflection components of textured surfaces using a single image. IEEE Trans. Pattern Anal. Mach. Intell. **27**(2), 178–193 (2005)
37. Van De Weijer, J., Schmid, C., Verbeek, J., Larlus, D.: Learning color names for real-world applications. IEEE Trans. Image Process. **18**(7), 1512–1523 (2009)

38. Vazquez, E., Baldrich, R., Van de Weijer, J., Vanrell, M.: Describing reflectances for color segmentation robust to shadows, highlights, and textures. IEEE Trans. Pattern Anal. Mach. Intell. **33**(5), 917–930 (2011)
39. Yan, S., Xu, D., Zhang, B., Zhang, H.J., Yang, Q., Lin, S.: Graph embedding and extensions: a general framework for dimensionality reduction. IEEE Trans. Pattern Anal. Mach. Intell. **29**(1), 40–51 (2007)
40. Yang, Y., Yang, J., Yan, J., Liao, S., Yi, D., Li, S.Z.: Salient color names for person re-identification. In: Fleet, D., Pajdla, T., Schiele, B., Tuytelaars, T. (eds.) ECCV 2014. LNCS, vol. 8689, pp. 536–551. Springer, Heidelberg (2014). doi:10. 1007/978-3-319-10590-1_35
41. Yu, Q., Liu, F., Song, Y., Xiang, T., Hospedales, T.M., Loy, C.C.: Sketch me that shoe. In: Conference on Computer Vision and Pattern Recognition (CVPR) (2016)
42. Zhao, R., Ouyang, W., Wang, X.: Person re-identification by salience matching. In: International Conference on Computer Vision (ICCV) (2013)
43. Zheng, L., Shen, L., Tian, L., Wang, S., Wang, J., Tian, Q.: Scalable person re-identification: a benchmark. In: International Conference on Computer Vision (ICCV) (2015)
44. Zheng, W.S., Gong, S., Xiang, T.: Person re-identification by probabilistic relative distance comparison. In: Conference on Computer Vision and Pattern Recognition (CVPR) (2011)

Speed Invariance vs. Stability: Cross-Speed Gait Recognition Using Single-Support Gait Energy Image

Chi Xu[1,2], Yasushi Makihara[2(✉)], Xiang Li[1,2], Yasushi Yagi[2], and Jianfeng Lu[1]

[1] School of Computer Science and Engineering, Nanjing University
of Science and Technology, Nanjing, China
xuchisherry@gmail.com, lixiangmzlx@gmail.com, lujf@mail.njust.edu.cn
[2] Institute of Scientific and Industrial Research, Osaka University, Osaka, Japan
{makihara,yagi}@am.sanken.osaka-u.ac.jp

Abstract. Gait recognition has recently attracted much attention since it can identify person at a distance without subject cooperation. Walking speed changes, however, cause gait changes in appearance, which significantly drops performance of gait recognition. Considering a speed-invariant property at single-support phases where stride change due to speed changes are mitigated, and a stability against phase estimation error and segmentation noise by aggregating multiple phases inspired by gait energy image (GEI), we propose a speed-invariant gait representation called single-support GEI (SSGEI), which realizes a good trade-off between the speed invariance and the stability by combining single-support phases and GEI concept. For this purpose, we firstly find out the optimal duration around single support phases using a training set so as to well balance the speed invariance and the stability. We then extract SSGEI by aggregating multiple single-support frames. Finally, we combine the proposed SSGEI with subsequent Gabor filters and metric learning for better performance. Experiments on the publicly available OU-ISIR Treadmill Dataset A composed of the largest speed variations demonstrated that the proposed method yielded 99.33% rank-1 identification rate on average for cross-speed gait recognition, which outperforms the other state-of-the-arts, and realized a low computational cost as well.

1 Introduction

Biometric person authentication has recently gained a growing demand in many applications, such as border control at an airport, access control to an amusement park, owner authentication for a bank card. Compared with physiological biometric cues such as DNA, fingerprint, iris, and face, gait has advantages in terms that it is difficult to be obscured and imitated. Moreover, it is possible to identifying a person from his/her gait at a large distance from a camera (*e.g.*, CCTV installed in the street) without subject cooperation, since gait recognition works even with relatively low-resolution images [1]. Gait recognition has therefore attracted considerable attention as a unique cue to authenticate a person from CCTV footage for surveillance and forensics [2–4].

© Springer International Publishing AG 2017
S.-H. Lai et al. (Eds.): ACCV 2016, Part II, LNCS 10112, pp. 52–67, 2017.
DOI: 10.1007/978-3-319-54184-6_4

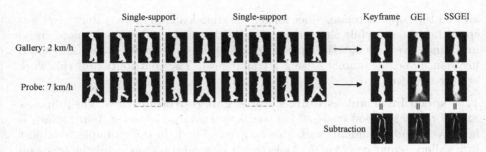

Fig. 1. Comparison of Keyframe, GEI and SSGEI. We choose 9 frames from a period evenly in both gallery (2 km/h) and probe (7 km/h) sequence. The corresponding single Keyframe, GEI and SSGEI features are shown in the right. The subtraction image for each feature are shown in the bottom. They illustrate that SSGEI can reduce such appearance differences caused by posture change, phase difference and speed change in the same time.

Involving uncooperative subjects in gait recognition, means that gait may be affected by various covariates, including but not limited to views, shoes, surfaces, clothing, carriages, and walking speed [5,6]. Among these covariates, walking speed is one of the most common challenging factors and also often observed in real scenes (*e.g.*, a perpetrator running out of a criminal scene). Since the change of walking speed causes the change of gait features in particular in dynamic ones like gait period, arm swing, and stride length, which may significantly drop the performance of gait recognition. In fact, many of popular gait descriptors such as gait energy image (GEI) [7], frequency-domain feature [8], chrono-gait image [9], gait flow image [10], does not work for cross-speed gait recognition if they are directly applied.

Hence, cross-speed gait recognition enjoy a rich body of literatures [11–20]. While they successfully mitigate the speed effect to some extent, yet most of them do not work well for larger speed changes, or suffer from high computational cost, which is an important problem in real-world scenarios.

Among them, use of single support phase [16] worth investigating more details. This is because the change of walking speed mainly affects the dynamic parts such as arm swing and stride length, which are the most outstanding at double support phases, and hence such effects are considerably mitigated at the single support phases where the limbs are the most closed as shown in Fig. 1. In other words, the single-support phases provide promising keyframes for speed invariance. A single keyframe at a single support phase itself may, however, be easily affected by phase (gait stance) estimation error, silhouette segmentation noises, and temporary posture changes, which also drops the gait recognition performance.

To overcome these defects, we propose a speed-invariant as well as stable gait representation called single-support GEI (SSGEI) for cross-speed gait recognition. Inspired by an idea of aggregating multiple frames for silhouette noise reduction in GEI [7], we also aggregate multiple frames of a certain duration around the support phase. Since longer duration leads to more stability but less speed invariance, while shorter duration leads to less stability but more speed invariance, we find out the optimal duration so as to well balance the speed invariance and the stability using a training set. The contribution of this work are three-folded.

1. A speed invariant as well as stable gait representation. The proposed SSGEI realizes a good trade-off between speed invariance and stability, which is intuitively understandable with an example in Fig. 1. In this example, a subject in a gallery sequence (2 km/h) looks down in several frames, while he keeps on walking normally in a probe sequence (7 km/h). In addition, selected keyframe of single support phase have a slight phase difference. Affected by such temporary posture change and phase difference, the difference of keyframes becomes large. On the other hand, GEI can mitigate such temporary posture change and phase difference, although it directly affected by speed variation in particular in dynamic parts in stride and arm swings. Compared these two, we observe that the cross-speed difference of the proposed SSGEI is well suppressed by balancing the speed invariance and stability, which are derived from concepts of keyframes at single support phase and aggregation in GEI, respectively.

2. State-of-the-art accuracy for cross-speed gait recognition. The proposed SSGEI in conjunction with Gabor filters and a standard metric learning technique yielded the best accuracy both in terms of verification and identification scenarios, compared with other state-of-the-arts approaches to cross-speed gait recognition, through experiments on publicly available OU-ISIR Treadmill Dataset A containing the largest speed variations.

3. Low computational cost. The proposed method is also executable with a low computational cost due to its simplicity, which is more applicable in real-world surveillance applications, while the state-of-the-art requires relatively high computational cost.

2   Gait Recognition Using SSGEI

2.1   Representation

As a preprocess, given input images, gait silhouettes have been extracted by background subtraction-based graph-cut segmentation [21], and then normalized by the height and registered by the region center to obtain size-normalized and registered silhouette sequences [8].

We then detect a gait period from lower body parts of the size-normalized silhouette sequence. Given a body height H, the vertical position of knee was suggested to be set to $0.285H^1$ in [22] based on statistics of anatomical data. We then compute a temporal series of the width of lower body from the foot bottom to the knee and then find the local maxima and minima as double support

---

[1] The vertical positions of the foot bottom and the head top are represented as 0 and H, respectively, in this coordinate system.

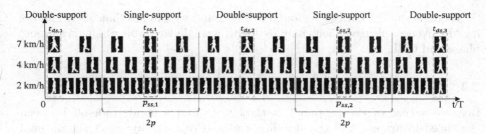

Fig. 2. Gait period setting and duration for SSGEI along with the size-normalized and registered silhouette sequence from three different walking speeds. The horizontal axis t/T means non-dimensional time normalized by the gait period T. Note that frame intervals are different among the walking speeds due to gait period difference. The non-dimensional time of two keyframes at single-support phases are represented by $p_{ss,k}(k = 1, 2)$. Two parts of multiple single-support phases within the range $[p_{ss,k} - p, p_{ss,k} + p](k = 1, 2)$ are selected to compose subsequences for constructing SSGEI, where p is a hyper parameter for duration selection.

phases and single support phases, respectively. Thus, we can set a gait period T [frames] so at that it starts from a double support phase ($t = t_{ds,1} = 0$), then goes through two single-support phases ($t = t_{ss,1}$ and $t = t_{ss,2}$) and another in-between double-support phase ($t = t_{ds,2}$), and finally ends with the third double-support phase ($t = t_{ds,3} = T$), as shown in Fig. 2.

In addition, we convert a time $t \in \mathbb{Z}$ [frames] into a non-dimensional time $p = t/T \in \mathbb{R}$ normalized with the period T so as that the duration around single support phases can be defined in a rate-invariant way against walking speeds. Assume that we take $2p$ duration around the single support phases $p_{ss,k}(k = 1, 2)$ in the non-dimensional time domain, the duration around the k-th single support phase is defined as $[p_{ss,k} - p, p_{ss,k} + p]$. Note that the duration parameter p is subject to $0 < p \leqslant 1/4$ (the duration will cover the whole period in case of $p = 1/4$).

Once we define the durations, we can convert them back to the original time domain and obtain the starting and ending frames for the k-th duration as $t_{ss,k}^s(p) = \lceil (p_{ss,k} - p)T \rceil$ and $t_{ss,k}^e(p) = \lfloor (p_{ss,k} + p)T \rfloor$, respectively, where $\lceil \cdot \rceil$ and $\lfloor \cdot \rfloor$ are ceiling and floor functions.

Now, we can define SSGEI based on the durations. Let a binary silhouette value at the position (x, y) at the t-th frame in the size-normalized and registered silhouette sequence, be $I(x, y, t)$, where 0 and 1 indicate background and foreground, respectively. We then compute SSGEI $S(x, y; p)$ with the duration parameter p as

$$S(x, y; p) = \frac{1}{2} \sum_{k=1}^{2} \frac{1}{t_{ss,k}^e(p) - t_{ss,k}^s(p) + 1} \sum_{t=t_{ss,k}^s(p)}^{t_{ss,k}^e(p)} I(x, y, t). \qquad (1)$$

Examples of SSGEI can be found in Fig. 1 and we can see that SSGEI shows its effectiveness clearly when compared with a single keyframe at single support phase and GEI.

2.2   The Optimal Duration Estimation

Because a core of the proposed method is to find a good trade-off between the speed invariance and the stability, we need to carefully select the optimal duration parameter p. For this purpose, we introduce a well-know criterion for discrimination capability, *i.e.*, Fisher ratio of between-class distance and within-class distance using a training set including speed variations.

Suppose that the training set is composed of a set of SSGEIs $\{S(p)_{i,j} \in \mathbb{R}^{H_S \times W_S}\}(i = 1, \ldots, N_c, j = 1, \ldots, n_i)$, where N_c and n_i are the number of training subjects and the number of training samples for the i-th training subject, and W_S and H_S are the width and the height of the SSGEI. We then compute summations of within-class distances and between-class distances as

$$D_W(p) = \sum_{i=1}^{N_c} \sum_{j=1}^{n_i} \|S_{i,j}(p) - \bar{S}_i(p)\|_F^2 \tag{2}$$

$$D_B(p) = \sum_{i=1}^{N_c} n_i \|\bar{S}_i(p) - \bar{S}(p)\|_F^2, \tag{3}$$

where $\| \cdot \|_F$ is Frobenius norm for a matrix, $\bar{S}_i(p)$ and $\bar{S}(p)$ are the i-th class mean and total mean which are given as

$$\bar{S}_i(p) = \frac{1}{n_i} \sum_{j=1}^{n_i} S_{i,j}(p) \tag{4}$$

$$\bar{S}(p) = \frac{\sum_{i=1}^{N_c} n_i \bar{S}_i(p)}{\sum_{i=1}^{N_c} n_c}. \tag{5}$$

Consequently, the optimal duration parameter p^* is obtained so as to make Fisher ratio of the between-class distances and the within-class distances be maximized as

$$p^* = \arg\max_p \frac{D_B(p)}{D_W(p)}. \tag{6}$$

2.3   Filtering as Postprocess

Recently, the Gabor-based feature has been demonstrated to be very effective for gait recognition [20, 23, 24], since Gabor-functions-based image decomposition is biologically relevant to image understanding and recognition as reported in [23, 25]. We therefore also introduce the Gabor filtering as a postprocess for the proposed SSGEI (referred to as Gabor-SSGEI later).

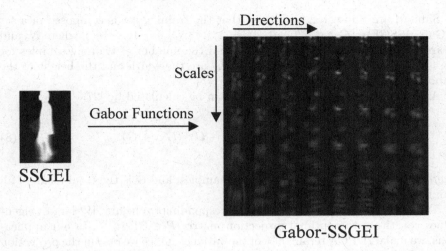

Gabor-SSGEI

Fig. 3. Example of Gabor-SSGEI. Here we choose Gabor kernel functions of 8 directions and 5 scales.

The Gabor wavelets can be defined as [23]

$$\psi_{s,d}(\boldsymbol{z}) = \frac{|\bar{k}_{s,d}|^2}{\delta^2} e^{-\frac{|\bar{k}_{s,d}|^2\|\boldsymbol{z}\|^2}{2\delta^2}} [e^{\boldsymbol{k}(i\bar{k}_{s,d})\cdot\boldsymbol{z}} - e^{-\frac{\delta^2}{2}}], \tag{7}$$

where $\boldsymbol{z} = [x, y]^T$ is a vector representing the spatial location in Gabor kernel window, i is an imaginary unit. $\boldsymbol{k}(\cdot)$ is a function to transform a complex number to a two-dimensional real vector. Moreover, $\bar{k}_{s,d} = k_s e^{i\phi_d}$ determines the scale and direction of Gabor functions, where $k_s = k_{max}/f^s$, with $k_{max} = \pi/2$, and k_{max} is the maximum frequency, and f is the spacing factor between kernels in the frequency domain [26]. Consequently, the Gabor kernels in Eq. (7) are self-similar and each kernel is a product of a Gaussian envelope and a complex plane wave.

After acquiring Gabor kernel functions of s scales and d directions, we convolve the SSGEI with Gabor functions. Similar to [24], we downsample each Gabor-filtered image from $M \times N$ to $\lfloor M/2 \times N/2 \rfloor$ for lower computational cost. Afterwards, all the Gabor-filtered images are aligned to represent the final feature Gabor-SSGEI, with rows show different scales and the columns show different directions. The example can be found in Fig. 3.

2.4   Metric Learning

Because direct matching in the original high dimensional feature space often leads to accuracy degradation as well as high computational cost, we employ two-dimensional principle component analysis (2DPCA) to reduce the feature dimension in the column direction.

Similarly to Subsect. 2.2, suppose that the training set is composed of a set of Gabor-SSGEIs $\{G_{i,j} \in \mathbb{R}^{H_G \times W_G}\}(i = 1, \ldots, N_c, j = 1, \ldots, n_i)$, where N_c and n_i are the number of training subjects and the number of training samples for the i-th training subject, and W_G and H_G are the width and the height of the Gabor-SSGEI.

A covariance matrix $S_T \in R^{W_G \times W_G}$ can be calculated by [27]

$$S_T = \frac{1}{N} \sum_{i=1}^{N_c} \sum_{j=1}^{n_i} (G_{i,j} - \bar{G})^T (G_{i,j} - \bar{G}), \tag{8}$$

where N is the total number of training samples, and $\bar{G}$ is the total mean of all training samples.

The orthogonal eigenvectors of S_T corresponding to the first W' largest eigenvalues constitute the optimal projection matrix $P \in \mathbb{R}^{W_G \times W'}$. In our applications, we make 2DPCA retain 99% of the variance. Once we obtain the projection matrix P, a dimension reduced feature matrix $Y_{i,j} \in \mathbb{R}^{H \times W}$ is computed as

$$Y_{i,j} = (G_{i,j} - \bar{G})P. \tag{9}$$

We then try finding a discriminative projection using two-dimensional linear discriminant analysis (2DLDA) in the row direction after the projection by 2DPCA. For this purpose, we consider a within-class scatter matrix $S_W \in \mathbb{R}^{H_G \times H_G}$ and a between-class scatter matrix $S_B \in \mathbb{R}^{H_G \times H_G}$ [28], which are computed as

$$S_W = \sum_{i=1}^{N_c} \sum_{j=1}^{n_i} (Y_{i,j} - \bar{Y}_i)(Y_{i,j} - \bar{Y}_i)^T \tag{10}$$

$$S_B = \sum_{i=1}^{N_c} n_i (\bar{Y}_i - \bar{Y})(\bar{Y}_i - \bar{Y})^T, \tag{11}$$

where $\bar{Y}_i$ and $\bar{Y}$ are the i-th class means and total mean in the 2DPCA space.

Finally, a projection q for 2DLDA is obtained so as to maximize the following criterion defined as a ratio of between-class scatter and within-class scatter as

$$J(q) = \frac{q^T S_B q}{q^T S_W q}. \tag{12}$$

The optimal projection is chosen when the $J(q)$ is maximized, and this problem can be solved by the generalized eigenvalue problem [28]. Similar to 2DPCA, the eigenvectors corresponding to the first H' largest eigenvalues make up the optimal projection matrix $Q \in \mathbb{R}^{H_G \times H'}$.

Consequently, a dimension reduced matrix $Y_{i,j} \in \mathbb{R}^{H_G \times W'}$ in the 2DPCA space is further transformed to $Z_{i,j} \in \mathbb{R}^{H' \times W'}$ in the 2DLDA space as

$$Z_{i,j} = Q^T Y_{i,j}. \tag{13}$$

3   Experiments

In this section, we first describe the datasets and parameter settings in Subsect. 3.1, then we design three experiments in Subsects. 3.2, 3.3 and 3.4 respectively, as follows:

1. Analyse the duration parameter p.
2. Compare five features (Keyframe, GEI, SSGEI, Gabor-GEI, Gabor-SSGEI) w/ and w/o metric learning in both verification (one-to-one matching) and identification (one-to-many matching) scenarios under speed variations, in order to confirm the proposed method realizes a good tradeoff between the speed invariance and the stability as well as confirming the contributions of individual components. Here, Keyframe is encoded as an average of two single support phases. In verification scenarios, we adopt an receiver operating characteristics (ROC) curve which shows a relation between false rejection rate (FRR) and false acceptance rate (FAR) when an acceptance threshold changes. In identification scenarios, as a performance evaluation measure, a cumulative matching characteristics (CMC) curve is used, which indicates rates that the genuine subjects are included within each of rank [29].
3. Compare the proposed method with the state-of-the-arts by rank-1 identification rate.

Finally, we evaluate the computational cost in Subsect. 3.5.

3.1   Datasets and Parameter Settings

For this experiments, we adopted the OU-ISIR Treadmill Dataset A [30], which contains image sequences of 34 subjects and speed variation ranging from 2 km/h to 10 km/h at 1 km/h interval to evaluate our method. In this paper, we focus on speed changes while walking (from 2 km/h to 7 km/h). Nice subjects were used for training the parameter p as well as 2DPCA and 2DLDA, and the other disjoint 25 subjects were used for testing.

As for parameter setting in Gabor functions, we set $f = \sqrt{2}$, and used five scale parameters (i.e., $s = 0, 1, 2, 3, 4$), and eight orientation parameters $\phi_d = \pi d/8$ for $d = 0, 1 \ldots 7$, following [23,24,26], which summed up to 40 Gabor functions in total. The number of oscillations under the Gaussian envelope is determined by $\delta = 2\pi$. The window size of Gabor filter is 45×45 pixels in our applications.

We empirically set the dimensions of 2DLDA to 90 for the feature whose dimension is 128×88 (Keyframe, GEI, SSGEI) and 110 for the feature whose dimension is 320×352 (Gabor-GEI, Gabor-SSGEI), respectively.

3.2   Analysis on the Optimal Duration Parameter

As described in Subsect. 2.1, we select the optimal duration parameter p within $0 < p \leqslant 1/4$. Concretely speaking, we empirically prepared a discrete set of

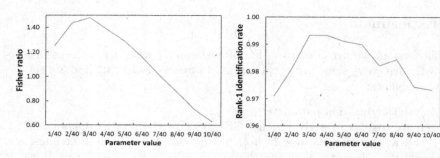

(a) Fisher ratio using training set   (b) Rank-1 identification on testing set

Fig. 4. Duration parameter analysis.

parameter candidates as $p \in \{i/40\}(i = 1, 2, \ldots, 10)$ at $1/40$ interval (when $p = 10/40$, the whole period is included in the duration). We report the Fisher ratio (Eq. (6)) corresponding to each parameter candidate p in Fig. 4(a). As a result, Fisher ratio is the largest for $p = 3/40$, and hence we adopted $p^* = 3/40$ in our experiments.

As for reference, we made sensitivity analysis of the duration parameter p on rank-1 identification rate for the testing set in order to investigate the generalization capability. Although the rank-1 identification rates over duration parameter p is not so smooth due to limited number of testing subjects (*i.e.*, 25 subjects), it is still worth mentioning to that the best rank-1 identification rate is obtained at the same optimal duration $p^* = 3/40$, which shows the generality of the duration parameter p.

3.3   Feature Comparison

In this section, five features (Keyframe, GEI, SSGEI, Gabor-GEI, Gabor-SSGEI) were tested w/o and w/ metric learning. For this purpose, we choose a pair of galleries at $4 \, \mathrm{km/h}$ and probes at each speed from $2 \, \mathrm{km/h}$ to $7 \, \mathrm{km/h}$ as examples.

Firstly, the accuracy in verification scenarios was evaluated with ROC curves in Fig. 5. When the differences of walking speeds between gallery and probe are small (see Fig. 5(b) for example), all the features get relatively good results. However, since Keyframe aggregates only two frames at single support phases in a period, it performs the worst when gallery and probe are the same speed (see Fig. 5(c)), where the stability is more meaningful than the speed invariance.

On the other hand, when the differences of walking speeds between gallery and probe becomes larger (see Fig. 5(f) for example), it is clearly seen that the result of GEI becomes worse as it is very sensitive to the walking speed change. In contrast, the proposed SSGEI yielded the better results in both cases of small

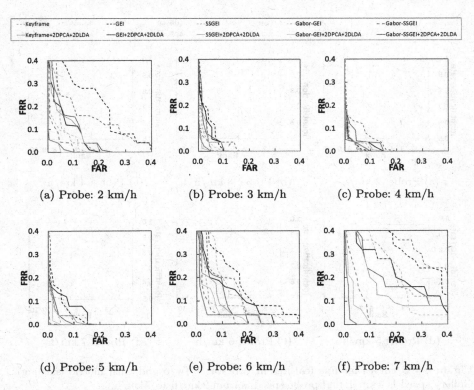

Fig. 5. ROC curves for five features in two stages (w/o and with metric learning). Gallery speed is 4 km/h and probe speed is from 2 km/h to 7 km/h.

and large speed change. What is more, when combined with Gabor filtering and 2DPCA, 2DLDA, the propose SSGEI achieved the best accuracy as a whole, and successfully suppress EER to 4.0% even at the worst case.

Next, we evaluate the accuracy in identification scenarios. Similarly, CMCs show performance of each feature w/o and with metric learning for each probe in Fig. 6. The results were basically consistent with those in the verification scenarios. Obviously, the proposed method (Gabor-SSGEI + 2DPCA + 2DLDA) yielded the highest accuracy, with 100% rank-1 identification rates in all the six pairs.

For a clearer and more intuitive explanation, we give examples of five features in Fig. 7 with a pair of true and false matches and also their corresponding subtraction and Euclidean distance. The subtraction images and Euclidean distances illustrate that, Keyframe, GEI, SSGEI, and Gabor-GEI all result in a false match since Euclidean distances for the false match is smaller than the true match. On the other hand, the proposed Gabor-SSGEI results in the true matches due to the good tradeoff between speed invariance and stability, as well as effectiveness of Gabor filtering.

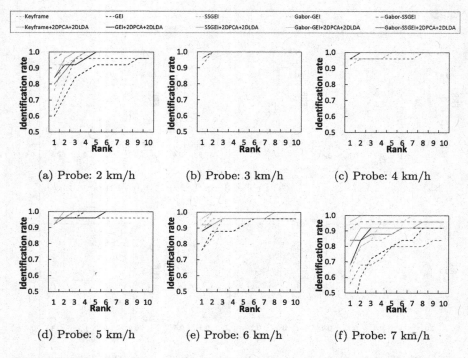

Fig. 6. CMC curves for five features in two stages (w/o and with metric learning). Gallery speed is 4 km/h and probe speed is from 2 km/h to 7 km/h.

Finally, for a comprehensive evaluation, we also give rank-1 identification rates of the five features averaged over all of the $36\,(= 6 \times 6)$ combinations of walking speeds in probe and gallery in Table 1.

As a result, the performance of SSGEI w/o and w/ metric learning are both better than Keyframe and GEI, which shows that the proposed SSGEI realizes a good tradeoff between the speed invariance and the stability as feature representation, and it is consistent with the results in Figs. 5, 6 and 7. In addition, if we exclude one of individual components SSGEI, Gabor filtering, and metric learning, from the full proposed method, the rank-1 identification rates drop from the best one, 99.33% for the full proposed method, *i.e.*, Gabor-SSGEI w/ metric learning, to 96.89% for Gabor-GEI w/ metric learning, 87.67% for SSGEI w/ metric learning, and 95.11% for Gabor-SSGEI w/o metric learning, respectively, which indicates individual components substantially contribute to the proposed method.

3.4   Comparison with State-of-the-arts

In this section, the proposed method is compared with the state-of-the-arts of cross-speed gait recognition, *i.e.*, hidden Markov model (HMM)-based

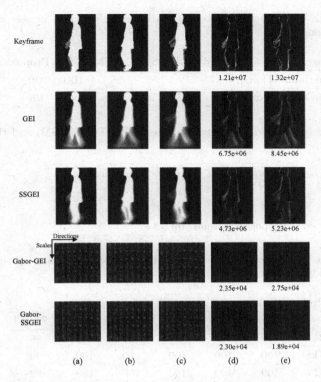

Fig. 7. Comparison examples of five features. Gallery speed is 4 km/h and probe speed is 2 km/h. (a) Probe. (b) False match in gallery (imposter). (c) True match in gallery (genuine). (d) Subtraction and corresponding Euclidean distance for false match. (e) Subtraction and corresponding Euclidean distance for true match.

Table 1. Rank-1 identification rates [%] of the five features w/o and w/ metric leaning averaged over all the 36 combinations of walking speeds in probe and gallery.

	Keyframe	GEI	SSGEI	Gabor-GEI	Gabor-SSGEI
w/o metric learning	74.89	62.56	80.33	*84.00*	**95.11**
w/ metric learning	84.44	85.89	87.67	*96.89*	**99.33**

approach [17], stride normalization (SN) [16], speed transformation model (STM) [15], differential composition model (DCM) [19], random subspace method (RSM) [20]. Following these works, we also report rank-1 identification rates to measure accuracy on cross-speed gait recognition. Although it is naturally preferable to evaluate the benchmarks using the same database under the same protocol, some of the benchmarks employed different databases (the number of subjects is almost consistent across the database used) and hence we set up similar experimental setup as much as possible as also doing the same thing in [15,20].

Table 2. Rank-1 identification rate [%] of different algorithms in case of small and large speed changes.

Speed change	HMM	SN	STM	DCM	RSM	Proposed method
Small (3 km/h and 4 km/h)	84	-	90	98	**100**	**100**
Large (2 km/h and 6 km/h)	-	35	58	82	*95*	**98**

Table 3. Averaged rank-1 identification rates [%] of DCM, RSM, and the proposed method.

Algorithms	Rank-1 identification rate
DCM	92.44
RSM	98.07
Proposed method	**99.33**

Table 4. Rank-1 identification rate (%) of the proposed method in all the 36 combinations of walking speeds.

Probe	Gallery					
	2 km/h	3 km/h	4 km/h	5 km/h	6 km/h	7 km/h
2 km/h	100	100	100	100	96	96
3 km/h	100	100	100	100	100	92
4 km/h	100	100	100	100	100	92
5 km/h	100	100	100	100	100	100
6 km/h	100	100	100	100	100	100
7 km/h	100	100	100	100	100	100

For example, SN [16] was evaluated with a different gait database whose walking speed differences were 2.5 km/h and 5.8 km/h, and hence we compared it with the matching results between 2 km/h and 6 km/h by the other methods. Moreover, HMM [17] also employed a different gait database whose walking speed difference is 3.3 km/h and 4.5 km/h, and hence we compared it with the matching results between 3 km/h and 4 km/h by the other methods.

Results are shown in Table 2. In addition, the rank-1 identification rates averaged over all the 36 combinations of walking speeds for the best three methods in Table 2, *i.e.*, DCM [19], RSM [20], and the proposed method, are listed in Table 3. Moreover, rank-1 identification rates of 36 individual combinations of walking speeds for the proposed method are reported in Table 4.

From Tables 2, 3 and 4, the proposed method clearly outperforms the other algorithms, in particular in case of large speed changes.

Table 5. Running time [s] of the proposed method.

Running stage	Time cost
Training time in optimizing duration parameter	0.009
Training time in 2DPCA and 2DLDA	0.115
Query time of each sequence	0.003

3.5   Evaluation of Running Time

To test the computational cost, Matlab code of the proposed method wad run on a PC with Intel Core i7 4.00 GHz processor and 32 GB RAM. The training time of parameter for optimizing duration and metric learning method, as well as the query time of each sequence are listed in Table 5. The result demonstrates the computational cost of the proposed method is very low and suitable for real applications, while some of the benchmarks requires high computational cost such as model fitting in [15] and substantial number of random projections in [20].

4   Conclusion

This paper presents a speed invariant as well as stable gait representation called SSGEI to cope with cross-speed gait recognition. In order to realize a good trade-off between the speed-invariance and the stability, we choose the optimal duration around single support phases so as to maximize Fisher ratio using a training set. SSGEI is then computed by aggregating multiple frames for the optimal duration and is further combined with Gabor filters and metric learning for better performance. Comprehensive experiments illustrated the effectiveness of the proposed method, which outperformed other state-of-the-art methods, with a low time consuming as well.

Since we focused on the cross-speed gait recognition within walking style in this work, a future research avenue is speed-invariant gait recognition across different modes, *i.e.*, walking and running, which may often the case with real-world scenes.

Acknowledgement. This work was supported by JSPS Grants-in-Aid for Scientific Research (A) JP15H01693, by Jiangsu Provincial Science and Technology Support Program (No. BE2014714), by the 111 Project (No. B13022), and by the Project Funded by the Priority Academic Program Development of Jiangsu Higher Education Institutions.

References

1. Nixon, M.S., Tan, T.N., Chellappa, R.: Human Identification Based on Gait. International Series on Biometrics. Springer, New York (2005)

2. Bouchrika, I., Goffredo, M., Carter, J., Nixon, M.: On using gait in forensic biometrics. J. Forensic Sci. **56**, 882–889 (2011)
3. Iwama, H., Muramatsu, D., Makihara, Y., Yagi, Y.: Gait verification system for criminal investigation. IPSJ Trans. Comput. Vis. Appl. **5**, 163–175 (2013)
4. Lynnerup, N., Larsen, P.: Gait as evidence. IET Biom. **3**, 47–54 (2014)
5. Sarkar, S., Phillips, J., Liu, Z., Vega, I., Grother, P., Bowyer, K.: The humanid gait challenge problem: data, sets, performance and analysis. IEEE Trans. Pattern Anal. Mach. Intell. **27**, 162–177 (2005)
6. Bouchrika, I., Nixon, M.: Exploratory factor analysis of gait recognition. In: Proceedings of 8th IEEE International Conference on Automatic Face and Gesture Recognition, Amsterdam, The Netherlands, pp. 1–6 (2008)
7. Han, J., Bhanu, B.: Individual recognition using gait energy image. IEEE Trans. Pattern Anal. Mach. Intell. **28**, 316–322 (2006)
8. Makihara, Y., Sagawa, R., Mukaigawa, Y., Echigo, T., Yagi, Y.: Gait recognition using a view transformation model in the frequency domain. In: Proceedings of 9th European Conference on Computer Vision, Graz, Austria, pp. 151–163 (2006)
9. Wang, C., Zhang, J., Wang, L., Pu, J., Yuan, X.: Human identification using temporal information preserving gait template. IEEE Trans. Pattern Anal. Mach. Intell. **34**, 2164–2176 (2012)
10. Lam, T.H.W., Cheung, K.H., Liu, J.N.K.: Gait flow image: a silhouette-based gait representation for human identification. Pattern Recogn. **44**, 973–987 (2011)
11. Boulgouris, N., Plataniotis, K., Hatzinakos, D.: Gait recognition using dynamic time warping. In: Proceedings of IEEE 6th Workshop on Multimedia Signal Processing, pp. 263–266 (2004)
12. Veeraraghavan, A., Roy-Chowdhury, A.K., Chellappa, R.: Matching shape sequences in video with applications in human movement analysis. IEEE Trans. Pattern Anal. Mach. Intell. **27**, 1896–1909 (2005)
13. Boulgouris, N., Plataniotis, K., Hatzinakos, D.: Gait recognition using linear time normalization. Pattern Recogn. **39**, 969–979 (2006)
14. Veeraraghavan, A., Srivastava, A., Roy-Chowdhury, A.K., Chellappa, R.: Rate-invariant recognition of humans and their activities. IEEE Trans. Image Process. **18**, 1326–1339 (2009)
15. Makihara, Y., Tsuji, A., Yagi, Y.: Silhouette transformation based on walking speed for gait identification. In: Proceedings of 23rd IEEE Conference on Computer Vision and Pattern Recognition, San Francisco, CA, USA (2010)
16. Tanawongsuwan, R., Bobick, A.: Modelling the effects of walking speed on appearance-based gait recognition. In: Proceedings of 17th IEEE Computer Society Conference on Computer Vision and Pattern Recognition, vol. 2, pp. 783–790 (2004)
17. Liu, Z., Sarkar, S.: Improved gait recognition by gait dynamics normalization. IEEE Trans. Pattern Anal. Mach. Intell. **28**, 863–876 (2006)
18. Kusakunniran, W., Wu, Q., Zhang, J., Li, H.: Speed-invariant gait recognition based on procrustes shape analysis using higher-order shape configuration. In: 18th IEEE International Conference on Image Processing, pp. 545–548 (2011)
19. Kusakunniran, W., Wu, Q., Zhang, J., Li, H.: Gait recognition across various walking speeds using higher order shape configuration based on a differential composition model. IEEE Trans. Syst. Man Cybern. Part B: Cybern. **42**, 1654–1668 (2012)
20. Guan, Y., Li, C.T.: A robust speed-invariant gait recognition system for walker and runner identification. In: Proceedings of 6th IAPR International Conference on Biometrics, pp. 1–8 (2013)

21. Makihara, Y., Yagi, Y.: Silhouette extraction based on iterative spatio-temporal local color transformation and graph-cut segmentation. In: Proceedings of 19th International Conference on Pattern Recognition, Tampa, Florida, USA (2008)
22. Hossain, M.A., Makihara, Y., Wang, J., Yagi, Y.: Clothing-invariant gait identification using part-based clothing categorization and adaptive weight control. Pattern Recogn. **43**, 2281–2291 (2010)
23. Tao, D., Li, X., Wu, X., Maybank, S.J.: General tensor discriminant analysis and gabor features for gait recognition. IEEE Trans. Pattern Anal. Mach. Intell. **29**, 1700–1715 (2007)
24. Xu, D., Huang, Y., Zeng, Z., Xu, X.: Human gait recognition using patch distribution feature and locality-constrained group sparse representation. IEEE Trans. Image Process. **21**, 316–326 (2012)
25. Lee, T.S.: Image representation using 2D Gabor wavelets. IEEE Trans. Pattern Anal. Mach. Intell. **18**, 959–971 (1996)
26. Liu, C., Wechsler, H.: Gabor feature based classification using the enhanced fisher linear discriminant model for face recognition. IEEE Trans. Image Process. **11**, 467–476 (2002)
27. Yang, J., Zhang, D., Frangi, A.F., Yang, J.Y.: Two-dimensional PCA: a new approach to appearance-based face representation and recognition. IEEE Trans. Pattern Anal. Mach. Intell. **26**, 131–137 (2004)
28. Li, M., Yuan, B.: 2D-LDA: a statistical linear discriminant analysis for image matrix. Pattern Recogn. Lett. **26**, 527–532 (2005)
29. Phillips, P., Blackburn, D., Bone, M., Grother, P., Micheals, R., Tabassi, E.: Face recogntion vendor test (2002). http://www.frvt.org
30. Makihara, Y., Mannami, H., Tsuji, A., Hossain, M., Sugiura, K., Mori, A., Yagi, Y.: The ou-isir gait database comprising the treadmill dataset. IPSJ Trans. Comput. Vis. Appl. **4**, 53–62 (2012)

Parametric Image Segmentation of Humans with Structural Shape Priors

Alin-Ionut Popa[2] and Cristian Sminchisescu[1,2(✉)]

[1] Department of Mathematics, Faculty of Engineering, Lund University,
Lund, Sweden
`cristian.sminchisescu@math.lth.se`
[2] Institute of Mathematics of the Romanian Academy,
Bucharest, Romania
`alin.popa@imar.ro`

Abstract. The figure-ground segmentation of humans in images captured in natural environments is an outstanding open problem due to the presence of complex backgrounds, articulation, varying body proportions, partial views and viewpoint changes. In this work we propose class-specific segmentation models that leverage parametric max-flow image segmentation and a large dataset of human shapes. Our contributions are as follows: (1) formulation of a sub-modular energy model that combines class-specific structural constraints and data-driven shape priors, within a parametric max-flow optimization methodology that systematically computes all breakpoints of the model in polynomial time; (2) design of a data-driven class-specific fusion methodology, based on matching against a large training set of exemplar human shapes (100,000 in our experiments), that *allows the shape prior to be constructed on-the-fly, for arbitrary viewpoints and partial views.*

1 Introduction

Detecting and segmenting people in real-world environments are central problems with applications in indexing, surveillance, 3D reconstruction and action recognition. Prior work in 3D human pose reconstruction from monocular images [1–3], as well as more recent, successful RGB-D sensing systems based on Kinect [4] have shown that the availability of a figure-ground segmentation opens paths towards robust and scalable systems for human sensing. Despite substantial progress, the figure-ground segmentation in RGB images remains extremely challenging, because people are observed from a variety of viewpoints, have complex articulated skeletal structure, varying body proportions and clothing, and are often partially occluded by other people or objects in the scene. The complexity of the background further complicates matters, particularly as any limb decomposition of the human body leads to parts that are relatively regular but not sufficiently distinctive even when spatial connectivity constraints are enforced [5]. Set aside appearance inhomogeneity and color variability due to clothing, which can overlap the background distribution significantly, it is well known that

© Springer International Publishing AG 2017
S.-H. Lai et al. (Eds.): ACCV 2016, Part II, LNCS 10112, pp. 68–83, 2017.
DOI: 10.1007/978-3-319-54184-6_5

many of the generic, parallel line (ribbon) detectors designed to detect human limbs, fire at high false positive rates in the background. This has motivated work towards detecting more distinctive part configurations, without restrictive assumptions on part visibility (e.g. full or upper view of the person), for which poselets [6] have been a successful example. However, besides relatively high false positive rates typical in detection, the transition from a bounding box of the person to a full segmentation of the human body is not straightforward. The challenge is to balance, on one hand, sufficient flexibility towards representing variability due to viewpoint, partial views and articulation, and, on the other hand, sufficient constraints in order to obtain segmentations that correspond to meaningful human shapes, all relying on region or structural human body part detectors that may only be partial or not always spatially accurate.

In this work we attempt to connect two relevant, recent lines of work, for the segmentation of people in real images. We rely on bottom-up figure-ground generation methods and region-level person classifiers in order to identify promising hypotheses for further processing. In a second pass, we set up informed constraints towards (human) class-specific figure-ground segmentation by leveraging skeletal information and data-driven shape priors computed on-the-fly by matching region candidates against exemplars of a large, recently introduced human motion capture dataset containing 3D and 2D semantic skeleton information of people, as well images and figure-ground masks from background subtraction (Human3.6M [7]). By exploiting globally optimal parametric max-flow energy minimization solvers, this time, based on a class dependent (as opposed to generic and regular) foreground seeding process [8–10], we show that we can considerably improve the quality of competitive object proposal generators. To our knowledge, this is one of the first formulations for class-specific segmentation that in principle can handle multiple viewpoints and any partial view of the person. It is also one of the first to leverage a large dataset of human shapes, together with semantic structural information, which until recently, have not been available. We show that such constraints are critical for accuracy, robustness, and computational efficiency.

1.1 Related Work

The literature on segmentation is huge, even when considering only subcategories like top-down (class-specific) and bottom-up segmentation. Humans are of significant interest to be devoted special methodology, and that proves to be effective [5,6,11–18]. One approach is to consider shape as category-specific property and integrate it within models that are driven by bottom-up processing [19–24]. Pishchulin *et al.* [23] develop pictorial structure formulations constrained by poselets, focusing on improving the response quality of an articulated part-based human model. The use of priors based on exemplars has also been explored, in a data-driven process. Both [25,26] focus on a matching process in order to identify exemplars that correspond to similar scene or object layouts, then used in a graph cut process that enforces spatial smoothness and provides a global solution. Our approach is related to such methods, but we use a novel data-driven prior construction, enforce structural constraints adapted to humans, and search the state space exhaustively by means of parametric max-flow. In contrast to priors used in [25, 26], which require a more repeatable scene layout, we focus on a prior generation process that can handle a diverse set of viewpoints and arbitrary partial views, not known a-priori, and different across the detected instances. Recently, there has been a rapid development of deep learning techniques towards the scene understanding task with focus on semantic segmentation [27–29], including humans.

Methods like [30] resemble ours in their reliance on a detection stage and the principle of matching that window representation against a training set where figure-ground segmentations are available, then optimizing an energy function via graph-cuts. Our window representation contains additional detail and this makes it possible to match exemplars based on the identified semantic content. Our matching and shape prior construction are optimized for humans, in contrast to the generic ones used in [30] (which can however segment any object, not just people, as is our focus here[1]). We use a large prior set of structurally annotated human shapes, and search the state space using a different, parametric multiple hypotheses scheme. Our prior construction uses, among other elements, a Procrustes alignment similar to [31] but differently: (1) we use it for shape prior construction (input dependent, on-the-fly) within the energy optimizer as opposed to object detection (classification, construction per class) as in [31], (2) we only use instances that align well with the query, thus reflecting accurate shape models, as opposed to fusing top-k instances to capture class variability in [31]. An alternative, interesting formulation for object segmentation with shape priors is branch-and-mincut [32], who propose a branch and bound procedure in the compound space of binary segmentations and hierarchically organized shapes. However, the bounding process used for efficient search in shape space would rely on knowledge of the type of shapes expected and their full visibility. We focus on a different optimization and modeling approach that can handle arbitrary occlusion patterns of shape. Our prior constraint for optimization is generated on-the-fly by fusing the visible exemplar components, following a structural alignment scheme.

Recently, there has been a resurrection of bottom-up segmentation methods based on multiple proposal generation, with surprisingly good results considering the low-level processing involved. Some of these methods generate segment hypotheses either by combining the superpixels [33] of a hierarchical clustering method [34–37], by varying the segmentation parameters [38] or by searching an energy model, parametrically, using graph cuts [10, 38–42]. Most of the latter techniques use mid-level shape priors for selection, either following hypotheses

---

[1] Notice, however, that the methodology we propose is also applicable to other categories than people. Here we focus on humans because for now, large training sets of segmented shapes with structural annotations are available only for them, through Human3.6M [7]. But, as large datasets for other object categories emerge, we expect our methodology to generalize well. In this respect, our results on a challenging visual category, humans, are indicative of the performance bounds one can expect.

generation [10,38,40] or during the process. Some methods provide a ranking, diversification and compression of hypotheses, using e.g. Maximal Marginal Relevance (MMR) diversification [10,38], whereas others report an unordered set [39,40]. Hypotheses pool sizes in the order of 1,000–10,000 range in the expansionary phase, and compressed models of 100–1,000 hypotheses following the application of trained rankers (operating on mid-level features extracted from segments) with diversification, are typical, with variance due to image complexity and edge structure. While prior work has shown that such hypotheses pools can contain remarkably good quality segments (60 − 80% intersection over union, IoU, scores are not uncommon) this leaves sufficient space for improvement particularly since sooner or later, one is inevitably facing the burden of decision making: *selecting one hypothesis to report*. It is then not uncommon for performance to sharply drop to 40%. This indicates that constraints and prior selection methods towards more compact, better quality hypotheses sets are necessary. Such issues are confronted in the current work.

2 Methodology

We consider an image as $I : \mathcal{V} \rightarrow R^3$, where $\mathcal{V}$ represents the set of nodes, each associated with a pixel in the image, and the range is the associated intensity (RGB) vector. The image is modeled as a graph $G = (\mathcal{V}, \mathcal{E})$. We partition the set of nodes $\mathcal{V}$ into two disjoint sets, corresponding to foreground and background, represented by labels 1 and 0, respectively. Seed pixels $\mathcal{V}_f$ and $\mathcal{V}_b$ are subsets of $\mathcal{V}$ and they are constrained to foreground and background, respectively. $\mathcal{E}$ is the subset of edges of the graph G which reflect the connections between adjacent pixels. The formulation we propose will rely on object (or foreground) structural skeleton constraints obtained from person detection and 2D localization (in particular the identification of keypoints associated with the joints of the human body, and the resulting set of nodes corresponding to the human skeleton, obtained by connecting keypoints, $T \subseteq \mathcal{V}$), as well as a data-driven, human shape fusion prior $S : \mathcal{V} \rightarrow [0, 1]$, constructed ad-hoc by fusing similar configurations with the one detected, based on a large dataset of human shapes with associated 2D skeleton semantics (see Sect. 2.1 for details). The energy function defined over the graph G, with $X = \cup\{x_u\}$ being the set of all image pixels, is:

$$E_\lambda(X) = \sum_{u \in \mathcal{V}} U_\lambda(x_u) + \sum_{(u,v) \in \mathcal{E}} V_{uv}(x_u, x_v) \tag{1}$$

where

$$U_\lambda(x_u) = D_\lambda(x_u) + S(x_u)$$

with $\lambda \in \mathbb{R}$, and unary potentials given by semantic foreground constraints $\mathcal{V}_f \leftarrow T$:

$$D_\lambda(x_u) = \begin{cases} 0 & \text{if } x_u = 1, \, u \notin \mathcal{V}_b \\ \infty & \text{if } x_u = 1, \, u \in \mathcal{V}_b \\ \infty & \text{if } x_u = 0, \, u \in \mathcal{V}_f \\ f(x_u) + \lambda & \text{if } x_u = 0, \, u \notin \mathcal{V}_f \end{cases} \tag{2}$$

The foreground bias is implemented as a cost incurred by the assignment of non-seed pixels to background, and consists of a pixel-dependent value $f(x_u)$ and an uniform offset λ. Two different functions $f(x_u)$ are used alternatively. The first is constant and equal to 0, resulting in a uniform (variable) foreground bias. The second function uses color. Specifically, RGB color distributions $p_f(x_u)$ on seed $\mathcal{V}_f$ and $p_b(x_u)$ on seed $\mathcal{V}_b$ are estimated and derive $f(x_u) = \ln \frac{p_f(x_u)}{p_b(x_u)}$. The probability distribution of pixel j belonging to the foreground is defined as $p_f(i) = \exp(-\gamma \cdot \min_j(\|I(i) - I(j)\|))$, with γ a scaling factor, and j indexes representative pixels in the seed region, selected as centers resulting from a k-means algorithm (k is set to 5 in all of our experiments). The background probability is defined similarly.

The pairwise term V_{uv} penalizes the assignment of different labels to similar neighboring pixels:

$$V_{uv}(x_u, x_v) = \begin{cases} 0 & \text{if } x_u = x_v \\ g(u, v) & \text{if } x_u \neq x_v \end{cases} \tag{3}$$

with similarity between adjacent pixels given by $g(u, v) = \exp\left[-\frac{\max(Gb(u), Gb(v))}{\sigma^2}\right]$. Gb returns the output of the multi-cue contour detector [43,44] at a pixel location. The *boundary sharpness* parameter σ controls the smoothness of the pairwise term.

The function $f(x_u)$ is the same as in the CPMC [10] algorithm. It takes two forms, the first is constant and acts as a foreground bias and the second uses color information, particularly the color distribution of the seed pixels computed with k-means algorithm. The energy function defined by (1) is submodular and can be optimized using parametric max-flow, in order to obtain all breakpoints of $E_\lambda(X)$ as a function of (λ, X) in polynomial time. The advantage of our approach is that it can be used with any object proposal generator based on graph-cut energy minimization.

Given the general formulation in (1) and (2), the key problems to address are: **(a)** the identification of a putative set of person regions and structural constraints hypotheses T; **(b)** the construction of an effective, yet flexible data-driven human shape prior S, based on a sufficiently diverse dataset of people shapes and skeletal structure, given estimates for T. **(c)** minimization of the resulting energy model (1). We address (a) without loss of generality, using a human region classifier (any other set of structural, problem dependent detectors can be used, here e.g. face and hand detectors based on skin color models or poselets). We address (b) using methodology that combines a large dataset of human pose shapes and body skeletons, collected from Human3.6M [7] with shape matching, alignment and fusion analysis, in order to construct the prior on-the-fly, for the instance being analyzed. We refer to a model that leverages both problem-dependent structural constraints T and a data-driven shape prior S, in a *single joint optimization problem*, as *Constraint Parametric Problem Dependent Cuts with Shape Matching. Alignment and Fusion (CPDC-MAF)*. The integration of bottom-up region detection constraints with a shape prior construction is described in Sect. 2.1. The CPDC-MAF model can be optimized in polynomial time using parametric max-flow, in order to obtain all breakpoints of the associated energy model (addressing c).

2.1   Data-Driven Shape Matching, Alignment and Fusion (MAF)

We aim to obtain an improved figure-ground segmentation for persons by combining bottom-up and top-down, class specific information. We initialize our proposal set using CPMC. While any figure-ground segmentation proposal method can be employed in principle, we choose CPMC due to its performance and because our method can be viewed as a generalization with problem dependent seeds and shape priors. We filter the top N segment candidates using an O2P-region classifier [45] trained to respond to humans, using examples from Human3.6M, to obtain $\mathcal{D} = \{d_i = \{\mathbf{z}, \mathbf{b}\}, |i = 1, \ldots N\}$. Each candidate segment is represented by a binary mask $\mathbf{z}$, where 1 stands for foreground and 0 stands for background, and a bounding box $\mathbf{b} \in \mathbb{R}^4$ where $\mathbf{b} = (m, n, w, h)$; m and n represent the image coordinates of the bottom left corner of the bounding box, w and h represents its width and its height.

We use the set of human region candidates in order to match against a set of human shapes and construct a shape prior. There are challenges however, particularly being able to: **(1)** access a sufficiently representative set of human shapes to construct the prior, **(2)** be sufficiently flexible so that human shapes from the dataset, which are very different from the shape being analyzed, would not negatively impact estimates, **(3)** handle partial views—while we rely on bottom-up proposals that can handle partial views, the use, in contrast, of a shape prior that can only represent, e.g. full or upper-body views, would not be effective.

We address (1) by employing a dataset of 100,000 human shapes together with the corresponding skeleton structure, sub-sampled from the recently created Human3.6M dataset [7]; (2) by employing a matching, alignment and fusion technique between the current segment and the individual exemplar shapes in the dataset. Shapes and structures which cannot be matched and aligned properly are discarded. (3) is adressed by leveraging the implicit correspondences available across training shapes, at the level of local shape matches, by only aligning and warping those components of the exemplar shapes that can be matched to the query. A sample flow of our entire method can be visualized in Figs. 1 and 2.

Boundary Point Sampling: Given a bottom-up figure-ground proposal represented as a binary mask $\mathbf{z} \in \mathcal{D}$, we sample through the image coordinates of the boundary points of the foreground segment. Thus we obtain a set of 2D points $\mathbf{p}_j, j = 1, \ldots, K$ with $\mathbf{p}_j \in \mathbb{R}^2$ where $\mathbf{p}_j = (x_j, y_j)$. We loop through the shapes of our human shape dataset Human3.6M and, for each shape, we rotate and

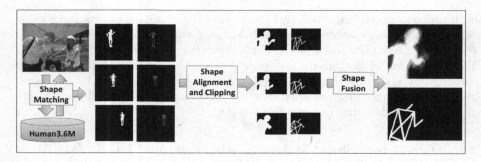

Fig. 1. Our Shape Matching Alignment Fusion (MAF) construction based on semantic matching, structural alignment and clipping, followed by fusion, to reflect the partial view. Notice that the prior construction allows us to match partial views of a putative human detected segment to fully visible exemplars in Human3.6M. This allows us to handle arbitrary patterns of occlusion. We can thus create a well adapted prior, on-the-fly, given a candidate segment.

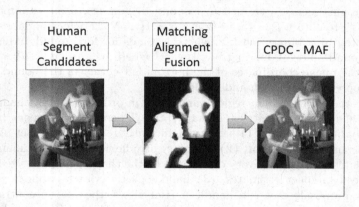

Fig. 2. Processing steps of our segmentation methods based on Constrained Parametric Problem Dependent Cuts (CPDC) with shape Matching, Alignment and Fusion (MAF).

scale it so that it has the same orientation and scale as the foreground candidate segment and sample through its boundary points. Thus we obtain a set of 2D points $\mathbf{q}_{jl}, j = 1, \ldots, K$, with $l = 1, \ldots, L$, where L represents the number of poses in the shape-pose dataset, in our case $L = 100,000$.

Shape Matching and Transform Matrix: We employ the shape context descriptor [46] at each position $\mathbf{p}_j$ from the candidate segment and at each position $\mathbf{q}_{jl}$. We evaluate a χ^2 distance [47] on the resulting descriptors to select the indexes l with sufficient good matches, such that we estimate an affine transformation.

We apply a 2D Procrustes transform with 5 degrees of freedom (rotation, anisotropic scaling including reflections, and translation) on each $\mathbf{q}_{jl}$ in order to align each shape in the dataset with the corresponding boundary point. This results in a 3×3 transformation matrix $\mathbf{W}_l$ and an error for the transform, e_l (average over $e_{jl}, j = 1, \ldots, K$) which represents the Euclidean distance between the boundary points $\mathbf{p}_j$ and the Procrustes transformed ones, $\mathbf{W}_l \cdot \mathbf{q}_{jl}$, in the image plane.

Prior Shape Selection and Warping: In order to determine which prior shapes are relevant for the current detected query, we identify the subset of indexes in the dataset $\mathcal{T}$ which correspond to transformation errors that are smaller than a given threshold ϵ. Thus, we obtain the corresponding figure-ground masks $\mathbf{m}_t, t \in \mathcal{T}$. For each mask $\mathbf{m}_t$, we select the coordinates of fore-ground pixels and warp them using the transform matrix computed using the 2D joint coordinates transformation. We apply the same procedure to the attached skeleton configuration of the corresponding mask. Thus, we obtain the coordinates of the foreground pixels for the transformed mask, Φ_t and the transformed skeleton coordinates Ψ_t.

Prior Shape Fusion: We compute the mean of the entire set of transformed masks, Φ_t, obtaining a MAF prior, S, corresponding to the detection d as seen in Fig. 1. The values of the shape prior mask range from 0 to 1, background and foreground probabilities, respectively. Also, we compute the mean of the entire set of transformed skeletons Ψ_t, obtaining a configuration of keypoints $\mathbf{B} \in \mathbb{R}^{3 \times 15}$ with $\mathbf{B}_j = (x, y, 1)$ where x and y represent the image coordinates of the warped joint from Human3.6M. This can be used to obtain a problem dependent mask $\mathbf{m}$ as follows. Initially we set the mask to have the same dimension as the entire image, filled with 0. We use Bresenham's algorithm to draw a line between the semantically adjacent joints, for example: left elbow - left wrist, right hip - right knee, and so on. We assign the set of skeleton nodes to the foreground as $T = \{i \in \mathcal{V} | \mathbf{m}(i) = 1\}$. This entire procedure of obtaining the shape prior information (mask and skeleton) is illustrated in Algorithm 1.

3   Experiments

We test our methodology on two challenging datasets: H3D [48] which contains 107 images and MPII [49] with 3799 images. We have figure-ground segmentation annotations available for all datasets. For the MPII dataset, we generate figure-ground human segment annotations ourselves. Both the H3D and the MPII datasets contain both full and partial views of persons with self-occlusion which makes them extremely challenging.

We run several segmentation algorithms including CPMC [10] as well as our proposed CPDC-MAF, where we use bottom-up person region detectors trained on Human3.6M, and region descriptors based on O2P [45]. We also constructed a model referred to as CPDC-MAF-POSELETS, built using problem dependent seeds based on a 2D pose detector instead of the proposed segments of a figure-ground segmentation algorithm. While any methodology that provides body keypoints (parts or articulations) is applicable, we choose the poselet detector

---

Algorithm 1. Calculate S and $\mathbf{B}$ (Shape Matching, Alignment and Fusion, **MAF**)

Require:
$d_i = \{\mathbf{z}, \mathbf{b}\}$
$\mathbf{d}_l, l = 1, \ldots, L$ - 2D joint positions (Human3.6M)
$\mathbf{m}_l, l = 1, \ldots, L$ - figure-ground masks (Human3.6M)
L - number of poses (Human3.6M, use $L = 100,000$)
ϵ - threshold value for transform error
$\mathbf{f}(\cdot)$ - shape context descriptor
μ - threshold value for χ^2 for shape context descriptors
Ensure: $S, \mathbf{B}$
  Sample boundary points $\mathbf{p}_j, j = 1, \ldots, K$ on $\mathbf{z}$
  for $l \in \mathcal{L}$ **do**
    Sample K boundary points $\mathbf{q}_{jl}, j = 1, \ldots, K$ on $\mathbf{m}_l$
    $J = \{(x, y) \in \mathbb{N}^2 | \chi^2(\mathbf{f}(\mathbf{q}_{xl}), \mathbf{f}(\mathbf{p}_y)) < \mu\}$
    if $|J| > 2$ **then**
      $\mathbf{a}_{jl}(\mathbf{W}) = \mathbf{p}_j - \mathbf{W} \cdot \mathbf{q}_{jl}$
      $\mathbf{W}_l = \underset{\mathbf{W}}{\operatorname{argmin}} \frac{1}{|K|} \sum_{j \in K} \mathbf{a}_{jl}(\mathbf{W})^\top \mathbf{a}_{jl}(\mathbf{W})$
      $e_l = \frac{1}{|K|} \sum_{j \in K} \mathbf{a}_{jl}(\mathbf{W}_l)^\top \mathbf{a}_{jl}(\mathbf{W}_l)$
    else
      $e_l = \infty$
    end if
  end for
  $\mathcal{T} = \{l \in \mathcal{L} | e_l < \epsilon\}$
  for $t \in \mathcal{T}$ **do**
    $\mathcal{V}_f$ - foreground pixels of $\mathbf{m}_t$; $\mathcal{V}_b$ - background pixels of $\mathbf{m}_t$
    $\mathcal{V} = \mathcal{V}_b \cup \mathcal{V}_f$
    for $\mathbf{u} \in \mathcal{V}$ **do**
      if $\mathbf{u} \in \mathcal{V}_f$ **then**
        $\Phi_t(\mathbf{W}_t \cdot \mathbf{u}) = 1$
      else
        $\Phi_t(\mathbf{W}_t \cdot \mathbf{u}) = 0$
      end if
    end for
    $\Psi_t = \mathbf{W}_t \cdot \mathbf{d}_l$
  end for
  $S = \frac{1}{|\mathcal{T}|} \sum_{t \in \mathcal{T}} \Phi_t$; $\mathbf{B} = \frac{1}{|\mathcal{T}|} \sum_{t \in \mathcal{T}} \Psi_t$

---

because it provides results under partial views of the body, or self occlusions of certain joints together with joint position estimates. Conditioned on a detection, we apply the same idea as in our CPDC-MAF, except that we use the detected skeletal keypoints to match against the exemplars in the Human3.6M dataset. A matching process based on semantic keywords (the body joints) is explicit, immediate (since joints are available both for the putative poselet detector and for the exemplar shapes in Human3.6M) and arguably simpler than matching shapes in the absence of skeletal information. The downside is that when the poselet detection is incorrect, the matching will also be (notice that alignments with high score following matching are nevertheless discarded within the MAF process).

We initialize CPDC-MAF, bottom-up, by using candidate segments from CPMC pool, selected based on their **person** ranking score after applying the O2P classifier. This is followed by a non-maximum suppression step were we remove the pair of segments with an overlap above 0.25. We use the MAF process to reject irrelevant candidates and to build shape prior masks and skeleton configuration seeds for the segments with good matching produced by shape context descriptors. On each resulting shape prior and skeleton seeds, we run the CPDC-MAF model with the resulting pools from each candidate segment merged to obtain the human region proposals for an entire image.

Table 1. Accuracy and pool size statistics for different methods, on data from H3D and MPII. We report average IoU over test set for the first segment of the ranked pool and the ground-truth figure-ground segmentation (*First*), the average IoU over test set of the segment with the highest IoU with the ground-truth figure-ground segmentation (*Best*) and average pool size (*Pool Size*).

Method	H3D test set [48]			MPII test set [49]		
	First	Best	Pool size	First	Best	Pool size
CPMC [10]	0.54	0.72	783	0.29	0.73	686
CPDC - MAF	0.60	0.72	77	0.55	0.71	102
CPDC - MAF - POSELETS	0.53	0.6	98	0.43	0.58	116

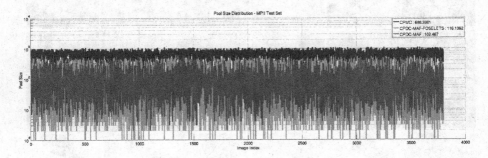

Fig. 3. Dimension of segmentation pool for MPII and various methods along with average pool size (in legend). Notice significant difference between the pool size values of CPDC-MAF-POSELETS and CPDC-MAF compared to the ones of CPMC. CPMC pool size values maintain an average of 700 units, whereas the pool sizes of CPDC-MAF and CPDC-MAF-POSELETS are considerably smaller, around 100 units.

For each testing setup, we report the mean values (computed over the entire testing dataset) of the intersection over union (IoU) scores for the first segment in the ranked pool and the ground-truth figure-ground segmentation for each

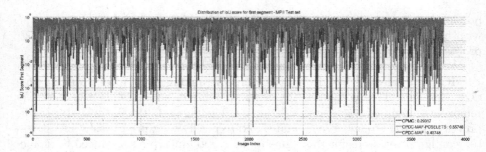

Fig. 4. IoU for the first segment from the ranked pool in MPII. The values for CPMC and CPDC-MAF-POSELETS have higher variance compared to CPDC-MAF resulting in the performance drop illustrated by their average.

Fig. 5. Sample of generated segments for images from MPII. From left to right, original image, top 5 first ranked segments of the CPDC-MAF generated pool of segments.

image. We also report the mean values of the IoU scores for the pool segment with the best IoU score with the ground-truth figure ground segmentation.

Results for different datasets can be visualized in Table 1. In turn, Figs. 3, 4 show plots for the size of the segment pools and IoU scores for highest ranked

Fig. 6. Segmentation examples for various methods. From left to right, original image, CPMC with default settings on person's bounding box, CPDC-MAF-POSELETS and CPDC-MAF. See also Table 1 for quantitative results.

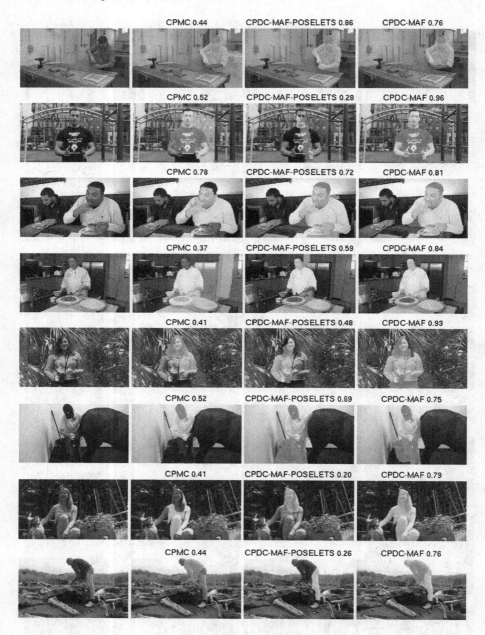

Fig. 7. Segmentation examples for difficult cases including partial views and occlusions. From left to right, original image, CPMC with default settings on person's bounding box, CPDC-MAF-POSELETS and CPDC-MAF.

segments generated by different methods, with image indexes sorted according to the best performing method (CPDP-MAF). Qualitative segmentation results for the various methods tested are given in Figs. 6 and 7. Also, we illustrate sample results with top ranked pool of segments in Fig. 5.

4   Conclusions

We have presented class-specific image segmentation models that leverage human body part detectors based on bottom-up figure-ground proposals, parametric max-flow solvers, and a large dataset of human shapes. Our formulation leads to a sub-modular energy model that combines class-specific structural constraints and data-driven shape priors, within a parametric max-flow optimization methodology that systematically computes all breakpoints of the model in polynomial time. We also propose a data-driven class-specific prior fusion methodology, based on shape matching, alignment and fusion, that *allows the shape prior to be constructed on-the-fly, for arbitrary viewpoints and partial views.* We demonstrate competitive results in two challenging datasets: H3D [48] and MPII [49], where we improve the first ranked hypothesis estimates of mid-level segmentation methods by 20%, *with pool sizes that are up to one order of magnitude smaller.* In future work we will explore additional class-dependent seed generation mechanisms and plan to study the extension of the proposed framework to video.

Acknowledgments. This work was supported in part by CNCS-UEFISCDI under PCE-2011-3-0438, JRP-RO-FR-2014-16, and NVIDIA through a GPU card donation.

References

1. Urtasun, R., Darrell, T.: Sparse probabilistic regression for activity-independent human pose inference. In: CVPR (2008)
2. Ionescu, C., Li, F., Sminchisescu, C.: Latent structured models for human pose estimation. In: ICCV (2011)
3. Ionescu, C., Carreira, J., Sminchisescu, C.: Iterated second-order label sensitive pooling for 3D human pose estimation. In: CVPR (2014)
4. Shotton, J., Fitzgibbon, A., Cook, M., Sharp, T., Finocchio, M., Moore, R., Kipman, A., Blake, A.: Real-time human pose recognition in parts from single depth images. In: CVPR (2011)
5. Yang, Y., Ramanan, D.: Articulated human detection with flexible mixtures of parts. PAMI **35**, 2878–2890 (2013)
6. Bourdev, L., Maji, S., Brox, T., Malik, J.: Detecting people using mutually consistent poselet activations. In: Daniilidis, K., Maragos, P., Paragios, N. (eds.) ECCV 2010. LNCS, vol. 6316, pp. 168–181. Springer, Heidelberg (2010). doi:10.1007/978-3-642-15567-3_13
7. Ionescu, C., Papava, D., Olaru, V., Sminchisescu, C.: Human3.6m: large scale datasets and predictive methods for 3D human sensing in natural environments. PAMI **7**, 1325–1339 (2014)

8. Gallo, G., Grigoriadis, M.D., Tarjan, R.E.: A fast parametric maximum flow algorithm and applications. SIAM J. Comput. **18**, 30–55 (1989)
9. Kolmogorov, V., Boykov, Y., Rother, C.: Applications of parametric maxflow in computer vision. In: ICCV (2007)
10. Carreira, J., Sminchisescu, C.: CPMC: automatic object segmentation using constrained parametric min-cuts. In: PAMI (2012)
11. Ladicky, L., Torr, P.H.S., Zisserman, A.: Human pose estimation using a joint pixel-wise and part-wise formulation. In: CVPR (2013)
12. Wang, H., Koller, D.: Multi-level inference by relaxed dual decomposition for human pose segmentation. In: CVPR (2011)
13. Ghiasi, G., Yang, Y., Ramanan, D., Fowlkes, C.C.: Parsing occluded people. In: CVPR (2014)
14. Xia, W., Song, Z., Feng, J., Cheong, L.-F., Yan, S.: Segmentation over detection by coupled global and local sparse representations. In: Fitzgibbon, A., Lazebnik, S., Perona, P., Sato, Y., Schmid, C. (eds.) ECCV 2012. LNCS, vol. 7576, pp. 662–675. Springer, Heidelberg (2012). doi:10.1007/978-3-642-33715-4_48
15. Ferrari, V., Marin, M., Zisserman, A.: Pose search: retrieving people using their pose. In: CVPR (2009)
16. Andriluka, M., Roth, S., Schiele, B.: Pictorial structures revisited: people detection and articulated pose estimation. In: CVPR (2009)
17. Zuffi, S., Freifeld, O., Black, M.J.: From pictorial structures to deformable structures. In: CVPR (2012)
18. Zuffi, S., Romero, J., Schmid, C., Black, M.J.: Estimating human pose with flowing puppets. In: ICCV (2013)
19. Boussaid, H., Kokkinos, I.: Fast and exact: ADMM-based discriminative shape segmentation with loopy part models. In: CVPR (2014)
20. Alpert, S., Galun, M., Basri, R., Brandt, A.: Image segmentation by probabilistic bottom-up aggregation and cue integration. In: CVPR (2007)
21. Kumar, M.P., Torr, P., Zisserman, A.: OBJCUT: efficient segmentation using top-down and bottom-up cues. PAMI **32**, 530–545 (2010)
22. Leibe, B., Leonardis, A., Schiele, B.: Robust object detection with interleaved categorization and segmentation. IJCV **77**, 259–289 (2008)
23. Pishchulin, L., Andriluka, M., Gehler, P., Schiele, B.: Poselet conditioned pictorial structures. In: CVPR (2013)
24. Flohr, F., Gavrila, D.M.: PedCut: an iterative framework for pedestrian segmentation combining shape models and multiple data cues. In: BMVC (2013)
25. Russell, B.C., Efros, A., Sivic, J., Freeman, W.T., Zisserman, A.: Segmenting scenes by matching image composites. In: NIPS (2009)
26. Rosenfeld, A., Weinshall, D.: Extracting foreground masks towards object recognition. In: ICCV (2011)
27. Long, J., Shelhamer, E., Darrell, T.: Fully convolutional networks for semantic segmentation. In: CVPR (2015)
28. Ren, S., He, K., Girshick, R., Sun, J.: Faster R-CNN: towards real-time object detection with region proposal networks. In: NIPS (2015)
29. Lin, G., Shen, C., Ian, R., van dan Hengel, A.: Efficient piecewise training of deep structured models for semantic segmentation. In: CVPR (2016)
30. Kuettel, D., Ferrari, V.: Figure-ground segmentation by transferring window masks. In: CVPR (2012)
31. Gu, C., Arbeláez, P., Lin, Y., Yu, K., Malik, J.: Multi-component models for object detection. In: Fitzgibbon, A., Lazebnik, S., Perona, P., Sato, Y., Schmid, C. (eds.) ECCV 2012. LNCS, vol. 7575, pp. 445–458. Springer, Heidelberg (2012). doi:10.1007/978-3-642-33765-9_32

32. Lempitsky, V., Blake, A., Rother, C.: Image segmentation by branch-and-mincut. In: Forsyth, D., Torr, P., Zisserman, A. (eds.) ECCV 2008. LNCS, vol. 5305, pp. 15–29. Springer, Heidelberg (2008). doi:10.1007/978-3-540-88693-8_2

33. Ren, X., Malik, J.: Learning a classification model for segmentation. In: ICCV (2003)

34. Arbelaez, P., Maire, M., Fowlkes, C., Malik, J.: Contour detection and hierarchical image segmentation. PAMI (2010)

35. Malisiewicz, T., Efros, A.: Improving spatial support for objects via multiple segmentations. In: BMVC (2007)

36. van de Sande, K.E., Uijlings, J.R., Gevers, T., Smeulders, A.W.: Segmentation as selective search for object recognition. In: ICCV (2011)

37. Brox, T., Bourdev, L., Maji, S., Malik, J.: Object segmentation by alignment of poselet activations to image contours. In: CVPR (2011)

38. Endres, I., Hoiem, D.: Category independent object proposals. In: Daniilidis, K., Maragos, P., Paragios, N. (eds.) ECCV 2010. LNCS, vol. 6315, pp. 575–588. Springer, Heidelberg (2010). doi:10.1007/978-3-642-15555-0_42

39. Kim, J., Grauman, K.: Shape sharing for object segmentation. In: Fitzgibbon, A., Lazebnik, S., Perona, P., Sato, Y., Schmid, C. (eds.) ECCV 2012. LNCS, vol. 7578, pp. 444–458. Springer, Heidelberg (2012). doi:10.1007/978-3-642-33786-4_33

40. Levinshtein, A., Sminchisescu, C., Dickinson, S.: Optimal contour closure by superpixel grouping. In: Daniilidis, K., Maragos, P., Paragios, N. (eds.) ECCV 2010. LNCS, vol. 6312, pp. 480–493. Springer, Heidelberg (2010). doi:10.1007/978-3-642-15552-9_35

41. Maire, M., Yu, S.X., Perona, P.: Object detection and segmentation from joint embedding of parts and pixels. In: ICCV (2011)

42. Dong, J., Chen, Q., Yan, S., Yuille, A.: Towards unified object detection and semantic segmentation. In: Fleet, D., Pajdla, T., Schiele, B., Tuytelaars, T. (eds.) ECCV 2014. LNCS, vol. 8693, pp. 299–314. Springer, Heidelberg (2014). doi:10.1007/978-3-319-10602-1_20

43. Maire, M., Arbelaez, P., Fowlkes, C., Malik, J.: Using contours to detect and localize junctions in natural images. In: CVPR (2008)

44. Leordeanu, M., Sukthankar, R., Sminchisescu, C.: Efficient closed-form solution to generalized boundary detection. In: Fitzgibbon, A., Lazebnik, S., Perona, P., Sato, Y., Schmid, C. (eds.) ECCV 2012. LNCS, vol. 7575, pp. 516–529. Springer, Heidelberg (2012). doi:10.1007/978-3-642-33765-9_37

45. Carreira, J., Caseiro, R., Batista, J., Sminchisescu, C.: Semantic segmentation with second-order pooling. In: Fitzgibbon, A., Lazebnik, S., Perona, P., Sato, Y., Schmid, C. (eds.) ECCV 2012. LNCS, vol. 7578, pp. 430–443. Springer, Heidelberg (2012). doi:10.1007/978-3-642-33786-4_32

46. Belongie, S., Malik, J., Puzicha, J.: Shape matching and object recognition using shape contexts. PAMI **24**, 509–522 (2002)

47. Ryabko, B.Y., Stognienko, V., Shokin, Y.I.: A new test for randomness and its application to some cryptographic problems. J. Stat. Plan. Infer. **123**, 365–376 (2004)

48. Bourdev, L., Malik, J.: Poselets: body part detectors trained using 3D human pose annotations. In: ICCV (2009)

49. Andriluka, M., Pishchulin, L., Gehler, P., Schiele, B.: 2D human pose estimation: new benchmark and state of the art analysis. In: CVPR (2014)

Faces

Lip Reading in the Wild

Joon Son Chung[✉] and Andrew Zisserman

Visual Geometry Group, Department of Engineering Science,
University of Oxford, Oxford, England
joon@robots.ox.ac.uk

Abstract. Our aim is to recognise the words being spoken by a talking face, given only the video but not the audio. Existing works in this area have focussed on trying to recognise a small number of utterances in controlled environments (*e.g.* digits and alphabets), partially due to the shortage of suitable datasets.

We make two novel contributions: first, we develop a pipeline for fully automated large-scale data collection from TV broadcasts. With this we have generated a dataset with over a million word instances, spoken by over a thousand different people; second, we develop CNN architectures that are able to effectively learn and recognize hundreds of words from this large-scale dataset.

We also demonstrate a recognition performance that exceeds the state of the art on a standard public benchmark dataset.

1 Introduction

Lip-reading, the ability to understand speech using only visual information, is a very attractive skill. It has clear applications in speech transcription for cases where audio is not available, such as for archival silent films or (less ethically) off-mike exchanges between politicians and celebrities (the visual equivalent of open-mike mistakes). It is also complementary to the audio understanding of speech, and indeed can adversely affect perception if audio and lip motion are not consistent (as evidenced by the McGurk [23] effect). For such reasons, lip-reading has been the subject of a vast research effort over the last few decades. It has also been the subject of excellent comedy sketches, e.g. Seinfeld "The Lip Reader", and its ambiguity and challenge can be exploited to replace/overdub actual speech, *e.g.* in the YouTube channel "Bad Lip Reading".

Our objective in this work is a scalable approach to large lexicon *speaker independent* lip-reading. Furthermore, we aim to recognize words from *continuous speech*, where words are not segmented, and there may be co-articulation of the lips from preceding and subsequent words.

In lip-reading there is a fundamental limitation on performance due to *homophemes*. These are sets of words that sound different, but involve identical movements of the speaker's lips. Thus they cannot be distinguished using visual information alone. For example, in English the phonemes 'p' 'b' and 'm' are visually identical, and consequently the words *mark*, *park* and *bark*, are homophemes (as

© Springer International Publishing AG 2017
S.-H. Lai et al. (Eds.): ACCV 2016, Part II, LNCS 10112, pp. 87–103, 2017.
DOI: 10.1007/978-3-319-54184-6_6

are *pat*, *bat* and *mat*) and so cannot be distinguished by lip-reading. This problem has been well studied and there are lists of ambiguous phonemes and words available [8,21]. It is worth noting that the converse problem also applies: for example '*m*' and '*n*' are easily confused as audio, but are visually distinct. We take account of such homopheme ambiguity in assessing the performance of our methods.

Apart from this limitation, lip-reading is a challenging problem in any case due to intra-class variations (such as accents, speed of speaking, mumbling), and adversarial imaging conditions (such as poor lighting, strong shadows, motion, resolution, foreshortening, etc.).

The usual approach to inference for temporal sequences is to employ sequence models such as Hidden Markov Models or Recurrent Neural Networks (e.g. LSTMs). For lip-reading such models can be employed for predicting individual characters or phonemes. In contrast, we investigate using Convolutional Neural Networks (CNNs) for directly recognizing individual *words* from a sequence of lip movements.

Clearly, visual registration is an important element to consider in the design of the networks. Typically, the imaged head will move in the video, either due to actual movement of the head or due to camera motion. One approach would be to tightly register the mouth region (including lips, teeth and tongue, that all contribute to word recognition), but another is to develop networks that are tolerant to some degree of motion jitter. We take the latter approach, and do not enforce tight registration.

We make contributions in two areas: first, we develop a pipeline for automated large scale data collection, including visual and temporal alignment. With this we are able to obtain training data for hundreds of distinct words, thousands of instances for each word, and over a thousand speakers (Sect. 2); second, we develop CNN architectures for classifying multi-frame time series of lips. In particular we propose and compare different input and temporal fusion architectures, and discuss their pros and cons (Sect. 3). We analyse the performance and ambiguity of the resulting classifications in Sect. 4.

As discussed in the related work below, in these three aspects: speaker independence, learning from continuous speech, and lexicon (vocabulary) size, we go far beyond the current state of the art. We also exceed the state of the art in terms of performance, as is also shown in Sect. 4 by comparisons on the standard OuluVS benchmark dataset [1,43].

1.1   Related Work

Research on lip reading (*a.k.a.* visual speech recognition) has a long history. A thorough survey of shallow (*i.e.* not deep learning) methods is given in the recent review [45], and will not repeated in detail here. Many of the existing works in this field have followed similar pipelines which first extract spatio-temporal features around the lips (either motion-based, geometric-feature based or both), and then align these features with respect to a canonical template. For example, Pei *et al.* [28], which holds state-of-the-art on many datasets, extracts the patch trajectory as a spatiao-temporal feature, and then aligns these features to reference motion patterns.

A number of recent papers have used deep learning methods to tackle problems related to lip reading. Koller *et al.* [16] train an image classifier CNN to discriminate *visemes* (mouth shapes, visual equivalent of *phonemes*) on a sign language dataset where the signers mouth words. Similar CNN methods have been performed by [25] to predict *phonemes* in spoken Japanese. In the context of word recognition, [33] has used deep bottleneck features (DBF) to encode *shallow* input features such as LDA and GIF [36]. Similarly [29] uses DBF to encode the image for every frame, and trains a LSTM classifier to generate a word-level classification.

One of the major obstacle to progress in this field has been the lack of suitable datasets [45]. Table 1 gives a summary of existing datasets. The amount of available data is far from sufficient to train scalable and representative models that will be able to generalise beyond the controlled environments and the very limited domains (*e.g.* digits and the alphabet).

Table 1. Existing lip reading datasets. **I** for **I**solated (one word, letter or digit per recording); **C** for **C**ontinuous recording. The reported performance is on speaker-independent experiments. (* For GRID [4], there are 51 classes in total, but the first word in a phrase is restricted to 4, the second word 4, etc. 8.5 is the average number of possible classes at each position in the phrase.)

Name	Env.	Output	I/C	# class	# subj.	Best perf.
AVICAR [19]	In-car	Digits	C	10	100	37.9% [7]
AVLetter [22]	Lab	Alphabet	I	26	10	43.5% [43]
CUAVE [27]	Lab	Digits	I	10	36	83.0% [26]
GRID [4]	Lab	Words	C	8.5*	34	79.6% [39]
OuluVS1 [43]	Lab	Phrases	I	10	20	89.7% [28]
OuluVS2 [1]	Lab	Phrases	I	10	52	73.5% [44]
OuluVS2 [1]	Lab	Digits	C	10	52	-
BBC TV	TV	Words	C	333/500	1000+	-

Word classification with large lexicons has not been attempted in lip reading, but [11] has tackled a similar problem in the context of text spotting. Their work shows that it is feasible to train a general and scalable word recognition model for a large pre-defined dictionary, as a multi-class classification problem. We take a similar approach.

Of relevance to the architectures and methods developed in this paper are ConvNets for action recognition that learn from multiple-frame image sequences such as [12,13,35], particularly the ways in which they capture spatio-temporal information in the image sequence using temporal pooling layers and 3D convolutional filters.

2   Building the Dataset

This section describes our multi-stage pipeline for automatically collecting and processing a very large-scale visual speech recognition dataset, starting from

British television programs. Using this pipeline we have been able to extract 1000s of hours of spoken text covering an extensive vocabulary of 1000s of different words, with over 1M word instances, and over 1000 different speakers.

Fig. 1. A sample of speakers in our dataset.

The key ideas are to: (i) obtain a temporal alignment of the spoken audio with a text transcription (broadcast as subtitles with the program). This in turn provides the time alignment between the visual face sequence and the words spoken; (ii) obtain a spatio-temporal alignment of the lower face for the frames corresponding to the word sequence; and, (iii) determine that the face is speaking the words (*i.e.* that the words are not being spoken by another person in the shot). The pipeline is summarised in Fig. 2 and the individual stages are discussed in detail in the following paragraphs.

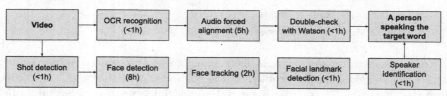

Fig. 2. Pipeline to generate the text and visually aligned dataset. Timings are for a one-hour video.

Stage 1. Selecting Program Types. We require programs that have a changing set of talking heads, so choose news and current affairs, rather than dramas with a fixed cast. Table 2 lists the programs. There is a significant variation of format across the programs – from the regular news where a single speaker is talking directly at the camera, to panel debate where the speakers look at each other and often shifts their attention. There are a few people who appear repeatedly in the videos (*e.g.* news presenter in BBC News or the host in the others), but the large majority of participants change every episode (Fig. 1).

Table 2. Video statistics. The yield is the proportion of useful face appearance relative to the total length of video. A useful face appearance is one that appears continuously for at least 5 s, with the face being that of the speaker.

Channel	Series name	Description	# vid.	Length	Yield
BBC 1 HD	News at 1	Regular news	1242	30 mins	39.9%
BBC 1 HD	News at 6	Regular news	1254	30 mins	33.9%
BBC 1 HD	News at 10	Regular news	1301	30 mins	32.9%
BBC 1 HD	Breakfast	Regular news	395	Varied	39.2%
BBC 1 HD	Newsnight	Current affairs debate	734	35 mins	40.0%
BBC 2 HD	World news	Regular news	376	30 mins	31.9%
BBC 2 HD	Question time	Current affairs debate	353	60 mins	48.8%

Fig. 3. Subtitles on BBC TV. **Left**: 'Question Time', **Right**: 'BBC News at One'.

Stage 2. Subtitle Processing and Alignment. We require the alignment between the audio and the subtitle in order to get a timestamp for every word that is being spoken in the videos. The BBC transmits subtitles as bitmaps rather than text, therefore subtitle text is extracted from the broadcast video using standard OCR methods [2,6]. The subtitles are not time-aligned, and also not verbatim as they are generated live. The Penn Phonetics Lab Forced Aligner [9,41] (based on the open-source HTK toolbox [40]) is used to force-align the subtitle to the audio signal. The aligner uses the Viterbi algorithm to compute the maximum likelihood alignment between the audio (modelled by PLP features [30]) and the text. This method of obtaining the alignment has significant performance benefits over regular speech recognition methods that do not use prior knowledge of what is being said. The alignment result, however, is not perfect due to: (1) the method often misses words that are spoken too quickly; (2) the subtitles are not verbatim; (3) the acoustic model is only trained to recognise American English. The noisy labels are filtered by double-checking against the commercial IBM Watson Speech to Text service. In this case, the only remaining label noise is where an interview is dubbed in the news, which is rare.

Stage 3. Shot Boundary Detection, Face Detection, and Tracking. The shot boundaries are determined to find the within-shot frames for which face tracking is to be run. This is done by comparing color histograms across consecutive frames [20]. The HOG-based face detection method of [15] is performed on every frame of the video (Fig. 4 left). As with most face detection methods, this results in many false positives and some missed detections. In a similar manner to [6], all face detections of the same person are grouped across frames using a KLT tracker [34] (Fig. 4 middle). If the track overlaps with face detections on the majority of frames, it is assumed to be correctly tracking the face.

Fig. 4. Left: face detections; **Middle:** KLT features and the tracked bounding box (in yellow); **Right:** facial landmarks. (Color figure online)

Stage 4. Facial Landmark Detection and Speaker Identification. Facial landmarks are needed to (1) determine the mouth position for cropping; and (2) for speaker/non-speaker classification. Facial landmarks are determined in every frame of the face track using the method of [14] (Fig. 4 right). To identify who is speaking, we assume that a person speaking will have lip movements that fall within a particular frequency range that is different to that arising from tracking noise. The 'openness' of the mouth is measured on every frame using the distance between the top and the bottom lip, normalised with respect to the size of the face in the video. For a speaking face, the openness signal contains the actual lip motion as well as the tracking noise, whereas for a non-speaking face (*e.g.* reaction shot, etc.), the only observed movement is the noise. A simple method of taking the Fourier transform of the mouth 'openness' temporal signal is performed to separate the lip movements that fall into different frequencies bins. A linear SVM classifier is trained on the frequency spectrum to make the distinction between a face that is speaker from a face that is not.

Stage 5. Compiling the Training and Test Data. The training, validation and test sets are disjoint in time. The dates of videos corresponding to each set is shown in Table 3. Note that we leave a week's gap between the test set and the rest in case any news footage is repeated. The lexicon is obtained by selecting the 500 most frequently occurring words between 5 and 10 characters in length (Fig. 6 gives the word duration statistics). This word length is chosen such that the speech duration does not exceed the fixed one-second bracket that is used in the recognition architecture, whilst shorter words are not included because there are too many ambiguities due to homophemes (*e.g.* 'bad', 'bat', 'pat', 'mat', etc. are all visually identical), and sentence-level context would be needed to disambiguate these.

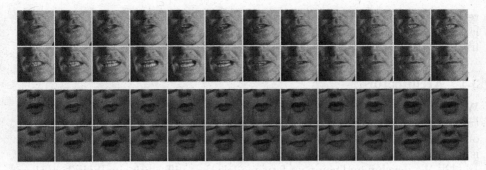

Fig. 5. One-second clips that contain the word '*about*'. Top: male speaker, bottom: female speaker.

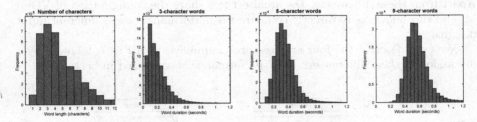

Fig. 6. Word statistics. Regardless of the actual duration of the word, we take a 1-second clip for training and test.

These 500 words occur at least 800 times in the training set, and at least 40 times in each of the validation and test sets. For each of the occurrences, the one-second clip is taken, and the face is cropped with the mouth centered using the registration found in Stage 4. The words are *not* isolated, as is the case in other lip-reading datasets; as a result, there may be co-articulation of the lips from preceding and subsequent words. The *test* set is manually checked for errors.

Table 3. Dataset statistics.

Set	Dates	# class	#/class
Train	01/01/2010–28/02/2015	500	800+
Val	01/03/2015–25/07/2015	500	50
Test	01/08/2015–31/03/2016	500	50

3 Network Architecture and Training

The task for the network is to predict which words are being spoken, given a video of a talking face. The input format to the network is a sequence of mouth regions, as shown in Fig. 5. Previous attempts at visual speech recognition have relied on very precise localisation of the facial landmarks (the mouth in particular); our aim is learn from from more noisy data, and tolerate some localisation irregularities both in position and in time.

3.1   Architecture

We cast the problem as one of multi-way classification, and so base our architecture on ones designed for image classification [3,18,32]. In particular, we build on the VGG-M model [3] since this has a good classification performance, but is much faster to train and experiment on than deeper models, such as VGG-16 [32]. We develop and compare four models that differ principally in how they 'ingest' the T input frames (where here T = 25 for a 1 s interval). These variations take inspiration from previous work on human action classification [12,13,35,42]. Apart from these differences, the architectures share the configuration of VGG-M, and this allows us to directly compare the performance across different input designs.

We next describe the four architectures, summarized in Fig. 7, followed by a discussion of their differences. Their performance is compared in Sect. 4.

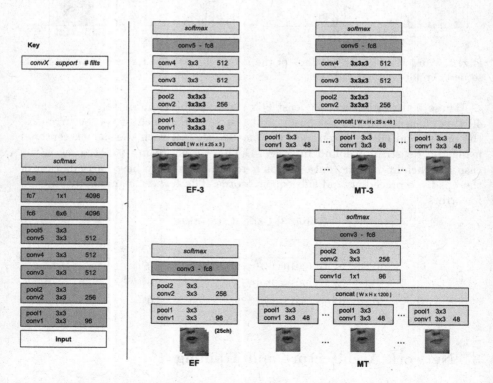

Fig. 7. CNN architectures. **Left:** VGG-M architecture that is used as a base. **Right:** **EF-3:** 3D convolution with early fusion; **MT-3:** 3D convolution with multiple towers; **EF:** early fusion; **MT:** multiple towers.

3D Convolution with Early Fusion (EF-3). This architecture is inspired by the work of [12] on human action recognition using 3D ConvNets. The general structure resembles that of an ordinary CNN used for image classification, but instead of taking H×W×3 input, it takes H×W×T×3 input. The convolutional and pooling filters operate and move along all three dimensions.

3D Convolution with Multiple Towers (MT-3). The model shares its basic design principles with the architecture of **EF-3**, however there is no explicit time-domain connectivity between frames before *conv2*. There are T = 25 towers with common *conv1* layers (with shared weights), each of which takes an input frame. Here, the activations at *pool1* are concatenated along a new dimension, and the 3D convolutions from *conv2* are performed in the same manner as [12] and **EF-3**.

Early Fusion (EF). The network ingests a T-channel image, where each of the channels encode an individual frame in *greyscale*. The layer structure for the subsequent layers is identical to that of the regular VGG-M network. This method is related to the Early Fusion model in [13], which takes *colour* images and uses a T × 3-channel convolutional filter at *conv1*. We did experiment with 25 × 3-channel colour input, but found that the increased number of parameters at *conv1* made training difficult due to overfitting (resulting in validation performance that is around 5% weaker; not quoted in Sect. 4).

Multiple Towers (MT). There are T = 25 towers with common *conv1* layers (with shared weights), each of which takes an input frame. The activations from the towers are concatenated channel-wise after *pool1*, producing an output activation with 1200 channels. The subsequent 1 × 1 convolution is performed to reduce this dimension, to keep the number of parameters at *conv2* at a manageable level. The rest of the network is the same as the regular VGG-M.

Discussion. There are two basic divisions of the architectures: between early fusion and multiple towers, and between 2D and 3D convolutions. We will discuss these in turn. The early fusion architectures, **EF-3** and **EF**, share similarities with previous work on human action recognition using ConvNets [12,13,42] in the way that they assume registration between frames. The models perform time-domain operations beginning from the first layer to precisely capture local motion direction and speed [13]. For these methods to capture useful information, good registration of details between frames is critical. However, we are not imposing strict registration, and in any case it goes slightly against the signal (lip motion and mouth region deformation) that we are trying to capture.

In contrast, the multiple towers architectures, **MT-3** and **MT**, both delay all time-domain registrations (and operations) until after the first set of convolutional and pooling layers. This gives tolerance against minor registration errors (the receptive field size at *conv2* is 11 pixels). Note, the common *conv1* layers of the multiple towers ensures that the same filter weights are used for all frames, whereas in the early fusion architecture **EF** it is possible to learn different weights for each frame. The experimental results show that these registration-tolerant models gives a modest improvement over their counterparts, and the performance improvement is likely to be more significant where the tracking quality is less ideal.

The reason for including 3D convolutions (the architectures **EF-3** and **MT-3**) is that intuitively a 3D convolution (that can have small spatial and *temporal* kernel size) should be able to match well a spatio-temporal feature, such as a particular lip shape over a particular sub-sequence. In contrast the 2D convolutions extend over the entire temporal range, and thus might be thought to waste parameters or require redundancy when trying to respond to such spatio-temporal features. Despite this intuition, the experimental results show that the 2D convolutions are superior to their 3D counterparts.

One other design choice is the size of the input images. This was chosen as 112×112 pixels, which is smaller than that typically used in image classification networks. The reason is that the size of the cropped mouth images are rarely larger than 112×112 pixels, and this smaller choice means that smaller filters can be used at *conv1* (than those used in VGG-M) without sacrificing receptive fields, but at a gain in avoiding unnecessary parameters being learnt.

3.2   Training

Data Augmentation. Data augmentation often helps to improve validation performance by reducing overfitting in ConvNet image classification tasks [18]. We apply the augmentation techniques used on the ImageNet classification task by [18,32] (*e.g.* random cropping, flipping, colour shift), with a consistent transformation applied to all frames of a single clip. To further augment the training data, we make random shifts in time by up to 0.2 s, which improves the *top-1* validation error by 3.5% compared to the standard ImageNet augmentation methods. It was not feasible to scale in the time-domain as this results in artifacts being shown due to the relatively low video refresh rate of 25 fps.

Details. Our implementation is based on the MATLAB toolbox MatConvNet [37] and trained on a NVIDIA Titan X GPU with 12GB memory. The network is trained using SGD with momentum 0.9 and batch normalisation [10], but without dropout. The training was stopped after 20 epochs, or when the validation error did not improve for 3 epochs, whichever is sooner. The learning rate of 10^{-2} to 10^{-4} was used, decreasing on log scale.

4   Experiments

In this section we evaluate and compare the several proposed architectures, and discuss the challenges arising from the visual ambiguities between words. We then compare to the state of the art on a public benchmark.

4.1   Comparison of Architectures

Evaluation Protocol. The models are evaluated on the independent test set (Sect. 2). We report *top-1* and *top-10* accuracies, as well as recall against rank curves. Here, the *'Recall@K'* is the proportion of times that the correct class is found in the top-K predictions for the word. We also report the character-level edit distance [17], which is the minimum number of character-level operations required to convert the predicted string to the ground truth. This metric imposes smaller penalties where the predicted string is similar to the ground truth (*e.g.* 'concerned' and 'concerns' have an edit distance of 2) and larger penalties where the words are very different (*e.g.* 'concerned' and 'company' have an edit distance of 6).

Results. As discussed in Sect. 3.1, the **MT-3** and **MT** variants have the advantage of being more tolerant to registration errors compared to their early fusion counterparts. The results in Table 4 and Fig. 8 confirm this, where we see a modest (3.2% on average for *top-1*) but consistent improvement in performance across the experiments. The performance of 3D ConvNets fall short of the 2D architectures by an average of around 14%.

The recall curves in Fig. 8 rise sharply for all models at low-K; the *top-10* figure for the **EF** and **MT** models being over 85%, despite the modest *top-1* figure of around 60%. This is a result of ambiguities in lip reading, which we will discuss next.

Table 4. Word classification results. **Left:** on the BBC data for the four different architectures. **ED** is the edit distance. **Right:** on OuluVS1 and OuluVS2 (short phrases, frontal view).

Net	500-class			333-class			OuluVS1	OuluVS2
	Top-1	Top-10	ED	Top-1	Top-10		Top-1	Top-1
EF-3	43.9%	81.0%	3.13	55.7%	87.9%	[29]	81.8%	-
MT-3	46.2%	82.4%	2.97	56.8%	88.7%	[44]	85.6%	73.5%
EF	57.0%	88.8%	2.32	63.2%	91.8%	[28]	89.7%	-
MT	**61.1%**	**90.4%**	**2.06**	**65.4%**	**92.3%**	MT	**91.4%**	**93.2%**

4.2 Analysis of Confusions

Here, we examine the classification results, in particular, the scenarios in which the network fails to correctly classify the spoken word. Table 5 shows the most common confusions between words in the test set. This is generated by taking the largest off-diagonal values in the word confusion matrix. This result confirms our prior knowledge about the challenges in visual speech recognition – almost all of the top confusions are either (i) a plural of the original word (*e.g.* 'report' and 'reports') which is ambiguous because one word is a subset of the other, and the words are not isolated in our dataset so this can be due to co-articulation; or (ii) a known homopheme visual ambiguity (explained in Sect. 1) where the words cannot be distinguished using visual information alone (*e.g.* 'billion' and 'million', 'worse' and 'worst').

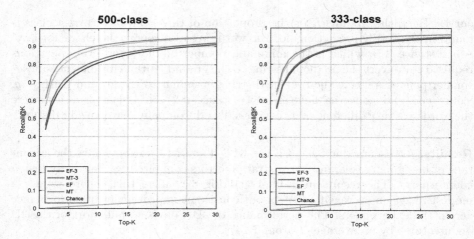

Fig. 8. Recall vs rank curves for the word classification.

Table 5. Most frequently confused word pairs.

500-class			333-class	
0.32	BENEFITS	BENEFIT	0.30	BORDER	IMPORTANT
0.31	QUESTIONS	QUESTION	0.29	PROBABLY	PROBLEM
0.31	REPORT	REPORTS	0.27	TAKING	TAKEN
0.31	BORDER	IMPORTANT	0.25	PERSONAL	PERSON
0.31	AMERICA	AMERICAN	0.23	CLAIMS	GAMES
0.29	GROUND	AROUND	0.22	AROUND	GROUND
0.28	RUSSIAN	RUSSIA	0.21	TONIGHT	NIGHT
0.28	FIGHT	FIGHTING	0.21	PROBLEM	PROBABLY
0.26	FAMILY	FAMILIES	0.19	SEVERAL	SEVEN
0.26	AMERICAN	AMERICA	0.19	CHALLENGE	CHANGE
0.26	BENEFIT	BENEFITS	0.18	PRICES	PERSON
0.25	ELECTIONS	ELECTION	0.18	WARNING	MORNING
0.24	WANTS	WANTED	0.18	CAPITAL	HAPPENED
0.24	HAPPEN	HAPPENED	0.18	OTHER	ANOTHER
0.24	FORCE	FORCES	0.17	AHEAD	AGAIN
0.23	HAPPENED	HAPPEN	0.16	WORKERS	WORDS
0.23	SERIOUS	SERIES	0.16	MEDIA	MEETING
0.23	TROOPS	GROUPS	0.16	UNITED	NIGHT
0.22	QUESTION	QUESTIONS	0.16	NEVER	SEVEN
0.21	PROBLEM	PROBABLY	0.15	WORLD	WORDS

Therefore, we generate a second test set where we eliminate these two types of known ambiguities. We first group the words according to the aforementioned criteria (*e.g.* 'billion', 'million' and 'millions' would form a single group), and keep only the most frequently occuring word in the training set for each group, eliminating the ambiguous words for that group. This process produces a new balanced test set containing a lexicon of 333 word-classes.

The network is finetuned on this new vocabulary for 1 epoch, before being re-evaluated. The results reported in Table 4 and Fig. 8 that are labelled '333-word' are evaluated on this vocabulary. The *top-10* performance increases from 90.4% (for the 500 word-class test set) to 92.3% (for the 333 word-class test set). This is an improvement, but still not perfect. The reason is that even excluding the known homopheme and plural ambiguities does not remove all confusion. Table 5 shows the common errors remaining, and these are phonetically understandable. For example, some of the most common confusions, e.g. 'claims' which is phonetically (K L EY M Z) and 'games' (G EY M Z), 'probably' (P R AA B AH B L IY) and 'problem' (P R AA B L AH M), actually share most of the phonemes.

Apart from these difficulties, the failure cases are typically for extreme samples. For example, due to strong international accents, or poor quality/low bandwidth location reports and Skype interviews, where there are motion compression artifacts or frames dropped from the transmission.

4.3   Visualisation of Salient Mouth Shapes

Our aim here is to visualize the frames of the temporal sequence that are most discriminative for the word. Simonyan *et al.* [31] have shown that it is possible to infer the localization of visual objects in an image as a saliency map for a network trained to classify images. We adapt this method to find the salient temporal information in a time-sequence.

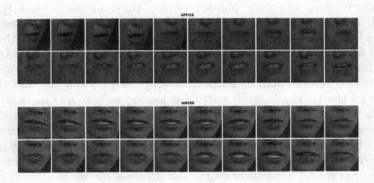

Fig. 9. Salient visual features of sequences 'office' and 'water' are highlighted in red. (Color figure online)

The method approximates the relation between the class score S and the input image I (represented as a vector) as $S(I) = w^T I + b$. The vector w is the same size as the input image, and the magnitude of its elements signify the influence of the corresponding elements of the image on the class score. Hence the magnitude of w determines a saliency map on the image. The vector w can be obtained as $w = \frac{\partial S_c}{\partial I}\big|_{I_0}$ and this derivative is obtained by back-prop from the class score $S_0(I_0)$ to the image.

The resulting salient regions are shown in Fig. 9. For example, the most distinctive mouth shape for 'office' (AO F AH S) is the 'AH' with the mouth open and 'F' with the top teeth biting the bottom lip.

4.4 Comparison to State of the Art

It is worth noting that the *top-1* classification accuracy of 65%, shown in Table 4, is comparable to that of many of the recent works [7,24,29] performed on lexicon sizes that are orders of magnitude smaller (Table 1).

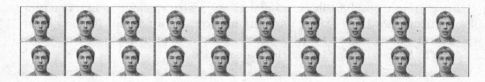

Fig. 10. Original video frames for 'hello' on OuluVS. Compare this to the our original input frames in Fig. 3.

OuluVS. We evaluate our method on the OuluVS datasets. OuluVS1 [43] consists of 20 subjects uttering 10 phrases (*e.g.* 'thank you', 'hello', etc.), and has been widely used in previous works. OuluVS2 [1] (short phrases) consists of 52 subjects uttering the same phrases as [43]. Here, we assess on a speaker-independent experiment, where some of the subjects are reserved for testing.

To apply our method on this dataset, we pre-train the model on the BBC data, and fine-tune the fully-connected layers. Training from scratch on OuluVS underperforms as the size of this dataset is insufficient to train a deep network. If the phrase is shorter than 25 frames, we simply repeat the first and the last frames to fill the 1-second clip. If the clip is longer, we take a random crop.

As can be seen in Table 4 our method achieves a strong performance, and sets the new state-of-the-art. Note that, without retraining the convolutional part of the network, we achieve these strong results on videos that are very different to ours in terms of lighting, background, camera perspective, etc. (Fig. 10), which shows that our model generalises well across different formats.

5 Summary and Extensions

We have shown that CNN architectures can be used to classify temporal sequences with excellent results. On the 333-word test set, we achieve *top-1* accuracy of 65.4%, which exceeds state-of-the-art [7,43] on multiple datasets [19,22] that have lexicon sizes that are orders of magnitude smaller, and a *top-10* accuracy of 92.3%. We also demonstrate a recognition performance that exceeds the state of the art on a standard public benchmark dataset, OuluVS.

Next steps include extending to lip reading of profile views, and combining the CNNs pre-trained using this approach with LSTMs trained with a language model [5,38], in order to recognize sentences rather than individual words. Of course, the visual only speech recognition method developed here can also be combined with audio only speech recognition to both their benefits.

Acknowledgements. Funding for this research is provided by the EPSRC Programme Grant Seebibyte EP/M013774/1. We are very grateful to Rob Cooper and Matt Haynes at BBC Research for help in obtaining the dataset.

References

1. Anina, I., Zhou, Z., Zhao, G., Pietikäinen, M.: OuluVS2: a multi-view audiovisual database for non-rigid mouth motion analysis. In: 2015 11th IEEE International Conference and Workshops on Automatic Face and Gesture Recognition (FG), vol. 1, pp. 1–5. IEEE (2015)
2. Buehler, P., Everingham, M., Zisserman, A.: Learning sign language by watching TV (using weakly aligned subtitles). In: Proceedings of CVPR (2009)
3. Chatfield, K., Simonyan, K., Vedaldi, A., Zisserman, A.: Return of the devil in the details: delving deep into convolutional nets. In: Proceedings of BMVC (2014)
4. Cooke, M., Barker, J., Cunningham, S., Shao, X.: An audio-visual corpus for speech perception and automatic speech recognition. J. Acoust. Soc. Am. **120**(5), 2421–2424 (2006)
5. Donahue, J., Anne Hendricks, L., Guadarrama, S., Rohrbach, M., Venugopalan, S., Saenko, K., Darrell, T.: Long-term recurrent convolutional networks for visual recognition and description. In: Proceedings of the IEEE Conference on Computer Vision and Pattern Recognition, pp. 2625–2634 (2015)
6. Everingham, M., Sivic, J., Zisserman, A.: "Hello! My name is.. Buffy" - automatic naming of characters in TV video. In: Proceedings of BMVC (2006)
7. Fu, Y., Yan, S., Huang, T.S.: Classification and feature extraction by simplexization. IEEE Trans. Inf. Forensics Secur. **3**(1), 91–100 (2008)
8. Goldschen, A.J., Garcia, O.N., Petajan, E.D.: Rationale for phoneme-viseme mapping and feature selection in visual speech recognition. In: Stork, D.G., Hennecke, M.E. (eds.) Speechreading by Humans and Machines, pp. 505–515. Springer, Heidelberg (1996)
9. Hermansky, H.: Perceptual linear predictive (PLP) analysis of speech. J. Acoust. Soc. Am. **87**(4), 1738–1752 (1990)
10. Ioffe, S., Szegedy, C.: Batch normalization: accelerating deep network training by reducing internal covariate shift. arXiv preprint arXiv:1502.03167 (2015)
11. Jaderberg, M., Simonyan, K., Vedaldi, A., Zisserman, A.: Synthetic data and artificial neural networks for natural scene text recognition. In: Workshop on Deep Learning, NIPS (2014)
12. Ji, S., Xu, W., Yang, M., Yu, K.: 3D convolutional neural networks for human action recognition. IEEE PAMI **35**(1), 221–231 (2013)
13. Karpathy, A., Toderici, G., Shetty, S., Leung, T., Sukthankar, R., Fei-Fei, L.: Large-scale video classification with convolutional neural networks. In: Proceedings of the IEEE conference on Computer Vision and Pattern Recognition, pp. 1725–1732 (2014)
14. Kazemi, V., Sullivan, J.: One millisecond face alignment with an ensemble of regression trees. In: Proceedings of the IEEE Conference on Computer Vision and Pattern Recognition, pp. 1867–1874 (2014)
15. King, D.E.: Dlib-ml: a machine learning toolkit. J. Acoust. Soc. Am. **10**, 1755–1758 (2009)
16. Koller, O., Ney, H., Bowden, R.: Deep learning of mouth shapes for sign language. In: Proceedings of the IEEE International Conference on Computer Vision Workshops, pp. 85–91 (2015)

17. Kondrak, G.: A new algorithm for the alignment of phonetic sequences. In: Proceedings of the 1st North American chapter of the Association for Computational Linguistics conference, pp. 288–295. Association for Computational Linguistics (2000)
18. Krizhevsky, A., Sutskever, I., Hinton, G.E.: ImageNet classification with deep convolutional neural networks. In: NIPS, pp. 1106–1114 (2012)
19. Lee, B., Hasegawa-Johnson, M., Goudeseune, C., Kamdar, S., Borys, S., Liu, M., Huang, T.S.: AVICAR: audio-visual speech corpus in a car environment. In: INTERSPEECH. Citeseer (2004)
20. Lienhart, R.: Reliable transition detection in videos: a survey and practitioner's guide. Images Graph. **1**, 469–486 (2001)
21. Lucey, P., Martin, T., Sridharan, S.: Confusability of phonemes grouped according to their viseme classes in noisy environments. In: Proceedings of Australian International Conference on Speech Science & Technical, pp. 265–270 (2004)
22. Matthews, I., Cootes, T.F., Bangham, J.A., Cox, S., Harvey, R.: Extraction of visual features for lipreading. IEEE Trans. Pattern Anal. Mach. Intell. **24**(2), 198–213 (2002)
23. McGurk, H., MacDonald, J.: Hearing lips and seeing voices. Nature **264**, 746–748 (1976)
24. Ngiam, J., Khosla, A., Kim, M., Nam, J., Lee, H., Ng, A.Y.: Multimodal deep learning. In: Proceedings of the 28th International Conference on Machine Learning (ICML 2011), pp. 689–696 (2011)
25. Noda, K., Yamaguchi, Y., Nakadai, K., Okuno, H.G., Ogata, T.: Lipreading using convolutional neural network. In: INTERSPEECH, pp. 1149–1153 (2014)
26. Papandreou, G., Katsamanis, A., Pitsikalis, V., Maragos, P.: Adaptive multimodal fusion by uncertainty compensation with application to audiovisual speech recognition. IEEE Trans. Audio Speech Lang. Process. **17**(3), 423–435 (2009)
27. Patterson, E.K., Gurbuz, S., Tufekci, Z., Gowdy, J.N.: CUAVE: a new audio-visual database for multimodal human-computer interface research. In: 2002 IEEE International Conference on Acoustics, Speech, and Signal Processing (ICASSP), vol. 2, pp. II-2017. IEEE (2002)
28. Pei, Y., Kim, T.K., Zha, H.: Unsupervised random forest manifold alignment for lipreading. In: Proceedings of the IEEE International Conference on Computer Vision, pp. 129–136 (2013)
29. Petridis, S., Pantic, M.: Deep complementary bottleneck features for visual speech recognition. ICASSP, pp. 2304–2308 (2016)
30. Rubin, S., Berthouzoz, F., Mysore, G.J., Li, W., Agrawala, M.: Content-based tools for editing audio stories. In: Proceedings of the 26th Annual ACM Symposium on User Interface Software and Technology, pp. 113–122. ACM (2013)
31. Simonyan, K., Vedaldi, A., Zisserman, A.: Deep inside convolutional networks: visualising image classification models and saliency maps. In: Workshop at International Conference on Learning Representations (2014)
32. Simonyan, K., Zisserman, A.: Very deep convolutional networks for large-scale image recognition. In: International Conference on Learning Representations (2015)
33. Tamura, S., Ninomiya, H., Kitaoka, N., Osuga, S., Iribe, Y., Takeda, K., Hayamizu, S.: Audio-visual speech recognition using deep bottleneck features and high-performance lipreading. In: 2015 Asia-Pacific Signal and Information Processing Association Annual Summit and Conference (APSIPA), pp. 575–582. IEEE (2015)
34. Lucas, B.D., Kanade, T.: An iterative image registration technique with an application to stereo vision, Vancouver, BC, Canada (1981)

35. Tran, D., Bourdev, L., Fergus, R., Torresani, L., Paluri, M.: Learning spatiotemporal features with 3D convolutional networks (2015)
36. Ukai, N., Seko, T., Tamura, S., Hayamizu, S.: GIF-LR: GA-based informative feature for lipreading. In: Signal & Information Processing Association Annual Summit and Conference (APSIPA ASC), pp. 1–4. IEEE (2012)
37. Vedaldi, A., Lenc, K.: Matconvnet - convolutional neural networks for matlab. CoRR abs/1412.4564 (2014)
38. Vinyals, O., Toshev, A., Bengio, S., Erhan, D.: Show and tell: a neural image caption generator. In: Proceedings of the IEEE Conference on Computer Vision and Pattern Recognition, pp. 3156–3164 (2015)
39. Wand, M., Koutník, J., Schmidhuber, J.: Lipreading with long short-term memory. arXiv preprint arXiv:1601.08188 (2016)
40. Woodland, P.C., Leggetter, C., Odell, J., Valtchev, V., Young, S.J.: The 1994 HTK large vocabulary speech recognition system. In: 1995 International Conference on Acoustics, Speech, and Signal Processing, 1995. ICASSP-95, vol. 1, pp. 73–76. IEEE (1995)
41. Yuan, J., Liberman, M.: Speaker identification on the scotus corpus. IEEE Trans. Audio Speech Lang. Process. **123**(5), 3878 (2008)
42. Yue-Hei Ng, J., Hausknecht, M., Vijayanarasimhan, S., Vinyals, O., Monga, R., Toderici, G.: Beyond short snippets: deep networks for video classification. In: Proceedings of the IEEE Conference on Computer Vision and Pattern Recognition, pp. 4694–4702 (2015)
43. Zhao, G., Barnard, M., Pietikäinen, M.: Lipreading with local spatiotemporal descriptors. IEEE Trans. Audio Speech Lang. Process. **11**(7), 1254–1265 (2009)
44. Zhou, Z., Hong, X., Zhao, G., Pietikäinen, M.: A compact representation of visual speech data using latent variables. IEEE Trans. Audio Speech Lang. Process. **36**(1), 1–1 (2014)
45. Zhou, Z., Zhao, G., Hong, X., Pietikäinen, M.: A review of recent advances in visual speech decoding. IEEE Trans. Audio Speech Lang. Process. **32**(9), 590–605 (2014)

Recurrent Convolutional Face Alignment

Wei Wang[✉], Sergey Tulyakov, and Nicu Sebe

University of Trento, Trento, Italy
`wei.wang@unitn.it`

Abstract. Mainstream direction in face alignment is now dominated by cascaded regression methods. These methods start from an image with an initial shape and build a set of shape increments by computing features with respect to the current shape estimate. These shape increments move the initial shape to the desired location. Despite the advantages of the cascaded methods, they all share two major limitations: (i) shape increments are learned separately from each other in a cascaded manner, (ii) the use of standard generic computer vision features such SIFT, HOG, does not allow these methods to learn problem-specific features. In this work, we propose a novel Recurrent Convolutional Face Alignment method that overcomes these limitations. We frame the standard cascaded alignment problem as a recurrent process and learn all shape increments jointly, by using a recurrent neural network with the gated recurrent unit. Importantly, by combining a convolutional neural network with a recurrent one we alleviate hand-crafted features, widely adopted in the literature and thus allowing the model to learn task-specific features. Moreover, both the convolutional and the recurrent neural networks are learned jointly. Experimental evaluation shows that the proposed method has better performance than the state-of-the-art methods, and further support the importance of learning a single end-to-end model for face alignment.

1 Introduction

Face alignment methods trace their lineage from Active Shape Models [1,2] and Active Appearance Models (AAM) [3], developed a couple of decades ago. These works first build a statistical shape and appearance models of the face, and during testing use numerical optimization techniques to find a set of parameters of the statistical model that could have generated the query face. Todays mainstream face alignment methods belong to Cascaded Regression Methods (CRM) group [4–9]. These methods operate in a cascaded fashion, *i.e.* starting from an initial shape and producing several shape increments that move the initial shape closer to the desired location. Shape increments are learned in a supervised manner during training stage. Formally CRMs operate in the following fashion:

$$\Delta\mathbf{S}_{t+1} = \mathbf{R}_t(\mathbf{F}_t(\mathbf{I}, \hat{\mathbf{S}}_t)), \tag{1}$$

$$\hat{\mathbf{S}}_{t+1} = \hat{\mathbf{S}}_t + \Delta\mathbf{S}_{t+1}, \tag{2}$$

© Springer International Publishing AG 2017
S.-H. Lai et al. (Eds.): ACCV 2016, Part II, LNCS 10112, pp. 104–120, 2017.
DOI: 10.1007/978-3-319-54184-6_7

where $\mathbf{I}$ denotes a 2D image, $\mathbf{F}_t(\mathbf{I}, \hat{\mathbf{S}}_t)$ represents the feature values extracted using the previous shape estimate $\hat{\mathbf{S}}_t$, $\Delta\mathbf{S}_{t+1}$ is a shape update produced by the t-th regressor $\mathbf{R}_t$ in the cascade. To initialize the pipeline the average face shape over all images in the training set $\bar{\mathbf{S}}$ is taken. The feature extraction function $(\mathbf{F}_t(\cdot, \cdot))$ and a set of regressors $(\mathbf{R}_t(\cdot))$ constitute the main ingredients of a CRM framework. The final outcome of the CRMs writes as:

$$\hat{\mathbf{S}}(T) = \bar{\mathbf{S}} + \sum_{t=1}^{T} \Delta\mathbf{S}_t, \tag{3}$$

where T is the total number of layers in the cascade. In order to frame a task at hand as a cascaded regression problem, one has to decide upon the feature extraction function $(\mathbf{F}_t(\cdot, \cdot))$, as well as to select a proper regression function $(\mathbf{R}_t(\cdot))$. Various features have been explored by the community e.g. HoG [9], SIFT [5,7], pixel differences [10–12], local binary features [13], as well as different regression functions have been tried: linear regression [5,12], random ferns [14], regression trees [10,11]. This brings to light two major limitations of the CRMs, that we are going to remove in this work: (i) manually designed features and (ii) relative independence of the regressors at the different layers in the cascade.

Hand-crafted computer vision features, such as HoG features for pedestrian detection [15], SIFT features for object recognition [16], attribute detection [17,18] have played an important role in many application domains for a long time since they offer illumination, rotation and scaling invariance. These features, however, represent a generic image transformation that lacks any domain specific knowledge. Many works, have tackled this problem by *selecting* best features out of an overcomplete set [10,11,13]. However, this feature selection is suboptimal, since it is still performed on a generated set. Recently, it has been shown for object detection [19], tracking [20], image labeling [21] and other fields that features learned for a specific problem using deep convolutional neural networks show much better performance. Moreover, features learned for image classification often generalize well for different tasks, showing the ability of CNNs, such as AlexNet [19], VGGNet [22] and GoogleNet [23], to learn a generic image representation.

The second limitation of the CRMs is the independence of the regressors at every level of the cascade. One can argue the regressor at time t is learned by using the output of the previous regressor at time $t - 1$, with the final prediction given by Eq. 3. This however, affects only the feature computation (see Eq. 1), while the regressors themselves are learned independently. It has been shown in [5] that a single regressor is not capable of arriving at the desired location in a single step. As shown in Eq. 3 the final prediction of the cascade $\hat{\mathbf{S}}(T)$ is a function of the number of layers in the cascade T. One can think of $\hat{\mathbf{S}}(T)$ as a sequence of measurements of some stochastic process. It has been recently shown that Recurrent Neural Network are extremely powerful in modeling the sequential inputs and outputs [24]. In order to model long time-varying sequences, various RNN units have been proposed. In particular, long-short term memory cells and later Gated Recurrent Units have proven to be efficient in modeling time-varying processes and sequence-to-sequence learning [25]. Additionally, it has been shown that using a CNN for feature extraction and an RNN for classification brings extra advantages [26–28].

This discussion naturally brings us to the main contribution of this work. We present a unified face alignment framework that features end-to-end learning starting from raw pixel values. We replace the manually hand-crafted features $\mathbf{F}_t(\cdot, \cdot)$ by learning a patch-based CNN. In contrast with boosted regression methods, where one has a sequence of regressors $\{\mathbf{R}_1(\cdot), \mathbf{R}_2(\cdot), \ldots, \mathbf{R}_T(\cdot)\}$, our method learns a single recurrent module trained jointly with the CNN which can generate the regressor $\mathbf{R}(\cdot)$ recursively based on the input data and the memory of the recurrent module. We would like to highlight for the reader, that the parameters of both the CNN module and the RNN module are learned jointly. Additionally, we show that our model is capable of generalizing beyond the learned number of recurrent iterations, being able to automatically decide when to stop iterating. The experimental evaluation we detail in Sect. 4 proves that learning a task-specific end-to-end model brings higher accuracy than that of the available state-of-the-art.

2   Related Work

In this section we review relevant works in face alignment as well as discuss recent advances in the neural-network learning important to formulate our Recurrent Convolutional Face Alignment (RFCA) method.

2.1   Face Alignment

According to the widely accepted classification, methods for face alignment can be grouped into three broad categories [29]: Active Appearance Models (AAM), Constrained Local Models (CLM) [30–32], and Cascaded Regression Methods (CRM). Initial works on face alignment such as ASMs [1,2] and AAMs [3,33], build a parametric statistical shape and appearance models from a set of training faces. These methods show reasonable accuracy when the testing image is close to the training distribution. However, they fail to generalize to an unseen subject [34]. Although such methods still attract the attention of researchers [35,36], the more recent Cascaded Regression Methods have shown higher accuracy at impressive frame rates [11,13]. In the following we will mostly detail this latter group of works.

Initially CRMs were introduced in the medical image processing community for anatomic structure prediction [37]. Since then they have been extensively exploited by the computer vision community with many seminal works proposed in the literature. Currently this avenue of research represents the mainstream direction of the deformable shape fitting. In [14] a method for cascaded pose regression was introduced. The authors used a *pose-indexed features* and learned a sequence of weak-regressors (random ferns in their case) to regress a deformable shape from an image. In order to compute pose-indexed features one has to provide the current belief regarding the shape. This naturally brings some form of pose invariance to the framework. Later, these ideas were extended to regress the whole face shape [38]. Importantly, it was shown that regressing the whole shape imposes the result to lie in the space constructed by all the training images.

The supervised descent method (SDM) [5] further extends the cascaded framework to generic non-linear optimization problems: face alignment, template tracking and camera calibration. SDM learns a sequence of descent directions that applied sequentially solve the optimization problem. The authors replace the feature extraction part with SIFT [39] and achieve impressive results by using linear regressors in the layers of the cascade. A downside of SDM, is its inability to generalize well to non-frontal poses, requiring to train separate regressors depending on the detected head pose. This constraint is relaxed in [9] by introducing a global SDM to automatically learn several descent maps at every level of the cascade to handle complex cost-functions. These ideas were extended in [7], where the authors learn both the Jacobian and the Hessian matrices, in a manner inspired by the Gauss-Newton optimization method. Similarly to the original SDM, the authors use hand-crafted SIFT features extracted around the keypoints locations. SDM-based methods have become popular in various applications of face analysis [40] and are used in several commercially available face alignment systems[1]. A different strategy for feature extraction is presented in [10, 11, 13]. Instead of employing hand-crafted features (e.g., HoG, SIFT), they perform feature selection using a framework of regression trees. Alleviating the need to compute hand-crafted features, these works reach impressive processing speed.

Multiple CRM-based 3D methods have been proposed. In [41], an extension of [38] is introduced to fit a 2D-3D parametric shape model. Similar ideas were explored in [42], where a cascaded coupled regressor is introduced to obtain the camera projection matrix and the 3D landmarks of the face. The work in [10] proposes to include the third dimension directly into the learning pipeline. They used a large generated set of training faces and showed that considering a face shape as a 3D object gives better results. An interesting work in [12] uses binary features to track a large number of points on the face, with subsequent 3D deformable model fitting to obtain a 3D mesh of the face. Notably, this work shows impressive frame rates for the whole pipeline as well as the tracking accuracy comparable to purely 3D methods [43].

From a higher perspective the aforementioned methods have two independent steps: (i) feature extraction and (ii) applying a sequence of regressors. Typically the first step is performed by using some hand-crafted features such as SIFT [5, 7], HOG [9]. Some form of feature learning is employed in [10, 11, 13], while the levels of the cascade in the second step still remain independent. This requires a researcher to use a trial-and-fail approach in selecting which features and which regressors work the best.

In contrast, the method presented in our study is end-to-end. By learning convolutional filters, RCFA does not require manual supervision in defining

---

feature extraction functions. Additionally, our method replaces a cascade of independent regressors by a single recurrent model, where all iterations are leaned jointly. This formulation merges the two steps of the typical CRM pipeline into a single unified framework, simultaneously trained using the available data.

2.2 Recurrent and Convolutional Neural Networks

Recurrent Neural Networks (RNN) have become increasingly popular to learn complex dynamic systems, because of their impressive capability to recurrently operate with sequential input. During each recurrence of the traditional RNN [44], an input signal is mapped to the hidden state, which is passed forward to the next recurrence. This way, the information of the previous states is memorized and persists during the whole process. Therefore, RNNs have proven to have an advantage in modeling sequences with long-term dependencies. During the last decade, we have seen a lot of success in applying RNNs to various application domains, such as generating text description of videos [45], image caption generation [46], face aging [47], machine translation [24] and speech recognition [48].

Given the success of RNNs, a lot of RNN variants have been explored, such as the Long Short Term Memory (LSTM) [49] networks, Gated Recurrent Unit (GRU) [25], and Clockwork RNN [50]. All these architectures consist of a chain of repeated modules, where each module contains several gates, controlling the information flow in the network and states, memorizing necessary information for future recurrences. Although the combination of gates/states varies depending on the selected architectures, each subsequent iteration is performed similarly, by processing a new input using the memory of the current state. These architectures show varying performances for different tasks. In [51] it was shown that, in general, GRU-based models feature superior performance compared with other architectures.

Convolutional Neural Networks (CNN) have recently demonstrated notable success in multiple tasks, such as image classification [19], super-resolution [52], as well as image segmentation [53]. One of the main advantages of CNNs, is that they do not require human supervision to design feature transformation. Their feature representations have shown to provide significantly higher performance, compared to commonly adopted hard-crafted features, in numerous application domains. Thus, it is very promising to combine the RNN architecture together with the CNN architecture into a hybrid architecture. This hybrid architecture has been successfully applied to many tasks, such as scene labeling [26], object recognition [27], and text classification [28].

3   Method

The overview of the proposed Recurrent Convolutional Face Alignment method is given in Fig. 1. The framework mainly consists of two parts, the *recurrent module* and the *convolutional module*. During each recurrent iteration t the current shape

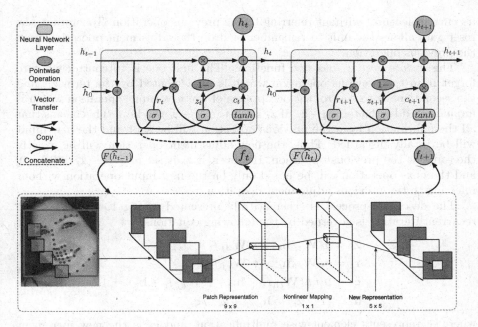

Fig. 1. The overview of the proposed approach. Top: the RNN with gated recurrent units unrolled in time. Bottom: the CNN architecture used for feature extraction. Note that feature extraction is performed at every recurrent iteration.

estimate $\mathbf{h}_t$ is imposed onto the image and the convolutional neural network is applied to the patches extracted around the points of the shape. The output of the last layer of the CNN is passed to the RNN as an input. During the first iteration, the average shape of all the images in the training set is set to the initial shape estimate: $\widehat{\mathbf{h}}_0 = \bar{\mathbf{S}}$.

3.1   Recurrent Module

In the current study we use an RNN with GRU module for its simplicity and superior performance as compared to other RNN types [51]. The structure of two recurrent iterations is given in the top row of Fig. 1. A GRU contains two gates and one state. The gates are the *reset* gate and the *update* gate. The *hidden* state $\mathbf{h}_t$ represents the relative movement or increment of the landmark positions after the adjustment in t-th iteration. Then the predicted position after t iterations is $\widehat{\mathbf{h}}_t = \widehat{\mathbf{h}}_0 + \mathbf{h}_t$.

The feature extraction function $f(t) = \mathbf{F}(\mathbf{I}, \widehat{\mathbf{h}}_t)$ is performed using a super resolution convolutional neural network (SRCNN), described in Sect. 3.

The *reset* gate $\mathbf{r}_t$ controls whether the adjustment from the previous recurrence should be ignored, *i.e.* if $\mathbf{r}_t$ is close to 0, the information of the previous adjustment operation will be forced to be discarded. Then the unit will focus on its current features without referring to the previous operation. To sum up, the reset gate allows the unit to remember or drop the adjustment operation from the previous operation.

The *update* gate $\mathbf{z}_t$ has two functions. The first one is to control what to forget from the previous operation which is implemented by the term $\mathbf{z}_t$, and the second one is to control the acceptance of the new input operation which is implemented by the term $1 - \mathbf{z}_t$. If $\mathbf{z}_t$ is set to 0, $1 - \mathbf{z}_t$ will be 1. This means that all the information from the previous operation will be kept and the new input will be totally discarded. Thus, the new adjustment operation will be exactly the same as the previous operation. However, if $\mathbf{z}_t$ is set to 1, $1 - \mathbf{z}_t$ will be 0, and the next operation will be based only on the new input operation without referring to the previous adjustment operation.

The described process is schematically presented in Fig. 1, where a single recurrent iteration is governed by the following equations:

$$\begin{aligned}
\mathbf{z}_t &= \sigma(\mathbf{W}_{zh}\mathbf{h}_{t-1} + \mathbf{W}_{zf}f_t + \mathbf{b}_z) \\
\mathbf{r}_t &= \sigma(\mathbf{W}_{rh}\mathbf{h}_{t-1} + \mathbf{W}_{rf}f_t + \mathbf{b}_r) \\
\mathbf{c}_t &= tanh(\mathbf{W}_{ch}\mathbf{r}_t \odot \mathbf{h}_{t-1} + \mathbf{W}_{cf}f_t + \mathbf{b}_c) \\
\mathbf{h}_t &= (1 - \mathbf{z}_t) \odot \mathbf{h}_{t-1} + \mathbf{z}_t \odot \mathbf{c}_t
\end{aligned} \tag{4}$$

where $\odot$ represents element-wise multiplication, and $\mathbf{c}_t$ is the new increment candidate created by the *tanh* layer that could be added to the current shape increment using the following rule:

$$\mathbf{h}_t = (1 - \mathbf{z}_t) \odot \mathbf{h}_{t-1} + \mathbf{z}_t \odot \mathbf{c}_t. \tag{5}$$

If the reset gate is always activated, the system will have only the short-term memory, since the calculation of the new increment candidate ignores the previous increments and focuses on the current input only. If the update gate is not activated, the system can have the long-term memory and the previous increments will be memorized.

Within this framework, the RNN acts as a refinement process which tries to find the optimal shape increment by *gradually* changing the previous shape. We use T recurrent steps to train RCFA. In order for the RNN to focus on the later iterations we define a series of weights $\mathbf{w} = [w_1, w_2, \ldots w_T]$ each one for a single recurrent iteration. These weights increase monotonically, therefore forcing the recurrent network to adjust the shape slowly and penalizing the model more for the error during the later recurrent steps. Formally, the loss writes as:

$$J = \sum_{i=1}^{n} \sum_{t=1}^{T} w_t \|(\widehat{\mathbf{h}}_0 + \mathbf{h}_t^i) - \mathbf{h}_*^i\|_F^2, \tag{6}$$

where $\widehat{\mathbf{h}}_0$ is the initial shape estimate, *i.e.* the average shape, $\mathbf{h}_t^i$ is the predicted shape increment after t iterations, $\mathbf{h}_*^i$ is the target shape, the superscript i defines the ith image in the mini-batch of n images. The final shape after t steps is obtained as $\widehat{\mathbf{h}}_0 + \mathbf{h}_t^i$. During training, for each face image $\mathbf{I}^i$, the initial shape $\widehat{\mathbf{h}}_0$ is sampled several times by adding noise to the mean shape.

3.2   Convolutional Module

We employ the super resolution convolutional neural network (SRCNN) for feature extraction [52]. We apply the SRCNN to the pixel values around the landmarks position Fig. 1. We denote the patch around a landmark location as $\mathbf{Y}$, and use it as an input for the SRCNN. The SRCNN consists of three convolution layers, formulated as the following operations:

$$F_1(\mathbf{Y}) = \max(0, \mathbf{W}_1 * \mathbf{Y} + \mathbf{B}_1)$$
$$F_2(\mathbf{Y}) = \max(0, \mathbf{W}_2 * F_1(\mathbf{Y}) + \mathbf{B}_2) \tag{7}$$
$$F_3(\mathbf{Y}) = \mathbf{W}_3 * F_2(\mathbf{Y}) + \mathbf{B}_3$$

where $\mathbf{W}_1$, $\mathbf{W}_2$, $\mathbf{W}_3$ and $\mathbf{b}_1$, $\mathbf{b}_2$, $\mathbf{b}_3$ represent the filters and biases respectively. The Rectified Linear Unit (ReLU) is employed as the activation function for the first two convolution layers. The dimensions of $\mathbf{W}_1$ are set to $c \times f_1 \times f_1 \times n_1 = [1 \times 9 \times 9 \times 64]$, where c is the number of channels of the input image, f_1 is the filter size, and n_1 is the number of filters which also corresponds to the number of feature maps. $\mathbf{W}_2$ is of the size $n_1 \times 1 \times 1 \times n_2 = [64 \times 1 \times 1 \times 32]$ and $\mathbf{W}_3$ has the size of $n_2 \times f_3 \times f_3 \times c = [32 \times 5 \times 5 \times 3]$. The first layer can be regarded as PCA where each filter works as a basis and projects the input $\mathbf{Y}$ to a high-dimension vector. The second layer has the filter size of 1×1, and this layer can be understood as a non-linear mapping operation which maps an $n-1$ dimensional vector to a n_2 dimensional vector. Originally, the last layer in the SRCNN works as an averaging filter which projects the n_2 dimensional vector to a high-resolution patch, and take the average of the overlapping high-resolution patches. However, instead of projecting the n_2 dimensional vector to a high-resolution patch, the last layer in our network will project the n_2 dimensional vector to a feature space which can is then passed to the recurrent module.

3.3   Supervised Descent Method as GRU

In this section we show that the proposed RCFA method is a generalization of the widely adopted Supervised Descent Method [5]. Given a set of images $[\mathbf{I}^1, \mathbf{I}^2, \dots \mathbf{I}^i, \dots, \mathbf{I}^n]$, $\mathbf{h}^i$ denotes the positions of the landmarks in image $\mathbf{I}^i$. $\mathbf{F}$ is a feature extraction function, and $\mathbf{F}(\mathbf{h}^i)$ represent the extracted features. Let $\mathbf{y}_*^i = \mathbf{F}(\mathbf{h}_*^i)$ be the ground-truth features extracted at the manually labeled landmark positions $\mathbf{h}_*^i$. Then we have the following objective function for face alignment with respect to image $\mathbf{I}^i$,

$$\min \|\mathbf{F}(\mathbf{h}^i) - \mathbf{y}_*^i\|_2^2. \tag{8}$$

SDM applies the gradient descent rule to Eq. 8, and yields the following discrete update equation:

$$\begin{aligned}
\mathbf{h}_t^i &= \mathbf{h}_{t-1}^i - \mathbf{R}_{t-1}(\mathbf{F}(\mathbf{h}_{t-1}^i) - \mathbf{y}_*^i) \\
&= \mathbf{h}_{t-1}^i - \mathbf{R}_{t-1}\mathbf{F}(\mathbf{h}_{t-1}^i) + \mathbf{R}_{t-1}\mathbf{y}_*^i,
\end{aligned} \tag{9}$$

where $\mathbf{R}_{t-1} = \alpha \mathbf{F}'(\mathbf{x}_{t-1}^i)$, and $\mathbf{R}_{t-1}$ is regarded as a regressor. Thus, instead of calculating the derivatives, the SDM learns a descend direction from the available training data.

However, Eq. 9 has an inconsistency problem, *i.e.* $\mathbf{y}_*^i$ is only available in the training phase and it is unknown in the testing phase. Therefore, Eq. 9 could not be used to calculate the position of the landmarks. To solve this inconsistency problem, $\mathbf{y}_*^i$ is replaced by $\overline{\mathbf{y}}_* = (\sum_i \mathbf{y}_*^i)/n$. By defining $\mathbf{b}_{t-1} = \mathbf{R}_{t-1}\overline{\mathbf{y}}_*$ we obtain the new update equation:

$$\mathbf{h}_t^i = \mathbf{h}_{t-1}^i - \mathbf{R}_{t-1}\mathbf{F}(\mathbf{h}_{t-1}^i) + \mathbf{b}_{t-1}, \tag{10}$$

which solves the inconsistency problem. During the training phase, $\mathbf{h}_t^i$ is set to $\mathbf{h}_*^i$ as our goal is to make $\mathbf{h}_t^i$ equal to the target $\mathbf{h}_*^i$. The loss is defined as:

$$\sum_i \|\mathbf{h}_*^i - \mathbf{h}_{t-1}^i + \mathbf{R}_{t-1}\mathbf{F}(\mathbf{h}_{t-1}^i) - \mathbf{b}_{t-1}\|^2 \tag{11}$$

where $\mathbf{h}_0^i$ is obtained using Monte Carlo integration.

Thus, Eq. 11 can be considered as a special case of Eq. 6, making the SDM a special case of our GRU network. As shown in Fig. 2(b), the traditional linear regressor is equivalent to GRU if the *update* gate and *reset* gate are removed. Finally, if we replace the *tanh* layer with the regressor $\mathbf{R}$, we obtain the formula for the shape increment $\mathbf{h}_t$ for the image $\mathbf{I}^i$, as follows:

$$\mathbf{h}_t^i = \mathbf{h}_{t-1}^i - \mathbf{R}_{t-1}\mathbf{F}(\mathbf{h}_{t-1}^i) \tag{12}$$

Equation 12 is a recurrent version of Eq. 10 except for the term $\mathbf{b}_{t-1}$ which can be implemented by expanding the feature space by several columns set to 1.

As shown in Fig. 2, the traditional regressors at different time steps $(\mathbf{R}_1, \mathbf{R}_2, \ldots)$ are trained independently, relying only on the input features while totally lacking memory regarding previous states, since as it is shown in Eq. 11,

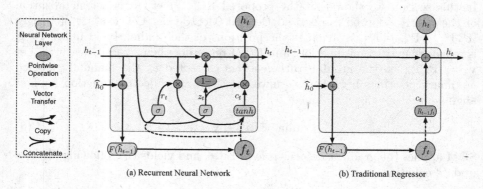

(a) Recurrent Neural Network   (b) Traditional Regressor

Fig. 2. Differences in the architecture of the proposed recurrent regressor (a) compared to the traditional regressor (b).

every step has a separate loss function. In contrast, for our model the overall loss over all the recurrent steps is defined and learned jointly by summing the steps up (Eq. 6). Another way of thinking about our recurrent module, is to treat it not as a regressor, but rather as a way of generating unique regressors at every recurrent step with respect to the memory and the input features.

4 Experiments

Datasets. We evaluate the performance of our algorithm using the widely adopted 300-W dataset [54]. This dataset is a combination of several in-the-wild datasets, including AFW [55], LFPW [56], HELEN [57] and XM2VTS [58], that are annotated with 68-point markup in a consistent manner. Similarly to previous works [8,13], for training the model we use the training samples from LFPW, HELEN and the whole AFW dataset, which makes 3148 images in total. Testing is performed on three different sets of images: (i) the *common* set includes the testing images from the LPFW and HELEN, (ii) the *challenging* set includes recently released 135 images also known as the IBUG set, and (iii) the *full* set is a combination of the first two. We do not report the results for the original annotations for HELEN and LFPW, since the accuracy of the state-of-the-art methods has saturated.

Evaluation Metrics. To evaluate the performance of our method, we follow the widely adopted evaluation metric [8,10,13], which is the **average error** of the point-to-point Euclidean distance, normalized by the distance between the outer corners of the eyes. This metric has been adopted for the 300-W challenge.

4.1 Implementation

For the CNN module, we follow the settings of the SRCNN framework in [52]. This module will extract features for all the image patches. After obtaining the features for each image patch, we concatenate the feature vectors of the 68 landmarks and feed the features to the RNN module. For the RNN module, we set the total number of recurrent iterations T to be 5. The weights in Eq. 6 are set to powers of 10: $\mathbf{w} = [10^{-2}, 10^{-1}, \ldots, 10^2]$. Powers of 2 and 5 showed slightly inferior performance to the powers of 10, while equal weights $[1, 1, 1, 1, 1]$ showed the worst performance.

To augment the size of the training data, we duplicate the images by adding the mirrored examples, and we also replicate the training data 3 times by adding noise to the bounding boxes. In the training phase, the batch size is set to 204 images. The learning rate is set to 0.01. The decay rate is set to 0.5, and the learning rate will be decayed after every 10 epochs and the training process is terminated after 200 epochs. After we obtain the model, we generate another three replicates in the same manner and fine tune the network with the new replicates for another 200 epochs.

4.2   Understanding When to Stop Iterating

One of the further advantages of our RCFA is that the model can be easily extended beyond the learned number of recurrent iterations without the need of retraining the whole pipeline. Importantly, there is no upper bound on the number of recurrent iterations one can perform. This, however, requires devising a strategy to stop iterating. During training the RNN performed 5 recurrent iterations. Intuitively, the model should require less iterations for a simple image, while difficult examples may need additional recurrences. In order to define a stopping strategy, we show the relationship between the average error and the recurrent steps as shown in Fig. 3.

As it is seen from the left graph in Fig. 3 for easy images from the common set additional iterations are redundant and do more harm, while the hard cases from the challenging set benefit from iterating further (see Fig. 3, right). Typically, for hard cases the error still continues decreasing when the number of iterations is more than 15, while for the easy ones it remains stable between 5-th and 9-th iterations and then goes up. This suggests a simple and efficient stopping criteria that was used to generate the results for the RCFA adaptive in the Table 1. If the difference between the previous landmark positions and the current landmark positions is smaller than a threshold, we stop iterating. To set the threshold value, we take the average difference between the 4-th and 5-th recurrence of all the images in the training data. This simple stopping strategy allows our model to automatically decide whether any additional iterations are necessary.

Figure 4 shows several qualitative examples of different number of recurrent iterations required for different testing examples. The first two rows show that 4 steps are not sufficient to localize the landmarks, as one can easily see that the landmarks on the jawline in the first row do not fit perfectly until the fifth recurrence. A similar observation can be made for the subject shown in the second row. The last two rows show cases, when even 5 iterations are not sufficient for the method

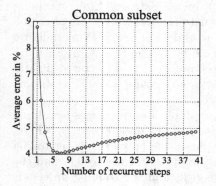
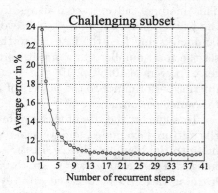

Fig. 3. Average error vs the number of recurrent iterations

Initial shape	Step 1	Step 2	Step 3	Step 4	Step 5

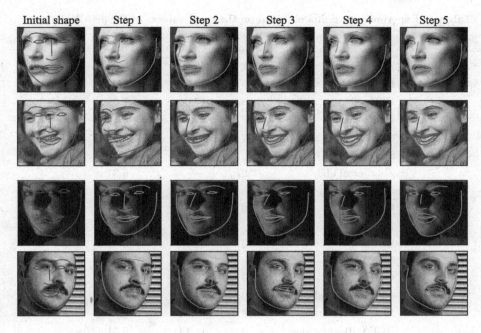

Fig. 4. Landmark localization for 5 recurrent steps. The top two rows show examples, for which 5 iterations is sufficient, while the examples in the last two rows require additional iterations.

to converge, due to difficult illumination conditions and extreme head poses. This further supports the importance of the adopted stopping strategy.

4.3 Experimental Results

We report evaluation results on the three subsets of the 300-W dataset in Table 1. It compares three different result of the same RCFA model against best performing state-of-the-art methods. The reported RCFA results are obtained using 5, 10 recurrent iterations and the proposed stopping strategy. We would like to highlight, that due to the end-to-end structure, our model shows better performance than the up-to-date face alignment methods regardless of the number of iterations for the common set and the full set. Notably, when the proposed stopping strategy is used, the proposed method outperforms other works by a large margin for all three testing sets.

Interestingly, RCFA outperforms CFSS [8] by a margin of 16% on the common subset, while showing a little bit lower performance gain for the challenging set and the full set (10% and 9% correspondingly). There are mainly two reasons for this. Firstly, the commonly accepted evaluation metric is severely affected by a small portion of hard examples. Secondly, these hard examples are not

Table 1. Experimental results obtained on the three subsets of the 300-W dataset.

Method	Common	Challenging	Full
Zhu and Ramanan [55]	8.22	18.33	10.20
DFMF [59]	6.65	19.79	9.22
ESR [60]	5.28	17.00	7.58
RCPR [61]	6.18	17.26	8.35
SDM [5]	5.57	15.40	7.50
Smith et al. [62]	-	13.30	—
Zhao et al. [63]	-	—	6.31
GN-DPM [64]	5.78	—	—
CFAN [65]	5.50	—	—
ERT [11]	-	—	6.40
LBF [13]	4.95	11.98	6.32
LBF fast [13]	5.38	15.50	7.37
CFSS [8]	4.73	9.98	5.76
CFSS practical [8]	4.79	10.92	5.99
RCFA 5 iterations	4.08	12.81	5.81
RCFA 10 iterations	4.13	11.14	5.51
RCFA adaptive	**4.03**	**9.85**	**5.32**

evenly presented in the 300-W training/testing sets. The dataset is rather biased towards having less extreme head poses, facial expressions and poor illumination conditions.

Figure 5 shows the qualitative results for the images taken from the full set. Clearly, due to end-to-end learning our framework handles even challenging face images, such as facial expressions, extreme head poses, difficult lighting conditions. It is also very interesting to observe that RCFA can handle faces with severe occlusions, including sun-glasses, hands and hats. The reason why our framework can work well for these images is because our RNN network can not only learn the dependencies between each regressor, it also learns the location dependencies between the landmarks. Thus, even though parts of the face is occluded, our framework can still predict the location of the occluded landmarks based on other landmarks.

In the current implementation, a single forward pass through the pipeline takes around 10ms on average on Tesla K40, making it possible to apply the proposed model for real-time video processing at 100 frames per second. We would like to note, that no specific performance optimizations were used, therefore, we believe the running time can be decreased dramatically.

Fig. 5. Selected qualitative examples taken from the full set of the 300-W dataset.

5   Conclusions

In this paper, we reformulate the classical cascaded regression face alignment problem as a recurrent process, alleviating the two major limitations of the CRMs. The proposed recurrent framework features end-to-end learning, starting from the raw pixel data, removing the previously used hand-crafted features. Replacing a standard cascade of independently learned shape regressors by a single recurrent regressor brings further advantage of iterating beyond the learned limit, making it possible to automatically decide when to stop.

The proposed RFCA method has room for further improvements. In our experiments an average shape is used to initialize the pipeline, while it has been shown that selecting a proper starting shape brings extra benefits [8]. Additionally, more rigorous data augmentation can alleviate the bias of the training set and can make the data more uniform. Furthermore, we believe similar recurrent-convolutional shape regression models can be employed to various other tasks such as action recognition [66] and human pose estimation.

References

1. Cootes, T.F., Taylor, C.J.: Active shape models - 'smart snakes'. In: Hogg, D., Boyle, R. (eds.) BMVC 1992. Springer, Heidelberg (1992)
2. Cootes, T.F., Taylor, C.J.: Active shape model search using local grey-level models: a quantitative evaluation. In: BMVC (1993)
3. Cootes, T.F., Edwards, G.J., Taylor, C.J.: TPAMI. Active appearance models **23**, 681–685 (2001)
4. Cao, C., Weng, Y., Lin, S., Zhou, K.: 3D shape regression for real-time facial animation. In: SIGGRAPH (2013)
5. Xiong, X., De La Torre, F.: Supervised descent method and its applications to face alignment. In: CVPR (2013)
6. Yang, H., Patras, I.: Sieving regression forest votes for facial feature detection in the wild. In: ICCV, pp. 1936–1943 (2013)
7. Tzimiropoulos, G.: Project-out cascaded regression with an application to face alignment. In: CVPR (2015)
8. Zhu, S., Li, C., Change, C., Tang, X.: Face alignment by coarse-to-fine shape searching. In: CVPR (2015)
9. Xiong, X., Torre, F.D.: Global supervised descent method. In: CVPR (2015)
10. Tulyakov, S., Sebe, N.: Regressing a 3D face shape from a single image. In: ICCV (2015)
11. Kazemi, V., Josephine, S.: One millisecond face alignment with an ensemble of regression trees. In: CVPR (2014)
12. Jeni, L.A., Cohn, J.F., Kanade, T.: Dense 3D face alignment from 2D videos in real-time. In: FG (2015)
13. Ren, S., Cao, X., Wei, Y., Sun, J.: Face alignment at 3000 FPS via regressing local binary features. In: CVPR (2014)
14. Doll, P., Pietro, W., Perona, P.: Cascaded pose regression. In: CVPR (2010)
15. Dalal, N., Triggs, B.: Histograms of oriented gradients for human detection. In: CVPR (2005)
16. Lowe, D.G.: Object recognition from local scale-invariant features. In: ICCV (1999)
17. Wang, W., Yan, Y., Winkler, S., Sebe, N.: Category specific dictionary learning for attribute specific feature selection. TIP **25**, 1465–1478 (2016)
18. Wang, W., Yan, Y., Sebe, N.: Attribute guided dictionary learning. In: ICMR (2015)
19. Krizhevsky, A., Sutskever, I., Hinton, G.E.: Imagenet classification with deep convolutional neural networks. In: NIPS (2012)
20. Wang, N., Yeung, D.Y.: Learning a deep compact image representation for visual tracking. In: NIPS (2013)
21. Simonyan, K., Zisserman, A.: Very deep convolutional networks for large-scale image recognition. arXiv (2014)
22. Chatfield, K., Simonyan, K., Vedaldi, A., Zisserman, A.: Return of the devil in the details: delving deep into convolutional nets. In: BMVC (2014)
23. Szegedy, C., Liu, W., Jia, Y., Sermanet, P., Reed, S., Anguelov, D., Erhan, D., Vanhoucke, V., Rabinovich, A.: Going deeper with convolutions. In: CVPR (2015)

24. Auli, M., Galley, M., Quirk, C., Zweig, G.: Joint language and translation modeling with recurrent neural networks. In: EMNLP (2013)
25. Cho, K., Van Merriënboer, B., Gulcehre, C., Bahdanau, D., Bougares, F., Schwenk, H., Bengio, Y.: Learning phrase representations using RNN encoder-decoder for statistical machine translation. arXiv (2014)
26. Pinheiro, P.H.O., Collobert, R.: Recurrent convolutional neural networks for scene parsing. arXiv (2013)
27. Liang, M., Hu, X.: Recurrent convolutional neural network for object recognition. In: CVPR (2015)
28. Lai, S., Xu, L., Liu, K., Zhao, J.: Recurrent convolutional neural networks for text classification. In: AAAI (2015)
29. Wang, N., Gao, X., Tao, D., Li, X.: Facial feature point detection: a comprehensive survey. arXiv (2014)
30. Saragih, J.M., Lucey, S., Cohn, J.F.: Deformable model fitting by regularized landmark mean-shift. IJCV **91**, 200–215 (2011)
31. Baltrusaitis, T., Robinson, P., Morency, L.P.: 3D constrained local model for rigid and non-rigid facial tracking. In: CVPR (2012)
32. Yu, X., Huang, J., Zhang, S., Yan, W., Metaxas, D.N.: Pose-free facial landmark fitting via optimized part mixtures and cascaded deformable shape model. In: ICCV (2013)
33. Cootes, T.F., Edwards, G.J., Taylor, C.J.: Active appearance models. TPAMI **23**(6), 681–685 (2001)
34. Gross, R., Matthews, I., Baker, S.: Generic vs. person specific active appearance models. IVC **23**, 1080–1093 (2005)
35. Tzimiropoulos, G., Pantic, M.: Optimization problems for fast aam fitting in-the-wild. In: ICCV (2013)
36. Fanelli, G., Dantone, M., Van Gool, L.: Real time 3D face alignment with random forests-based active appearance models. In: FG (2013)
37. Zhou, S.K., Comaniciu, D.: Shape regression machine. In: Karssemeijer, N., Lelieveldt, B. (eds.) IPMI 2007. LNCS, vol. 4584, pp. 13–25. Springer, Heidelberg (2007). doi:10.1007/978-3-540-73273-0_2
38. Cao, X.: Face alignment by explicit shape regression. In: CVPR (2012)
39. Lowe, D.G.: Distinctive image features from scale-invariant keypoints. IJCV **60**, 91–110 (2004)
40. Tulyakov, S., Alameda-Pineda, X., Ricci, E., Yin, L., Cohn, J.F., Sebe, N.: Self-adaptive matrix completion for heart rate estimation from face videos under realistic conditions. In: CVPR (2016)
41. Cao, C., Weng, Y., Zhou, S., Tong, Y., Zhou, K.: Facewarehouse: a 3D facial expression database for visual computing. TVCG **20**, 413–425 (2014)
42. Jourabloo, A., Liu, X.: Pose-invariant 3D face alignment. In: ICCV (2015)
43. Tulyakov, S., Vieriu, R.L., Semeniuta, S., Sebe, N.: Robust real-time extreme head pose estimation. In: International Conference on Pattern Recognition (2014)
44. Schuster, M., Paliwal, K.K.: Bidirectional recurrent neural networks. TSP **45**, 2673–2681 (1997)
45. Venugopalan, S., Rohrbach, M., Donahue, J., Mooney, R., Darrell, T., Saenko, K.: Sequence to sequence-video to text. In: ICCV (2015)
46. Karpathy, A., Fei-Fei, L.: Deep visual-semantic alignments for generating image descriptions. In: CVPR (2015)
47. Wang, W., Cui, Z., Yan, Y., Feng, J., Yan, S., Shu, X., Sebe, N.: Recurrent face aging. In: CVPR, pp. 2378–2386 (2016)

48. Graves, A., Mohamed, A.R., Hinton, G.: Speech recognition with deep recurrent neural networks. In: ICASSP (2013)
49. Hochreiter, S., Schmidhuber, J.: Long short-term memory. Neural Comput. 9(8), 1735–1780 (1997)
50. Koutnik, J., Greff, K., Gomez, F., Schmidhuber, J.: A clockwork RNN. arXiv (2014)
51. Jozefowicz, R., Zaremba, W., Sutskever, I.: An empirical exploration of recurrent network architectures. In: ICML (2015)
52. Dong, C., Loy, C.C., He, K., Tang, X.: Learning a deep convolutional network for image super-resolution. In: Fleet, D., Pajdla, T., Schiele, B., Tuytelaars, T. (eds.) ECCV 2014. LNCS, vol. 8692, pp. 184–199. Springer, Heidelberg (2014). doi:10. 1007/978-3-319-10593-2_13
53. Liang, X., Liu, S., Shen, X., Yang, J., Liu, L., Dong, J., Lin, L., Yan, S.: Deep human parsing with active template regression. TPAMI 37, 2402–2414 (2015)
54. Sagonas, C., Tzimiropoulos, G., Zafeiriou, S., Pantic, M.: 300 faces in-the-wild challenge: the first facial landmark localization challenge. In: ICCV Workshops (2013)
55. Zhu, X., Ramanan, D.: Face detection, pose estimation, and landmark localization in the wild. In: CVPR (2012)
56. Belhumeur, P.N., Jacobs, D.W., Kriegman, D.J., Kumar, N.: Localizing parts of faces using a consensus of exemplars. TPAMI 35, 2930–2940 (2013)
57. Le, V., Brandt, J., Lin, Z., Bourdev, L., Huang, T.S.: Interactive facial feature localization. In: Fitzgibbon, A., Lazebnik, S., Perona, P., Sato, Y., Schmid, C. (eds.) ECCV 2012. LNCS, vol. 7574, pp. 679–692. Springer, Heidelberg (2012). doi:10.1007/978-3-642-33712-3_49
58. Messer, K., Matas, J., Kittler, J., Luettin, J., Maitre, G.: XM2VTSDB: the extended M2VTS database. In: Second International Conference on Audio and Video-based Biometric Person Authentication (1999)
59. Asthana, A., Zafeiriou, S., Cheng, S., Pantic, M.: Robust discriminative response map fitting with constrained local models. In: CVPR (2013)
60. Cao, X., Wei, Y., Wen, F., Sun, J.: Face alignment by explicit shape regression. IJCV 107, 177–190 (2014)
61. Burgos-Artizzu, X., Perona, P., Dollár, P.: Robust face landmark estimation under occlusion. In: ICCV (2013)
62. Smith, B., Brandt, J., Lin, Z., Zhang, L.: Nonparametric context modeling of local appearance for pose-and expression-robust facial landmark localization. In: CVPR (2014)
63. Zhao, X., Kim, T.K., Luo, W.: Unified face analysis by iterative multi-output random forests. In: CVPR (2014)
64. Tzimiropoulos, G., Pantic, M.: Gauss-Newton deformable part models for face alignment in-the-wild. In: CVPR (2014)
65. Zhang, J., Shan, S., Kan, M., Chen, X.: Coarse-to-fine auto-encoder networks (CFAN) for real-time face alignment. In: Fleet, D., Pajdla, T., Schiele, B., Tuytelaars, T. (eds.) ECCV 2014. LNCS, vol. 8690, pp. 1–16. Springer, Heidelberg (2014). doi:10.1007/978-3-319-10605-2_1
66. Wang, W., Yan, Y., Nie, L., Zhang, L., Winkler, S., Sebe, N.: Sparse code filtering for action pattern mining. In: ACCV (2016)

Continuous Supervised Descent Method
for Facial Landmark Localisation

Marc Oliu[1,4(✉)], Ciprian Corneanu[2,4], László A. Jeni[3], Jeffrey F. Cohn[3,5],
Takeo Kanade[3], and Sergio Escalera[2,4]

[1] Universitat Oberta de Catalunya, 156 Rambla del Poblenou,
Barcelona, Spain
moliusimon@gmail.com
[2] Universitat de Barcelona, 585 Gran Via de les Corts Catalanes,
Barcelona, Spain
[3] Robotics Institute, Carnegie Mellon University, Pittsburgh, PA, USA
[4] Computer Vision Center, O Building, UAB Campus, Bellaterra, Spain
[5] Department of Psychology, University of Pittsburgh, Pittsburgh, PA, USA

Abstract. Recent methods for facial landmark location perform well
on close-to-frontal faces but have problems in generalising to large head
rotations. In order to address this issue we propose a second order linear regression method that is both compact and robust against strong
rotations. We provide a closed form solution, making the method fast
to train. We test the method's performance on two challenging datasets.
The first has been intensely used by the community. The second has
been specially generated from a well known 3D face dataset. It is considerably more challenging, including a high diversity of rotations and more
samples than any other existing public dataset. The proposed method is
compared against state-of-the-art approaches, including RCPR, CGPRT,
LBF, CFSS, and GSDM. Results upon both datasets show that the proposed method offers state-of-the-art performance on near frontal view
data, improves state-of-the-art methods on more challenging head rotation problems and keeps a compact model size.

1 Introduction

Facial landmark location consists of detecting a set of particular points on the
face. Usually these points have semantic meaning, their location being in highly
distinctive places around the eyes, mouth or nose. A set of such points is useful
for expressing both the rigid and non-rigid deformations of the face geometry.
Because facial geometry changes with identity, facial expression and head pose,
it is an important step in many automatic facial analysis tasks such as face
recognition, face expression recognition, face synthesis and age or gender estimation [1].

A common approach for locating landmarks on the face is to model the
relation between the face appearance and its geometry. If we consider $\mathbf{X}^*$ to be
the ground truth geometry, and $\Phi(\mathbf{I}, \mathbf{X})$ a representation function of a geometry

© Springer International Publishing AG 2017
S.-H. Lai et al. (Eds.): ACCV 2016, Part II, LNCS 10112, pp. 121–135, 2017.
DOI: 10.1007/978-3-319-54184-6_8

$\mathbf{X}$ on an image $\mathbf{I}$, then starting from an initial estimation $\mathbf{X}^0$ landmark location can be formulated as an optimisation problem of the form:

$$\underset{\varDelta \mathbf{X}}{\arg \min} f(\mathbf{X}^0 + \varDelta \mathbf{X}) = ||\varPhi(\mathbf{I}, \mathbf{X}^0 + \varDelta \mathbf{X}) - \varPhi(\mathbf{I}, \mathbf{X}^*)||_2^2 \qquad (1)$$

Because $\varPhi$ is a highly non-linear function, f is non-convex and has many local minima, the problem becomes severe in the case of large variations of the texture which is normally the case with rotations of the head and strong non-rigid deformations. Additionally, successfully solving the optimisation problem is highly dependent on the initialisation.

Historically, Active appearance models (AAM) [2] are one of the most used methods for 2D face registration. They are an extension of active shape models (ASM) [3] which encode both geometry and intensity information. More recently, even though single step landmark location methods have been proposed [4,5], the most common approach is to model the relationship between texture and geometry with a cascade of regression functions [6–12]. Features are extracted from the current estimated geometry and passed to the learnt mapping in order to update the geometry. This process is repeated iteratively for each step of the cascade, applying a specific mapping to each. If we denote by $\mathbf{R}^i$ the regression function at the ith step of the cascade, by $\mathbf{\Phi}^i = \varPhi(\mathbf{I}, \mathbf{X}^i)$ the corresponding representation and by $\mathbf{b}^i$ a constant bias, then at every step of the cascade, the geometry $\mathbf{X}$ will be updated in the following way:

$$\mathbf{X}^{i+1} = \mathbf{X}^i + \mathbf{R}^i \mathbf{\Phi}^i + \mathbf{b}^i \qquad (2)$$

While most cascaded regression methods share this approach, considerable variation can be found in representation, regression functions and initialisation strategies. The simplest way to initialise the geometry is by starting with the mean [8,9,11]. For faces, this works well in close-to-frontal scenarios but proves inefficient when large pose variation occurs. A common solution is to try a set of random initialisations and consider the median of the predictions as the final solution [6,13]. Unfortunately, this considerably increases the computational cost. An alternative approach is to apply the initial part of the cascade and continue only if the variance of the regressed shapes is low, which is a strong predictor of convergence towards the global minimum [7]. If this is not the case then a different set of initial shapes is generated. Even so, all these methods are dependent on the initialisation and prove low generalisation to large head pose rotation. A coarse-to-fine searching approach was recently proposed to deal with the initialisation dependency problem [14]. A regression function is learnt from a set of shapes generated according to a probabilistic distribution on the shape space. A dominant set approach is used to eliminate outliers between the regressed shapes in an unsupervised manner. From the filtered subset the centre of a smaller region of the original space is computed and the process repeated until convergence. While it prevents locality of the solutions it improves robustness to large pose variation.

The work of Dollar et al. which proved influential in the field of facial landmark localisation, uses intensities of sparse sets of pixels at predefined locations to represent texture in a shape indexed fashion for learning a fixed linear sequence of weak regressors [13]. In this way, representation's output depends on both the image data and the current estimate of the geometry. Some of the methods propose to jointly learn the representation and the regression function [6–8,11]. In this sense, several shape indexed locations are randomly generated and then selected based on a certain optimisation criteria. Alternatively, local binary features are learnt for each landmark independently [8]. During test, very fast landmark localisation is obtained. In a recent method [15], *Difference of Gaussians* (DoG) features are selectively extracted from locations arranged in a pattern inspired by the human visual system [16]. Learnt trees at early stages tend to select indexed DoG features computed from distant sampling points while trees at later stages tend to use nearby sampling points. Finally, a very common problem of most of the proposed methods, the lack of sensitivity to occlusions is tackled in the work of Burgos-Artizzu et al. [7]. They propose a method that reduces exposure to outliers by detecting occlusions explicitly and using robust shape-indexed features. It incorporates occlusion directly during learning to improve shape estimation.

A distinct group of methods use predefined handcrafted representations while learning the regression function from the data. For example, to overcome the large computational time required by the regression of many generated shapes at each stage more simple descriptors are used in the initial stages when coarse localisation is performed. More complicated representations are used on final stages when fine localisation takes place [14]. A particularly important set of methods that use fixed representations are the ones derived from the *Supervised Descent Method* (SDM) [9]. SDM uses simplified SIFT features and linear regressors. As is the case of previous methods, SDM works well for near frontal faces but fails on strong rotations. To overcome this problem, *Global Supervised Descent Method* (GSDM) [10] introduced an approach which uses a sub-space defined by a set of directions of maximum variance of the training data to partition the original feature space. Each partition shares a similar descent direction for the training instances falling within it. A linear regressor is learnt for each partition. However, GSDM suffers from two main problems. Both the number of training instances and model size increase exponentially with the number of sub-space dimensions.

In order to perform landmark localisation under strong rotations while keeping a fast and compact model, this work proposes a continuous formulation of GSDM. Instead of using the sub-space to partition the feature space as GSDM does, it is used to describe a space of linear regressors. This is equivalent to proposing a regressor which estimates the second derivative of the gradient, instead of the first as a standard linear regressor would (e.g. in SDM). While this formulation may not be as expressive as GSDM, the amount of memory and training instances required increases linearly with the number of dimensions of the sub-space. Also, the proposed formulation defines a specific linear regressor for each instance.

In summary, our list of contributions is as follows:

- we present a method that improves state-of-the-art results on strongly rotated faces
- the trained models are small, the amount of memory and training instances required increase only linearly with the number of dimensions of the sub-spaces
- the method is fast to train due to its closed form solution
- we have synthesised largest 2D face dataset to date, with a challenging face rotation distribution

The rest of the paper is organised as follows: in Sect. 2 we formulate the proposed method, in Sect. 3 we present the experimental analysis and finally, in Sect. 4, we conclude the paper.

Notations. Vectors ($\mathbf{a}$) and matrices ($\mathbf{A}$) are denoted by bold letters. An $\mathbf{u} \in \mathbb{R}^d$ vector's Euclidean norm is $\|\mathbf{u}\|_2 = \sqrt{\sum_{i=1}^d u_i^2}$. $\mathbf{B} = [\mathbf{A}_1; \dots; \mathbf{A}_K] \in \mathbb{R}^{(d_1 + \dots + d_K) \times N}$ denotes the concatenation of matrices $\mathbf{A}_k \in \mathbb{R}^{d_k \times N}$.

2 Continuous Supervised Descent Method

2.1 Second Order Regressor

The original SDM method [9] is an exemplar-based method which learns a series of linear regressors approximating the data to the global optima in a cascaded manner. Lets consider $\mathbf{X}^i \in \mathbb{R}^{n \times m}$ the m targets for each of n samples at a given cascade step i, $\Delta\mathbf{\Phi}^i \in \mathbb{R}^{n \times (k+1)} = \mathbf{\Phi}^i - \overline{\mathbf{\Phi}^i}$ the difference of the feature vectors of length k from the mean, with a column vector of ones added in order to account for the bias, and $\mathbf{R}^i \in \mathbb{R}^{(k+1) \times m}$ the linear regressor for each of the m parameters. Then the update formula for SDM can be expressed as follows:

$$\mathbf{X}^{i+1} = \mathbf{X}^i + (\mathbf{\Phi}^i - \overline{\mathbf{\Phi}^i})\mathbf{R}^i = \mathbf{X}^i + \Delta\mathbf{\Phi}^i\mathbf{R}^i \qquad (3)$$

This can be seen as learning a linear approximation of the first-order partial derivatives for each parameter. These correspond to $\partial\Delta\mathbf{X}^{i+1}/\partial\Delta\mathbf{\Phi}^i_j = \Delta\mathbf{\Phi}^i_j\mathbf{R}^i_j$, with $\mathbf{R}^i$ being the Jacobian matrix, $\Delta\mathbf{\Phi}^i_j$ the jth column of $\Delta\mathbf{\Phi}^i$ and $\mathbf{R}^i_j$ the jth row of $\mathbf{R}^i$. To make this approximation, the slope is considered homogeneous for any point of the feature space. This assumption does not hold for most problems, where the gradient direction suffers from large variations on different locations of that space. On Global SDM [10] these variations are handled by partitioning the space into different regions and learning a linear regressor for each one. This approach can approximate with high accuracy the gradient variations at different regions of the space, but has the problem of doubling the amount of learnt regressors and required training data each time the space is divided.

Here we introduce a continuous formulation, where a set of bases are learnt for the regressors, effectively learning a linear approximation of the second derivative. To do so, first a set of main modes of variation are learnt from either $\Delta\mathbf{X}^*$ or $\Delta\mathbf{\Phi}^i$ using Principal Component Analysis (PCA):

$$\widetilde{\Delta\mathbf{\Phi}^i} = \left[\Delta\mathbf{\Phi}^i\mathbf{P}_{1:l}, \mathbf{1}_n\right], \qquad (4)$$

where l represents the number of bases to learn and $\mathbf{P}_{1:l}$ is the projection matrix. $\mathbf{1}_n \in \mathbb{R}^{n \times 1}$ denotes an all-ones vector. Given that the total number of learnable parameters for one of m targets equals $p = (k+1)(l+1)$, learning the second derivative for all parameters $(l = k)$ would drastically increase the problem dimensionality. Estimating the second derivative on the l main variation modes is a more treatable problem. Given one of the targets $\Delta\mathbf{X}_j^i \in \mathbb{R}^{n \times 1}$, its associated second order regressor is expressed as the solution to the following minimisation problem:

$$\underset{\mathbf{R}_j^i}{\arg\min} ||(\Delta\mathbf{\Phi}^i \circ (\Delta\widetilde{\mathbf{\Phi}}^i \mathbf{R}_j^i))\mathbf{1}_{(k+1)} - \Delta\mathbf{X}_j^i||_2^2 \tag{5}$$

Here, $\mathbf{R}_j^i \in \mathbb{R}^{(l+1) \times (k+1)}$ is the set of l bases (and baseline or bias regressor) describing the regressor for the jth target at the ith cascade step, and $\circ$ denotes the Hadamard product. Note that, according to Eq. 7, this formulation learns a linear approximation to the second order partial derivatives $\partial^2 \Delta\mathbf{X}_j^{i+1}/(\partial\Delta\mathbf{\Phi}_p^i \, \partial\Delta\widetilde{\mathbf{\Phi}}_q^i) = \Delta\mathbf{\Phi}_p^i \Delta\widetilde{\mathbf{\Phi}}_q^i (\mathbf{R}_j^i)_{pq}$. Thus $\mathbf{R}_j^i$ corresponds to a compact version of the Hessian matrix for target j at cascade step i, having the dimensionality of the feature space reduced before applying the second derivative. Equation 5 can be seen as a compact formulation defining a quadratic regressor for each target, which is known to be a linear problem, having a closed form solution. This minimisation problem can be expressed in a least squares form, providing a closed form solution, as follows:

$$\underset{\mathbf{R}_j^i}{\arg\min} ||(\Delta\widetilde{\mathbf{\Phi}}^i \odot \Delta\mathbf{\Phi}^i)vec(\mathbf{R}_j^{i\,\mathsf{T}}) - \Delta\mathbf{X}_j^i||_2^2 \tag{6}$$

Here $\odot$ denotes the Khatri–Rao product, considering each instance (row) on $\Delta\mathbf{\Phi}^i$ and $\Delta\widetilde{\mathbf{\Phi}}^i$ as a partition of the matrix, and $vec(\mathbf{R}_j^{i\,\mathsf{T}}) \in \mathbb{R}^{(kl+2) \times 1}$ is the vectorisation of the regressor bases. Thus, while the second derivative estimate is used for a subset of principal components, the regressor remains linear. This allows us to rapidly and directly find the optimal regression weights given the training instances. Note that this formulation could be extended to estimate higher order derivatives by applying the Khatri–Rao product multiple times. At test time, the parameters are updated with the following equation:

$$\mathbf{X}_j^{i+1} = \mathbf{X}_j^i + (\Delta\mathbf{\Phi}^i \circ (\Delta\widetilde{\mathbf{\Phi}}^i \mathbf{R}_j^i))\mathbf{1}_{(k+1)} \tag{7}$$

This formula estimates the regressor weights and bias for the current value of the principal components $\widetilde{\mathbf{\Phi}}^i$, and applies it to the features. This is more memory-efficient than performing the Khatri-Rao product of $\Delta\mathbf{\Phi}^i$ and $\Delta\widetilde{\mathbf{\Phi}}^i$ and then performing a linear regression. The bias for the regressor bases is the baseline regressor for an instance with the mean value for the l principal components (PCs) of the feature vector. Each of the l regression bases in $\mathbf{R}^i$ corresponds to the second derivative estimate wrt. a given PC. Note that when $l = 0$ the model is a standard linear regressor. Thus, SDM can be seen as a special case of our method where the second derivative is not taken into account for any PC.

The proposed approach estimates a standard linear regressor for each instance given the coordinates of the features sub-space $\Delta\widetilde{\mathbf{\Phi}}^i$. Global SDM assigns the same one to all instances falling into a given region of the partitioned sub-space. Another advantage of this approach is that the number of parameters learnt p at each cascade step increases linearly with the number of bases ($p = (k + 1)(l + 1)$). With Global SDM it increases quadratically ($p = (k + 1)min(1, l^2)$). These two factors make the proposed approach both more compact in terms of memory and more accurate, as shown in Sect. 3.3. Because the regression space is continuous, the weights of the linear regressor are adapted to each instance, providing more flexibility to the model. During training, this also implies that for the proposed approach all the training data is available for each base of the sub-space, helping to reduce over-fitting. GDSM distributes the data between quadrants, logarithmically reducing the available training data for each quadrant with the number of sub-space bases.

2.2   Implementation Details

As discussed in Sect. 2.1, the second derivative of the feature space is calculated over the l principal components. For this work, similarly to [9], a simplified SIFT descriptor is extracted from each landmark estimate. The descriptor has a fixed 32×32 window around the landmark, rotated according to the in-plane rotation of the current geometry relative to the mean facial shape. PCA is then applied in order to reduce its dimensionality. Thus, the feature vector for an instance j at the cascade step i is defined as $\mathbf{\Phi}_j^i = sift(\mathbf{I}_j, \mathbf{X}_j^i)^\intercal \mathbf{P}_{1:k}^i$, the k principal components of the extracted SIFT descriptors. This implicitly provides the l parameters for the regressor bases, being $\widetilde{\mathbf{\Phi}}_j^i = (\mathbf{\Phi}_j^i)_{1:l}$. The targets $\Delta\mathbf{X}^i$ are rotated in the same way as the descriptor windows in order to maintain a coherent update direction.

The feature vector length k and number of regression bases l may depend on the problem and are free parameters of the model. Still, there are two considerations to take into account. In a cascaded regression approach, the first steps of the cascade broadly approximate the face pose and general shape, while later steps tend to fine-tune the location of each landmark, working more locally. This implies that at the fist steps a smaller amount of the total descriptors variance may be enough. Conversely, a higher amount of regression bases would increase the adaptability to the descriptors main modes of variation, which are expected to be caused by pose/illumination variations. The feature vector length k is defined as a fixed percentage of the original SIFT features variance. While it may be possible to adjust the number of bases l at each cascade step (for instance with forward selection), in this work a global value is chosen for all cascade steps.

The initial shape at the first cascade step is the mean shape. It is calculated from the training instances ground truth shapes using Generalised Procrustes Analysis.

3 Experiments

This section is dedicated to the description and discussion of the experiments conducted to validate the proposed method. We begin in Sect. 3.1 by describing the two datasets we used, 300W a dataset intensely used by the community and BU4DFE-S, a dataset we have specially synthesised from BU4DFE, a 3D face dataset. In Sect. 3.2 we present the experimental setup and the methods used for comparison[1]. In Sect. 3.3 we discuss the results.

The objective of these experiments are two-fold. First we want to show that the proposed method achieves state-of-the-art results on close-to-frontal faces. For this purpose we use 300W, a well known public dataset which is the de-facto standard benchmarking dataset for facial landmark localisation. We then want to show that the method outperforms other methods when applied to heavily rotated faces. For this purpose we show results on the BU4DFE-S, a dataset specially synthesised for this purpose. The reader is referred to Table 1 for the overall results on the two considered datasets and to Fig. 3 for a comparative study of the robustness to rotation 3. Detection examples are presented in Fig. 4. The code for the experiments is made publicly available.

3.1 Data

In order to test the proposed method we used 300W, a well known facial expression dataset. We also designed a new dataset, which we called BU4DFE-S, consisting of 2D faces synthesised from BU4DFE, a public 3D dynamic facial expression dataset.

300W. The 300 Faces In-the-Wild (300W) [17] database is a compilation of six re-annotated datasets (68 landmarks). Following the same approach as in [8] [15], four of the six datasets are used: AFW [18], LFPW [19], Helen [20] and iBUG [17]. The test data for LFPW and Helen, along with iBug, are used as test. The rest of data is used for training. This provides a total of 3148 and 689 train and test instances. The data is captured outside the lab and it has balanced ethnic and gender distribution. While challenging and diverse, it does not contain far-from-frontal faces and its number of samples is rather low.

BU4DFE-S. While annotated face datasets have become more challenging and diverse in recent years, they still provide a low number of training instances with limited variation in rotation. In order to compare the robustness of the proposed method with state-of-the-art facial landmark localisation methods, we have created BU4DFE-S, a new large 2D dataset synthesised from the publicly available BU4DFE. BU4DFE is a high resolution dynamic 3D facial dataset [21]. 101 subjects of ages between 18 to 45 years old are captured while showing facial expressions in a controlled environment. The 3D facial expressions are captured at 25 frames per second. Each sequence begins with the neutral expression, proceeds to target emotion and then back to neutral. For creating the BU4DFE-S we sample 5 frames from each captured sequence. The sampled 3D frames

---

[1] Code and data generation script available at https://github.com/moliusimon/csdm.

are equally distributed along the sequence, portraying varying intensities of the same expression during onset, apex and offset.

We use the extracted 3D samples, to build 25 2D projections by rotating the 3D model in pitch and yaw. The projected images are generated as follows. The BU4DFE provides the 3D point cloud of the face and an RGB image. Additionally for each of the 3D points the mapping is provided to the corresponding position on the RGB image, making it possible to map the 3D geometry to the colour texture. We first homogeneously down-sample the 3D points set by 20 and build a triangle mesh from the remaining points. The down-sampling factor was heuristically found as a trade-off between the computational cost for generating the projected images and their quality. We consider an isometric projection to associate texture patches to the mesh triangles. The mesh is then rotated with the desired angle and the triangles are again projected to the 2D plane. The new face is built by affine piece-wise warping of the initial texture patches to the newly projections of the rotated triangles by taking into account self occlusions. Inpainting is used to fill warping holes or artifacts. Finally, the images are resized to a standard size of 200 × 200 pixels and a background is painted on the remaining regions.

We have used the test partition of the Places Dataset, a scene recognition dataset, to build the backgrounds [22]. It contains 41000 images of size 256 × 256 pixels. From every image we crop two 200 × 200 regions, one on the top-let corner and the other on the bottom-down corner. The former is flipped. We use these images to place a different background behind each of the generated faces. In Fig. 1(a) we provide a summarised depiction of the process. The rotation angles follow an inverse normal distribution for angles between ±90° in yaw and ±45° in pitch as shown in Fig. 1(b). In this way we obtain more highly rotated faces in all directions than close-to-frontal faces. The generated data contains a total of 75000 rotated images of 100 persons. Each person appears in 750 samples with 6 different facial expressions at 5 different intensities rotated 25 times. As the BU4DFE, the subjects are from different ethnicities and follow a balanced gender distribution. The generated dataset has more instances and rotation variation than any other existing public 2D dataset. We show some examples in Fig. 2.

Besides containing a larger number of samples (approximately 24 times more than 300W), BU4DFE-S has two more important characteristics. First, for each of the samples the pose is known which is not the case with most of the other 2D face datasets. There exist datasets containing captured faces under different angles in the lab, but the angle distribution is extremely skewed [23]. Another advantage of the BU4DFE-S is that we have total control over the pose distribution of the synthesised data. This makes possible benchmarking the robustness to pose rotation against state-of-the-art methods as shown in Fig. 3.

3.2   Experimental Settings

For the proposed method the parameter space is larger than for SDM, specially at the first cascade steps. In order to avoid over-fitting, the training data is augmented. For both the 300 W and BU4DFE-S datasets the images and geometries

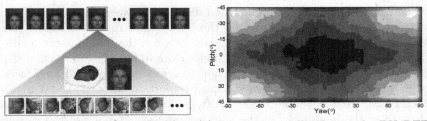

(a) Data synthesis for BU4DFE-S. (b) Pose rotation distribution for BU4DFE-S.

Fig. 1. BU4DFE-S contains 2D rotated faces synthesised from BU4DFE, a 3D dynamic facial expression dataset. In (a) we show how from a original sequence we sample a limited number of equally spaced frames. For each of these frames we use the provided 3D mesh and the texture to generate 25 rotated projections. The rotation angle distribution is shown in (b). We favour far-from-frontal faces with respect to close-to-frontal ones in order to make the data as challenging as possible.

are mirrored, doubling the number of training instances. In the case of 300W, which consists of only 3148 training images, the dataset is further augmented by providing 25 different initial geometries. These are generated by applying a random rotation between $[-\pi/4, \pi/4]$, a displacement between $[-5\%, 5\%]$ for both width and height, and a scaling factor between $[0.9, 1.1]$ to the mean shape.

Regarding the number of bases l and the captured feature space variance, the values have been manually chosen for each dataset. For 300W, 2 bases and 95% of variance are used, while for BU4DFE-S, 5 bases and 85% of variance are used. It is necessary to use fewer bases in 300 W in order to avoid over-fitting, since the number of training instances is smaller.

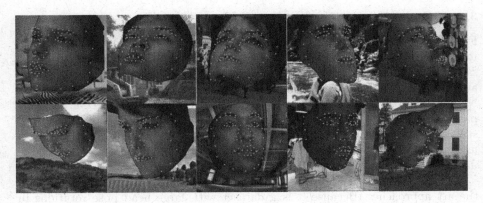

Fig. 2. BU4DFE-S dataset samples. Annotated face landmarks are shown in green. (Color figure online)

We compare the proposed method with the most important facial landmark localisation methods in recent years. This is done using the Normalised Mean Euclidean Error (NMEE) metric, a standard error metric in the literature

[7, 9]. It corresponds to the mean euclidean distance between the detected and ground truth landmarks, normalised by the inter-ocular distance. In the case of BU4DFE-S, where large head rotations are present, the 3D inter-ocular distance is used instead. Otherwise for yaw angles close to 90° the inter-ocular distance would tend to zero, giving more weight to errors on heavily rotated faces. For comparing results we considered the most important state-of-the-art methods [6–11, 14, 20]. RCPR is able to deal with occlusions by including occlusion ground-truth of the landmarks in the learning process. As none of the considered datasets has annotated occlusions we discarded this feature during training. For the ERT [11] and LBF [8], we compare with already published results for the 300W. For a fair comparison we compare the results for the SDM and the GSDM after training with the same number of steps as the proposed method. It is important to note that GSDM is a method oriented to tracking the facial geometry, but can be easily applied to the static case by modifying the definition of the subspace used to partition the feature space. Instead of using two principal components from $\Delta \mathbf{X}^i$ and one from $\Delta \mathbf{\Phi}^i$, all principal components are taken from $\Delta \mathbf{\Phi}^i$. For the proposed approach, a 2-dimensional subspace is used in the case of 300W, and a 5-dimensional one for BU4DFE-S. Finally, two recent methods, CFSS [14] and CGPRT [15], have been considered. In their paper, the authors of CGPRT publish two results, with different number of training steps. The result we have obtained was with the larger number of steps and by initialising with the mean shape. CFSS does a constraint search of the shape in a coarse-to-fine manner in subsequent finer shape subspaces. Even though a parallel training on the CPU was attempted, we found training to be very slow, which made impossible obtaining results for BU4DFE-S with the available hardware resources.

3.3 Results Discussion

For the 300 W dataset, the trained model has been fit to the test data both using mean shape initialisation and with 25 random initialisations sampled using the same criteria used during training (see Sect. 3.2). The results of both approaches are shown in Table 1. Without multiple test initialisations, the method has a NMEE lower than those achieved by ESR, RCPR, SDM and GSDM, also surpassing ERT when using multiple initialisations. Yet LBF, CGPRT and CFSS still have lower errors. Thus, the proposed approach surpasses, or is close, to most state-of-the-art methods in the near frontal view conditions of the 300-w dataset.

On BU4DFE-S the proposed method outperforms all considered state of the art approaches. Because it is a dataset with large head pose rotations in both pitch and yaw, this dataset better represents the strength of the proposed algorithm to better adjust to the main modes of variation of the data. This is analysed in Fig. 3. There, the NMEE is shown relative to the yaw rotation, for two ranges of pitch. Without using multiple test initialisations, the proposed method has an accuracy similar to that of the other state-of-the-art approaches for near-frontal faces, but is much more robust to pose variations. It works specially

Table 1. Our method compared with state-of-the-art methods in terms of mean landmark displacement as percentage of interocular distance (For the BU4DFE-S we compute the interocular distance in 3D.) without (CSDM) and with multiple test initialisations (CSDMa).

	ESR [6]	RCPR [7]	SDM [9]	ERT [11]	LBF [8]	CGPRT [15]	CFSS [14]	GSDM [10]	CSDM	CSDMa
300W	7.58	8.38	7.52	6.40	6.32	**5.71**	5.76	6.96	6.83	6.40
BU4 DFE-S	9.45	8.61	9.57	-	-	15.81	-	9.01	8.28	**7.62**

well for both large pitch and yaw rotations. This contrasts with CGPRT, which performed specially well for the 300 W dataset, but had problems with BU4DFE-S. The only method still far from, but approaching the accuracy obtained by the proposed approach is RCPR. It can be seen in Fig. 3 that while RCPR has the lowest NMEE for frontal faces, it one of the best approaches when dealing with large pose variations. When using multiple test initialisations, a much lower average error is obtained, achieving the same accuracy for near-frontal faces as ERT. This accuracy improvement is maintained regardless of the facial pose, except for large pose rotations in both pitch and yaw, where the yaw angle is close to 90°. For these extreme cases, the error is only slightly lower than CSDM without using multiple shape initialisations.

A breakdown of the NMEE by facial regions, as shown in Table 2, gives a better insight on the method performance. For far from frontal head poses, the proposed approach surpasses the state of the art accuracies on all facial regions, both with and without multiple test initialisations. In the case of near-frontal head poses, RCPR has a higher precision for the eyes and eyebrows. CSDM is better at localising landmarks at the nose, mouth and contour regions when using multiple shape initialisations. An interesting result is the error reduction when localising the contour landmarks with multiple shape initialisations. While the other facial regions reduce the RMSE by about 5%, in the case of the contour it is reduced by over 10%, both in close to and far from frontal head poses. This is likely caused by the lack of edges and strong gradients on this region. By averaging multiple predictions, the noise is reduced, obtaining a higher accuracy.

GSDM is another method that exploits the features main modes of variation to better approximate the descent direction at different regions of the feature space. Compared to it, the proposed method obtains better results for both 300 W and BU4DFE-S while also producing a more compact model. The memory required by GSDM increases quadratically with the number of considered bases, while the proposed approach does so linearly. Furthermore, each position of the subspace has a unique regressor assigned, while GSDM shares the same regression weights for a given partition of the subspace. One downside to the proposed approach is that the computational cost increases linearly with the number of bases, while for GSDM the cost remains constant.

Similarly to SDM and GSDM, the proposed method provides a closed-form solution. Compared to other state-of-the-art methods such as CFSS, CGPRT and

Table 2. Normalised Mean Euclidean Error (NMEE) for different landmark subsets corresponding to facial regions on BU4DFE-S. We group faces according to their pose. Close-to-frontal faces have an yaw angle between $\pm30°$ and pitch angle between $\pm15°$. Correspondingly far-from frontal faces have both yaw and pitch angles above $\pm30°$ and $\pm15°$ respectively.

	Close to frontal							Far from frontal						
	ESR	RCPR	SDM	CGPRT	GSDM	CSDM	CSDMa	ESR	RCPR	SDM	CGPRT	GSDM	CSDM	CSDMa
Eyes	3.92	**3.38**	4.02	10.53	3.92	4.04	3.82	6.94	6.11	6.76	14.29	6.25	5.55	**5.20**
Eyebrows	5.84	**5.17**	5.60	13.15	5.56	5.84	5.54	9.01	8.02	8.50	17.73	8.12	7.20	**6.77**
Nose	6.03	5.59	5.60	10.30	5.51	5.58	**5.27**	8.26	7.69	8.58	13.21	8.00	7.41	**6.99**
Mouth	5.46	4.28	4.47	10.91	4.27	4.40	**4.27**	8.20	6.70	8.18	14.52	6.72	6.17	**5.84**
Contour	12.59	12.11	13.26	17.49	13.52	13.27	**12.04**	17.30	17.19	18.54	22.43	18.53	17.20	**15.27**

LBF, which use stochastic processes when learning each regressor, the proposed approach ensures a consistent result on different training runs given the same data.

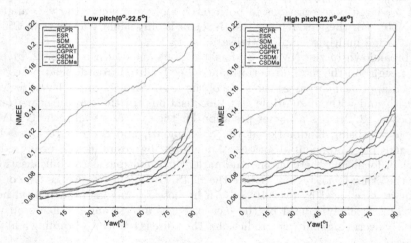

Fig. 3. Normalised Mean Euclidean Error (NMEE) as a function of yaw on two different pitch ranges on BU4DFE-S.

Multiple qualitative examples of faces from the BU4DFE-S dataset, with the landmark predictions for different methods, are shown in Fig. 4. From these examples it can be seen that SDM, CGPRT and RCPR struggle to correctly locate inner face landmarks for heavily rotated faces. Compared to all other considered methods, our proposal has a high accuracy on inner face landmarks even with highly rotated faces, followed by GSDM and ESR. The main weakness is the localisation of face contour landmarks, which is noisy due to the lack of edges and little texture information on that area, resulting in a lack of smoothness in the contour line. Even with this noise, as shown in Table 2, the proposed

ESR   RCPR   SDM   CGPRT   GSDM   CSDM

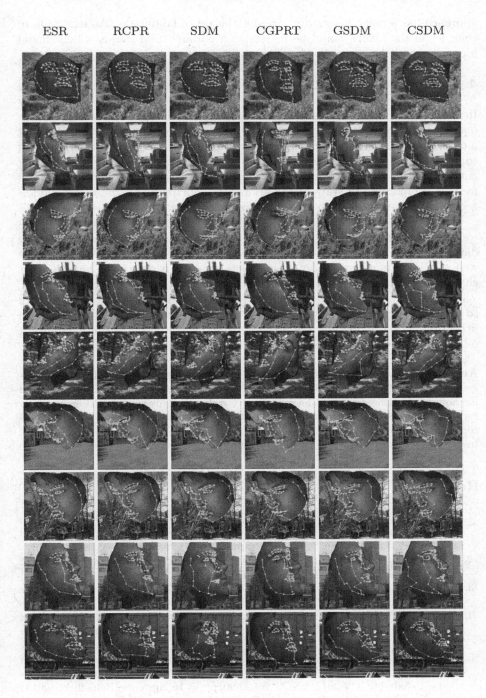

Fig. 4. Facial landmark localisation examples for BU4DFE-S.

approach has a much better precision for this set of landmarks. An extension to consider in the future would be regressing a parametrised shape, which should increase the accuracy for the face contours.

4   Conclusion

In this work we extended cascaded regression approaches by introducing the second order derivative over the main modes of variation of the features, presenting a closed-form solution to the face alignment problem. We showed that by doing so, the robustness to large head pose variations is greatly increased, surpassing current state of the art methods. At the same time, the accuracy for near-frontal faces is comparable to state of the art results. Furthermore, the learnt models are smaller than those from other similar approaches.

In order to prove the effectiveness of our method on heavily rotated faces we have built a new synthetic dataset based on a well known public 3D face dataset. It contains large variations in both head pose and facial expressions, as well as a large number of training instances, making it one of the largest, most challenging datasets for facial landmark localisation to date.

Several future improvements can be envisioned, like parameterizing the face to increase shape consistency especially for landmarks situated in regions with little texture and extending the method to 3D, which would make it useful for a larger number of applications.

Acknowledgement. The work of Marc Oliu is supported by the FI-DGR 2016 fellowship, granted by the Universities and Research Secretary of the Knowledge and Economy Department of the Generalitat de Catalunya. This work has been partially supported by the Spanish project TIN2013-43478-P, the European Comission Horizon 2020 granted project SEE.4C under call H2020-ICT-2015 and the U.S. National Institutes of Health under the grant MH096951.

References

1. Corneanu, C.A., Oliu, M., Cohn, J.F., Escalera, S.: Survey on RGB, 3D, thermal, and multimodal approaches for facial expression recognition: history, trends, and affect-related applications. Trans. Pattern Anal. Mach. Intell. Spec. Issue (2016)
2. Cootes, T.F., Edwards, G.J., Taylor, C.J.: Active appearance models. IEEE Trans. Pattern Anal. Mach. Intell. **23**, 681–685 (2001)
3. Cootes, T.F., Taylor, C.J., Cooper, D.H., Graham, J.: Active shape models-their training and application. Comput. Vis. Image Underst. **61**, 38–59 (1995)
4. Sun, Y., Wang, X., Tang, X.: Deep convolutional network cascade for facial point detection. In: 2013 IEEE Conference on Computer Vision and Pattern Recognition (CVPR), pp. 3476–3483. IEEE (2013)
5. Zhang, J., Shan, S., Kan, M., Chen, X.: Coarse-to-Fine Auto-Encoder Networks (CFAN) for real-time face alignment. In: Fleet, D., Pajdla, T., Schiele, B., Tuytelaars, T. (eds.) ECCV 2014. LNCS, vol. 8690, pp. 1–16. Springer, Heidelberg (2014)
6. Cao, X., Wei, Y., Wen, F., Sun, J.: Face alignment by explicit shape regression. Int. J. Comput. Vis. **107**, 177–190 (2014)

7. Burgos-Artizzu, X.P., Perona, P., Dollár, P.: Robust face landmark estimation under occlusion. In: 2013 IEEE International Conference on Computer Vision (ICCV), pp. 1513–1520. IEEE (2013)
8. Ren, S., Cao, X., Wei, Y., Sun, J.: Face alignment at 3000 FPS via regressing local binary features. In: 2014 IEEE Conference on Computer Vision and Pattern Recognition (CVPR), pp. 1685–1692. IEEE (2014)
9. Xiong, X., De la Torre, F.: Supervised descent method and its applications to face alignment. In: 2013 IEEE Conference on Computer Vision and Pattern Recognition (CVPR), pp. 532–539. IEEE (2013)
10. Xiong, X., De la Torre, F.: Global supervised descent method. In: Proceedings of the IEEE Conference on Computer Vision and Pattern Recognition, pp. 2664–2673 (2015)
11. Kazemi, V., Sullivan, J.: One millisecond face alignment with an ensemble of regression trees. In: 2014 IEEE Conference on Computer Vision and Pattern Recognition (CVPR), pp. 1867–1874. IEEE (2014)
12. Jeni, L.A., Cohn, J.F., Kanade, T.: Dense 3D face alignment from 2D videos in real-time. In: 2015 11th IEEE International Conference and Workshops on Automatic Face and Gesture Recognition (FG), vol. 1, pp. 1–8. IEEE (2015)
13. Dollár, P., Welinder, P., Perona, P.: Cascaded pose regression. In: 2010 IEEE Conference on Computer Vision and Pattern Recognition (CVPR), pp. 1078–1085. IEEE (2010)
14. Zhu, S., Li, C., Change Loy, C., Tang, X.: Face alignment by coarse-to-fine shape searching. In: Proceedings of the IEEE Conference on Computer Vision and Pattern Recognition, pp. 4998–5006 (2015)
15. Lee, D., Park, H., Yoo, C.D.: Face alignment using cascade Gaussian process regression trees. In: 2015 IEEE Conference on Computer Vision and Pattern Recognition (CVPR), pp. 4204–4212. IEEE (2015)
16. Alahi, A., Ortiz, R., Vandergheynst, P.: Freak: fast retina keypoint. In: 2012 IEEE Conference on Computer Vision and Pattern Recognition (CVPR), pp. 510–517. IEEE (2012)
17. Sagonas, C., Tzimiropoulos, G., Zafeiriou, S., Pantic, M.: 300 faces in-the-wild challenge: the first facial landmark localization challenge. In: Proceedings of the IEEE International Conference on Computer Vision Workshops, pp. 397–403 (2013)
18. Zhu, X., Ramanan, D.: Face detection, pose estimation, and landmark localization in the wild. In: 2012 IEEE Conference on Computer Vision and Pattern Recognition (CVPR), pp. 2879–2886. IEEE (2012)
19. Belhumeur, P.N., Jacobs, D.W., Kriegman, D.J., Kumar, N.: Localizing parts of faces using a consensus of exemplars. IEEE Trans. Pattern Anal. Mach. Intell. **35**, 2930–2940 (2013)
20. Le, V., Brandt, J., Lin, Z., Bourdev, L., Huang, T.S.: Interactive facial feature localization. In: Fitzgibbon, A., Lazebnik, S., Perona, P., Sato, Y., Schmid, C. (eds.) ECCV 2012. LNCS, vol. 7574, pp. 679–692. Springer, Heidelberg (2012)
21. Yin, L., Chen, X., Sun, Y., Worm, T., Reale, M.: A high-resolution 3D dynamic facial expression database. In: 8th IEEE International Conference On Automatic Face and Gesture Recognition, FG 2008, pp. 1–6. IEEE (2008)
22. Zhou, B., Lapedriza, A., Xiao, J., Torralba, A., Oliva, A.: Learning deep features for scene recognition using places database. In: Advances in Neural Information Processing Systems, pp. 487–495 (2014)
23. Gross, R., Matthews, I., Cohn, J., Kanade, T., Baker, S.: Multi-PIE. Image Vis. Comput. **28**, 807–813 (2010)

Modeling Stylized Character Expressions via Deep Learning

Deepali Aneja[1(✉)], Alex Colburn[2], Gary Faigin[3], Linda Shapiro[1], and Barbara Mones[1]

[1] Department of Computer Science and Engineering, University of Washington, Seattle, WA, USA
{deepalia,shapiro,mones}@cs.washington.edu
[2] Zillow Group, Seattle, WA, USA
alexco@cs.washington.edu
[3] Gage Academy of Art, Seattle, WA, USA
gary@gageacademy.org

Abstract. We propose **DeepExpr**, a novel expression transfer approach from humans to multiple stylized characters. We first train two Convolutional Neural Networks to recognize the expression of humans and stylized characters independently. Then we utilize a transfer learning technique to learn the mapping from humans to characters to create a shared embedding feature space. This embedding also allows human expression-based image retrieval and character expression-based image retrieval. We use our perceptual model to retrieve character expressions corresponding to humans. We evaluate our method on a set of retrieval tasks on our collected stylized character dataset of expressions. We also show that the ranking order predicted by the proposed features is highly correlated with the ranking order provided by a facial expression expert and Mechanical Turk experiments.

1 Introduction

Facial expressions are an important component of almost all human interaction and face-to-face communication. As such, the importance of clear facial expressions in animated movies and illustrations cannot be overstated. Disney and Pixar animators [1,2] have long understood that unambiguous expression of emotions helps convince an audience that an animated character has underlying cognitive processes. The viewer's emotional investment in a character depends on the clear recognition of the character's emotional state [3]. To achieve lifelike emotional complexity, an animator must be able to depict characters with clear, unambiguous expressions, while retaining the fine level control over intensity and expression mix required for nuance and subtlety [4]. However,

---

Electronic supplementary material The online version of this chapter (doi:10.1007/978-3-319-54184-6_9) contains supplementary material, which is available to authorized users.

S.-H. Lai et al. (Eds.): ACCV 2016, Part II, LNCS 10112, pp. 136–153, 2017.
DOI: 10.1007/978-3-319-54184-6_9

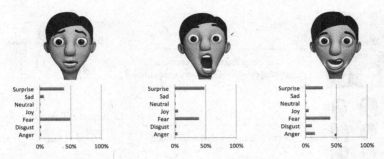

Fig. 1. Expressions are surprisingly difficult to create for professional animators. Three professional animators were asked to make the character appear as surprised as possible. None of the expressions achieved above 50% recognition on Mechanical Turk with 50 test subjects.

explicit expressions are notoriously difficult to create [5], as illustrated in Fig. 1. This difficulty is in part due to animators and automatic systems relying on geometric markers and features modeled for human faces, not stylized character faces.

We focus our efforts on *stylized* 3D characters, defined as characters that no human would mistake for another person, but would still be perceived as having human emotions and thought processes. Our goal is to develop a model of facial expressions that enables accurate retrieval of stylized character expressions given a human expression query.

To achieve this goal, we created DeepExpr, a perceptual model of stylized characters that accurately recognizes human expressions and transfers them to a stylized character without relying on explicit geometric markers. Figure 2 shows an overview of the steps to develop the framework of our model. We created a database of labeled facial expressions for six stylized characters as shown in Fig. 7. This database with expressions is created by facial expression artists and initially labeled via Mechanical Turk (MT) [6]. Images are labeled for each of six cardinal expressions: joy, sadness, anger, surprise, fear, disgust, and neutral. First, we trained a Convolutional Neural Network (CNN) on a large database of human expressions to input a human expression and output the probabilities of each of the seven classes. Second, we trained a similar character model on an artist-created character expression image database. Third, we learned a mapping between the human and character feature space using the transfer learning approach [7]. Finally, we can retrieve character expressions corresponding to a human using perceptual model mapping and human geometry.

We make the following contributions[1]:

1. A data-driven perceptual model of facial expressions.
2. A novel stylized character data set with cardinal expression annotations.
3. A mechanism to accurately retrieve plausible character expressions from human expression queries.

---

[1] Project page: http://grail.cs.washington.edu/projects/deepexpr/.

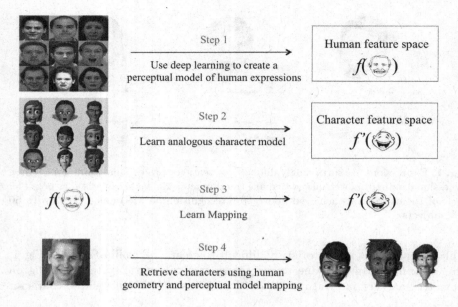

Fig. 2. Overview of our pipeline. Feature extraction using CNNs and transfer learning builds a model of expression mapping.

2   Related Work

There is a large body of literature classifying, recognizing, and characterizing human facial expressions. Notably, Paul Ekman's widely adopted Facial Action Coding System (FACS) [8] is used as a common basis for describing and communicating human facial expressions. The FACS system is often used as a basis for designing character animation systems [5,9] and for facial expression recognition on scanned 3D faces [10]. However, despite these advances, creating clear facial animations for 3D characters remains a difficult task.

2.1   Facial Expression Recognition and Perception

FACS for Animation. Though a reliable parameterization of emotion and expression remains elusive, the six cardinal expressions pervade stories and face-to-face interactions, making them a suitable focus for educators and facial expression researchers [11]. To guide and automate the process of expression animation, animators and researchers turn to FACS. For example, FACSGen [9] allows researchers to control action units on realistic 3D synthetic faces. Though Roesch et al. confirmed the tool's perceptual validity settings by asking viewers to rate the presence of emotions in faces developed using action unit combinations found in "real life situations", we found that their faces were unclear as demonstrated in Fig. 3.

HapFACS [5], an alternative to FACSGen, allows users to control facial movement at the level of both action units and whole expressions (EmFACS) according to Ekman's formulas. The strict use of anatomy-based and constrained motion by these systems limits their generalizability to characters with different anatomy and limits their application, because the most believable animation may require the violation of physical laws [1].

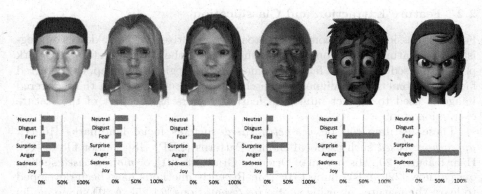

Fig. 3. DeepExpr yields clearer expressions than other approaches when tested on MT. From left to right each generated face was intended to clearly convey an expression: anger from MPEG-4 [12] scored 20% clarity for anger. Anger from HapFACS [5] scored 8% clarity for anger. Fear from HapFACS scored 20% clarity for fear. Fear using FAC-SGen [9] scored 6% clarity for fear. Anger and fear faces retrieved with our approach, both scored over 85% clarity.

Alternatively, the MPEG-4 standard [12] can describe motion in stylized faces by normalizing feature motion to a standard distance. The MPEG-4 standard provides users with archetypal expression profiles for the six cardinal expressions, but like the FACS-based systems does not give the user feedback on the perceptual validity of their expression, which may lead to unclear faces. As demonstrated in Fig. 3, anatomically valid faces generated by these systems did not consistently yield high recognition rates in MT with 50 test subjects.

Other Perceptual Models. The results shown in Fig. 3 support artists' intuition that anatomy based formulas for expressions must be tailored to each unique face, and necessitate a perceptually guided system to find the optimal configuration for a clear expression. Perceptual models such as Deng and Ma [13] have also been explored for realistic faces with promising results. Deng and Ma polled students' perceptions of the expression of different motion-captured facial configurations and ran Principal Component Analysis (PCA) [14] on the vertices of the meshes of these faces. Using these results, they developed a Support Vector Machine (SVM) [15] model for expression clarity as a function of PCA weights for different areas of the face. They also showed significantly increased expression clarity of generated speech animation by constraining the characters' motion to fit their model. However, the scalability of their procedure is limited by its reliance on on-site subjects and the size and specificity of the seeding dataset. We addressed these limitations by incorporating MT tests in our character expression data collection and training a deep learning model for expression clarity.

2.2 Feature Extraction and Classification

Facial expression recognition can be broadly categorized into face detection, registration, feature extraction, and classification. In the detection step, landmark points are used to detect a face in an image. In the registration step, the detected faces are geometrically aligned to match a template image. Then the registered image is used to extract numerical feature vectors as the part of the feature extraction step.

These features can be *geometry based* such as facial landmarks [16,17], *appearance based* such as Local Binary Patterns (LBP) [18], Gabor filters [19], Haar features [20], Histogram of Oriented Gradients [21], or *motion based* such as optical flow [22] and Volume LBP [23]. Recently, methods have been developed to learn the features by using sparse representations [24,25]. A 3D shape model approach has also been implemented to improve the facial expression recognition rate [26]. A variety of fusion of features has also been utilized to boost up the facial expression recognition performance [27,28]. They are mostly a combination of geometric and appearance based features. In the current practice of facial expression analysis, CNNs have shown the capability to learn the features that statistically allow the network to make the correct classification of the input data in various ways [29,30]. CNN features fused with geometric features for customized expression recognition [31] and Deep Belief Networks have also been utilized to solve the Facial Expression recognition (FER) problem. A recent approach [32] termed "AU (Action Unit)-Aware" Deep Networks demonstrated the effectiveness in classifying the six basic expressions. Joint Fine-Tuning in Deep Neural Networks [33] have also been used to combine temporal appearance features from image sequences and temporal geometry features from temporal facial landmark points to enhance the performance of the facial expression recognition. Along similar lines, we have utilized deep learning techniques as a tool to extract useful features from raw data for both human faces and stylized characters. We then deploy a transfer learning approach, where the weights of the stylized character are initialized with those from a network pre-trained on a human face data set, and then fine-tuned with the target stylized character dataset.

In the last step of classification, the algorithm attempts to classify the given face image into seven different classes of basic emotions using machine learning techniques. SVMs are most commonly used for FER tasks [18,34,35]. As SVMs treat the outputs as scores for each class which are uncalibrated and difficult to interpret, the softmax classifier gives a slightly more intuitive output with normalized class probabilities and also has a probabilistic interpretation. Based on that, we have used a softmax classifier to recognize the expressions in our classification task using the features extracted by the deep CNNs.

3 Methodology

We first describe the data collection approach and design of facial features that can capture the seven expressions: joy, sadness, anger, surprise, fear, disgust, and neutral. Then, we discuss our customized expression recognition and transfer learning framework using deep learning.

3.1 Data Collection and Pre-processing

To learn deep CNN models that generalize well across a wide range of expressions, we need sufficient training data to avoid over-fitting of the model. For human facial expression data collection, we combined publicly available annotated facial expression databases: extended CK+ [36], DISFA [37], KDEF [38] and MMI [39]. We also created a novel database of facial expressions for six stylized characters: the Facial Expression Research Group-Database (**FERG-DB**). Both the databases have labels for the six cardinal expressions and neutral.

Fig. 4. Examples of registered faces from CK+, DISFA, KDEF, and MMI databases showing disgust, joy, anger, and surprise emotion from left to right.

CK+: The Extended Cohn-Kanade database (CK+) includes 593 video sequences recorded from 123 subjects. Subjects portrayed the six cardinal expressions. We selected only the final frame of each sequence with the peak expression for our method, which resulted in 309 images.

DISFA: Denver Intensity of Spontaneous Facial Actions (DISFA) database consists of 27 subjects, each recorded while watching a four minutes video clip by two cameras. As DISFA is not emotion-specified coded, we used the EMFACS system [5] to convert AU FACS codes to expressions, which resulted in around 50,000 images using the left camera only.

KDEF: The Karolinska Directed Emotional Faces (KDEF) is a set of 4900 images of human facial expressions of emotion. This database consists of 70 individuals, each displaying 7 different emotional expressions. We used only the front facing angle for our method and selected 980 images.

MMI: The MMI database includes expression labeled videos for more than 20 subjects of both genders for which subjects were instructed to display 79 series of facial expressions. We extracted static frames from each corresponding sequence for the six cardinal emotions, resulting in 10,000 images.

We balanced out the final number of samples for each class for training our network to avoid any bias towards a particular expression.

Stylized Character Database. We created a novel database (**FERG-DB**) of labeled facial expressions for six stylized characters. The animator created the key poses for each expression, and they were labeled via MT to populate the database initially. The number of key poses created depends on the complexity of the expression for each character. We only used the expression key poses having 70% MT test agreement among 50 test subjects for the same pose. On average, 150 key poses (15–20 per expression) were created for each character. Interpolating between the key poses resulted in 50,000 images (around 8,000 images per character). The motivation behind the combination of different characters is to have a generalized feature space among various stylized characters.

Data Pre-processing. For our combined human dataset, Intraface [40] was used to extract 49 facial landmarks. We use these points to register faces to an average frontal face via an affine transformation. Then a bounding box around the face is considered to be the face region. Geometric measurements between the points are also taken to produce geometric features for refinement of expression retrieval results as described in Sect. 3.2. Once the faces are cropped and registered, the images are re-sized to 256×256 pixels for analysis. Figure 4 shows examples of registered faces from different databases using this method.

The corresponding 49 landmark points are marked on the neutral expression of the 3D stylized character rig. This supplementary information is saved along with each expression rendering and used later to perform geometric refinement of the result. This step is performed only once per character.

3.2   Network Training Using Deep Learning

With approximately 70,000 images of labeled samples of human faces and 50,000 images for stylized character faces, the datasets are smaller in comparison to other image classification datasets that have been trained from scratch in the past. Moreover, since we have to use a portion of this data set for validation, effectively only 80% of the data was available for training. We performed data augmentation techniques to increase the number of training examples. This step helps in reducing overfitting and improving the model's ability to generalize. During the training phase, we extracted 5 crops of 227×227 from the four corners and the center of the image and also used the horizontal mirror images for data augmentation.

Training Human and Character CNN Models. Our human expression network consists of three elements: multiple convolutional layers followed by max-pooling layers and fully connected layers as in [41]. Our character network is analogous to the human CNN architecture and does not require CONV4 for the recognition task as the character images are not very complex. Unlike the human dataset, there are fewer variations in the character dataset (light, pose, accessories, etc.). To avoid overfitting, we limited our model to a fewer number of convolutional parameters (until CONV3). Both networks are trained independently. The details of the network layers are shown in Fig. 5 and network parameters are given in the supplementary material.

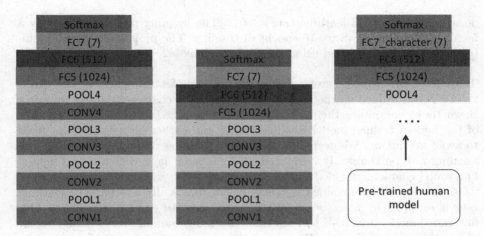

Fig. 5. Outline of the CNN architecture. The convolutional layers, max pooling layers and fully connected layers are denoted as CONV, POOL and FC followed by the layer number. Human expression image trained model (left), Stylized character expression image trained model (middle) and fine-tuned character trained model (right) are shown. In the transfer learning step, the last fully-connected layer (FC7_character) is fine-tuned using stylized character data.

All three color channels are processed directly by the network. Images are first rescaled to 256×256 and a crop of 227×227 is fed to the network. Finally, the output of the last fully connected layer is fed to a softmax layer that assigns a probability for each class. The prediction itself is made by taking the class with the maximal probability for the given test image.

In the forward propagation step, the CONV layer computes the output of neurons that are connected to local regions in the input (resized to 256×256 in the data pre-processing step), each computing a dot product between their weights and a small region they are connected to in the input volume, while the POOL layer performs a downsampling operation along the spatial dimensions. The output of each layer is a linear combination of the inputs mapped by an activation function given as:

$$h^{i+1} = f((W^{i+1})^T h^i) \tag{1}$$

where h^{i+1} is the ith layer output, W^i is the vector of weights that connect to each output node and $f(\cdot)$ is the non-linear activation function which is implemented by the RELU layer given as: $f(x) = max(0, x)$ where x is the input to the neuron. The back-propagation algorithm to used to calculate the gradient with respect to the parameters of the model. The weights of each layer are updated as:

$$\delta^i = (W^i)^T \delta^{i+1} . f'(h^i) \tag{2}$$

where δ^i is the increment of weights at layer i. We train our networks using stochastic gradient descent with hyperparameters (momentum $= 0.9$, weight decay = 0.0005, initial learning rate = 0.01). The learning rate is dropped by a factor of 10 following every 10 epochs of training. The proposed network architectures were implemented using the Caffe toolbox [42] on a Tesla k40c GPU.

Transfer Learning. To create a shared embedding feature space, we fine-tuned the CNN pre-trained on the human dataset with the character dataset for every character by continuing the backpropagation step. The last fully connected layer of the human trained model was fine-tuned, and earlier layers were kept fixed to avoid overfitting. We decreased the overall learning rate while increasing the learning rate on the newly initialized FC7_character layer which is highlighted fine-tuned character trained model in Fig. 5. We set an initial learning rate of 0.001, so that the pre-trained weights are not drastically altered. The learning rate is dropped by a factor of 10 following every 10 epochs of training. Our fine-tuned model used 38 K stylized character image samples for training, 6K for validation, and 6 K for test. The proposed architecture was trained for 50 epochs with 40 K iterations on batches of size 50 samples.

Distance Metrics. In order to retrieve the stylized character closest expression match to the human expression, we used the Jensen—Shannon divergence distance [43] for expression clarity and geometric feature distance for expression refinement. It is described by minimizing the distance optimization function in Eq. 3 given as:

$$\phi_d = \alpha \, |\text{JS Distance}| + \beta \, |\text{Geometric Distance}| \qquad (3)$$

where JS Distance is given as the Jensen—Shannon divergence distance between FC6 feature vectors of *human* and *character*, and Geometric distance is given as the L^2 norm distance between geometric features of *human* and *character*. Our implementation uses JS Distance as a retrieval parameter and then geometric distance as a sorting parameter to refine the retrieved results with α and β as relative weight parameters. Details of the computation are given as follows:

Expression Distance. For a given human expression query image, FC6 (512 outputs) features are extracted from the query image using the human expression trained model and for the test character images from the shared embedding feature space using the fine-tuned character expression model. The FC7 (7 outputs) layer followed by a softmax can be interpreted as the probability that a particular expression class is predicted for a given input feature vector. By normalizing each element of the feature vector by the softmax weight, the FC6 feature vectors are treated as discrete probability distributions. To measure the similarity between human and character feature probability distributions, we used the Jensen—Shannon divergence [43] which is symmetric and is computed as:

$$JSD(H||C) = \frac{1}{2}D(H||M) + \frac{1}{2}D(C||M) \qquad (4)$$

where $M = \frac{1}{2}(H + C)$, $D(H||M)$ and $D(C||M)$ represents the Kullback—Leibler divergence [44] which is given as:

$$D(X||M) = \sum_i X(i)log\frac{X(i)}{M(i)} \tag{5}$$

where X and M are discrete probability distributions.

We used this distance metric to order the retrievals from the closest distance to the farthest in the expression feature space. Our results show that the retrieval ordering matched the query image label, and retrievals were ordered in order of similarity to the query label. To choose the best match out of the multiple retrievals with the same label as shown in Fig. 6, we added a geometric refinement step as described in the next section.

Geometric Distance. The JS Divergence distance results in the correct expression match, but not always the closest geometric match to the expression. Figure 6 shows the retrieval of the correct label (joy). To match the geometry, we extract geometric distance vectors and use them to refine the result.

Fig. 6. Multiple retrieval results for the joy query image

We use the facial landmarks as described in Sect. 3.1, to extract the geometric features including the following measurements: the left/right eyebrow height (vertical distance between top of the eyebrow and center of the eye), left/right eyelid height (vertical distance between top of an eye and bottom of the eye), nose width (horizontal distance between leftmost and rightmost nose landmarks), mouth width (left mouth corner to right mouth corner distance), closed mouth measure (vertical distance between the upper and the lower lip), and left/right lip height (vertical distance between the lip corner from respective the lower eyelid). The geometric distance is a normalized space. Each of the distances between landmarks is normalized by the bounding box of the face. After normalization, we compute the L^2 norm distance between the human geometry vector and character geometry vectors with the correct expression label. Finally, we re-order the retrieved images within the matched label based on matched geometry.

4   Experimental Results

The combined DeepExpr features and geometric features produce significant performance enhancement in retrieving the stylized character facial expressions based on human facial expressions. The top results for all seven expressions on six stylized characters are shown in Fig. 7. Human expression-based image retrievals and character expression-based image retrievals are shown in the supplementary material.

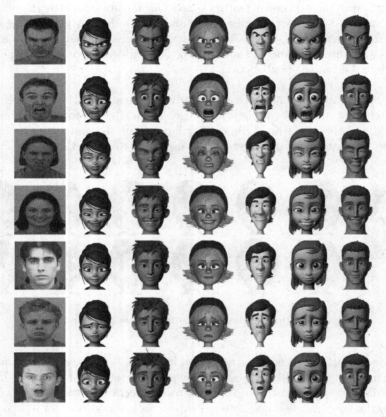

Fig. 7. Results from our combined approach - DeepExpr and geometric features. The leftmost image in each row is the query image and all six characters are shown portraying the top match of the same expression - anger, fear, joy, disgust, neutral, sad and surprise (top to bottom).

5   Evaluation

5.1   Expression Recognition Accuracy

For human facial expression recognition accuracy, we performed the subject independent evaluation, where the classifier is trained on the training set and evaluated on images in the same database (validation and test set) using K-fold cross-validation with K = 5. On average, we used 56 K samples for training in batches of 50 samples, 10 K samples for validation and 10 K for testing. The overall accuracy of human facial expression recognition was 85.27%. Similarly, for stylized character expression, we used 38 K character images for training in batches of 50 samples, 6 K for validation, and 6 K for testing, and achieved the recognition accuracy of 89.02%. Our aim with human expression accuracy was to achieve a good score on the expression recognition which is close to the state-of-the-art results in order to extract relevant features corresponding to a facial expression. The details of human expression recognition accuracy for each expression are given in the supplementary material.

5.2   Expression Retrieval Accuracy

We analyze our retrieval results by computing the retrieval score to measure how close is the retrieved character expression label is to the human query expression label. We also compare our results with a facial expression expert by choosing 5 random samples from the retrieved results with the same label and rank order them based on their similarity to the query image. The details of analysis are discussed as follows:

Retrieval Score. We measured the retrieval performance of our method by calculating the average normalized rank of relevant results (same expression label) [45]. The evaluation score for a query human expression image was calculated as:

$$score(q) = \frac{1}{1 - N \cdot N_{rel}} \left(\sum_{k=1}^{N_{rel}} R_k - \frac{N_{rel}(N_{rel} + 1)}{2} \right) \tag{6}$$

where N is the number of images in the database, N_{rel} the number of database images that are relevant to the query expression label q (all images in the character database that have the same expression label as the human query expression label), and R_k is the rank assigned to the kth relevant image. The evaluation score ranges from 0 to 1, where 0 is the best score as it indicates that all the relevant database images are retrieved before all other images in the database. A score that is greater than 0 denotes that some irrelevant images (false positives) are retrieved before all relevant images.

The retrieval performance was measured over all the images in the human test dataset using each test image in turn as a query image. The average retrieval score for each expression class was calculated by averaging the retrieval score for all test images in the same class. Table 1 shows the final class retrieval score, which was calculated by averaging the retrieval scores across all characters for each expression class using only geometry and DeepExpr expression features. The best match results in Fig. 8 confirm that the geometric measure is not sufficient to match the human query expression with clarity.

Comparison. In order to judge the effectiveness of our system, we compared DeepExpr to a human expert and MT test subjects. We asked the expert and

Table 1. Average retrieval score for each expression across all characters using only geometry and DeepExpr features.

Expression	Geometry	DeepExpr
Anger	0.384	0.213
Disgust	0.386	0.171
Fear	0.419	0.228
Joy	0.276	0.106
Neutral	0.429	0.314
Sad	0.271	0.149
Surprise	0.322	0.125

Query (Disgust)   Geometry   DeepExpr

Query (Fear)   Geometry   DeepExpr

Fig. 8. Best match results from our Deep-Expr approach compared to only geometric feature based retrieval for Disgust (top) and Fear (bottom).

the MT subjects to rank five stylized character expressions in order of decreasing expression similarity to a human query image. The facial expression expert, 50 MT test subjects and DeepExpr ranked the same 30 validation test sets. We aggregated the MT results into a single ranking using a voting scheme. We then compared the DeepExpr ranking to the results, measuring similarity with two measures. Both measures found a high correlation between DeepExpr ranking compared with the expert and the MT ranking results. The details of the ranking comparison tests are given in the supplementary material.

The **Spearman rank correlation coefficient** ρ measures the strength and direction of the association between two ranked variables [46]. The closer the ρ coefficient is to 1, the better the two ranks are correlated.

The average ρ coefficient for the expert rank orderings is 0.773 ± 0.336 and for MT tests is 0.793 ± 0.3561. The most relevant correlation coefficient is between the first rank chosen by the expert and the first rank chosen by DeepExpr as they represent the best match with the query image. The Spearman correlation with expert best rank is 0.934 and with MT best rank is 0.942, which confirms the agreement on selection of the closest match to the human expression.

The **Kendall τ test** is a non-parametric hypothesis test for statistical dependence based on the τ coefficient [47]. It is a pairwise error that represents how many pairs are ranked discordant. The best matching ranks receive a τ value of 1. The average τ coefficient for expert validation rank orderings is 0.706 ± 0.355, and the best rank correlation is 0.910. For the MT ranking, the average Kendall correlation coefficient is 0.716 ± 0.343 and 0.927 is the best rank correlation.

The Spearman and Kendall correlation coefficients of DeepExpr ranking with the expert ranking and MT test ranking for 30 validation experiments are shown in Fig. 9. Note that more than half the rankings are perfectly correlated, and most of them are above 0.8. Only two of the rankings had (small) negative correlations in both correlation experiments: the order was confusing because of very subtle difference in expressions (see supplementary material for details).

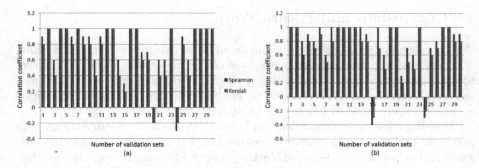

Fig. 9. Correlation rank order result charts with (a) expert and (b) MT tests

6   Comparison with Character Animator

Currently to our knowledge, no other system performs stylized character retrieval based on a learned feature set. The closest match to DeepExpr tool is Adobe Character Animator (Ch) [48] which creates 2.5-D animations for characters. We conducted an expression recognition experiment by creating a similar character in Ch with different expressions as layers. We queried three human expression images for each of the seven expressions. Then, we asked 50 MT test subjects to recognize the expression for best matches from DeepExpr retrieved images and Ch results. The results of the experiment are shown in Table 2. On an average, joy, neutral and surprise had comparable recognition performance. DeepExpr showed great improvement in recognition of fear and disgust. In Ch, fear was confused with surprise due to the dependence on geometric landmarks of the face showing an open mouth and disgust was most confused with anger. For anger and sad, the closed mouth was most confused with neutral in Ch. An example of a fear expression MT test is shown in Fig. 10. DeepExpr achieved higher (83%) expression recognition accuracy as compared to the Ch animator tool (41%).

Table 2. Average expression recognition accuracy (%) for each expression across all characters using Ch animator and DeepExpr.

Expression	Ch animator	DeepExpr
Anger	60	85
Disgust	47	86
Fear	42	81
Neutral	87	88
Joy	95	97
Sad	43	89
Surprise	93	95

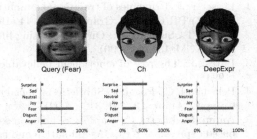

Fig. 10. Expression matching results for fear query image. Ch [48] result scored 41% clarity for fear and DeepExpr result scored 83% clarity for fear.

7 Conclusions and Future Work

We have demonstrated a perceptual model of facial expression clarity and geometry using a deep learning approach combined with artistic input and crowd-sourced perceptual data. Our results are highly correlated with a facial expression expert, in addition to MT subjects and have a higher expression recognition accuracy as compared to Character Animator.

DeepExpr has several practical applications in the field of storytelling, puppeteering and animation content development. For example, the system could assist animators during the initial blocking stage for 3D characters in any production pipeline. When there are multiple animators working on the same character during a production, using our expression recognition system will help enable a consistent approach to the personality for that character. More importantly, our approach provides a foundation for future facial expression studies. For example, our perceptual model could be used to evaluate existing FACS.

Our system demonstrates a perceptual model of facial expressions that provides insight into facial expressions displayed by stylized characters. It can be used to automatically create desired character expressions driven by human facial expressions. The model can also be incorporated into the animation pipeline to help animators and artists to better understand expressions, communicate how to create expressions to others, transfer expressions from humans to characters, and to provide a mechanism for animators/storytellers to more quickly and accurately create the expressions they intend.

Acknowledgements. We would like to thank Jamie Austad for creating our stylized character database. We would also like to thank the creators of the rigs we used in our project: Mery (www.meryproject.com), Ray (*CGTarian Online School*), Malcolm (www.animSchool.com), Aia & Jules (www.animationmentor.com), and Bonnie (*Josh Sobel Rigs*).

References

1. Lasseter, J.: Principles of traditional animation applied to 3D computer animation. SIGGRAPH Comput. Graph. **21**, 35–44 (1987)
2. Porter, T., Susman, G.: On site: creating lifelike characters in pixar movies. Commun. ACM **43**, 25 (2000)
3. Bates, J.: The role of emotion in believable agents. Commun. ACM **37**, 122–125 (1994)
4. Pelachaud, C., Poggi, I.: Subtleties of facial expressions in embodied agents. J. Vis. Comput. Anim. **13**, 301–312 (2002)
5. Amini, R., Lisetti, C.: HapFACS: an open source API/Software to generate FACS-based expressions for ECAs animation and for corpus generation. In: 2013 Humaine Association Conference on Affective Computing and Intelligent Interaction (ACII), pp. 270–275 (2013)
6. Buhrmester, M., Kwang, T., Gosling, S.D.: Amazon's Mechanical Turk: a new source of inexpensive, yet high-quality, data? Perspect. Psychol. Sci. **6**, 3–5 (2011)

7. Oquab, M., Bottou, L., Laptev, I., Sivic, J.: Learning and transferring mid-level image representations using convolutional neural networks. In: Proceedings of the IEEE Conference on Computer Vision and Pattern Recognition, pp. 1717–1724 (2014)
8. Ekman, P., Friesen, W.: Facial Action Coding System: A Technique for the Measurement of Facial Movement. Consulting Psychologists Press, Palo Alto (1978)
9. Roesch, E.B., Tamarit, L., Reveret, L., Grandjean, D., Sander, D., Scherer, K.R.: FACSGen: a tool to synthesize emotional facial expressions through systematic manipulation of facial action units. J. Nonverbal Behav. **35**, 1–16 (2011)
10. Sandbach, G., Zafeiriou, S., Pantic, M., Yin, L.: Static and dynamic 3D facial expression recognition: a comprehensive survey. Image Vis. Comput. **30**, 683–697 (2012)
11. Adolphs, R.: Recognizing emotion from facial expressions: psychological and neurological mechanisms. Behav. Cogn. Neurosci. Rev. **1**(1), 21–62 (2002)
12. Pereira, F.C., Ebrahimi, T.: The MPEG-4 Book. Prentice Hall PTR, Upper Saddle River (2002)
13. Deng, Z., Ma, X.: Perceptually guided expressive facial animation. In: Proceedings of the 2008 ACM SIGGRAPH/Eurographics Symposium on Computer Animation, SCA 2008 Eurographics Association (2008)
14. Jolliffe, I.: Principal Component Analysis. Wiley, Hoboken (2002)
15. Cortes, C., Vapnik, V.: Support-vector networks. Mach. Learn. **20**, 273–297 (1995)
16. Kobayashi, H., Hara, F.: Facial interaction between animated 3D face robot and human beings. In: 1997 IEEE International Conference on Systems, Man, and Cybernetics, Computational Cybernetics and Simulation, vol. 4, pp. 3732–3737. IEEE (1997)
17. Dibeklioglu, H., Salah, A., Gevers, T.: Like father, like son: facial expression dynamics for kinship verification. In: Proceedings of the IEEE International Conference on Computer Vision, pp. 1497–1504 (2013)
18. Shan, C., Gong, S., McOwan, P.W.: Facial expression recognition based on local binary patterns: a comprehensive study. Image Vis. Comput. **27**, 803–816 (2009)
19. Liu, C., Wechsler, H.: Independent component analysis of Gabor features for face recognition. IEEE Trans. Neural Netw. **14**, 919–928 (2003)
20. Whitehill, J., Littlewort, G., Fasel, I., Bartlett, M., Movellan, J.: Toward practical smile detection. IEEE Trans. Pattern Anal. Mach. Intell. **31**, 2106–2111 (2009)
21. Shu, C., Ding, X., Fang, C.: Histogram of the oriented gradient for face recognition. Tsinghua Sci. Technol. **16**, 216–224 (2011)
22. Kenji, M.: Recognition of facial expression from optical flow. IEICE Trans. Inf. Syst. **74**, 3474–3483 (1991)
23. Zhao, G., Pietikainen, M.: Dynamic texture recognition using local binary patterns with an application to facial expressions. IEEE Trans. Pattern Anal. Mach. Intell. **29**, 915–928 (2007)
24. Mahoor, M.H., Zhou, M., Veon, K.L., Mavadati, S.M., Cohn, J.F.: Facial action unit recognition with sparse representation. In: 2011 IEEE International Conference on Automatic Face & Gesture Recognition and Workshops (FG 2011), pp. 336–342. IEEE (2011)
25. Lin, Y., Song, M., Quynh, D.T.P., He, Y., Chen, C.: Sparse coding for flexible, robust 3D facial-expression synthesis. IEEE Comput. Graph. Appl. **32**, 76–88 (2012)

26. Jeni, L.A., Lőrincz, A., Szabó, Z., Cohn, J.F., Kanade, T.: Spatio-temporal event classification using time-series kernel based structured sparsity. In: Fleet, D., Pajdla, T., Schiele, B., Tuytelaars, T. (eds.) ECCV 2014. LNCS, vol. 8692, pp. 135–150. Springer, Heidelberg (2014). doi:10.1007/978-3-319-10593-2_10

27. Tan, X., Triggs, B.: Fusing Gabor and LBP feature sets for kernel-based face recognition. In: Zhou, S.K., Zhao, W., Tang, X., Gong, S. (eds.) AMFG 2007. LNCS, vol. 4778, pp. 235–249. Springer, Heidelberg (2007). doi:10.1007/978-3-540-75690-3_18

28. Ying, Z.-L., Wang, Z.-W., Huang, M.-W.: Facial expression recognition based on fusion of sparse representation. In: Huang, D.-S., Zhang, X., Reyes García, C.A., Zhang, L. (eds.) ICIC 2010. LNCS (LNAI), vol. 6216, pp. 457–464. Springer, Heidelberg (2010). doi:10.1007/978-3-642-14932-0_57

29. Mollahosseini, A., Chan, D., Mahoor, M.H.: Going deeper in facial expression recognition using deep neural networks. In: 2016 IEEE Winter Conference on Applications of Computer Vision (WACV). IEEE (2016)

30. Yu, Z., Zhang, C.: Image based static facial expression recognition with multiple deep network learning. In: Proceedings of the 2015 ACM on International Conference on Multimodal Interaction, pp. 435–442. ACM (2015)

31. Yu, X., Yang, J., Luo, L., Li, W., Brandt, J., Metaxas, D.: Customized expression recognition for performance-driven cutout character animation. In: Winter Conference on Computer Vision (2016)

32. Liu, M., Li, S., Shan, S., Chen, X.: Au-aware deep networks for facial expression recognition. In: 2013 10th IEEE International Conference and Workshops on Automatic Face and Gesture Recognition (FG), pp. 1–6. IEEE (2013)

33. Jung, H., Lee, S., Yim, J., Park, S., Kim, J.: Joint fine-tuning in deep neural networks for facial expression recognition. In: Proceedings of the IEEE International Conference on Computer Vision, pp. 2983–2991 (2015)

34. Zhong, L., Liu, Q., Yang, P., Liu, B., Huang, J., Metaxas, D.N.: Learning active facial patches for expression analysis. In: 2012 IEEE Conference on Computer Vision and Pattern Recognition (CVPR), pp. 2562–2569. IEEE (2012)

35. Dumas, M.: Emotional expression recognition using support vector machines. In: Proceedings of International Conference on Multimodal Interfaces. Citeseer (2001)

36. Lucey, P., Cohn, J.F., Kanade, T., Saragih, J., Ambadar, Z., Matthews, I.: The extended Cohn-Kanade dataset (CK+): a complete dataset for action unit and emotion-specified expression. In: 2010 IEEE Computer Society Conference on Computer Vision and Pattern Recognition Workshops (CVPRW), pp. 94–101. IEEE (2010)

37. Mavadati, S.M., Mahoor, M.H., Bartlett, K., Trinh, P., Cohn, J.F.: Disfa: a spontaneous facial action intensity database. IEEE Trans. Affect. Comput. 4, 151–160 (2013)

38. Lundqvist, D., Flykt, A., Öhman, A.: The Karolinska directed emotional faces-KDEF. CD-ROM from department of clinical neuroscience, psychology section, Karolinska Institutet, Stockholm, Sweden. Technical report (1998). ISBN 91-630-7164-9

39. Pantic, M., Valstar, M., Rademaker, R., Maat, L.: Web-based database for facial expression analysis. In: 2005 IEEE International Conference on Multimedia and Expo, ICME 2005, p. 5. IEEE (2005)

40. Xiong, X., Torre, F.: Supervised descent method and its applications to face alignment. In: Proceedings of the IEEE Conference on Computer Vision and Pattern Recognition, pp. 532–539 (2013)

41. Krizhevsky, A., Sutskever, I., Hinton, G.E.: Imagenet classification with deep convolutional neural networks. In: Advances in Neural Information Processing Systems, pp. 1097–1105 (2012)
42. Jia, Y., Shelhamer, E., Donahue, J., Karayev, S., Long, J., Girshick, R., Guadarrama, S., Darrell, T.: Caffe: convolutional architecture for fast feature embedding. In: Proceedings of the ACM International Conference on Multimedia, pp. 675–678. ACM (2014)
43. Lin, J.: Divergence measures based on the Shannon entropy. IEEE Trans. Inf. Theor. **37**, 145–151 (1991)
44. Kullback, S., Leibler, R.A.: On information and sufficiency. Ann. Math. Stat. **22**, 79–86 (1951)
45. Müller, H., Marchand-Maillet, S., Pun, T.: The truth about corel - evaluation in image retrieval. In: Lew, M.S., Sebe, N., Eakins, J.P. (eds.) CIVR 2002. LNCS, vol. 2383, pp. 38–49. Springer, Heidelberg (2002). doi:10.1007/3-540-45479-9_5
46. Spearman, C.: The proof and measurement of association between two things. Am. J. Psychol. **15**, 72–101 (1904)
47. Kendall, M.G.: A new measure of rank correlation. Biometrika **30**, 81–93 (1938)
48. Character Animator: Adobe After Effects CC 2016. Adobe Systems Incorporated, San Jose, CA 95110–2704 (2016)

Variational Gaussian Process Auto-Encoder for Ordinal Prediction of Facial Action Units

Stefanos Eleftheriadis[1]([⊠]), Ognjen Rudovic[1], Marc Peter Deisenroth[1],
and Maja Pantic[1,2]

[1] Department of Computing, Imperial College London,
London, UK
{stefanos,orudovic,m.deisenroth,m.pantic}@imperial.ac.uk
[2] EEMCS, University of Twente, Enschede, The Netherlands

Abstract. We address the task of simultaneous feature fusion and modeling of discrete ordinal outputs. We propose a novel Gaussian process (GP) auto-encoder modeling approach. In particular, we introduce GP *encoders* to project multiple observed features onto a latent space, while GP *decoders* are responsible for reconstructing the original features. Inference is performed in a novel variational framework, where the recovered latent representations are further constrained by the ordinal output labels. In this way, we seamlessly integrate the ordinal structure in the learned manifold, while attaining robust fusion of the input features. We demonstrate the representation abilities of our model on benchmark datasets from machine learning and affect analysis. We further evaluate the model on the tasks of feature fusion and joint ordinal prediction of facial action units. Our experiments demonstrate the benefits of the proposed approach compared to the state of the art.

1 Introduction

Automated analysis of facial expressions has attracted significant attention because of its practical importance in psychology studies, human-computer interfaces, marketing research, and entertainment, among others [1]. The most objective way to describe facial expressions is by means of the facial action coding system (FACS) [2]. This is the most comprehensive anatomically-based system that can be used to describe virtually all possible facial expressions in terms of 30+ facial muscle movements, named action units (AUs). FACS also defines rules for scoring the intensity of each AU in the range from absent to maximal intensity on a six-point ordinal scale. Therefore, FACS is critical for high-level interpretation of facial expressions. For instance, the high intensity of AU12 (lip corner puller), as in full-blown smiles, may indicate joy. Conversely, its low intensity may indicate fake smiles as in the case of sarcasm.

The machine analysis of AU intensities is challenging mainly due to the complexity and subtlety of human facial behavior as well as individual differences in expressiveness and variations in head-pose, illumination, occlusions, etc. [3]. These sources of variation are typically accounted for at the feature level by

© Springer International Publishing AG 2017
S.-H. Lai et al. (Eds.): ACCV 2016, Part II, LNCS 10112, pp. 154–170, 2017.
DOI: 10.1007/978-3-319-54184-6_10

means of geometric- and appearance-based features, capturing the geometry and texture changes in a face, respectively. Furthermore, some AUs usually appear in combination with other AUs. For instance, the criteria for intensity scoring of AU7 (lid tightener) are changed significantly if AU7 appears with a maximal intensity of AU43 (eye closure) since this combination changes the appearance as well as timing of these AUs [4]. Furthermore, co-occurring AUs can be non-additive, *e.g.*, if one AU masks another a new and distinct set of appearances is created [2]. Thus, combining different facial features while accounting for AU co-occurrences in a common framework is expected to result in a robust and more accurate estimation of target AUs intensity.

Most existing approaches to AU intensity estimation model each AU independently and cast it as a classification [4–8] or regression [9–12] task. While classification seems to be a natural choice to handle the problem, the related literature fails to account for the ordinal nature of the target intensity levels (misclassification of different levels is equally penalized). The regression-based approaches model the intensity levels on a continuous scale, which is sub-optimal when dealing with discrete outputs. Similarly, the models that do attempt multiple AU intensity estimation (*e.g.*, [13–17]) adopt the same sub-optimal approach to deal with the nature of the output as the independent methods. However, they have showed improved performance in the target task due to the modeling of AU co-occurrences. Apart from a few exceptions that treat each AU independently [7,9,10], none of the aforementioned approaches addresses the task of joint output modeling (*i.e.*, multiple AUs) while accounting for different modalities in the input (*i.e.*, fusion of geometric and appearance features). These limitations can naturally be addressed by following recent advances in manifold learning [18–20] and, in particular, using the framework of Gaussian processes (GPs) [21]. Within this framework, the problem of feature fusion is transformed to that of learning from multiple views, while continuous-valued predictions can be handled efficiently, for more than one output. However, as with the regression-based models described above, these models treat the ordinal labels as continuous values. This also limits their potential to unravel an 'ordinal' manifold, needed to facilitate estimation of target ordinal intensities.

In this work, we propose a novel manifold-based GP approach based on the Bayesian GP latent variable model (B-GPLVM) [22] that performs simultaneously the feature fusion and joint estimation of the AU ordinal intensity. Specifically, we propose the variational GP auto-encoder (VGP-AE), which is composed of a probabilistic *recognition* model, used to project the observed features onto the manifold, and a generative model, used for their reconstruction. This, in contrast to existing work (*e.g.*, [23]) that applies deterministic back-mappings, allows us to explicitly model the uncertainty in the projections onto the learned manifold. Additionally, we endow the proposed VGP-AE with the ordinal outputs [24]. The fusion of the information from the input features and learning of the joint ordinal output is performed simultaneously in a joint Bayesian framework. In this way, we seamlessly integrate the ordinal structure into the recovered manifold while attaining robust fusion of the target features.

To the best of our knowledge, this is the first approach that achieves simultaneous feature fusion and joint AU intensity estimation in the context of facial behavior analysis.

2   Related Work on AU Intensity Estimation

To date, most existing work on automated analysis focuses on the detection of AU activations [25–28]. The problem of AU intensity estimation is relatively new in the field. Most of the research in this area focuses on independent modeling of AU intensities [4–12]. Only recently, joint estimation of the intensity levels has been addressed [13–17]. This is motivated by the fact that intensity annotations are difficult to obtain (due to the tedious process of manually coding) and that AU levels are highly imbalanced. Thus, by imposing the structure on the output in terms of AU co-occurrences robust intensity estimation is expected.

Toward this direction, [13] proposed a two-stage learning strategy, where a multi-class support vector machine (SVM) is first trained for each AU independently. Then, the structure modeling is handled via a dynamic Bayesian network, which captures the semantic relationship among the AU-specific SVMs. In a similar fashion, [14] used support vector regressors (SVR) and a Markov random field (MRF). However, these two-stage approaches are sub-optimal for the target task as the regressors/classifiers and the AU relations are learned independently. To overcome this, [15] proposed to learn latent trees that encode both the input features and (multiple) output AU labels. The structure of the latent variables is modeled using a tree-like graph. However, in the presence of high-dimensional inputs and multiple AUs, this method becomes prohibitively expensive. Moreover, the authors show that with this approach the fusion of different features does not benefit the estimation of AU intensity, achieving similar performance to when individual modalities are used. More recently, [17] proposed a sparse learning approach that uses the notion of robust principal component analysis [29] to decompose expression from facial identity. Then, joint intensity estimation of multiple AUs is performed via a regression model based on dictionary learning. However, this approach can deal with a single modality only. [16] casts the joint AU intensity estimation as a multi-task learning problem based on kernel regression (MLKR). However, in their formulation of the model, the use of MLKR does not scale to high-dimensional features, let alone when using features of different modalities (*e.g.*, geometric and appearance).

The work presented in this paper advances the current state of the art in several aspects: (1) The proposed VGP-AE can efficiently perform the fusion of multiple modalities by means of a shared manifold; (2) Automatic feature selection is implicitly performed via the manifold. The recovered latent representations are used as input to multiple *ordinal* regressors [24], which are concurrently learned in a joint Bayesian framework; (3) GPs allow us to efficiently deal with high-dimensional input and output variables without significantly affecting the model's complexity.

3   Variational Gaussian Process Auto-Encoder

We assume that we have access to a training data set $\mathcal{D} = \{\boldsymbol{Y}, \boldsymbol{Z}\}$, which is comprised of V observed input channels $\boldsymbol{Y} = \{\boldsymbol{Y}^{(v)}\}_{v=1}^{V}$, and the associated output labels $\boldsymbol{Z}$. Each input channel consists of N i.i.d. samples $\boldsymbol{Y}^{(v)} = \{\boldsymbol{y}_i^{(v)}\}_{i=1}^{N}$, where $\boldsymbol{y}_i^{(v)} \in \mathbb{R}^{D_v}$ denotes corresponding facial features. $\boldsymbol{Z} = \{\boldsymbol{z}_i\}_{i=1}^{N}$ is the common label representation, where $z_{ic} \in \{1, \ldots, S\}$ denotes the discrete, ordinal state of the c-th output (*i.e.*, AU intensity level), $c = 1, \ldots, C$. We are interested in simultaneously addressing the tasks of feature fusion and ordinal prediction of the multiple outputs. For this purpose, we propose an approach that resembles recent work of generative models [30,31]. In these models, auto-encoders are employed to learn compact representations of the input data. In a standard auto-encoding setting, the encoding/decoding functions are modeled via neural networks. Here we replace these functions with probabilistic non-parametric mappings, significantly reducing the number of optimized parameters, and naturally modeling the uncertainty in the mappings. The proposed approach can be regarded as a B-GPLVM (generative model) with a fast inference mechanism based on the non-parametric, probabilistic mapping (recognition model). To achieve this, we impose GP priors on both models, and hence, obtain a well-defined GP-*encoder*, in accordance to the GP-*decoder*.

3.1   The Model

Within the above setting, we assume that the observed features $\boldsymbol{Y}^{(v)}$ are generated by a random process, involving a latent (unobserved) set of variables $\boldsymbol{X} = \{\boldsymbol{x}_i\}_{i=1}^{N}, \boldsymbol{x}_i \in \mathbb{R}^q$, with $q \ll D_v$. The data pairs $\mathcal{D} = \{\boldsymbol{Y}, \boldsymbol{Z}\}$ are assumed to be conditionally independent given the latent variables, *i.e.*, $\boldsymbol{Y} \perp\!\!\!\perp \boldsymbol{Z}|\boldsymbol{X}$. The random process of recovering the latent variables has two distinctive stages: (a) a latent variable $\boldsymbol{x}_i$ is generated from some general prior distribution

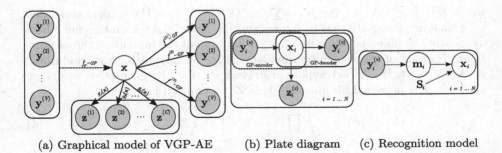

(a) Graphical model of VGP-AE   (b) Plate diagram   (c) Recognition model

Fig. 1. The proposed VGP-AE. (a) $f^{(v)}$ and f_r are the GP-decoder and GP-encoder, respectively. The projection of the latent variable $\boldsymbol{x}$ to the labels' ordinal plane is facilitated through the ordinal regression $g(\boldsymbol{x})$. (b) Compact representation of the model. (c) The proposed recognition model (GP-encoder) with the intermediate variable $\boldsymbol{m}$.

$p(\boldsymbol{x}) = \mathcal{N}(\boldsymbol{0}, \boldsymbol{I})$, and further projected to the labels' ordinal plane via $p(\boldsymbol{z}|\boldsymbol{x})$; (b) an observed input $\boldsymbol{y}_i^{(v)}$ is generated from the conditional distribution $p(\boldsymbol{y}^{(v)}|\boldsymbol{x})$. This process is described in Fig. 1(a),(b). Using this approach, we can now perform classification in the lower-dimensional space of $\boldsymbol{X}$. However, this requires access to the intractable true posterior $p(\boldsymbol{x}|\boldsymbol{y}^{(v)})$.

To constrain the distribution of the latent variables we follow [30,31] and introduce the *recognition* model $p_r(\boldsymbol{x}|\boldsymbol{y}^{(v)})$. Hence, we end up with a supervised auto-encoder setting

$$\boldsymbol{y}_i^{(v)}|\boldsymbol{x}_i = f^{(v)}(\boldsymbol{x}_i; \boldsymbol{\theta}^{(v)}) + \epsilon^{(v)}, \quad \boldsymbol{x}_i|\boldsymbol{y}_i^{(v)} = f_r(\boldsymbol{y}_i^{(v)}; \boldsymbol{\theta}_r) + \epsilon_r, \quad \boldsymbol{z}_i|\boldsymbol{x}_i = g(\boldsymbol{x}_i; \boldsymbol{W}), \quad (1)$$

where the latent space is further encouraged to reflect the structure of the output labels. Here, $\epsilon^{(v)} \sim \mathcal{N}(0, \sigma_v^2 \boldsymbol{I})$, $\epsilon_r \sim \mathcal{N}(0, \sigma_r^2 \boldsymbol{I})$. We place GP priors on $f^{(v)}, f_r$ with corresponding hyper-parameters $\boldsymbol{\theta}^{(v)}, \boldsymbol{\theta}_r$.[1] g denotes the ordinal regression that transforms the latent variables to the labels' ordinal plane, via $\boldsymbol{W} = \{\boldsymbol{w}_c\}_{c=1}^C, \boldsymbol{w}_c \in \mathbb{R}^q$.

In the following, we detail how to learn the GP auto-encoder in Eq. (1) by deriving a variational approximation to the log-marginal likelihood

$$\log p(\boldsymbol{Y}, \boldsymbol{Z}) = \log \int p(\boldsymbol{Z}|\boldsymbol{X}) \prod_v p(\boldsymbol{Y}^{(v)}|\boldsymbol{X}) p(\boldsymbol{X}) d\boldsymbol{X}. \quad (2)$$

3.2 Deriving the Lower Bound

We exploit the conditional independence property of $\boldsymbol{Y} \perp\!\!\!\perp \boldsymbol{Z}|\boldsymbol{X}$ and focus our analysis on the GP auto-encoder. The ordinal information from the labels is incorporated in the presented variational framework in Sect. 3.3. As in [28], we place GP priors on $f^{(v)}, f_r$, and after integrating out the mapping functions, we obtain the conditionals

$$p(\boldsymbol{Y}^{(v)}|\boldsymbol{X}) = \mathcal{N}(\boldsymbol{0}, \boldsymbol{K}^{(v)} + \sigma_v^2 \boldsymbol{I}), \qquad p_r(\boldsymbol{X}|\boldsymbol{Y}) = \mathcal{N}(\boldsymbol{0}, \boldsymbol{K}_r + \sigma_r^2 \boldsymbol{I}), \quad (3)$$

where $\boldsymbol{K}^{(v)} = k^{(v)}(\boldsymbol{X}, \boldsymbol{X})$ and $\boldsymbol{K}_r = \sum_v k_r^{(v)}(\boldsymbol{Y}^{(v)}, \boldsymbol{Y}^{(v)})$ are the kernels associated with each process. Note that in the recognition model the relevant kernel allows us to easily combine multiple features via the sum of the individual kernel functions. Training of the recognition model consists of maximizing the conditional $p_r(\boldsymbol{X}|\boldsymbol{Y})$ w.r.t. the kernel hyper-parameters $\boldsymbol{\theta}_r$. For the generative model we maximize the marginal likelihood (labels $\boldsymbol{Z}$ are omitted here)

$$p(\boldsymbol{Y}) = \int \prod_{v=1}^V p(\boldsymbol{Y}^{(v)}|\boldsymbol{X}) p(\boldsymbol{X}) d\boldsymbol{X}. \quad (4)$$

Since the above integral is intractable, we resort to approximations. Our main interest is to recover a Bayesian non-parametric solution for both the GP encoder and decoder. We first need to break the circular dependence between $\boldsymbol{Y}^{(v)}$ and $\boldsymbol{X}$ in order to train the two GPs simultaneously.

---

[1] The subscript r indicates that the process facilitates the recognition model.

GP-Encoder. We decouple X and Y by introducing an intermediate variable $M = \{m_i\}_{i=1}^{N}$, so that the recognition model becomes $y^{(v)} \to m \to x$. The GP operates on $y^{(v)}, m$, while x is the noisy observations of m. This process is described in Fig. 1(c). We follow a mean field approximation and introduce the variational distribution $q(X|M) = \prod_i q_i(x_i|m_i) = \prod_i \mathcal{N}(m_i, S_i)$. Here, $m_i, S_i \in \mathbb{R}^q$ are variational parameters[2] of q_i. We define M by employing the cavity distribution of the leave-one-out solution of GP [21]

$$p(M|Y) = \prod_i p(m_i|Y, M_{\setminus i}) = \prod_i \mathcal{N}(\hat{m}_i, \hat{\sigma}_i^2 I), \tag{5}$$

where the subscript $\setminus i$ means 'all datapoints except i', and the mean and variance of the Gaussian are given by [21]

$$\hat{m}_i = m_i - \left[K_r^{-1}M\right]_i / \left[K_r^{-1}\right]_{ii}, \qquad \hat{\sigma}_i^2 = 1 / \left[K_r^{-1}\right]_{ii}. \tag{6}$$

We now integrate out the intermediate layer and propagate the uncertainty of the GP mapping to the latent variable X, which yields the variational distribution

$$q(X|Y) = \prod_i \mathcal{N}(\hat{m}_i, S_i + \hat{\sigma}_i^2 I). \tag{7}$$

GP-Decoder. The proposed recognition model, *i.e.*, the variational distribution of Eq. (7), can be employed to approximate the intractable marginal likelihood of Eq. (4). By introducing the variational distribution as an approximation to the true posterior, and after applying the Jensen's inequality, we obtain the lower bound to the log-marginal likelihood (again, labels Z are omitted)

$$\log p(Y) \geq \mathcal{F}_1 = \sum_v \mathbb{E}_{q(X|Y)} \left[\log p(Y^{(v)}|X)\right] - KL(q(X|Y)\|p(X)). \tag{8}$$

Training our model consists of maximizing the lower bound of Eq. (8) w.r.t. the variational parameters M, S and the hyper-parameters of the kernels $K^{(v)}, K_r$. Further details are given in Sect. 3.4.

3.3   Incorporating Ordinal Variables

In the previous section, we presented the recognition model that we employ to learn a nonlinear manifold from the observed inputs. In the following, we further constrain this manifold by imposing an ordinal structure. This is attained by introducing ordinal variables that account for C ordinal levels of AUs. We use the notion of ordinal regression [24] and, in particular, the ordinal threshold model that imposes the monotonically increasing structure of the discrete output labels to the continuous manifold. Formally, the non-linear mapping between the manifold X and the ordinal outputs Z is modeled as

$$p(Z|g(X)) = \prod_{i,c} p(z_{ic}|g_c(x_i)), \ \ p(z_{ic} = s|g_c(x_i)) = \begin{cases} 1 & \text{if } g_c(x_i) \in (\gamma_{c,s-1}, \gamma_{c,s}] \\ 0 & \text{otherwise,} \end{cases}$$
$$\tag{9}$$

---

[2] For simplicity we assume an isotropic (diagonal) covariance across the dimensions.

where $i = 1, \ldots, N$ indexes the training data. $\gamma_{c,0} = -\infty \leq \cdots \leq \gamma_{c,S} = +\infty$ are the thresholds or cut-off points that partition the real line into $s = 1, \ldots, S$ contiguous intervals. These intervals map the real function value $g_c(\boldsymbol{x})$ into the discrete variable s, corresponding to each of S intensity levels of an AU, while enforcing the ordinal constraints. The threshold model $p(z_{ic} = s | g_c(\boldsymbol{x}_i))$ is used for ideally noise-free cases. Here, we assume that the latent functions $g_c(\cdot)^3$ are corrupted by Gaussian noise, leading to the following formulation

$$g_c(\boldsymbol{x}_i) = \boldsymbol{w}_c^T \boldsymbol{x}_i + \epsilon_g, \quad \epsilon_g \sim \mathcal{N}(0, \sigma_g^2). \tag{10}$$

By integrating out the noisy projections from Eq. (9) (see [32] for details), we arrive at the ordinal log-likelihood

$$\log p(\boldsymbol{Z}|\boldsymbol{X}, \boldsymbol{W}) = \sum_{i,c} \mathbb{I}(z_{ic} = s) \log \left(\Phi \left(\frac{\gamma_{c,s} - \boldsymbol{w}_c^T \boldsymbol{x}_i}{\sigma_g} \right) - \Phi \left(\frac{\gamma_{c,s-1} - \boldsymbol{w}_c^T \boldsymbol{x}_i}{\sigma_g} \right) \right), \tag{11}$$

where $\Phi(\cdot)$ is the Gaussian cumulative density function, and $\mathbb{I}(\cdot)$ is the indicator function. Finally, by using the ordinal likelihood defined in Eq. (11), we obtain the final lower bound of our log-marginal likelihood

$$\log p(\boldsymbol{Y}, \boldsymbol{Z}|\boldsymbol{W}) \geq \mathcal{F}_2 = \sum_v \mathbb{E}_{q(\boldsymbol{X}|\boldsymbol{Y})} \left[\log p(\boldsymbol{Y}^{(v)}|\boldsymbol{X}) \right] - KL(q(\boldsymbol{X}|\boldsymbol{Y}) || p(\boldsymbol{X}))$$
$$+ \sum_{i,c} \mathbb{I}(z_{ic} = s) \mathbb{E}_{q(\boldsymbol{X}|\boldsymbol{Y})} \left[\log \left(\Phi \left(\frac{\gamma_{c,s} - \boldsymbol{w}_c^T \boldsymbol{x}_i}{\sigma_g} \right) - \Phi \left(\frac{\gamma_{c,s-1} - \boldsymbol{w}_c^T \boldsymbol{x}_i}{\sigma_g} \right) \right) \right]. \tag{12}$$

3.4 Learning and Inference

Training our model consists of maximizing the lower bound of Eq. (12) w.r.t. the variational parameters $\{\boldsymbol{S}, \boldsymbol{M}\}$, the hyper-parameters $\{\boldsymbol{\theta}^{(v)}, \sigma_v, \boldsymbol{\theta}_r^{(v)}, \sigma_r\}$ of the GP mappings, and the parameters $\{\boldsymbol{W}, \gamma, \sigma_g\}$ of the ordinal classifier. For the kernel of the GP-decoder we use the radial basis function (RBF) with automatic relevance determination (ARD), which can effectively estimate the dimensionality of the latent space [18]. For the kernel of the GP-encoder we use the isotropic RBF for each observed input. To utilize a joint optimization scheme, we use stochastic backpropagation [30,31], where the re-parameterization trick is applied in Eq. (12). Thus, we can obtain the Monte Carlo estimate of the expectation of the GP auto-encoder from

$$\mathbb{E}_{q(\boldsymbol{X}|\boldsymbol{Y})} \left[\log p(\boldsymbol{Y}^{(v)}|\boldsymbol{X}) \right] = \sum_i \mathbb{E}_{\mathcal{N}(\boldsymbol{\xi}|\boldsymbol{0}, \boldsymbol{I})} \left[\log p(\boldsymbol{y}_i^{(v)} | \hat{\boldsymbol{m}}_i + (\boldsymbol{S}_i^{1/2} + \hat{\sigma}_i \boldsymbol{I})\boldsymbol{\xi}) \right]. \tag{13}$$

The expectation of the ordinal classifier is computed in a similar manner. The advantage of Eq. (13) is twofold: (i) It allows for an efficient computation of the lower bound even when using arbitrary kernel functions (in contrast to [18]);

---

^{3} Note that we adopt here a linear model for $g_c(\cdot)$ as it operates on a low-dimensional non-linear manifold $\boldsymbol{X}$, already obtained by the GP auto-encoder.

(ii) It provides an efficient, low-variance estimator of the gradient [30]. The extra approximation (via the expectation) in the gradient step requires stochastic gradient descent. We use AdaDelta [33] for this purpose.

Inference in the proposed method is straightforward: The test data $\boldsymbol{y}_*^{(v)}$, are first projected onto the manifold using the trained GP-encoder. In the second step, we apply the ordinal classifier to the obtained latent position.

3.5   Relation to Prior Work on Gaussian Processes

Our auto-encoder approach is inspired by neural-network counterparts proposed in [30,31], where probabilistic distributions are defined for the input and output mapping functions. In the GP literature, auto-encoders are closely related to the notion of 'back-constraints'. Back-constraints were introduced in [34] as a deterministic, parametric mapping (commonly a multi-layer perceptron (MLP)) that pairs the latent variables of the GPLVM [35] with the observations. This mapping facilitates a fast inference mechanism and enforces structure preservation in the manifold. The same mechanism has been used to constrain the shared GPLVM [36], from one view in [37] and multiple views in [38].

Back-constraints have been recently introduced to the B-GPLVM [22]. In [39] the authors proposed to approximate the true posterior of the latent space by introducing a variational distribution conditioned on some unobserved inputs. However, those inputs are not related to the observation space considered in this paper (*i.e.*, the outputs $\boldsymbol{Y}$ of the GPLVM). In [23] the variational posterior of the latent space is constrained by using the trick of the parametric deterministic mapping from [34]. Finally, in [28], the authors replaced the variational approximation with a Monte Carlo expectation-maximization algorithm. Samples were obtained from the GP mapping from the observed inputs to the manifold.

Our proposed VGP-AE advances the current literature in many aspects: (1) We introduce a GP mapping for our recognition model. Hence, can model different uncertainty levels per input, which allows us to learn more confident latent representations. (2) The use of the non-parametric GPs also allows us to model complex structures at a lesser expense than the MLP (fewer parameters). Thus, it is less prone to overfitting and scales better to high-dimensional data. (3) Compared to [39] our probabilistic recognition model facilitates a low-dimensional projection of our observed features, while the variational constraint in [39] does not constitute a probabilistic mapping. (4) We learn the GP encoders/decoders in a joint optimization, while [28] train the two models in an alternating scheme.

4   Experiments

We empirically assess the structure learning abilities of the proposed VGP-AE as well as its efficacy when dealing with data of ordinal nature.

4.1   Experimental Protocol

Datasets. We first show the qualitative evaluation of the proposed VGP-AE on the MNIST [40] benchmark dataset of images of handwritten digits. We use it to assess the properties of the auto-endoced manifold. We then show the performance of VGP-AE on two benchmark datasets of facial affect: DISFA [6], and BP4D [41] (using the publicly available data subset from the FERA2015 [8] challenge). Specifically, DISFA contains video recordings of 27 subjects while watching YouTube videos. Each frame is coded in terms of the intensity of 12 AUs, on a six-point ordinal scale. The FERA2015 database includes video of 41 participants. There are 21 subjects in the training and 20 subjects in the development partition. The dataset contains intensity annotations for 5 AUs.

Features. In the experiment on MNIST dataset, we use the normalized raw pixel intensities as input, resulting in a 784D feature vector. For DISFA and FERA2015, we use both geometric and appearance features. Specifically, DISFA and FERA2015 datasets come with frame-by-frame annotations of 66 and 49 facial landmarks, respectively. After removing the contour landmarks from DISFA annotations, we end up with the same set of 49 facial points. We register the images to a reference face using an affine transform based on these points. We then extract Local Binary Patterns (LBP) histograms [42] with 59 bins from patches centered around each registered point. Hence, we obtain 98D (geometric) and 2891D (appearance) feature vectors, commonly used in modeling of facial affect.

Evaluation. As evaluation measures, we use the negative log-predictive density (NLPD) to assess the generative ability (reconstruction part) of our model. For the task of ordinal classification, we report the mean squared error (MSE) and the intra-class correlation (ICC(3,1)) [43]. These are the standard measures for ordinal data. The MSE measures the classifier's consistency regarding the relative order of the classes. ICC is a measure of agreement between annotators (in our case, the ground truth of the AU intensity and the model's predictions). Finally, we adopt the subject-independent setting: for FERA2015 we report the results on the subjects of the development set, while for DISFA we perform a 9-fold (3 subjects per fold) cross-validation procedure.

Models. We compare the proposed VGP-AE to the state of the art GP manifold learning methods that perform multi-input multi-output inference. These include: (i) manifold relevance determination (MRD) [18], a regression model based on variational inference, (ii) variational auto-encoded deep GP (VAE-DGP) [23], which uses a recognition model based on an MLP to constrain the learning of MRD, and (iii) multi-task latent GP (MT-LGP) [19], which uses the same MLP-based recognition model and a maximum likelihood learning approach. We also compare to the variational GP for ordinal regression (vGPOR) [44]. As a baseline, we use the standard GP [21] with a shared covariance function among the multi-outputs. We also compare to the single-output ordinal threshold model (SOR) [24]. Finally, we compare to state of the art methods for joint estimation of AU intensity based on MRFs [14] and latent trees (LT) [15], respectively. For the single input (no fusion) methods (GP, vGPOR,

SOR, LT, MRF), we concatenate the two feature sets. The parameters of each method were tuned as described in the corresponding papers. For the GP subspace methods, we used the RBF kernel with ARD, and initialized with the 20D manifold. For the GP regression methods, we used the standard RBF. For the sparse variational GP methods (vGPOR, MRD, VAE-DGP) we used 200 inducing points, and 20 hidden units for the MLP in the recognition models of VAE-DGP and MT-LGP.

4.2   Assessing the Recognition Model

In the following, we qualitatively assess the benefits of the proposed recognition model in the task of manifold recovery from the MNIST dataset. We select an image depicting the digit '1' and rotate it around 360°. This results in a set of images of '1's rotated at a step of 1°. Our goal is to infer the true structure of the data, for which we know *a priori* that it should correspond to a diagonal-like kernel and a circular manifold. However, the challenge arises from the symmetry of digit '1', which is almost identical at opposite degrees (*e.g.*, 0° and 180°). The results are depicted in Fig. 2. Note that since we do not deal with the classification task we exclude the ordinal component in VGP-AE. We compare the learned manifold structure to the B-GPLVM [22], which does not model the back-projection to the latent space, and a single layer VAE-DGP, where the back-projections are modeled using MLP. In Fig. 2 (upper row), we see from the learned kernels that the B-GPLVM is unable to fully unravel the dissimilarity between the 'inverted' images, resulting also in a non-smooth kernel with a discontinuity at 180° and 270°. By contrast, the VAE-DGP benefits from the recognition model and manages to resolve this to some extent. Yet, the recovered kernel still suffers from a discontinuity around 180°. On the other hand, the proposed VGP-AE, by using the more general recognition model based on GPs (infinitely wide MLP), succeeds to accurately discover the true underlying manifold, also resulting in a more smooth, almost ideal kernel. These observations are further supported by the instances of the learned 2D manifolds in Fig. 2 (lower row). B-GPLVM learns a disconnected manifold with 'jumps' at 180° and 270°. However, both the VAE-DGP and proposed VGP-AE recover a circular manifold, with the manifold recovered by VGP-AE being more symmetric.

4.3   Convergence Analysis

We next demonstrate the convergence of VGP-AE in the task of AU intensity estimation on FERA2015. Figure 3(a) shows the effect of learning the ordinal classifier and the auto-encoded manifold within the joint optimization framework. It can be clearly seen from the recovered space that the information from the labels has been correctly encoded in the manifold, which now has an ordinal structure (the depicted coloring accounts for the 'ordinality' of AU12). As depicted in Fig. 3(b), we can accurately reconstruct face shapes with different AU intensities, by sampling from different regions of the space. Figure 3(c) shows the convergence of the proposed method when optimizing the lower bound $\mathcal{F}_2$ of

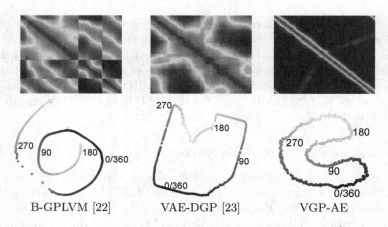

B-GPLVM [22]   VAE-DGP [23]   VGP-AE

Fig. 2. Recovering the structure of a rotated '1' from MNIST. The learned kernel matrices (upper row) and 2D manifolds (lower row) obtained from B-GPLVM (left), VAE-DGP (middle) and the proposed VGP-AE (right), initialized from the same random instance.

Eq. (12) for different batch sizes of the stochastic optimization. With a small batch size (100 datapoints) the model cannot estimate the structure of the inputs well. Hence, it approximates the log-marginal likelihood less accurately. By increasing the batch size to 500, the model converges to a better solution and optimization becomes more stable since the curve becomes smoother over the iterations. Further increase of the batch size does not have a considerable effect.

In Fig. 3(d)–(e) we evaluate the generative part of the auto-encoder by measuring the model's ability to reconstruct both input features (points and LBPs) in terms of NLPD. First of all, it is clear that our Bayesian training prevents the model from overfitting, since the NLPD of the test data follows the trend of the training data. Furthermore, we can see that the model can reconstruct the geometric features better than the appearance, which is evidenced by the lower NLPD (around −50 for points and 1500 for LBPs). We partly attribute this to the fact that the LBPs are of higher dimension and therefore more difficult to reconstruct. Another reason for this difference is that the model learns to reconstruct the part of the features that enclose the more relevant information regarding the task of classification. The latter is further supported by Fig. 3(e), where we see the progress of the average ICC during the optimization. In the beginning, the model has no information since the latent space is initialized randomly. As we progress the model fuses the information of the input features in the latent space and unravels the structure of the data. Thus, ICC starts rising and reaches its highest value, .65 on the test data. After that point the model does no longer benefit from the appearance features: it has reached the plateau.

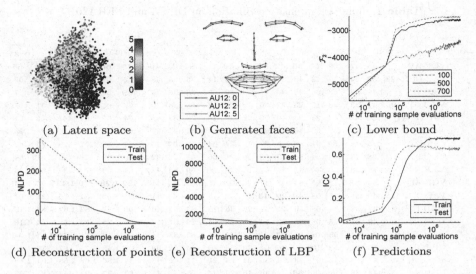

(a) Latent space   (b) Generated faces   (c) Lower bound

(d) Reconstruction of points (e) Reconstruction of LBP   (f) Predictions

Fig. 3. Convergence analysis of the proposed method on FERA2015. (a) The recovered latent space with ordinal information from AU12, and (b) reconstructed face shapes sampled from different regions of the manifold. (c) the estimated average variational lower bound, $\mathcal{F}_2$, per datapoint, for different batch sizes. The model's reconstruction capacity for the points (d) and LBP (e) features, measured by the NLPD. (f) the average ICC for the joint AU intensity estimation. The horizontal axis corresponds to the amount of training points evaluated after 1500 epochs of the stochastic optimization. (Color figure online)

4.4   Model Comparisons on Spontaneous Data of Facial Expressions

We compare the proposed approach to several methods on the spontaneous data from the DISFA and FERA2015 datasets. Table 1 summarizes the results. First, we observe that all methods perform significantly better (in terms of ICC) on the data from FERA2015 than on DISFA. This is mainly due to the fact that FERA2015 contains a much more balanced set of AUs (in terms of activations), and hence, all models (single- and multi-output) can learn the classifiers for the target task better. Furthermore, our proposed approach performs significantly better than the compared GP manifold learning methods, which treat the output labels as continuous variables. MRD lacks the modeling of back-projections. This results in learning a less smooth manifold of facial expressions, which affects its representation abilities, and hence, its predictions. On the other hand, the VAE-DGP learns explicitly the mapping from the observed features to the latent space in a deterministic and parametric fashion. Although this strategy is proven to be superior to unconstrained learning, it can be severely affected in cases where we have access to noisy and high-dimensional features. MT-LGP also models the back-mappings. However, it reports worse results, especially on DISFA. This drop in the performance is accounted to the non-Bayesian learning of the manifold, which constitutes the model more prone to overfitting.

Table 1. Joint AU intensity estimation on DISFA and FERA2015

Dataset		DISFA													FERA2015					
AU		1	2	4	5	6	9	12	15	17	20	25	26	Avg.	6	10	12	14	17	Avg.
ICC	VGP-AE	.48	.47	.62	.19	**.50**	**.42**	**.80**	.19	**.36**	.15	**.84**	.53	**.46**	**.75**	**.66**	**.88**	.47	**.49**	**.65**
	VAE-DGP [23]	.39	.34	.46	.13	.40	.31	.75	.14	.23	.14	.75	.45	.38	.72	.61	.82	.40	.38	.59
	MRD [18]	.46	.39	.43	.09	.28	.34	.71	.09	.30	.09	.73	.36	.36	.68	.59	.80	.38	.38	.57
	MT-LGP [19]	.41	.33	.28	.10	.23	.22	.56	.13	.26	.18	.65	.23	.30	.67	.61	.80	.37	.41	.57
	vGPOR [44]	**.53**	**.49**	.54	.21	.35	.40	.75	.18	.30	.16	.79	.39	.42	.74	.62	.84	**.48**	.35	.61
	GP [21]	.28	.13	.42	.03	.13	.23	.62	.08	.26	.19	.67	.23	.27	.69	.58	.81	.35	.38	.56
	SOR [24]	.25	.18	**.65**	.08	.46	.15	.77	.14	.24	.04	.82	**.57**	.36	.61	.50	.77	.28	.45	.52
	LT [15]	.28	.26	.44	.24	.50	.13	.69	.06	.21	.06	.62	.37	.32	.70	.59	.76	.30	.31	.53
	MRF [14]	.46	.38	.50	**.37**	.41	.34	.67	**.32**	.29	**.20**	.69	.46	.42	.64	.53	.79	.34	.46	.55
MSE	VGP-AE	.51	**.32**	1.13	.08	.56	.31	.47	.20	**.28**	.16	**.49**	**.44**	.41	**.82**	**1.28**	**.70**	1.43	**.77**	**1.00**
	VAE-DGP [23]	.40	.36	.95	.08	.48	.29	.43	.19	.32	.16	.76	**.44**	.41	.91	1.33	.81	1.46	.86	1.07
	MRD [18]	.42	.38	1.31	.08	.56	**.27**	.47	.20	.36	.18	.82	.53	.46	1.00	1.39	.83	1.64	.88	1.15
	MT-LGP [19]	.40	.35	1.25	.08	.60	.30	.73	.18	.36	.16	1.19	.67	.52	.97	1.31	.81	1.58	.84	1.10
	vGPOR [44]	.38	.34	.95	**.06**	.57	**.27**	**.43**	.18	.33	.18	.65	.53	.41	1.00	1.54	.76	1.78	1.11	1.24
	GP [21]	.52	.51	1.13	.13	.65	.36	.61	.23	.38	.20	.94	.66	.53	.94	1.40	.76	1.62	.88	1.12
	SOR [24]	.47	.40	1.13	.07	.63	.37	.55	.21	.35	.21	.71	.61	.48	1.44	1.82	1.08	2.58	1.01	1.59
	LT [15]	.44	.38	**.93**	**.06**	**.36**	.32	.46	.16	.29	**.15**	.97	**.44**	.41	.89	1.33	.91	1.48	.85	1.09
	MRF [14]	**.37**	.35	.94	**.06**	.45	.29	.46	**.13**	.32	.16	.77	**.44**	**.40**	1.20	1.66	.86	2.19	.92	1.37

Regarding the sparse ordinal regression instance of GPs, *i.e.*, vGPOR, we see that it manages to learn relatively accurate mappings between features and labels, and thus, performs close to our proposed method. However, it reports worse results since it cannot achieve the desirable fusion of the features without learning an intermediate latent space. The baseline methods, *i.e.*, GP and SOR, report lower results. The GP attains low scores due to handling the ordinal outputs in a continuous manner while the ordinal modeling helps SOR to report consistently better.

Finally, the proposed approach significantly outperforms the state of the art methods in the literature of AU intensity estimation, *i.e.*, LT and MRF. LT learns the label information in a generative manner, and treats them as extra feature dimensions. Although this approach can be beneficial in the presence of noisy features [15], it suffers from learning complicated and large tree structures when falsely detecting connections between features and AUs. Hence, it performs worse. The MRF performs on par to the proposed method on DISFA and achieves the best average MSE, but it is consistently worse on FERA2015. This inconsistency is due to its two-step learning strategy, which results in unraveling a graph that cannot explain simultaneously all different features and AUs.

In Fig. 4 we evaluate the attained fusion between the best performing methods on FERA2015, *i.e.*, the proposed VGP-AE, VAE-DGP [23] and vGPOR [44]. As we can see, the proposed approach (solid line, first tuple) manages to accurately fuse the information from the two input features in the learned manifold. Thus, it achieves higher ICC on all AUs compared to when the two modalities are used individually as input features. On the other hand, although vGPOR (third tuple, dotted line) reports also high ICC scores, it does not benefit from

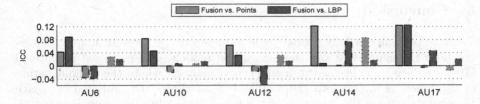

Fig. 4. Demonstration of the gain/loss from feature fusion for joint AU intensity estimation on FERA2015. Within each AU the first tuple (solid line) corresponds to the proposed VGP-AE, the second tuple (dashed line) to the VAE-DGP [23], and the third tuple (dotted line) to the vGPOR [44].

the presence of the two features: In most cases it cannot achieve a significant increase compared to the individual inputs. Finally, VAE-DGP (middle tuple, dashed line) consistently attains better performance on all AUs with a single feature as input. This can be attributed to modeling the recognition model via the parametric MLP. The latter affects the learning of the manifold, especially when dealing with the high-dimensional noisy appearance features.

The above mentioned difference between our approach and the VAE-DGP is further evidenced in Fig. 5. The proposed fusion along with the novel non-parametric, probabilistic recognition model in our auto-encoder leads to less confusion between the ordinal states across all AUs. We further attribute this to the ordinal modeling of outputs in our VGP-AE, contrary to VAE-DGP that treats the output as continuous variables. This is especially pronounced in the case of the subtle AUs 14 & 17, where examples of high intensity levels are scarce.

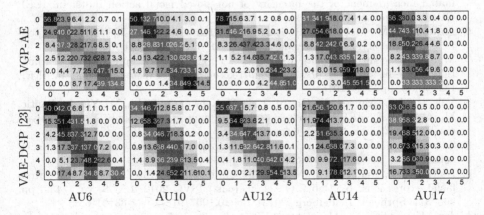

Fig. 5. Confusion matrices for predicting the 0–5 intensity of all AUs on FERA2015, when performing fusion with VGP-AE (upper row) and VAE-DGP [23] (lower row).

5 Conclusion

We have presented a fully probabilistic auto-encoder, where GP mappings govern both the generative and the recognition models. The proposed variational GP auto-encoder is learned in a supervised manner, where the ordinal nature of the labels is imposed to the manifold. This allows the proposed approach to accurately learn the structure of the input data, while also remain competitive in the classification task. We have empirically evaluated our model on the task of facial feature fusion for joint intensity estimation of facial action units. The proposed model outperforms related GP methods and the state of the art approaches for the target task.

Acknowledgement. This work has been funded by the European Community Horizon 2020 under grant agreement No. 645094 (SEWA), and No. 688835 (DE-ENIGMA). MPD has been supported by a Google Faculty Research Award.

References

1. Bartlett, M., Whitehill, J.: Automated facial expression measurement: recent applications to basic research in human behavior, learning, and education. In: Handbook of Face Perception. Oxford University Press, USA (2010)
2. Ekman, P., Friesen, W.V., Hager, J.C.: Facial action coding system. UT: A Human Face, Salt Lake City (2002)
3. Pantic, M.: Machine analysis of facial behaviour: naturalistic and dynamic behaviour. Philos. Trans. Roy. Soc. B: Biol. Sci. **364**, 3505–3513 (2009)
4. Rudovic, O., Pavlovic, V., Pantic, M.: Context-sensitive dynamic ordinal regression for intensity estimation of facial action units. IEEE TPAMI **37**, 944–958 (2015)
5. Mahoor, M.H., Cadavid, S., Messinger, D.S., Cohn, J.F.: A framework for automated measurement of the intensity of non-posed facial action units. In: IEEE CVPR-W, pp. 74–80 (2009)
6. Mavadati, S.M., Mahoor, M.H., Bartlett, K., Trinh, P., Cohn, J.F.: DISFA: a spontaneous facial action intensity database. IEEE TAC **4**, 151–160 (2013)
7. Ming, Z., Bugeau, A., Rouas, J.L., Shochi, T.: Facial action units intensity estimation by the fusion of features with multi-kernel support vector machine. In: IEEE FG, vol. 6, pp. 1–6 (2015)
8. Valstar, M.F., Almaev, T., Girard, J.M., McKeown, G., Mehu, M., Yin, L., Pantic, M., Cohn, J.F.: FERA 2015 - second facial expression recognition and analysis challenge. In: IEEE FG, vol. 6, pp. 1–8 (2015)
9. Savran, A., Sankur, B., Bilge, M.T.: Regression-based intensity estimation of facial action units. Image Vis. Comput. **30**, 774–784 (2012)
10. Kaltwang, S., Rudovic, O., Pantic, M.: Continuous pain intensity estimation from facial expressions. In: Bebis, G., et al. (eds.) ISVC 2012. LNCS, vol. 7432, pp. 368–377. Springer, Heidelberg (2012). doi:10.1007/978-3-642-33191-6_36
11. Jeni, L.A., Girard, J.M., Cohn, J.F., De La Torre, F.: Continuous AU intensity estimation using localized, sparse facial feature space. In: IEEE FG, pp. 1–7 (2013)
12. Kaltwang, S., Todorovic, S., Pantic, M.: Doubly sparse relevance vector machine for continuous facial behavior estimation. IEEE TPAMI **38**, 1748–1761 (2015)

13. Li, Y., Mavadati, S.M., Mahoor, M.H., Ji, Q.: A unified probabilistic framework for measuring the intensity of spontaneous facial action units. In: IEEE FG (2013)
14. Sandbach, G., Zafeiriou, S., Pantic, M.: Markov random field structures for facial action unit intensity estimation. In: IEEE ICCV-W, pp. 738–745 (2013)
15. Kaltwang, S., Todorovic, S., Pantic, M.: Latent trees for estimating intensity of facial action units. In: IEEE CVPR, pp. 296–304 (2015)
16. Nicolle, J., Bailly, K., Chetouani, M.: Facial action unit intensity prediction via hard multi-task metric learning for kernel regression. In: IEEE FG, pp. 1–6 (2015)
17. Mohammadi, M.R., Fatemizadeh, E., Mahoor, M.H.: Intensity estimation of spontaneous facial action units based on their sparsity properties. IEEE TCYB **46**, 817–826 (2016)
18. Damianou, A., Ek, C.H., Titsias, M., Lawrence, N.: Manifold relevance determination. In: ICML, pp. 145–152 (2012)
19. Urtasun, R., Quattoni, A., Lawrence, N., Darrell, T.: Transferring nonlinear representations using Gaussian processes with a shared latent space. Technical report MIT-CSAIL-TR-08-020 (2008)
20. Calandra, R., Peters, J., Rasmussen, C.E., Deisenroth, M.P.: Manifold Gaussian processes for regression. In: IJCNN (2016)
21. Rasmussen, C., Williams, C.: Gaussian Processes for Machine Learning, vol. 1. MIT Press, Cambridge (2006)
22. Titsias, M.K., Lawrence, N.D.: Bayesian Gaussian process latent variable model. In: AISTATS, pp. 844–851 (2010)
23. Dai, Z., Damianou, A., González, J., Lawrence, N.: Variational auto-encoded deep Gaussian processes. In: ICLR (2016)
24. Agresti, A.: Analysis of Ordinal Categorical Data. Wiley, Hoboken (2010)
25. Mahoor, M.H., Zhou, M., Veon, K.L., Mavadati, S.M., Cohn, J.F.: Facial action unit recognition with sparse representation. In: IEEE FG, pp. 336–342 (2011)
26. Chu, W.S., Torre, F.D.L., Cohn, J.F.: Selective transfer machine for personalized facial action unit detection. In: IEEE CVPR, pp. 3515–3522 (2013)
27. Zhao, K., Chu, W.S., De la Torre, F., Cohn, J.F., Zhang, H.: Joint patch and multi-label learning for facial action unit detection. In: IEEE CVPR (2015)
28. Eleftheriadis, S., Rudovic, O., Pantic, M.: Multi-conditional latent variable model for joint facial action unit detection. In: IEEE ICCV, pp. 3792–3800 (2015)
29. Candès, E.J., Li, X., Ma, Y., Wright, J.: Robust principal component analysis? J. ACM **58**, 11 (2011)
30. Kingma, D.P., Welling, M.: Auto-encoding variational Bayes. In: ICLR (2013)
31. Rezende, D.J., Mohamed, S., Wierstra, D.: Stochastic backpropagation and approximate inference in deep generative models. In: ICML, pp. 1278–1286 (2014)
32. Chu, W., Ghahramani, Z.: Gaussian processes for ordinal regression. JMLR **6**, 1019–1041 (2005)
33. Zeiler, M.D.: ADADELTA: an adaptive learning rate method. arXiv preprint arXiv:1212.5701 (2012)
34. Lawrence, N.D., Candela, J.Q.: Local distance preservation in the GP-LVM through back constraints. In: ICML, vol. 148, pp. 513–520 (2006)
35. Lawrence, N.: Probabilistic non-linear principal component analysis with Gaussian process latent variable models. JMLR **6**, 1783–1816 (2005)
36. Shon, A., Grochow, K., Hertzmann, A., Rao, R.: Learning shared latent structure for image synthesis and robotic imitation. NIPS **18**, 1233–1240 (2006)

37. Ek, C.H., Torr, P.H.S., Lawrence, N.D.: Gaussian process latent variable models for human pose estimation. In: Popescu-Belis, A., Renals, S., Bourlard, H. (eds.) MLMI 2007. LNCS, vol. 4892, pp. 132–143. Springer, Heidelberg (2008). doi:10. 1007/978-3-540-78155-4_12
38. Eleftheriadis, S., Rudovic, O., Pantic, M.: Discriminative shared Gaussian processes for multiview and view-invariant facial expression recognition. IEEE TIP **24**, 189–204 (2015)
39. Damianou, A., Lawrence, N.: Semi-described and semi-supervised learning with Gaussian processes. In: UAI (2015)
40. LeCun, Y., Cortes, C., Burges, C.J.: The MNIST database of handwritten digits (1998)
41. Zhang, X., Yin, L., Cohn, J.F., Canavan, S., Reale, M., Horowitz, A., Liu, P., Girard, J.M.: BP4D-spontaneous: a high-resolution spontaneous 3D dynamic facial expression database. Image Vis. Comput. **32**, 692–706 (2014)
42. Ojala, T., Pietikainen, M., Maenpaa, T.: Multiresolution gray-scale and rotation invariant texture classification with local binary patterns. IEEE TPAMI **24**, 971–987 (2002)
43. Shrout, P.E., Fleiss, J.L.: Intraclass correlations: uses in assessing rater reliability. Psychol. Bull. **86**, 420 (1979)
44. Sheth, R., Wang, Y., Khardon, R.: Sparse variational inference for generalized GP models. In: ICML, pp. 1302–1311 (2015)

Multi-Instance Dynamic Ordinal Random Fields for Weakly-Supervised Pain Intensity Estimation

Adria Ruiz[1](✉), Ognjen Rudovic[2], Xavier Binefa[1], and Maja Pantic[2,3]

[1] DTIC, Universitat Pompeu Fabra, Barcelona, Spain
adria.ruiz@upf.edu
[2] Department of Computing, Imperial College London, London, UK
[3] EEMCS, University of Twente, Enschede, The Netherlands

Abstract. In this paper, we address the Multi-Instance-Learning (MIL) problem when bag labels are naturally represented as ordinal variables (Multi-Instance-Ordinal Regression). Moreover, we consider the case where bags are temporal sequences of ordinal instances. To model this, we propose the novel Multi-Instance Dynamic Ordinal Random Fields (MI-DORF). In this model, we treat instance-labels inside the bag as latent ordinal states. The MIL assumption is modelled by incorporating a high-order cardinality potential relating bag and instance-labels, into the energy function. We show the benefits of the proposed approach on the task of weakly-supervised pain intensity estimation from the UNBC Shoulder-Pain Database. In our experiments, the proposed approach significantly outperforms alternative non-ordinal methods that either ignore the MIL assumption, or do not model dynamic information in target data.

1 Introduction

Multi-Instance-Learning (MIL) is a popular modelling framework for addressing different weakly-supervised problems [1–3]. In traditional Single-Instance-Learning (SIL), the fully supervised setting is assumed with the goal to learn a model from a set of feature vectors (instances) each being annotated in terms of target label y. By contrast, in MIL, the weak supervision is assumed, thus, the training set is formed by bags (sets of instances), and only labels at bag-level are provided. Furthermore, MIL assumes that there exist an underlying relation between the bag-label (e.g., video) and the labels of its constituent instances (e.g., image frames). In standard Multi-Instance-Classification (MIC) [4], labels are considered binary variables $y \in \{-1, 1\}$ and negative bags are assumed to contain only instances with an associated negative label. In contrast, positive bags must contain at least one positive instance. Another MIL assumption is related to the Multi-Instance-Regression (MIR) problem [5], where $y \in R$ is a real-valued variable and the maximum instance-label within the bag is usually assumed to be equal to y. Note, however, that none of these assumptions accounts for structure in the bag labels. Yet, this can be important in case when the bag labels are ordinal, i.e., $y \in \{0 \prec ... \prec l \prec L\}$, as in the case of various ratings or

© Springer International Publishing AG 2017
S.-H. Lai et al. (Eds.): ACCV 2016, Part II, LNCS 10112, pp. 171–186, 2017.
DOI: 10.1007/978-3-319-54184-6_11

intensity estimation tasks. In this work, we focus on the novel modelling task to which we refer as Multi-Instance-Ordinal Regression (MIOR). Similar to MIR, in MIOR we assume that the maximum instance ordinal value within a bag is equal to its label.

To demonstrate the benefits of the proposed approach to MIOR, we apply it to the task of automatic pain estimation [6]. Pain monitoring is particularly important in clinical context, where it can provide an objective measure of the patient's pain level (and, thus, allow for proper treatment) [7]. The aim is to predict pain intensity levels from facial expressions (in each frame in a video sequence) of a patient experiencing pain. To obtain the labelled training data, the pain level is usually manually coded on an ordinal scale from low to high intensity [8]. To estimate the pain, several SIL methods have been proposed [9,10]. Yet, the main limitation of these approaches is they require the frame-based pain level annotations to train the models, which can be very expensive and time-consuming. To reduce the efforts, MIL approaches have recently been proposed for automatic pain detection [3,11,12]. Specifically, a weak-label is provided for the whole image sequence (in terms of the maximum observed pain intensity felt by the patient). Then, a video is considered as a bag, and image frames as instances, where the pain labels are provided per bag. In contrast to per-frame annotations, the bag labels are much easier to obtain. For example, using patients self-reports or external observers [6]. Yet, existing MIL approaches for the task focus on the MIC setting, i.e., pain intensities are binarized and model predicts only the presence or absence of pain. Consequently, these approaches are unable to deal with Ordinal Regression problems, and, thus, estimate different intensity levels of pain – which is critical for real-time pain monitoring.

In this paper, we propose Multi-Instance Dynamic Ordinal Random Fields (MI-DORF) for MIL with ordinal bag labels. We build our approach using the notion of Hidden Conditional Ordinal Random Fields framework (HCORF) [13], for modeling of linear-chains of ordinal latent variables. In contrast to HCORF that follows the Single-Instance paradigm, the energy function employed in MI-DORF is designed to model the MIOR assumption relating instance and bag labels. In relation to static MIL methods, our MI-DORF also incorporates dynamics within the instances, encoded by transitions between ordinal latent states. This information is useful when instances (frames) in a bag are temporally correlated, as in pain videos. The main contributions of this work can be summarised as follows:

- To the best our knowledge, the proposed MI-DORF is the first MIL approach that imposes ordinal structure on the bag labels. The proposed method also incorporates dynamic information that is important when modeling temporal structure in instances within the bags (i.e., image sequences). While modeling the temporal structure has been attempted in [11,14], there are virtually no works that account for both ordinal and temporal data structures within MIL framework.
- We introduce an efficient inference method in our MI-DORF, which has a similar computational complexity as the forward-backward algorithm [15] used in standard first-order Latent-Dynamic Models (e.g. HCORF). This is despite

the fact that we model high-order potentials modelling the Multi-Instance assumption.

- We show in the task of automated pain intensity estimation from the UNBC Shoulder-Pain Database [6] that the proposed MI-DORF outperforms significantly existing related approaches applicable to this task. We show that due to the modeling of the ordinal and temporal structure in the target data, we can infer instance-level pain intensity levels that largely correlate with manually obtained frame-based pain levels. Note that we do so by using only the bag labels for learning, that are easy to obtain. To our knowledge, this has not been attempted before.

2 Related Work

Multi-Instance-Learning. Existing MIC/MIR approaches usually follow the bag-based or instance-based paradigms [16]. In bag-based methods, a feature vector representation for each bag is first extracted. Then, these representations are used to train standard Single-Instance Classification or Regression methods, used to estimate the bag labels. Examples include Multi-Instance Kernel [17], MILES [18], MI-Graph [19] and MI-Cluster Regression [20]. The main limitation of these approaches is that the learned models can only make predictions at the bag-level. However, these methods cannot work in the weakly-supervised settings, where the goal is to predict instance-labels (e.g., frame-level pain intensity) from a bag (e.g., a video). In contrast, instance-based methods directly learn classifiers which operate at the instance level. For this, MIL assumptions are incorporated into the model by considering instance-labels as latent variables. Examples include Multi-Instance Support Vector Machines [21] (MI-SVM), MIL-Boost [22], and Multi-Instance Logistic Regression [23]. The proposed MI-DORF model follows the instance-based paradigm by treating instance-labels as ordinal latent states in a Latent-Dynamic Model. In particular, it follows a similar idea to that in the Multi-Instance Discriminative Markov Networks [24]. In this approach, the energy function of a Markov Network is defined by using cardinality potentials modelling the relation between bag and instance labels. MI-DORF also make use of cardinality potentials, however, in contrast to the works described above, it accounts for the ordinal structure at both the bag and instance level, while also accounting for the dynamics in the latter.

Latent-Dynamic Models. Popular methods for sequence classification are Latent-Dynamic Models such as Hidden Conditional Random Fields (HCRFs) [25] or Hidden-Markov-Models (HMMs) [26]. These methods are variants of Dynamic Bayesian Networks (DBNs) where a set of latent states are used to model the conditional distribution of observations given the sequence label. In these approaches, dynamic information is modelled by incorporating probabilistic dependence between time-consecutive latent states. MI-DORF builds upon the HCORF framework [13] which considers latent states as ordinal variables. However, HMM and HCRF/HCORF follow the SIL paradigm where the main goal is to predict sequence labels. In contrast, in MI-DORF, we define a novel energy function that encodes the MI relationship between the bag labels, and also their latent ordinal states. Note also that the recent works (e.g., [11,14]) extended HMMs/HCRFs, respectively, for MIC. The reported results in this work suggested that modeling dynamics in MIL can be beneficial when bag-instances exhibit temporal structure. However, these methods limit their consideration to the case where bag labels are binary and, therefore, are unable to solve the MIOR problem.

MIL for Weakly-Supervised Pain Detection. Several works attempted pain detection in the context of the weakly-supervised MIL. As explained in Sect. 1, these approaches adopt the MIC framework where pain intensities are binarized. For instance, [12] proposed to extract a Bag-of-Words representation from video segments and treat them as bag-instances. Then, MILBoosting [22] was applied to predict sequence-labels under the MIC assumption. Following the bag-based paradigm, [3] developed the Regularized Multi-Concept MIL method capable of discovering different discriminative pain expressions within an image sequence. More recently, [11] proposed MI Hidden Markov Models, an adaptation of standard HMM to the MIL problem. The limitation of these approaches is that they focus on the binary detection problem, and, thus, are unable to deal with (ordinal) multi-class problems (i.e., pain intensity estimation). This is successfully attained by the proposed MI-DORF.

3   Multi-Instance Dynamic Ordinal Random Fields (MI-DORF)

3.1   Multi Instance Ordinal Regression (MIOR)

In the MIOR weakly-supervised setting, we are provided with a training set $\mathcal{T} = \{(\mathbf{X}_1, y_1), (\mathbf{X}_2, y_2), ..., (\mathbf{X}_N, y_N)\}$ formed by pairs of structured-inputs $X \in \mathcal{X}$ and labels $y \in \{0 \prec ... \prec l \prec L\}$ belonging to a set of L possible ordinal values. In this work, we focus on the case where $\mathbf{X} = \{\mathbf{x}_1, \mathbf{x}_2, ..., \mathbf{x}_T\}$ are temporal sequences of T observations $\mathbf{x} \in R^d$ in a d-dimensional space [1]. Given the training-set $\mathcal{T}$, the goal is to learn a model $\mathcal{F} : \mathcal{X} \rightarrow \mathcal{H}$ mapping sequences $\mathbf{X}$ to an structured-output $\mathbf{h} \in \mathcal{H}$. Concretely, $\mathbf{h} = \{h_1, h_2, ..., h_T\}$ is a sequence of variables $h_t \in \{0 \prec ... \prec l \prec L\}$ assigning one ordinal value for each observation $\mathbf{x}_t$. In order to learn the model $\mathcal{F}$ from $\mathcal{T}$, MIOR assumes that the maximum ordinal value in $\mathbf{h}_n$ must be equal to the label y_n for all sequences $\mathbf{X}_n$:

$$\mathcal{F}(\mathbf{X}_n) = \mathbf{h}_n \quad s.t \quad y_n = \max_h(\mathbf{h}_n) \quad \forall \quad (\mathbf{X}_n, y_n) \in \mathcal{T} \tag{1}$$

3.2   MI-DORF: Model Overview

We model the structured-output $\mathbf{h} \in \mathcal{H}$ as a set of ordinal latent variables. We then define the conditional distribution of y given observations $\mathbf{X}$. Formally, $P(y|\mathbf{X}; \theta)$ is assumed to follow a Gibbs distribution as:

---

[1] Total number of observations T can vary across different sequences.

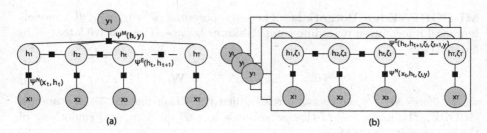

Fig. 1. (a) Graphical representation of the proposed MI-DORF model. Node potentials Ψ^N model the compatibility between a given observation $\mathbf{x}_t$ and a latent ordinal value h_t. Edge potentials Ψ^E take into account the transition between consecutive latent ordinal states h_t and h_{t+1}. Finally, the high-order cardinality potential Ψ^M models the MIOR assumption relating all the latent ordinal states $\mathbf{h}_t$ with the bag-label y. (b) Equivalent model defined using the auxiliary variables ζ_t for each latent ordinal state. The use of these auxiliary variables and the redefinition of node and edge potentials allows us to perform efficient inference over the MI-DORF model (see Sect. 3.4).

$$P(y|\mathbf{X};\theta) = \frac{\sum_h e^{-\Psi(\mathbf{X},\mathbf{h},y;\theta)}}{\sum_{y'}\sum_h e^{-\Psi(\mathbf{X},\mathbf{h},y';\theta)}}, \tag{2}$$

where θ is the set of the model parameters. As defined in Eq. 3, the energy function Ψ defining the Gibbs distribution is composed of the sum of three different types of potentials. An overview of the model is shown in Fig. 1(a).

$$\Psi(\mathbf{X},\mathbf{h},y;\theta) = \sum_{t=1}^{T} \Psi^N(\mathbf{x}_t,h_t;\theta^N) + \sum_{t=1}^{T-1} \Psi^E(h_t,h_{t+1};\theta^E) + \Psi^M(\mathbf{h},y,\theta^M), \tag{3}$$

MI-DORF: Ordinal Node Potentials. The node potentials $\Psi^N(\mathbf{x},h;\theta^N)$ aim to capture the compatibility between a given observation $\mathbf{x}_t$ and the latent ordinal value h_t. Similar to HCORF, it is defined using the ordinal likelihood model [27]:

$$\Psi^N(\mathbf{x},h=l;\theta^N) = \log\left(\Phi\left(\frac{b_l - \beta^T\mathbf{x}}{\sigma}\right) - \Phi\left(\frac{b_{(l-1)} - \beta^T\mathbf{x}}{\sigma}\right)\right), \tag{4}$$

where $\Phi(\cdot)$ is the normal cumulative distribution function (CDF), and $\theta^N = \{\beta,\mathbf{b},\sigma\}$ is the set of potential parameters. Specifically, the vector $\beta \in R^d$ projects observations $\mathbf{x}$ onto an ordinal line divided by a set of cut-off points $b_0 = -\infty \leq \cdots \leq b_L = \infty$. Every pair of contiguous cut-off points divide the projection values into different bins corresponding to the different ordinal states $l = 1, ..., L$. The difference between the two CDFs provides the probability of the latent state l given the observation $\mathbf{x}$, where σ is the standard deviation of a Gaussian noise contaminating the ideal model (see [13] for details). In our case, we fix $\sigma = 1$, to avoid model over-parametrization.

MI-DORF: Edge Potentials. The edge potential $\Psi^E(h_t, h_{t+1}; \theta^E)$ models temporal information regarding compatibilities between consecutive latent ordinal states as:

$$\Psi^E(h_t = l, h_{t+1} = l'; \theta^E) = \mathbf{W}_{l,l'}, \tag{5}$$

where $\theta^E = \mathbf{W}^{L \times L}$ represents a real-valued transition matrix, as in standard HCORF. The main goal of this potential is to perform temporal smoothing of the instance intensity levels.

MI-DORF: Multi-Instance-Ordinal Potential. In order to model the MIOR assumption (see Eq. 1), we define a high-order potential $\Psi^M(\mathbf{h}, y; \theta^M)$ involving label y and all the sequence latent variables $\mathbf{h}$ as:

$$\Psi^M(\mathbf{h}, y; \theta^M) = \begin{cases} w \sum_{t=1}^{T} \mathbf{I}(h_t == y) & \text{iff} \quad \max(\mathbf{h}) = y \\ -\infty & \text{otherwise} \end{cases}, \tag{6}$$

where $\mathbf{I}$ is the indicator function, and $\theta^M = w$. Note that when the maximum value within $\mathbf{h}$ is not equal to y, the energy function is equal to $-\infty$ and, thus, the probability $P(y|\mathbf{X}; \theta)$ drops to 0. On the other hand, if the MI assumption is fulfilled, the summation $w \sum_{t=1}^{T} \mathbf{I}(h_t == y)$ increases the energy proportionally to w and the number of latent states $\mathbf{h} \in h_t$ that are equal to y. This is convenient since, in sequences annotated with a particular label, it is more likely to find many latent ordinal states with such ordinal level. Therefore, the defined MI potential does not only model the MI-OR assumption but also provides mechanisms to learn how important is the proportion of latent states $\mathbf{h}$ that are equal to the label. Equation 6 is a special case of cardinality potentials [28] also employed in binary Multi-Instance Classification [24].

3.3 MI-DORF: Learning

Given a training set $\mathcal{T} = \{(\mathbf{X}_1, y_1), (\mathbf{X}_2, y_2), ..., (\mathbf{X}_N, y_N)\}$, we learn the model parameters θ by minimizing the regularized log-likelihood:

$$\min_\theta \quad \sum_{i=1}^{N} \log P(y|\mathbf{X}; \theta) + \mathcal{R}(\theta), \tag{7}$$

where the regularization function $\mathcal{R}(\theta)$ over the model parameters is defined as:

$$\mathcal{R}(\theta) = \alpha(||\beta||_2^2 + ||\mathbf{W}||_F^2) \tag{8}$$

and α is set via a validation procedure. The objective function in Eq. 7 is differentiable and standard gradient descent methods can be applied for optimization. To this end, we use the L-BFGS Quasi-Newton method [29]. The gradient evaluation involves marginal probabilities $p(h_t|\mathbf{X})$ and $p(h_t, h_{t+1}|\mathbf{X})$ which can be efficiently computed using the proposed algorithm in Sect. 3.4.

3.4   MI-DORF: Inference

The evaluation of the conditional probability $P(y|\mathbf{X};\theta)$ in Eq. 2 requires computing $\sum_h e^{-\Psi(\mathbf{X},\mathbf{h},y;\theta)}$ for each label y. Given the exponential number of possible latent states $\mathbf{h} \in \mathcal{H}$, efficient inference algorithms need to be used. In the case of Latent-Dynamic Models such as HCRF/HCORF, the forward-backward algorithm [15] can be applied. This is because the pair-wise linear-chain connectivity between latent states $\mathbf{h}$. However, in the case of MI-DORF, the inclusion of the cardinality potential $\Psi^M(\mathbf{h}, y; \theta^M)$ introduces a high-order dependence between the label y and all the latent states in $\mathbf{h}$. Inference methods with cardinality potentials has been previously proposed in [28,30]. However, these algorithms only consider the case where latent variables are independent and, therefore, they cannot be applied in MI-DORF. For these reasons, we propose an specific inference method. The idea behind it is to apply the standard forward-backward algorithm by converting the energy function defined in Eq. 3 into an equivalent one preserving the linear-chain connectivity between latent states $\mathbf{h}$.

To this end, we introduce a new set of auxiliary variables $\boldsymbol{\zeta} = \{\zeta_1, \zeta_2, ..., \zeta_T\}$, where each $\zeta_t \in \{0, 1\}$ takes a binary value denoting whether the sub-sequence $\mathbf{h}_{1:t}$ contains at least one ordinal state h equal to y. Now we redefine the MI-DORF energy function in Eq. 3 as:

$$\Psi(\mathbf{X}, \mathbf{h}, \boldsymbol{\zeta}, y; \theta) = \sum_{t=1}^{T} \Psi^N(\mathbf{x}_t, h_t, \zeta_t, y; \theta^N) + \sum_{t=1}^{T-1} \Psi^E(h_t, h_{t+1}, \zeta_t, \zeta_{t+1}, y; \theta^E), \quad (9)$$

where the new node and edge potentials are given by:

$$\Psi^N(\mathbf{x}_t, h_t, \zeta_t, y; \theta^N) = \begin{cases} \Psi^N(\mathbf{x}_t, h_t; \theta^N) + w\mathbf{I}(h_t == y) & \text{iff} \quad h_t <= y \\ -\infty & \text{otherwise} \end{cases}, \quad (10)$$

$$\Psi^E(h_t, h_{t+1}, \zeta_t, \zeta_{t+1}, y; \theta^E) = \begin{cases} \mathbf{W}_{h_t, h_{(t+1)}} & \text{iff} \quad \zeta_t = 0 \wedge \zeta_{t+1} = 0 \wedge h_{t+1} \neq y \\ \mathbf{W}_{h_t, h_{(t+1)}} & \text{iff} \quad \zeta_t = 0 \wedge \zeta_{t+1} = 1 \wedge h_{t+1} = y \\ \mathbf{W}_{h_t, h^{(t+1)}} & \text{iff} \quad \zeta_t = 1 \wedge \zeta_{t+1} = 1 \\ -\infty & \text{otherwise} \end{cases}$$

$$(11)$$

Note that Eq. 9 does not include the MIO potential and, thus, the high-order dependence between the label y and latent ordinal-states $\mathbf{h}$ is removed. The graphical representation of MI-DORF with the redefined energy function is illustrated in Fig. 1(b). In order to show the equivalence between energies in Eqs. 3 and 9, we explain how the original Multi-Instance-Ordinal potential Ψ^M is incorporated into the new edge and temporal potentials. Firstly, note that Ψ^N now also takes into account the proportion of ordinal variables h_t that are equal to the sequence label. Moreover, it enforces $\mathbf{h}$ not to contain any h_t greater than y, thus aligning the bag and (max) instance labels. However, the original Multi-Instance-Ordinal potential also constrained $\mathbf{h}$ to contain at least one h_t with the same ordinal value than y. This is achieved by using the set of auxiliary variables

ζ_t and the re-defined edge potential Ψ^E. In this case, transitions between latent ordinal states are modelled but also between auxiliary variables ζ_t. Specifically, when the ordinal state in h_{t+1} is equal to y, the sub-sequence $\mathbf{h}_{1:t+1}$ fulfills the MIOR assumption and, thus, ζ_{t+1} is forced to be 1. By defining the special cases at the beginning and the end of the sequence ($t = 1$ and $t = T$):

$$\Psi^N(\mathbf{x}_1, h_1, , \zeta_1, y) = \begin{cases} \Psi^N(\mathbf{x}_1, h_1) + w\mathbf{I}(h_1 == y) & \text{iff} \quad \zeta_1 = 0 \wedge l_1 < y \\ \Psi^N(\mathbf{x}_1, h_1) + w\mathbf{I}(h_1 == y) & \text{iff} \quad \zeta_1 = 1 \wedge l_1 = y \\ -\infty & \text{otherwise} \end{cases}, \quad (12)$$

$$\Psi^N(\mathbf{x}_T, h_T, \zeta_T, y) = \begin{cases} \Psi^N(\mathbf{x}_T, h_T) + w\mathbf{I}(h_T == y) & \text{iff} \quad \zeta_T = 1 \wedge h_T <= y \\ -\infty & \text{otherwise} \end{cases}, \quad (13)$$

we can see that the energy is $-\infty$ when the MIOR assumption is not fulfilled. Otherwise, it has the same value than the one defined in Eq. 3 since no additional information is given. The advantage of using this equivalent energy function is that the standard forward-backward algorithm can be applied to efficiently compute the conditional probability:

$$P(y|\mathbf{X}; \theta) = \frac{\sum_{\mathbf{h}} \sum_{\zeta} e^{-\Psi(\mathbf{X}, \mathbf{h}, \zeta, y; \theta)}}{\sum_{y'} \sum_{\mathbf{h}} \sum_{\zeta} e^{-\Psi(\mathbf{X}, \mathbf{h}, \zeta, y'; \theta)}}, \quad (14)$$

The proposed procedure has a computational complexity of $\mathcal{O}(T \cdot (2L)^2)$ compared with $\mathcal{O}(T \cdot L^2)$ using standard forward-backward in traditional linear-chain latent dynamical models. Since typically $L << T$, this can be considered a similar complexity in practice. The presented algorithm can also be applied to compute the marginal probabilities $p(h_t|\mathbf{X})$ and $p(h_t, h_{t+1}|\mathbf{X})$. This probabilities are used during training for gradient evaluation and during testing to predict ordinal labels at the instance and bag level.

4 Experiments

4.1 Baselines and Evaluation Metrics

The introduced MI-DORF approach is designed to address the Multi-Instance-Ordinal Regression when bags are structured as temporal sequences of ordinal states. Given that this has not been attempted before, we compare MI-DORF with different approaches that either ignore the MIL assumption (Single-Instance) or do not model dynamic information (Static):

Single-Instance Ordinal Regression (SIL-OR): MIL can be posed as a SIL problem with noisy labels. The main assumption is that the majority of instances will have the same label than their bag. In order to test this assumption, we train standard Ordinal Regression [27] at instance-level by setting all their labels to the same value as their corresponding bag. During testing, bag-label is set to the maximum value predicted for all its instances. Note that this baseline can be considered an Static-SIL approach.

Static Multi-Instance Ordinal Regression (MI-OR): Given that no MIOR methods have previously been proposed for this task, we implemented this static approach following the MIOR assumption. This method is inspired by MI-SVM [21], where instance labels are considered latent variables and are iteratively optimized during training. To initialize the parameters of the ordinal regressor, we follow the same procedure as described above in SIL-OR. Then, ordinal values for each instance are predicted and modified so that the MIOR assumption is fulfilled for each bag. Note that if all the predictions within a bag are lower than its label, the instances with the maximum value are set to the bag-label. On the other hand, all the predictions greater than the bag-label are decreased to this value. With this modified labels, Ordinal Regression is applied again and this procedure is applied iteratively until convergence.

Multi-Instance-Regression (MIR): Several methods have been proposed in the literature to solve the MIL problem when bags are real-valued variables. In order to evaluate the performance of this approach in MIOR, we have implemented a similar method as used in [23]. Specifically, a linear regressor at the instance-level is trained by optimizing a loss function over the bag-labels. This loss models the MIR assumption by using a soft-max function which approximates the maximum instance label within a bag predicted by the linear regressor. Note that a similar approach is also applied in Multi-Instance Logistic Regression [31]. In these works, a logistic loss is used because instance labels take values between 0 and 1. However, we use a squared-error loss to take into account the different ordinal levels.

Multi-Instance HCRF (MI-HCRF): This approach is similar to the proposed MI-DORF. However, MI-HCRF ignores the ordinal nature of labels and models them as nominal variables. For this purpose, we replace the MI-DORF node potentials by a multinomial logistic regression model[2]. Inference in MI-HCRF is performed by using the algorithm described in Sect. 3.4.

Single-Instance Latent-Dynamic Models (HCRF/HCORF): We also evaluate the performance of HCRF and HCORF. For this purpose, the Mutli-Instance-Ordinal potential in MI-DORF is replaced by the employed in standard HCRF [25]. This potential models the compatibility of hidden state values $\mathbf{h}$ with the sequence-label y but ignores the Multi-Instance assumption. For HCRF, we also replace the node potential as in the case of MI-HCRF. Inference is performed using the standard forward-backward algorithm.

Evaluation Metrics: In order to evaluate the performance of MI-DORF and the compared methods, we report results in terms of instance and bag-labels

---

[2] The potential with the Multinomial Logistic Regession model is defined as $\log(\frac{\exp(\beta_l^T x)}{\sum_{l' \in L} \exp(\beta_{l'}^T x)})$. Where all β_l defines a linear projection for each possible ordinal value l [32].

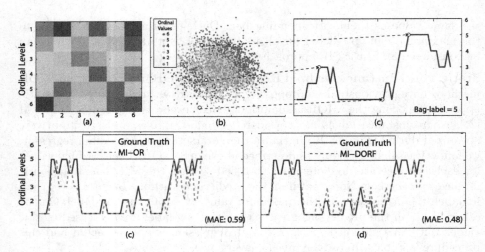

Fig. 2. Description of the procedure used to generate synthetic sequences. (a) A random matrix modelling transition probabilities between consecutive latent ordinal values. (b) Ordinal levels assigned to the random feature vectors according to the ordinal regressor. (c) Example of a sequence of ordinal values obtained using the generated transition matrix. The feature vector representing each observation is randomly chosen between the samples in (b) according to the probability for each ordinal level. (c–d) Examples of instance-level predictions in a sequence for MI-OR and MI-DORF.

predictions. Note that in the MIL literature, results are usually reported only at bag-level. However, in problems such as weakly-supervised pain detection, the main goal is to predict instance labels (frame-level pain intensities). Given the ordinal nature of the labels, the reported metrics are the Pearson's Correlation (CORR), Intra-Class-Correlation (ICC) and Mean-Average-Error (MAE). For bag-label predictions, we also report the Accuracy and average F1-score as discrete metrics.

4.2   Synthetic Experiments

Synthetic Data: Given that no standard benchmarks are available for MIOR, we have generated synthetic data. To create the synthetic sequences, we firstly generated a sequence of ordinal values using a random transition matrix. It represents the transition probabilities between temporally-consecutive ordinal levels. The first value for the sequence is randomly chosen with equal probability among all possible ordinal levels. Secondly, we generate random parameters of an Ordinal Regressor as defined in Eq. 4. This regressor is used to compute the probabilities for each ordinal level in a set o feature-vectors randomly sampled from a Gaussian distribution. Thirdly, the corresponding sequence observation for each latent state in the sequence is randomly chosen between the sampled feature vectors according to the obtained probability for each ordinal value. Finally, the sequence-label is set to the maximum ordinal state within the sequence following the MIOR assumption, and Gaussian noise ($\sigma = 0.25$) is added to the feature

Table 1. The performance of different methods obtained on the synthetic data.

	Frame-level			Sequence-level				
	CORR	MAE	ICC	CORR	MAE	ICC	ACC	F1
SIL-OR	0.77	1.40	0.71	0.85	1.43	0.57	0.26	0.19
MI-OR	0.82	0.54	0.80	0.92	0.58	0.91	0.48	0.44
HCORF [13]	0.81	1.33	0.80	0.94	0.28	0.94	0.74	0.74
HCRF [25]	0.49	1.41	0.42	0.93	0.36	0.92	0.67	0.66
MIR [23]	0.79	0.58	0.78	0.92	0.42	0.91	0.61	0.61
MI-HCRF	0.77	0.75	0.67	0.93	0.43	0.93	0.59	0.58
MI-DORF	0.86	0.39	0.85	0.96	0.20	0.96	0.80	0.80

vectors. Figure 2(a–c) illustrates this procedure. Following this strategy, we have generated ten different data sets by varying the ordinal regressor parameters and transition matrix. Concretely, each dataset is composed of 100 sequences for training, 150 for testing and 50 for validation. The last set is used to optimize the regularization parameters of each method. The sequences have a variable length between 50 and 75 instances. The dimensionality of the feature vectors was set to 10 and the number of ordinal values to 6.

Results and Discussion: Table 1 shows the results computed as the average performance over the ten datasets. SIL methods (SIL-OR, HCRF and HCORF) obtain worse performance than their corresponding MI versions (MI-OR, MI-HCRF and MI-DORF) in most of the evaluated metrics. This is expected since SIL approaches ignore the Multi-Instance assumption. Moreover, HCORF and MI-DORF obtain better performance compared to HCRF and MI-HCRF. This is because the former model the latent states as nominal variables, thus, ignoring their ordinal nature. Finally, note that MI-DORF outperforms the static methods MI-OR and MIR. Although these approaches use the Multi-Instance assumption and incorporate the labels ordering, they do not take into account temporal information. In contrast, MI-DORF is able to model the dynamics of latent ordinal states and use this information to make better predictions when sequence observations are noisy. As Fig. 2(c–d) shows, MI-OR predictions tends to be less smooth because dynamic information is not taken into account. In contrast, MI-DORF better estimate the actual ordinal levels by modelling transition probabilities between consecutive ordinal levels.

4.3 Weakly-Supervised Pain Intensity Estimation

In this experiment, we test the performance of the proposed model for weakly-supervised pain intensity estimation. To this end, we use the UNBC Shoulder-Pain Database [6]. This dataset contains recordings of different subjects performing active and passive arm movements during rehabilitation sessions. Each video is annotated according to the maximum pain felt by the patient during the recording in an ordinal scale between 0 (no pain) and 5 (strong pain). These annotations are used as the bag label in the MIOR task. Moreover, pain intensities are also annotated at frame-level in terms of the PSPI scale [33]. This ordinal scale ranges from 0 to 15. Frame PSPI annotations are normalized between 0 and 5, in order to align the scale with the one provided at the sequence level. Furthermore, we used a total of 157 sequences from 25 subjects. The remaining 43 were removed because a high discrepancy between sequence and frame-level annotations was observed. Concretely, we do not consider the cases where the sequence label is 0 and frame annotations contains higher pain levels. Similarly, we also remove sequences with a high-discrepancy in the opposite way. Given the different scales used in frame and sequence annotations, we computed the agreement between them. For this purpose, we firstly obtained the maximum pain intensities at frame-level for all the used sequences. Then, we computed the CORR and ICC between them and their corresponding sequence labels. The results were 0.83 for CORR, and 0.78 in the case of ICC. This high agreement indicates that predictions in both scales are comparable. More importantly, this supports our hypothesis that sequence labels are highly correlated with frame labels; thus, the used bag labels provide sufficient information for learning the instance labels in our weakly-supervised setting.

Facial-Features: For each video frame, we compute a geometry-based facial-descriptor as follows. Firstly, we obtain a set of 49 landmark facial-points with the method described in [34]. Then, the obtained points locations are aligned with a mean-shape using Procrustes Analysis. Finally, we generate the facial descriptor by concatenating the x and y coordinates of all the aligned points. According to the MIL terminology, these facial-descriptors are considered the instances in the bag (video).

Experimental Setup: We perform Leave-One-Subject-Out Cross Validation similar to [12]. In each cycle, we use 20 subjects for training, 1 for testing and 4 for validation. This last subset is used to cross-validate the regularization parameters of each particular method. In order to reduce computational complexity and redundant information between temporal consecutive frames, we have segmented the sequences using non-overlapping windows of 0.5 s, similar to [12]. The instance representing each segment is computed as the mean of its corresponding facial-descriptors. Apart from the baselines described in Sect. 4.1, we also evaluate the performance of the MIC approach considering pain levels as binary variables. For this purpose, we have implemented the MILBoosting [22] method used in [12] and considered videos with a pain label greater than 0 as positive. Given that MI-Classification methods are only able to make binary predictions, we use the output probability as indicator of intensity levels, at bag and instance-level, i.e., the output probability is normalized between 0 and 5.

Results and Discussion: Table 2 shows the results obtained by the evaluated methods following the experimental setup previously described. By looking into the results of the compared methods, we can derive the following conclusions. Firstly, SI approaches (SIL-OR, HCORF and HCRF) obtain worse performance than MI-OR and MIR. This is because pain events are typically

Table 2. The performance of different methods obtained on the UNBC Database.

	Frame-level			Sequence-level				
	CORR	MAE	ICC	CORR	MAE	ICC	ACC	F1
SIL-OR	0.31	1.67	0.22	0.59	1.52	0.56	0.19	0.16
MI-OR	0.39	0.76	0.28	0.64	1.01	0.63	0.39	0.31
HCORF [13]	0.24	1.92	0.12	0.30	1.36	0.30	0.39	0.19
HCRF [25]	0.09	2.29	0.05	0.26	1.52	0.26	0.29	0.13
MIR [23]	0.35	0.84	0.24	0.63	0.94	0.63	0.41	0.30
MILBoost [12]	0.28	1.77	0.11	0.38	1.7	0.38	0.3	0.2
MI-HCRF	0.17	1.45	0.11	0.26	1.69	0.26	0.28	0.21
MI-DORF	0.40	0.19	0.40	0.67	0.80	0.66	0.52	0.34

very sparse in these sequences and most frames have intensity level 0 (neutral). Therefore, the use of the MIL assumption has a critical importance in this problem. Secondly, poor results are obtained by HCRF and MI-HCRF. This can be explained because these approaches consider pain levels as nominal variables and are ignorant of the ordering information of the different pain intensities. Finally, MILBoost trained with binary labels also obtains low performance compared to the MI-OR and MIR. This suggest that current approaches posing weakly-supervised pain detection as a MIC are suboptimal, thus, unable to predict accurately the target pain intensities. By contrast, MI-DORF obtains the best performance across all the evaluated metrics at both the sequence and frame-level. We attribute this to the fact the MI-DORF models the MIL assumption with ordinal variables. Moreover, the improvement of MI-DORF compared to the static approaches, such as MI-OR and MIR, suggests that modelling dynamic information is beneficial in this task. To get better insights into the performance of our weakly supervised approach, we compare its results (in terms of ICC) to those obtained by the fully supervised (at the frame level) state-of-the-art approach to pain intensity estimation - Context-sensitive Dynamic Ordinal Regression [35]. While this approach achieves an ICC of 0.67/0.59, using context/no-context features, respectively, our MI-DORF achieves an ICC of 0.40 without ever seeing the frame labels. This is a good trade-off between the need for the "very-expensive-to-obtain" frame-level annotation, and the model's performance.

Finally, in Fig. 3, we show more qualitative results comparing predictions of MI-OR, MIR and MI-DORF. The shown example sequences depict image frames along with the per-frame annotations and those obtained by compared models, using the adopted weakly-supervised setting (thus, only bag labels are provided). First, we note that all methods succeed to capture the segments in the sequences where the intensity changes occur, as given by the frame-level ground truth. However, note that MI-DORF achieves more accurate localization of the pain activations and prediction of their actual intensity. This is also reflected in terms of the MAE depicted, showing clearly that the proposed outperforms the competing methods on target sequences.

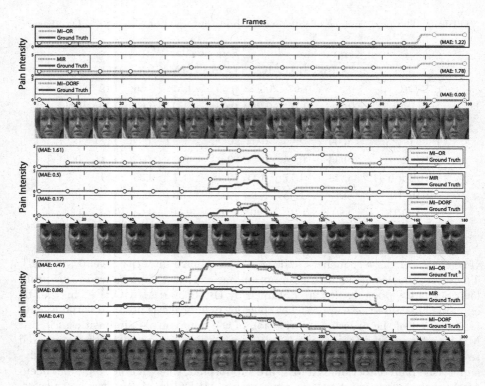

Fig. 3. Visualization of the pain intensity predictions at frame-level for MI-OR, MIR and the proposed MI-DORF method. From top to bottom, three sequences with ground-truth where MI-DORF predicted the sequence labels: 0, 3 & 5 respectively.

5 Conclusions

In this work, we introduced MI-DORF for the task of Multi-Instance-Ordinal Regression. This is the first MI approach that imposes an ordinal structure on the bag labels, and also attains dynamic modeling of temporal sequences of corresponding ordinal instances. In order to perform inference in the proposed model, we have developed an efficient algorithm with similar computational complexity to that of the standard forward-backward method - despite the fact that we model high-order potentials modelling the MIOR assumption. We demonstrated on the task of weakly supervised pain intensity estimation that the proposed model can successfully unravel the (ordinal) instance labels by using only the (ordinal) bag labels. We also showed that this approach largely outperforms related MI approaches – all of which fail to efficiently account for either temporal or ordinal, or both types of structure in the target data.

Acknowledgement. This paper is part of a project that has received funding from the European Union's Horizon 2020 research and innovation programme under grants agreement no. 645012 (KRISTINA), no. 645094 (SEWA) and no. 688835 (DE-ENIGMA). Adria Ruiz would also like to acknowledge Spanish Government to provide support under grant FPU13/01740.

References

1. Babenko, B., Yang, M.H., Belongie, S.: Robust object tracking with online multiple instance learning. IEEE Trans. Pattern Anal. Mach. Intell. **33**, 1619–1632 (2011)
2. Wu, J., Zhao, Y., Zhu, J.Y., Luo, S., Tu, Z.: Milcut: a sweeping line multiple instance learning paradigm for interactive image segmentation. In: Proceedings of the Computer Vision and Pattern Recognition (2014)
3. Ruiz, A., Van de Weijer, J., Binefa, X.: Regularized multi-concept MIL for weakly-supervised facial behavior categorization. In: Proceedings of the British Machine Vision Conference (2014)
4. Maron, O., Lozano-Pérez, T.: A framework for multiple-instance learning. In: Advances in Neural Information Processing Systems (1998)
5. Ray, S., Page, D.: Multiple instance regression. In: Proceedings of the International Conference on Machine Learning (2001)
6. Lucey, P., Cohn, J.F., Prkachin, K.M., Solomon, P.E., Matthews, I.: Painful data: the UNBC-McMaster shoulder pain expression archive database. In: International Conference on Automatic Face and Gesture Recognition (2011)
7. Aung, M.S., Kaltwang, S., Romera-Paredes, B., Martinez, B., Singh, A., Cella, M., Valstar, M.F., Meng, H., Kemp, A., Shafizadeh, M., Elkins, A.C., Kanakam, N., rothschild, A.D., Tyler, N., Watson, P.J., Williams, A.C., Pantic, M., Bianchi-berthouze, N.: The automatic detection of chronic pain-related expression: requirements, challenges and a multimodal dataset. IEEE Trans. Affect. Comput. (2015, to appear)
8. Hjermstad, M.J., Fayers, P.M., Haugen, D.F., Caraceni, A., Hanks, G.W., Loge, J.H., Fainsinger, R., Aass, N., Kaasa, S., EPCRC, E.P.C.R.C., et al.: Studies comparing numerical rating scales, verbal rating scales, and visual analogue scales for assessment of pain intensity in adults: a systematic literature review. J. Pain Symptom Manag. **41**, 1073–1093 (2011)
9. Rudovic, O., Pavlovic, V., Pantic, M.: Automatic pain intensity estimation with heteroscedastic conditional ordinal random fields. In: Bebis, G., et al. (eds.) ISVC 2013. LNCS, vol. 8034, pp. 234–243. Springer, Heidelberg (2013). doi:10.1007/978-3-642-41939-3_23
10. Kaltwang, S., Rudovic, O., Pantic, M.: Continuous pain intensity estimation from facial expressions. In: Bebis, G., et al. (eds.) ISVC 2012. LNCS, vol. 7432, pp. 368–377. Springer, Heidelberg (2012). doi:10.1007/978-3-642-33191-6_36
11. Wu, C., Wang, S., Ji, Q.: Multi-instance hidden Markov model for facial expression recognition. In: International Conference on Automatic Face and Gesture Recognition (2015)
12. Sikka, K., Dhall, A., Bartlett, M.: Weakly supervised pain localization using multiple instance learning. In: International Conference on Automatic Face and Gesture Recognition (2013)
13. Kim, M., Pavlovic, V.: Hidden conditional ordinal random fields for sequence classification. In: Balcázar, J.L., Bonchi, F., Gionis, A., Sebag, M. (eds.) ECML PKDD 2010. LNCS (LNAI), vol. 6322, pp. 51–65. Springer, Heidelberg (2010). doi:10.1007/978-3-642-15883-4_4
14. Liu, J., Chen, C., Zhu, Y., Liu, W., Metaxas, D.N.: Video classification via weakly supervised sequence modeling. Comput. Vis. Image Underst. **152**, 79–87 (2015)
15. Barber, D.: Bayesian Reasoning and Machine Learning. Cambridge University Press, Cambridge (2012)

16. Amores, J.: Multiple instance classification: review, taxonomy and comparative study. Artif. Intell. **201**, 81–105 (2013)
17. Gärtner, T., Flach, P.A., Kowalczyk, A., Smola, A.J.: Multi-instance kernels. In: Proceedings of the International Conference on Machine Learning (2002)
18. Chen, Y., Bi, J., Wang, J.Z.: Miles: Multiple-instance learning via embedded instance selection. IEEE Trans. Pattern Anal. Mach. Intell. **28**, 1931–1947 (2006)
19. Zhou, Z.H., Sun, Y.Y., Li, Y.F.: Multi-instance learning by treating instances as non-I.I.D. samples. In: Proceedings of the International Conference on Machine Learning (2009)
20. Wagstaff, K.L., Lane, T., Roper, A.: Multiple-instance regression with structured data. In: International Conference on Data Mining (2008)
21. Andrews, S., Tsochantaridis, I., Hofmann, T.: Support vector machines for multiple-instance learning. In: Advances in Neural Information Processing Systems (2002)
22. Zhang, C., Platt, J.C., Viola, P.A.: Multiple instance boosting for object detection. In: Advances in Neural Information Processing Systems (2005)
23. Hsu, K.J., Lin, Y.Y., Chuang, Y.Y.: Augmented multiple instance regression for inferring object contours in bounding boxes. IEEE Trans. Image Process. **23**, 1722–1736 (2014)
24. Hajimirsadeghi, H., Li, J., Mori, G., Zaki, M., Sayed, T.: Multiple instance learning by discriminative training of Markov networks. In: Uncertainty in Artificial Intelligence (2013)
25. Quattoni, A., Wang, S., Morency, L.P., Collins, M., Darrell, T.: Hidden conditional random fields. IEEE Trans. Pattern Anal. Mach. Intell. **29**, 1848–1852 (2007)
26. Rabiner, L.R., Juang, B.H.: An introduction to hidden Markov models. ASSP Mag. **3**, 4–16 (1986)
27. Winkelmann, R., Boes, S.: Analysis of Microdata. Springer Science Business Media, Berlin (2006)
28. Gupta, R., Diwan, A.A., Sarawagi, S.: Efficient inference with cardinality-based clique potentials. In: Proceedings of the International Conference on Machine Learning (2007)
29. Byrd, R.H., Nocedal, J., Schnabel, R.B.: Representations of quasi-newton matrices and their use in limited memory methods. Math. Program. **63**, 129–156 (1994)
30. Tarlow, D., Swersky, K., Zemel, R.S., Adams, R.P.: Fast exact inference for recursive cardinality models. In: Conference on Uncertainty in Artificial Intelligence (2012)
31. Ray, S., Craven, M.: Supervised versus multiple instance learning: an empirical comparison. In: Proceedings of the International Conference on Machine Learning (2005)
32. Walecki, R., Rudovic, O., Pavlovic, V., Pantic, M.: Variable-state latent conditional random fields for facial expression recognition and action unit detection. In: International Conference on Automatic Face and Gesture Recognition (2015)
33. Prkachin, K.M.: The consistency of facial expressions of pain: a comparison across modalities. Pain **51**, 297–306 (1992)
34. Xuehan-Xiong, D., la Torre, F.: Supervised descent method and its application to face alignment. In: Proceedings of the Computer Vision and Pattern Recognition (2013)
35. Rudovic, O., Pavlovic, V., Pantic, M.: Context-sensitive dynamic ordinal regression for intensity estimation of facial action units. IEEE Trans. Pattern Anal. Mach. Intell. **37**, 944–958 (2015)

Deep Learning

Analysis on the Dropout Effect in Convolutional Neural Networks

Sungheon Park and Nojun Kwak[✉]

Graduate School of Convergence Science and Technology,
Seoul National University, Seoul, Korea
{sungheonpark,nojunk}@snu.ac.kr

Abstract. Regularizing neural networks is an important task to reduce overfitting. Dropout [1] has been a widely-used regularization trick for neural networks. In convolutional neural networks (CNNs), dropout is usually applied to the fully connected layers. Meanwhile, the regularization effect of dropout in the convolutional layers has not been thoroughly analyzed in the literature. In this paper, we analyze the effect of dropout in the convolutional layers, which is indeed proved as a powerful generalization method. We observed that dropout in CNNs regularizes the networks by adding noise to the output feature maps of each layer, yielding robustness to variations of images. Based on this observation, we propose a stochastic dropout whose drop ratio varies for each iteration. Furthermore, we propose a new regularization method which is inspired by behaviors of image filters. Rather than randomly drop the activation, we selectively drop the activations which have high values across the feature map or across the channels. Experimental results validate the regularization performance of selective max-drop and stochastic dropout is competitive to the dropout or spatial dropout [2].

1 Introduction

Convolutional neural networks (CNNs) have been widely used for many computer vision tasks such as image classification, segmentation, and detection in recent years, mainly due to their high representation power and superior performance. Since deep neural networks are involved with a large number of parameters, regularization is a critical task to reduce overfitting. Other than a weight decay term, many algorithms have been presented to regularize neural networks. Dropout [1] is the most commonly used technique for regularization. For CNNs, stochastic pooling [3] or maxout networks [4] are well known techniques to regularize convolutional layers. Though dropout has shown its effectiveness in convolutional layers in some cases [1,5,6], it is still rarely used with convolutional layers in practice. Moreover, the effect of dropout in convolutional layers has not been studied thoroughly. Different from the fully connected layers, convolutional layers have smaller number of parameters compared to the size of feature maps. Hence, it is believed that convolutional layers suffer less from overfitting.

In this paper, we analyze the effect of dropout in convolutional layers. We found that dropout in convolutional layers as well as the fully connected layers

© Springer International Publishing AG 2017
S.-H. Lai et al. (Eds.): ACCV 2016, Part II, LNCS 10112, pp. 189–204, 2017.
DOI: 10.1007/978-3-319-54184-6_12

are effective for regularization. The generalization effect of dropout in convolutional layers is due to the enhanced robustness by adding noise to the inputs of convolutional layers, not due to the model averaging in the case of fully connected layers. Based on this observation, we propose two variants of dropout which is suited for convolutional layers of CNNs. Similar to dropout [1], the proposed methods turn off the activations of convolutional layers. While dropout turns off the activations randomly, the first variant, max-drop, selectively drops the activation which is the maximum value within each feature map or within the same spatial position of feature maps. Since the neurons with high activation values contain key information about the problem at hand, dropping the maximum activation probabilistically can grant generalization power to the networks. The other variant, stochastic dropout, varies the dropout probability based on the probability distribution which makes the network robust to inputs with different levels of noise. Experimental results show that the proposed method effectively regularizes convolutional layers and shows competitive performance against dropout. This result indicates that unlike dropout, only dropping a small portion of activations in the network can lead to a powerful generalization performance.

The rest of the papers will be presented as follows. Related works are introduced in Sect. 2, and we analyze the effect of dropout in convolutional layers in Sect. 3. Based on the analysis, two variants of dropout, max-drop and stochastic dropout, are proposed in Sects. 4 and 5 respectively. Experiments on various datasets are conducted to compare the generalization performance of proposed methods with dropout, and the results are illustrated in Sect. 6. Finally, conclusions are made in Sect. 7.

2   Related Work

Many efforts have been made for regularizing neural networks. Dropout [1] is the most popular method for network regularization. It randomly drops the pre-designed portion of activations at each iteration to regularize the network. Dropout can be viewed as cooperation of multiple models trained on different subsets of data. From similar inspiration, DropConnect [7] drops the connections of the network instead of activations. It showed comparable generalization performance with dropout.

Dropout works well in practice especially with fully connected layers. However, when applied to convolutional layers in a deep CNN, the performance of dropout has been thought to be questionable. It is argued that convolutional layers does not suffer from overfitting because the number of parameters for the convolutional layers is small relative to the number of activations. Nevertheless, dropout in convolutional layer is proven to improve generalization performance in some extent by adding noise to the activations [1]. Network-in-Network [8] efficiently integrated dropout in convolutional layer by using 1×1 convolutional layer followed by dropout, which enhances both representation and generalization power. On the other hand, spatial dropout [2] has been suggested to consider the correlated activations in convolutional layers. The method drops the entire feature maps rather than individual activations. Since spatially close activations in the same feature map are tend to be correlated, the paper argues that dropout does not suitably applied to volumetric feature map since it assumes independence between the activations.

Various pooling methods have been proposed to regularize CNNs. Stochastic pooling [3] determines the elements to pool probabilistically based on the input activation values. Generalized pooling functions [9] learn parameters to combine average and max pooling. Strided convolution [5] can also be viewed as generalization of pooling operations. Wu and Gu [10] proposed probabilistic weighted pooling which combines dropout in convolutional layers and max pooling together.

Adding noise in the training step or to the activation function also helps enhancing generalization power. Neelakantan et al. [11] found that adding noise to the gradient during backpropagation helps deep networks converge faster and prevent overfitting. Audhkhasi et al. [12] and Gulcehre et al. [13] showed that adding carefully chosen noise to the activation can speed up training procedure. Maxout networks [4] regularize networks by propagating only maximum activations. Huang et al. [14] proposed the regularization techniques which combine maxout and dropout. Opposed to the maxout networks, our method prohibits maximum activations from forward and backward propagation.

Among the numerous regularization methods, dropout is still used in most neural networks due to its simplicity and reasonable performance. Though Srivastava et al. [1] empirically proved the effectiveness of dropout in the convolutional layers, dropout is not preferable to apply every layer in a deep convolutional neural network since the scale of backpropagated error drops whenever it passes the layer with dropout, which slows down the learning speed in the lower convolutional layers. Therefore, dropout has been applied only to the fully connected networks in most cases.

3 Effectiveness of Dropout in Convolutional Layer

We first investigate the effect of dropout in convolutional layers of CNNs. Dropout is interpreted as bagging of different models which is trained on different subsets of data. On the other hand, it is believed that the regularization effect of dropout in convolutional layers is mainly obtained from the robustness to noisy inputs. To analyze the characteristics of dropout that actually help generalizing convolutional layers, we scrutinized the distribution of activations in a CNN trained on the CIFAR-10 dataset [15] with and without dropout in convolutional layers. The network model used in this section consists of 10 convolutional layers and 4 pooling layers. All convolutional layers have kernels of 3×3 size, and inputs to the convolutional layers are padded by 1 pixels for both sides. All pooling layers use 2×2 max pooling with stride of 2 except the last layer for which we used 4×4 mean pooling. Dropout after *pool4* with probability of 0.5 is applied regardless of using dropout in convolutional layers or not. The number of filters is doubled after each pooling layer, which is a similar approach to the VGGnet [16]. Rectified linear unit (ReLU) is used as a activation function in all layers. Detailed configuration is illustrated in Fig. 1(a). While the CNN that does not use dropout achieved 83.16% accuracy, when dropout is applied to the output of every convolutional layer except the last *conv4_3* layer with ratio of 0.1, the network achieved 87.78% accuracy. We analyzed the reason of accuracy improvement by looking into the behavior of the activated neurons in the convolutional layers.

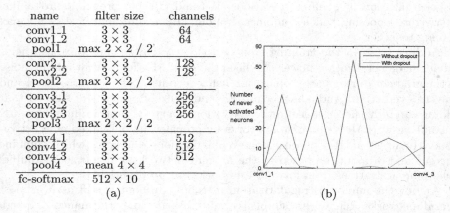

name	filter size	channels
conv1_1	3 × 3	64
conv1_2	3 × 3	64
pool1	max 2 × 2 / 2	
conv2_1	3 × 3	128
conv2_2	3 × 3	128
pool2	max 2 × 2 / 2	
conv3_1	3 × 3	256
conv3_2	3 × 3	256
conv3_3	3 × 3	256
pool3	max 2 × 2 / 2	
conv4_1	3 × 3	512
conv4_2	3 × 3	512
conv4_3	3 × 3	512
pool4	mean 4 × 4	
fc-softmax	512 × 10	

(a)   (b)

Fig. 1. (a) Structure of CNN used in the experiments. (b) Number of neurons that never activated in each layer.

First, we investigated that every neuron in CNNs are working effectively, which means that the filters in CNNs do not learn redundant or useless information. One of the difficulties for training deep CNNs is that there exist dead neurons in the convolutional layer that are never activated. Using variants of ReLU activation functions such as leaky ReLU [17] or parametric ReLU [18] is one of the solutions to avoid dead neurons. We verify that dropout is also useful for avoiding dead neurons while the network still uses ReLU activation function. We counted the number of neurons that are not activated at the test time for each layer. The portions of never activated neurons with and without dropout are shown in Fig. 1(b). Large number of dead neurons are observed in most layers when dropout is not applied. On the other hand, when dropout is applied to the convolutional layers, almost all neurons are activated. Therefore, we verify that dropout in convolutional layers helps filters to learn informative features of images, which improves representation power of the network and classification performance as a consequence.

Next, as discussed in [1], we compared the sparsity of the activations. It is verified from [1] that in the fully connected layer, the activations are sparser when dropout is used. To confirm that this statement also holds for the convolutional layers in both lower and higher layers, we calculated the mean activation of

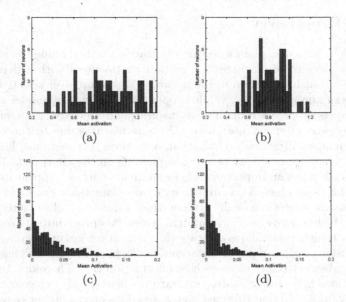

Fig. 2. Histogram of mean activation of (a) conv1_1 layer without dropout. (b) conv1_1 layer with dropout. (c) conv4_2 layer without dropout. (d) conv4_2 layer with dropout.

all neurons in each layer. The mean activation of lower convolutional layers, *conv*1_1 (Fig. 2(a) and (b)), and that of higher convolutional layers, *conv*4_2 (Fig. 2(c) and (d)) are shown. In the lower convolutional layers, the histogram is almost flat when dropout is not used while it is bell shape when dropout is applied. This indicates that some neurons are activated frequently or have larger activation values, and others are activated less frequently or have small activation values when dropout is not applied. Meanwhile, with dropout, every neuron has similar mean activation value, which means that every neuron is similarly activated. Since lower layers in CNN usually captures common features, it is preferable behavior that neurons have similar mean activation values. In the higher convolutional layer, we could verify the sparsity of activations with dropout. A high peak near zero value is observed, which implies that the mean activation of most neurons are concentrated at small values when dropout is applied.

Based on these observations, we conclude that dropout in convolutional layers helps filters to learn informative features. However, when dropout is applied to every convolutional layers in deep CNNs, training process can be slow since activation signals are dropped exponentially as dropout is applied repeatedly. If higher drop probability such as 0.5 is applied in convolutional layers, CNNs perform poor or cannot be trained at all. In the next sections, we propose two variants of dropout to deal with this problem while maintaining the competitive generalization power with dropout.

4   Max-Drop Layer

In this section, we will explain a new regularization method named as max-drop. Based on the information from Sect. 3, we note that neurons with high activation contain important information in the network. Max-drop layer selectively drops only the maximum activations. While dropout is motivated by model averaging, max-drop layers originate from different inspiration from CNNs. Different images of the same class often do not share the same features due to the occlusion, viewpoint changes, illumination variation, and so on. For instance, human face images may contain one eye or two eyes depending on the viewpoint. Therefore, a feature which plays an important role in an image may not appear in different images of the same class. Max-drop aims to simulate these cases by dropping high activations deterministically, rather than randomly select activations to drop off. In the lower layer of convolutional layers, this procedure of dropping the maximum activations simulates the case that important features are not present due to occlusion or other types of variations. In the higher layer, each feature map learns more abstracted and class-specific information [19]. Therefore, turning off high activations helps other neurons to learn the class-specific characteristics. It is intuitively uncertain that dropped high activations on the higher convolutional layers give generalization power, but we empirically prove that max-drop layers effectively regularize higher convolutional layers similar to dropout.

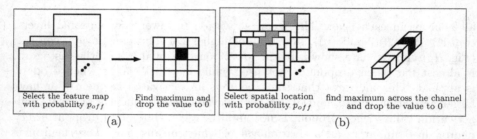

Fig. 3. Illustration of max-drop layer. Two different ways to find maximum value is proposed in this paper: (a) feature-wise max-drop finds maximum value within each feature map and drops the maximum values with probability p_{off}. (b) Channel-wise max-drop finds maximum value across each channel in the same spatial position and drops the maximum values with probability p_{off}.

In the max-drop layer, maximum element is found in the activations of convolutional layers and the maximum activation is dropped to zero with the probability of p_{off}. Max-drop can be applied to both outputs of convolutional layers or pooling layers as in the case of dropout. We propose two different strategies to find maximum value which is illustrated in Fig. 3. The first strategy is to find maximum within each feature map, which will be called as feature-wise max-drop. This scheme turns off the most informative feature within the feature map and drop the value to 0 with the probability p_{off}. The portion of dropped activations with respect to the entire activations of the convolutional layer is calculated as

$$p_f = \frac{1}{n_w \times n_f} p_{off}, \tag{1}$$

where n_w, n_h is the width and height of the feature map respectively. For instance, if convolutional layer outputs 4×4 size feature map, then the maximum probability of drop will be 0.0625, which is smaller than the typical dropout ratio.

Another strategy is to find maximum across the channels in the same position of feature map, which is denoted as channel-wise max-drop. This scheme prevents the highest activation to be propagated to the next layer on a certain spatial position of the feature map. The actual drop probability for the channel-wise max-drop is

$$p_c = \frac{1}{n_c} p_{off}, \tag{2}$$

where n_c is the number of channels in the convolutional layer outputs. With the same drop rate, channel-wise max-drop will drop smaller number of activations than the feature-wise max-drop in higher layers where the size of feature map is much smaller than the number of channels, and vice versa in the lower layers.

Dropping small number of neurons has an advantage over conventional dropout. Max-drop does not suffer from slow training since gradients are propagated through all activations except the maximum activations that are selected to turn off. Empirically, when max-drop is applied to every convolutional layer, test error decreases faster in the early stage of training than the case when dropout is applied. Moreover, with the same learning rate, the network with max-drop can be trained when high p_{off} (larger than 0.5) is used while the network with dropout usually failed to be trained when the drop probability exceeds 0.2.

5   Stochastic Dropout

If we interpret the effect of dropout as gaining robustness by putting in noisy inputs, giving different degrees of noise might be helpful. Also, it is hard to determine an appropriate drop rate for the convolutional layers in most cases. If dropout ratio is determined differently for every iteration, we believe that CNN can be learned to handle different amount of information. Based on this idea, we propose a stochastic dropout in which dropout ratio is determined from probability distribution. In stochastic dropout, probability of dropping neurons is drawn from the uniform distribution or normal distribution, i.e.,

$$p_{off} \sim N(\mu, \sigma) \quad \text{or} \quad p_{off} \sim U(a, b) \tag{3}$$

where $N(\mu, \sigma)$ is the normal distribution with mean μ and standard deviation σ, and $U(a, b)$ is the uniform distribution whose range is $[a, b]$. For our implementation, if $\mu = 0$ for normal distribution, we used the absolute value of drawn probability as p_{off}. When μ is non-zero, we set $p_{off} = 0$ if negative number is drawn.

We implemented max-drop and stochastic dropout using Caffe framework [20]. Like the dropout implementation, the activations scale up by inverse of drop probability when max-drop or stochastic dropout is applied.[1] Note that the scale factor is almost 1 for max-drop since the actual drop probability is near 0. We also found that the performance is almost the same for max-drop regardless of scale factor multiplication. GPU implementation of max-drop showed similar computation time for one iteration of backpropagation with dropout.

6   Experimental Results

We examined the regularization performance of various algorithms using MNIST [21], CIFAR-10, CIFAR-100 [15], and the street view house numbers (SVHN) [22] dataset. Max-drop and stochastic dropout are compared to dropout [1] and spatial dropout [2] to validate the generalization performance of the proposed methods against the conventional algorithms. To verify the regularization effect on the recently proposed very deep neural networks, we also conducted an experiment with ResNet [23] on CIFAR-10 dataset.

The baseline model structure of the CNN is the same as the model described in Fig. 1(a) except the MNIST dataset in Sect. 6.1 and ResNet [23] experiment in Sect. 6.2. For the MNIST dataset, the number of channels for the network is reduced from 64, 128, 256, 512 to 64, 96, 128, 256. Also, $pool3$ layer has 3×3 kernels with a stride of 2, and $pool4$ has 3×3 kernels to deal with the 28×28 input size. For ResNet experiment, we used the same 32-layer model suggested in [23] except that the number of feature maps in every convolutional layer is doubled. Mean substraction is the only preprocessing for the whole experiments.

For fair comparison, we searched the best parameter (e.g. drop probability) for each method. To ease the parameter tuning process, we applied the regularization algorithms for every convolutional layer with the same parameters. For all models in the experiments, dropout with probability of 0.5 is applied after $pool4$ and before the softmax. Dropout, spatial dropout, max-drop, or stochastic dropout is applied after every convolution layers except the last $conv4\_3$ layer. When batch normalization [24] is applied, dropout after $pool4$ is removed and the regularization methods are applied after $conv4\_3$. Since drop probability of max-drop has large values, we searched the parameter for max-drop with the interval of 0.1, ranging from 0.1 to 0.9, and we used the interval of 0.05 for dropout and spatial dropout, ranging from 0.05 to 0.5. We reported the selected parameter together with the regularization method.

6.1   MNIST Dataset

As a sanity check, we experimented the proposed max-drop and stochastic dropout on the MNIST dataset. MNIST has 60,000 training images and 10,000

---

[1] Caffe implementation of dropout scales up the activations at training time instead of scaling down them at test time unlike the original dropout paper [1].

test images with 28×28 size. We trained CNNs for 60 epochs with the initial learning rate of 0.01 and the batch size of 128. The learning rate is decreased by 0.1 for every 20 epochs. Since MNIST classification is an easy task, and the performance is highly saturated, we conducted training 5 times for each model. We reported the average classification error for each method with the standard deviation as well as the classification of the ensemble classification error by averaging the predictions of 5 models. The results are shown in Table 1.

Table 1. Classification error on MNIST

Method	Classification error (%)	
	Average of 5 models	Ensemble of 5 models
Baseline (without dropout)	0.604 ± 0.0829	0.57
Dropout ($p = 0.2$)	0.430 ± 0.0212	**0.38**
Spatial dropout ($p = 0.1$)	0.504 ± 0.0493	0.42
Feature-wise max-drop ($p = 0.2$)	0.488 ± 0.0657	0.42
Channel-wise max-drop ($p = 0.5$)	0.502 ± 0.0148	0.40
Stochastic dropout ($N(0.2, 0.05)$)	$\mathbf{0.410 \pm 0.0122}$	**0.38**
Stochastic dropout ($U(0.1, 0.3)$)	0.448 ± 0.0363	0.42

It is shown that all regularization methods significantly improve the performance of the baseline. Dropout has higher improvement on classification accuracy than max-drop. For average performance, stochastic dropout also showed the best performance.

We also analyzed the effect of regularization methods with small amount of training data. We randomly select 20% of the training images from the MNIST dataset and trained with the small dataset. The performance is illustrated in Table 2, which shows similar tendency to Table 1.

Table 2. Classification error on MNIST with 20% of training data.

Method	Classification error (%)	
	Average of 5 models	Ensemble of 5 models
Baseline	1.126 ± 0.0802	0.92
Dropout ($p = 0.2$)	0.808 ± 0.0740	0.76
Spatial dropout ($p = 0.1$)	0.872 ± 0.0335	0.78
Feature-wise max-drop ($p = 0.4$)	0.882 ± 0.0676	0.83
Channel-wise max-drop ($p = 0.5$)	0.888 ± 0.0638	0.79
Stochastic dropout ($N(0.2, 0.05)$)	0.810 ± 0.0534	0.80
Stochastic dropout ($U(0.1, 0.3)$)	$\mathbf{0.802 \pm 0.0444}$	**0.75**

In general, regularization methods reduced the classification error by 20 ∼ 30%. Also, standard deviation has smaller values when regularization methods are applied, which means that regularization in convolutional layers provides stable results. Dropout shows superior performance to max-drop in MNIST dataset. Though stochastic dropout works slightly better than dropout with fixed probability, it seems that giving different levels of noise does not take much advantage against dropout. Spatial dropout showed inferior performance, which indicates that independence between feature map does not play an important role in regularization of convolutional layers.

6.2   CIFAR-10 and CIFAR-100 Dataset

CIFAR-10 and CIFAR-100 datasets are image classification dataset which consist of 10 and 100 classes respectively. Each dataset has 50,000 training images and 10,000 test images with 32×32 size. For the CIFAR datasets, we reported the classification error of a single model for each method. To ensure convergence of models, we trained CNNs for 250 epochs with the initial learning rate of 0.02 and the batch size of 128. The learning rate is decreased by 0.5 for every 25 epochs.

The classification accuracy is illustrated in Table 3. In CIFAR-10, channel-wise max-drop showed better result than dropout. Note that despite the high drop probability, the actual drop probability of channel-wise max-drop is very small, about 0.01 for the first convolutional layer and about 0.001 for the last convolutional layer. The result verifies that dropping only high activations results in similar regularization effect to random drop. Also, unlike MNIST experiment, stochastic dropout of normal distribution with zero mean showed best performance. One possible interpretation is that giving different levels of noise to the input of the convolutional layers makes the layers robust to intra-class variations, thus obtaining enhanced generalization power.

Table 3. Classification error on CIFAR-10 dataset.

Method	Classification error (%)
Baseline	16.84%
Dropout ($p = 0.1$)	12.22%
Spatial dropout ($p = 0.05$)	13.78%
Feature-wise max-drop ($p = 0.2$)	12.55%
Channel-wise max-drop ($p = 0.7$)	12.00%
Stochastic dropout ($N(0.0, 0.2)$)	**11.79%**
Stochastic dropout ($U(0.0, 0.4)$)	12.86%

Next, we compared the progress of training for baseline, dropout, and channel-wise max-drop models. The losses on the training set and the test set, and the accuracies on the test set is illustrated in Fig. 4. The training losses are plotted in log-scale (Fig. 4(a)). It is shown that the training loss of dropout and max-drop fluctuates heavily compared to the baseline model since each layer in those models takes noisy inputs. Dropout has larger variations than max-drop since the number of dropped activations is larger. These fluctuations does not affect the test loss or accuracy, as shown in Fig. 4(b). It is interesting that the test loss of max-drop is even higher than the baseline model while maintaining similar accuracy with dropout. Since max-drop drops the highest activation which contains important information, the model is learned to classify an image with less informative feature. This will increase the uncertainty of the prediction, which leads to high softmax loss values.

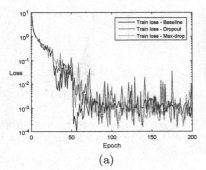 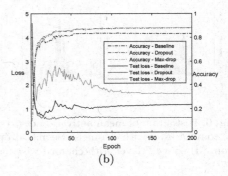

(a)   (b)

Fig. 4. (a) Training error of baseline, dropout ($p = 0.1$), and channel-wise max-drop ($p = 0.7$). (b) Test error and test set accuracy of the models.

We also analyzed the activation behavior of max-drop following the analysis of dropout in Sect. 3. The number of neurons that are never activated at the test time are counted for each regularization method and reported in Fig. 5. Similar to dropout, all regularization methods have little number of never activated neurons for all layers. Thus, it is verified that max-drop helps neurons to learn discriminative features as in the case of dropout.

The histogram of mean activation in the lower and higher convolutional layers for max-drop models are shown in Fig. 6. As observed in Fig. 6(a) and (b), the histogram is bell-shaped in the lower convolutional layer like dropout, which indicates that max-drop also make neurons evenly activated. Meanwhile, for the higher convolutional layer, number of neurons that has mean activation near zero is small unlike either dropout or no regularization case. Max-drop pushes neurons to have similar mean activations, but it does not prefer sparse activations.

To investigate the usefulness of the regularization methods in the specific layers, we trained the model by applying dropout and max-drop only to the lower layers (*conv*1_1 and *conv*1_2) and only to the higher layers (*conv*4_1 and *conv*4_2). The classification errors for both cases are shown in Table 4. Regularization methods improves the network in both lower and higher layers, but the regularization effect is more powerful in the higher layers. We found that high

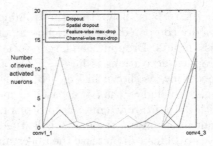

Fig. 5. Number of never activated neurons in the models with dropout, max-drop, and spatial dropout.

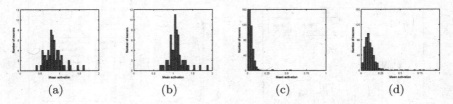

     (a)   (b)   (c)   (d)

Fig. 6. Histogram of mean activation of (a) conv1_1 layer with feature-wise max-drop. (b) conv1_1 layer with channel-wise max-drop. (c) conv4_2 layer with feature-wise max-drop. (d) conv4_2 layer with channel-wise max-drop.

drop ratio is preferable for higher layers, while low drop ratio showed better performance in lower layers. Feature-wise max-drop in higher layers and channel-wise max-drop in lower layers showed better regularization performance. Spatial dropout also proved its effectiveness in higher layers.

Table 4. Effect of regularization in lower and higher convolutional layers.

Method	$conv1$ regularization		$conv4$ regularization	
	Parameter	Classification err. (%)	Parameter	Classification err. (%)
Dropout	$p = 0.05$	15.69	$p = 0.3$	15.14
Spatial dropout	$p = 0.05$	16.16	$p = 0.25$	**14.48**
Feature-wise max-drop	$p = 0.1$	15.93	$p = 0.7$	15.06
Channel-wise max-drop	$p = 0.1$	**15.02**	$p = 0.4$	15.47

Next, we combined the regularization methods with other methods that improves generalization performance. Batch normalization [24] improves training speed and the performance of network by normalizing the activations of each layer in neural networks. We applied the regularization methods after batch normalization is performed. Data augmentation is also a simple way to grant generalization power to neural networks. Following the previous works [4,8,25], we applied data augmentation to training data by padding 4 pixels on all sides of images and by flipping images horizontally. The classification errors are shown in Table 5. With batch normalization, dropout showed the best performance. After activations are normalized, it seems that the importance of maximum value is decreased, which leads to the poor generalization performance of max-drop compared to dropout or spatial dropout. With data augmentation, spatial dropout showed the smallest error, but the improvement of all regularization methods from the baseline is very small. The result indicates that data augmentation imposes generalization power to the CNNs which make the regularization methods less effective.

Table 5. Effect of regularization when combined with batch normalization and data augmentation.

Method	With batch normalization		With data augmentation	
	Parameter	Classification err. (%)	Parameter	Classification err. (%)
Baseline	-	12.10	-	8.01
Dropout	$p = 0.1$	**9.85**	$p = 0.05$	8.54
Spatial dropout	$p = 0.15$	10.69	$p = 0.05$	**7.17**
Feature-wise max-drop	$p = 0.2$	10.67	$p = 0.2$	7.73
Channel-wise max-drop	$p = 0.2$	11.15	$p = 0.4$	7.49

Recently, deep residual learning [23] enabled training of very deep networks. To investigate the regularization performance in the very deep CNNs, we trained 32-layer ResNet on CIFAR-10 dataset. We followed the training procedure and hyper parameters selection from [23] without data augmentation. The result is illustrated in Table 6. All of the tested methods showed superior performance over the baseline with a margin of $2 \sim 4\%$ except spatial dropout. This indicates that dropout and max-drop is still effective for regularizing very deep networks.

Table 6. Classification error on CIFAR-10 dataset using ResNet-32.

Method	Classification error (%)
Baseline	12.84%
Dropout ($p = 0.1$)	9.14%
Spatial dropout ($p = 0.1$)	16.33%
Feature-wise max-drop ($p = 0.2$)	10.72%
Channel-wise max-drop ($p = 0.1$)	11.15%

Lastly, we evaluated the regularization methods on CIFAR-100 dataset. The dataset has much less training samples for each class than CIFAR-10. The classification errors without data augmentation are shown in Table 7. Regularization effect is much stronger than CIFAR-10 mainly due to the small amount of training samples, which reduced the classification error up to 15%. Without batch normalization, max-drop methods outperforms dropout. When batch normalization is used, dropout shows more improvements.

Table 7. Classification errors on CIFAR-100 dataset.

Method	W/O batch normalization		W/batch normalization	
	Parameter	Classification err. (%)	Parameter	Classification err. (%)
Baseline	-	50.26	-	38.84
Dropout	$p = 0.3$	37.23	$p = 0.15$	**32.46**
Spatial dropout	$p = 0.15$	42.07	$p = 0.1$	35.28
Feature-wise max-drop	$p = 0.4$	36.22	$p = 0.2$	34.27
Channel-wise max-drop	$p = 0.7$	**35.33**	$p = 0.3$	34.71

Table 8. Classification errors on SVHN dataset.

Method	W/O batch normalization		W/batch normalization	
	Parameter	Classification err. (%)	Parameter	Classification err. (%)
Baseline	-	2.46	-	2.34
Dropout	$p = 0.25$	2.46	$p = 0.1$	**2.02**
Spatial dropout	$p = 0.05$	2.58	$p = 0.15$	2.07
Feature-wise max-drop	$p = 0.4$	**2.29**	$p = 0.2$	2.14
Channel-wise max-drop	$p = 0.4$	2.30	$p = 0.7$	2.28

6.3 SVHN Dataset

SVHN dataset contains much more training samples than the previous datasets. The dataset consists of over 600,000 training images and 26,032 test images. We trained CNN for 15 epochs with the initial learning rate of 0.01 and the batch size of 128 for the experiments on SVHN dataset. The learning rate is decreased by 0.1 for every 5 epochs. Data augmentation is not applied. The classification errors are reported in Table 8. Huge number of training samples weakens the effect of regularization. Without batch normalization, max-drop methods showed a small improvement, while dropout and spatial dropout worsen the performance of the network. Dropout showed the best performance when batch normalization is applied.

7   Conclusion

We have investigated and verified the usefulness of dropout-like methods in convolutional layers. Usage of dropout in convolutional layers is justified by looking into the activation behavior of neurons. Regularization effect in the convolutional layers is strong when training samples are small and when data augmentation is not used. Also, newly-proposed max-drop and stochastic dropout methods showed competitive results to the conventional dropout, which implies that these methods can substitute dropout in convolutional layers of CNNs. Max-drop layer can be generalized such as dropping largest k activations or suppress the activations by multiplying constant value instead of dropping them to zero. We expect that carefully adjusted parameters may increase the performance.

Acknowledgement. This research was supported by Basic Research Program through the National Research Foundation of Korea (NRF-2016R1A1A1A05005442).

References

1. Srivastava, N., Hinton, G., Krizhevsky, A., Sutskever, I., Salakhutdinov, R.: Dropout: a simple way to prevent neural networks from overfitting. J. Mach. Learn. Res. **15**, 1929–1958 (2014)
2. Tompson, J., Goroshin, R., Jain, A., LeCun, Y., Bregler, C.: Efficient object localization using convolutional networks. In: Proceedings of the IEEE Conference on Computer Vision and Pattern Recognition, pp. 648–656 (2015)
3. Zeiler, M.D., Fergus, R.: Stochastic pooling for regularization of deep convolutional neural networks. arXiv preprint arXiv:1301.3557 (2013)
4. Goodfellow, I., Warde-Farley, D., Mirza, M., Courville, A., Bengio, Y.: Maxout networks. In: Proceedings of the 30th International Conference on Machine Learning, pp. 1319–1327 (2013)
5. Springenberg, J.T., Dosovitskiy, A., Brox, T., Riedmiller, M.: Striving for simplicity: the all convolutional net. arXiv preprint arXiv:1412.6806 (2014)
6. Graham, B.: Spatially-sparse convolutional neural networks. arXiv preprint arXiv:1409.6070 (2014)
7. Wan, L., Zeiler, M., Zhang, S., Cun, Y.L., Fergus, R.: Regularization of neural networks using dropconnect. In: Proceedings of the 30th International Conference on Machine Learning (ICML 2013), pp. 1058–1066 (2013)
8. Lin, M., Chen, Q., Yan, S.: Network in network. arXiv preprint arXiv:1312.4400 (2013)
9. Lee, C.Y., Gallagher, P.W., Tu, Z.: Generalizing pooling functions in convolutional neural networks: mixed, gated, and tree. arXiv preprint arXiv:1509.08985 (2015)
10. Wu, H., Gu, X.: Towards dropout training for convolutional neural networks. Neural Netw. **71**, 1–10 (2015)
11. Neelakantan, A., Vilnis, L., Le, Q.V., Sutskever, I., Kaiser, L., Kurach, K., Martens, J.: Adding gradient noise improves learning for very deep networks. arXiv preprint arXiv:1511.06807 (2015)
12. Audhkhasi, K., Osoba, O., Kosko, B.: Noise-enhanced convolutional neural networks. Neural Netw. **78**, 15–23 (2016)

13. Gulcehre, C., Moczulski, M., Denil, M., Bengio, Y.: Noisy activation functions. arXiv preprint arXiv:1603.00391 (2016)
14. Huang, Y., Sun, X., Lu, M., Xu, M.: Channel-max, channel-drop and stochastic max-pooling. In: Proceedings of the IEEE Conference on Computer Vision and Pattern Recognition Workshops, pp. 9–17 (2015)
15. Krizhevsky, A., Hinton, G.: Learning multiple layers of features from tiny images (2009)
16. Simonyan, K., Zisserman, A.: Very deep convolutional networks for large-scale image recognition. arXiv preprint arXiv:1409.1556 (2014)
17. Maas, A.L., Hannun, A.Y., Ng, A.Y.: Rectifier nonlinearities improve neural network acoustic models. In: Proceedings of the ICML, vol. 30, p. 1 (2013)
18. He, K., Zhang, X., Ren, S., Sun, J.: Delving deep into rectifiers: surpassing human-level performance on imagenet classification. In: Proceedings of the IEEE International Conference on Computer Vision, pp. 1026–1034 (2015)
19. Zeiler, M.D., Fergus, R.: Visualizing and understanding convolutional networks. In: Fleet, D., Pajdla, T., Schiele, B., Tuytelaars, T. (eds.) ECCV 2014. LNCS, vol. 8689, pp. 818–833. Springer, Heidelberg (2014). doi:10.1007/978-3-319-10590-1_53
20. Jia, Y., Shelhamer, E., Donahue, J., Karayev, S., Long, J., Girshick, R., Guadarrama, S., Darrell, T.: CAFFE: convolutional architecture for fast feature embedding. arXiv preprint arXiv:1408.5093 (2014)
21. LeCun, Y., Bottou, L., Bengio, Y., Haffner, P.: Gradient-based learning applied to document recognition. Proc. IEEE **86**, 2278–2324 (1998)
22. Netzer, Y., Wang, T., Coates, A., Bissacco, A., Wu, B., Ng, A.Y.: Reading digits in natural images with unsupervised feature learning. In: NIPS Workshop on Deep Learning and Unsupervised Feature Learning, Granada, Spain, vol. 2011, p. 4 (2011)
23. He, K., Zhang, X., Ren, S., Sun, J.: Deep residual learning for image recognition. In: The IEEE Conference on Computer Vision and Pattern Recognition (CVPR) (2016)
24. Ioffe, S., Szegedy, C.: Batch normalization: accelerating deep network training by reducing internal covariate shift. In: Proceedings of the 32nd International Conference on Machine Learning, pp. 448–456 (2015)
25. Lee, C.Y., Xie, S., Gallagher, P., Zhang, Z., Tu, Z.: Deeply-supervised nets. In: Proceedings of the Eighteenth International Conference on Artificial Intelligence and Statistics, pp. 562–570 (2015)

Efficient Model Averaging for Deep Neural Networks

Michael Opitz[✉], Horst Possegger, and Horst Bischof

Institute for Compute Graphics and Vision,
Graz University of Technology, Graz, Austria
{michael.opitz,possegger,bischof}@icg.tugraz.at

Abstract. Large neural networks trained on small datasets are increasingly prone to overfitting. Traditional machine learning methods can reduce overfitting by employing bagging or boosting to train several diverse models. For large neural networks, however, this is prohibitively expensive. To address this issue, we propose a method to leverage the benefits of ensembles without explicitly training several expensive neural network models. In contrast to Dropout, to encourage diversity of our sub-networks, we propose to maximize diversity of individual networks with a loss function: DivLoss. We demonstrate the effectiveness of DivLoss on the challenging CIFAR datasets.

1 Introduction

Ensemble methods such as bagging [1], boosting (e.g. [2]), or more specifically Random Forests [3], have shown great success in improving generalization performance of machine learning methods. They combine several diverse classifiers to a single predictor, e.g. by averaging their responses. This reduces the generalization error compared to the individual classifiers, since an ensemble of diverse classifiers reduces the variance term in the bias-variance trade-off.

Unfortunately, traditional ensembling methods such as bagging or boosting are prohibitively expensive for neural networks. Large neural networks need several days to train, e.g. [4,5]. Further, especially for real-world applications with real-time requirements, evaluating an ensemble of several networks at test time is computationally too expensive. Additionally, for systems with low memory capacity, such as embedded systems, employing large ensembles is infeasible.

Previous work [6] proposes Dropout to randomly omit neurons of the hidden layers to implement efficient model averaging for neural networks. This can be interpreted as an efficient combination of an exponential number of different neural networks. However, we found that individual sub-networks trained by Dropout have low diversity. This is due to the fact that these sub-networks share all their parameters with each other and only rely on random feature subsampling to encourage diversity. Further, Dropout is applied on the hidden layers of a neural network.

© Springer International Publishing AG 2017
S.-H. Lai et al. (Eds.): ACCV 2016, Part II, LNCS 10112, pp. 205–220, 2017.
DOI: 10.1007/978-3-319-54184-6_13

In contrast to this we focus on efficient model averaging on the output layer of a neural network. To this end, we divide the last hidden layer of a neural network into several, possibly overlapping, groups and optimize a loss function for each of the groups individually rather than over the full output layer. We group the neurons during training such that the ensemble can be mapped back to a regular neural network. By doing this no additional computational cost is incurred at runtime. Instead of relying on sub-sampling of training samples and features, as done by Dropout, we propose a loss function to maximize diversity of our individual network predictors. With this loss function we can effectively balance diversity and discriminativeness of our sub-networks and achieve competitive accuracy to Dropout. We name our method DivLoss.

As our experiments show, sub-networks trained with DivLoss have a larger diversity compared to sub-networks trained with Dropout. Further, we demonstrate that our method can outperform Dropout on the CIFAR-10 and CIFAR-100 datasets. Finally, we show that our method benefits from the decorrelation of hidden units, similar to [7].

The remainder of this paper is structured as follows. In Sect. 2 we discuss related work. Next, in Sect. 3 we review preliminaries on learning theory and introduce our DivLoss. In Sect. 4 we demonstrate effectiveness of our method in several experiments.

2   Related Work

Improving performance of neural networks for supervised learning problems has recently received a lot of attention from the research community. There is a lot of work which is complementary to our method.

A simple, yet effective way to improve accuracy is data augmentation, e.g. [4]. During training, before showing an input sample to the network, a transformation can be applied on the training sample, which preserves the label of the sample. For example mirroring, crops, affine transformations and photometric transformations can be used for image categorization.

Another way to improve neural networks are activation functions. Recently proposed activation functions are more expressive than standard activation functions such as sigmoid or tanh, or are presumably easier to optimize than standard regularization functions, e.g. [8–12].

Since deeper networks are exponentially more expressive than shallow networks, and training very deep networks is challenging due to exploding and vanishing gradients [13], there is a line of work which focuses on enabling training of deeper neural networks. These methods add residual connections or use gating functions from lower to higher layers to enable a better gradient flow in the network and reduce the vanishing and exploding gradient problem [14,15].

Further, some recent contributions focus on improving optimization algorithms for training deep neural networks. They propose accelerated first-order gradient methods specifically designed for neural networks, e.g. [16–19]. These methods focus on reducing the training time (i.e. fewer iterations), and presumably let the network converge to a better local minimum. Additionally, Ioffe et al. [20] leverage batch statistics to normalize inputs to activation functions. This reduces the internal covariate shift and significantly accelerates training.

Further, there are methods which aim to improve the weight initialization of neural networks. This is especially useful for training very large neural networks, as these models do not converge if the weight initialization is not carefully tuned, e.g. [9, 21–23].

Since networks presumably perform better if their hidden features are discriminative, several methods propose auxiliary loss layers on top of hidden layers to regularize neural networks [24, 25]. Further, Cogswell et al. [7] use auxiliary functions to decorrelate hidden neurons. This enables the network to learn more diverse features and reduces redundancy in the representation of deep networks.

Some methods change the structure of the networks, e.g. by adding additional 1×1 convolutions on top of convolutional layers [26], using layers of multiple scales [27], adding an "Inception" layer, consisting of convolutions of different sizes combined with max-pooling [24] or replacing 5×5 convolutions with 3×3 convolutions [5].

Closely related to our method are contributions which leverage the benefits of ensembles to improve generalization performance of neural networks. Recently, Hinton et al. [28] propose to leverage the "dark knowledge" of neural networks to train a network on the predictions of an ensemble to improve accuracy of the new model. The ensemble predictions are used as soft-labels in combination with the original labels to train a new network achieving better accuracy compared to individual networks of the ensemble. This idea is extended by Romero et al. [29] to train a wide teacher network and a smaller network, which mimics the predictions of the teacher network on the output and hidden layers. In contrast to this kind of work, we leverage the benefits of ensembles without explicitly training several full networks to improve performance of a single neural network. We argue that our method is complementary to these approaches, as better individual predictors result in better ensemble performance. This results in more accurate soft-labels which are useful for these methods.

Another promising line of research focuses on improving accuracy by efficient model averaging. The most prominent work is Dropout [6], which randomly omits hidden units from the network during training. Wan et al. [30] generalize this idea to randomly omit weights of the network during training. Stochastic Pooling [31] introduces a pooling method which samples the activations of the receptive fields, rather than just taking the max or the mean. These methods rely on random noise to increase diversity of neural networks. In contrast to these methods we propose a loss function to increase diversity.

Most closely related to our work is the pioneering work of negative correlation learning [32], which also uses a loss function to reduce correlation of different networks in an ensemble. However, networks trained with negative correlation learning do not share parameters, which is prohibitively expensive for training large neural networks. Further, negative correlation learning focuses on non-computer vision related regression problems and penalizes the correlation of predictions. We show that optimizing cross entropy can achieve better accuracy compared to negative correlation learning for computer vision related classification tasks. Further, we compare this method in a more modern setting, with larger networks, larger datasets and recent contributions such as ReLU activations or Dropout.

3   Towards Efficient Neural Network Ensembles

Given a fully annotated dataset, we want to efficiently train a neural network ensemble to obtain a single highly accurate neural network model. However, training and evaluating several independent neural networks on a dataset is computationally expensive, especially for very large networks. Hence, we propose to share most parameters between the individual models, as illustrated in Fig. 1. We divide the last hidden layer into several groups, which is indicated by the respective color. Groups might overlap and share parameters with each other. Further, in contrast to standard ensembles, we train our network ensemble jointly and not sequentially. With this strategy expensive computations for shared parameters can be re-used among different neural network models. Additionally, due to parameter sharing, we can map our networks back to a regular neural network at test time. Hence, DivLoss does not impose any additional computational cost at test time.

As we will discuss in Sect. 3.1, one key-requirement for ensembles is to reduce correlation among individual models and make them diverse. However, by sharing the feature representation as well as the training set, the individual classifiers

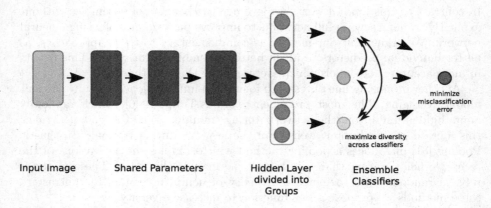

Fig. 1. We divide the last hidden layer into several possibly overlapping groups. Neurons of each group are combined into a classifier. For each of these classifiers we separately optimize a loss which minimizes the training error (e.g. cross-entropy). To increase diversity of the classifiers we add a separate loss between classifiers. (Color figure online)

will make highly correlated decisions. To address this issue, we propose to maximize the pairwise cross entropy between different classifiers of the ensemble. As we will see, this increases the diversity of classifiers and improves generalization performance.

3.1 Learning Theory

One well-known theoretical result in machine learning is the bias-variance trade-off, e.g. [33]. It states that the generalization error can be decomposed into a bias and variance term. Here, we briefly review the main results of Ueda et al. [34], which analyze the bias-variance trade-off in context of neural network ensembles. For the sake of clarity, we stick to the notation introduced by Ueda et al.

The purpose of learning methods is to construct a model $f(x; \theta)$ that approximates an unknown target function $g(x)$. θ is a parameter vector which is estimated by leveraging a set of i.i.d. samples $z^N = \{z_1, z_2, \ldots, z_N\}$, where $z_i = (x_i, y_i)$, $x_i \in \mathbb{R}^d$, $y_i \in \mathbb{R}$ and N is the total number of training samples. z^N is the realization of a random sequence $Z^N = \{Z_1, \ldots, Z_N\}$, whose ith component consists of a random vector $Z_i = (X_i, Y_i)$. Hence, each z_i is generated from an unknown joint probability function $p(x, y)$. The parameters θ of the neural network are estimated by an optimization algorithm given the dataset z^N:

$$\hat{\theta}(z^N) = \arg \min_{\theta} \sum_{i=1}^{N} (y_i - f(x_i; \theta))^2 / N. \tag{1}$$

Note that since $\hat{\theta}$ depends on a z^N, the estimated predictor $f(x; \hat{\theta}(z^N))$ is also a realization of a random variable $f(x; Z^N)$. Further, Ueda et al. [34] introduces a new random variable $Z_0 = (X_0, Y_0) \in \mathbb{R}^{d+1}$, which has a distribution identical to that of Z_i, but is independent of Z_i for all i. The generalization error of the estimator can then be defined as

$$GErr(f) = E_{Z^N} \left\{ E_{Z_0} \left\{ \left[Y_0 - f(X_0; Z^N) \right]^2 \right\} \right\}, \tag{2}$$

where $E_{Z_0}\{\cdot\}$ and $E_{Z^N}\{\cdot\}$ denotes expectation with respect to the distribution Z_0 and Z^N, respectively. This generalization error decomposes into the well-known bias and variance terms

$$GErr(f) = E_{X_0} \left\{ Var\{f|X_0\} + Bias\{f|X_0\}^2 \right\} + \sigma^2, \tag{3}$$

where σ^2 denotes the irreducible noise, $Var\{f|X_0 = x_0\}$ and $Bias\{f|X_0 = x_0\}$ are conditional variance and bias given $X_0 = x_0$

$$Var\{f|X_0\} = E_{Z^N} \left\{ \left(f(X_0; Z^N) - E_{Z^N}\{f(X_0; Z^N)\} \right)^2 \right\}, \tag{4}$$

$$Bias\{f|X_0\} = E_{Z^N} \left\{ f(X_0; Z^N) \right\} - g(X_0). \tag{5}$$

For an ensemble, let $f_1, f_2, \ldots, f_M$ denote M estimators, where the mth estimator is separately trained on $z_{(m)}^N$, i.e. the training set for the mth estimator, $m = 1, \ldots, M$. The output of the ensemble is the average of the estimators:

$$f_{ens}^{(M)}(x) = \frac{1}{M} \sum_{m=1}^{M} f_m(x; z_{(m)}^N) \tag{6}$$

Ueda et al. [34] derive the following generalization error of the ensemble estimator

$$GErr(f_{ens}^{(M)}) = E_{X_0} \left\{ \frac{1}{M} \overline{Var}(X_0) + \left(1 - \frac{1}{M} \right) \overline{Cov}(X_0) + \overline{Bias}(X_0)^2 \right\} + \sigma^2, \tag{7}$$

where $\overline{Var}(\cdot)$, $\overline{Bias}(\cdot)$ and $\overline{Cov}(\cdot)$ are variance, bias and covariance of the M estimators, defined as follows

$$\overline{Var}(X_0) = \frac{1}{M} \sum_{m=1}^{M} Var\{f_m | X_0\},$$

$$\overline{Cov}(X_0) = \frac{1}{M(M-1)} \sum_{m} \sum_{m' \neq m} Cov\{f_m, f_{m'} | X_0\},$$

$$\overline{Bias}(X_0) = \frac{1}{M} \sum_{m=1}^{M} Bias\{f_m | X_0\}. \tag{8}$$

Interestingly, the correlation between individual estimators $\overline{Cov}(X_0)$ is part of this generalization bound. Hence, low correlated and diverse classifiers are desirable to achieve good ensemble performance. Similar results were observed for other popular ensemble methods, such as Random Forests. For this specific learning method Breiman et al. [3] show an upper bound on the error which depends on the strength (i.e. inverse proportional to the bias) and correlation between individual models.

Motivated by these results, in addition to reducing the variance term, we aim to reduce $\overline{Cov}(X_0)$ of our ensemble. Unfortunately, directly minimizing $\overline{Cov}(X_0)$ is impossible, since the distribution of $p(x, y)$ is unknown so we cannot compute expectations over it. Ensemble methods typically subsample the training set or features to reduce correlation among estimators [3]. In the context of neural networks these ideas have been leveraged by Dropout [6], which subsamples different hidden features for each network for each training sample. In contrast to this work, we propose using a loss function to increase the diversity among several networks in the following section.

3.2   Efficient Model Averaging for Deep Neural Networks

To create our individual predictors we divide the last hidden layers into several, possibly overlapping, groups and optimize a loss function for each of these groups separately (recall Fig. 1). The architecture of our ensemble method allows mapping it back to a regular neural network at test time. Hence, by our ensemble method, no additional computational cost is incurred at runtime and negligible additional cost is incurred at training time. Training time is dominated by computing the forward and backward passes of the convolution layers.

For the sake of clarity, to avoid cluttering the notation, we here consider only non-overlapping groups of hidden units. To implement overlapping groups we simply share a subset of weights between classifiers. Let x_i denote the activations of the last hidden layer ($i \in \{1 \ldots H\}$) and W the output weight matrix $W \in \mathbb{R}^{H \times D}$ with entries w_{ij}, where H is the number of hidden neurons and D the number of outputs units of the neural network. We group C non-overlapping neurons in the hidden layer to classifiers. For classifiers with softmax activation, we define the logit (i.e. the inputs to the last softmax nonlinearity) c_{bj} of such a classifier as

$$c_{bj} = \sum_{i=(b-1)\cdot C}^{b\cdot C} x_i \cdot w_{ij} + b_{bj}, \tag{9}$$

where b_{bj} denotes the bias term, b is the block index and $j \in \{1, \ldots, C\}$ indicates the output class.

We define the ensemble logit as average of the $B = H/C$ individual classifiers

$$o_j = \frac{1}{B} \sum_{b=1}^{B} c_{bj}. \tag{10}$$

The final classifier output is defined as softmax function over o_j. By setting our method up this way, we can map it back to a regular neural network at test time, hence, imposing no additional runtime overhead. For non-overlapping groups we can push the scaling factor $\frac{1}{B}$ back into the last weight matrix W. For overlapping groups we have to scale weights which are used by multiple classifiers by an appropriate scaling factor. The ensemble prediction can then be computed by a simple forward pass. Note that by setting our network up this way, it corresponds to taking the geometric mean of the individual classifier softmax outputs and re-normalizing them to a probability distribution:

$$\sigma(o_j) = \frac{e^{\frac{1}{B}\cdot\sum_{b=1}^{B} c_{bj}}}{Z} = \frac{\left(\prod_{b=1}^{B} e^{c_{bj}}\right)^{\frac{1}{B}}}{Z} \tag{11}$$

$$= \frac{\left(\prod_{b=1}^{B} \sigma(c_{bj}) \cdot Z_b\right)^{\frac{1}{B}}}{Z} = \frac{\left(\prod_{b=1}^{B} \sigma(c_{bj})\right)^{\frac{1}{B}}}{\hat{Z}},$$

where Z_b denotes the normalization for the softmax activation of the bth classifier, Z denotes the normalization for the softmax of the classifier ensemble and

$$\hat{Z} = \frac{Z}{\left(\prod_{b=1}^{B} Z_b\right)^{\frac{1}{B}}} = \frac{\sum_{j=1}^{D}\left[\left(\prod_{b=1}^{B} Z_b\right)^{\frac{1}{B}}\left(\prod_{b=1}^{B}\sigma(c_{bj})\right)^{\frac{1}{B}}\right]}{\left(\prod_{b=1}^{B} Z_b\right)^{\frac{1}{B}}} \tag{12}$$

$$= \sum_{j=1}^{D}\left(\prod_{b=1}^{B}\sigma(c_{bj})\right)^{\frac{1}{B}}.$$

We see that the geometric mean of the normalizations of the individual classifier, i.e. $\left(\prod_{b=1}^{B} Z_b\right)^{\frac{1}{B}}$, is independent of j, can be pulled out of the sum and cancels with the denominator. Hence, the ensemble output is proportional to the geometric mean of the responses of the individual classifiers.

We want both, our final ensemble and our individual classifiers to be discriminative on our training set. To this end, we minimize the cross entropy on both, the ensemble predictions and the predictions of the individual classifiers by introducing the loss

$$\mathcal{L}_{discr} = \sum_{i=1}^{N}\left(\mathcal{H}(y^{(i)}, \sigma(o^{(i)})) + \lambda_{parts}\cdot\left(\frac{1}{B}\sum_{b=1}^{B}\mathcal{H}(y^{(i)}, \sigma(c_b^{(i)}))\right)\right), \tag{13}$$

where N is the total number of training samples and $y^{(i)}$ is the label of the ith training sample. With a slight abuse of notation $\sigma(o^{(i)})$ denotes the softmax activations of the full ensemble for the ith sample, $\sigma(c_b^{(i)})$ denotes the softmax activation for the ith sample of the bth classifier. The parameter λ_{parts} is a hyperparameter which balances the influence of the individual classifiers and the ensemble and is set by (cross-)validation. We typically sweep it out on a log scale, i.e. $2^{\{0,1,2,3\}}$. Finally, $\mathcal{H}(p, q)$ denotes the cross entropy between probability distributions p and q.

3.3   Enforcing Diversity

Naively applying Eq. (13) to a learning problem will result in several individual classifiers, which all have highly correlated predictions. Hence, according to the bias-variance-correlation trade-off there is no benefit in such a setup. To address this problem, we propose to *maximize* the cross entropy between all classifier pairs. Cross entropy is employed in logistic regression and in most neural networks for classification as loss function. It measures the dissimilarity between two probability distributions and is typically used to minimize dissimilarity between ground-truth label and predicted label in supervised learning problems.

In contrast to that, to encourage diversity for different classifiers, we propose to maximize the cross entropy (i.e. maximize dissimilarity or minimize similarity) between all pairs of classifiers. More formally, we define the following loss function:

$$\mathcal{L}_{diversity} = \frac{1}{B \cdot (B-1)} \sum_{i=1}^{N} \sum_{b=1}^{B} \sum_{b' \neq b} -\mathcal{H}(\sigma(c_b^{(i)}), \sigma(c_{b'}^{(i)})), \tag{14}$$

where N is the number of samples, B the number of classifiers, $\sigma(c_b^{(i)})$ denotes the output for the ith sample from the bth classifier and $\mathcal{H}$ is the cross entropy between the two classifiers.

Our final loss function $\mathcal{L}$ is a combination of $\mathcal{L}_{discr}$ and $\mathcal{L}_{diversity}$:

$$\mathcal{L} = \mathcal{L}_{discr} + \lambda_{diversity} \cdot \mathcal{L}_{diversity} \tag{15}$$

where $\lambda_{diversity}$ is a hyperparameter, balancing the influence of the diversity loss and the discriminative loss. The parameter is set by (cross-)validation on a log scale, i.e. $10^{\{2,3,4\}}$. We call this loss function DivLoss, as it encourages diversity between individual predictors of an ensemble.

3.4   Loss Function on Hidden Layers

Compared to Dropout, our method is applied only on the output layer of a neural network, and not on an arbitrary hidden layer. For very large networks, however, it might be beneficial to apply regularization already on top of hidden layers. To address this issue, inspired by deeply supervised networks [24,25], we propose to apply our ensemble layer on top of intermediate hidden layers as auxiliary layer (see Fig. 2). During training time, we can divide any hidden layer of a network into possibly overlapping groups, as indicated by the corresponding color, and optimize our loss on them. The next layer in the regular feed forward pass receives all neurons from this hidden layer as input (i.e. it does not operate on individual groups).

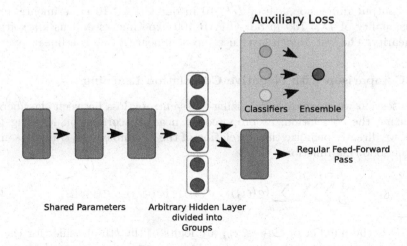

Fig. 2. We can apply our method on top of any hidden layer in a neural network.

Note that this setup does not introduce any additional computational cost during test time, since these auxiliary layers are just used during training time, and not during test time. In Sect. 4.3 we show that this setup can indeed improve accuracy.

4   Evaluation

In this section we provide a detailed evaluation of our method on CIFAR-10 and CIFAR-100 [35]. These datasets each consist of 50,000 training images and 10,000 test images of size 32×32. CIFAR-10 has 10 object classes, wheras CIFAR-100 has 100 object classes. Both datasets have a uniform class distribution, i.e. there are 6,000 images per class in CIFAR-10, from which 1,000 are in the test set, and 600 images per class in CIFAR-100, from which 100 are in the test set. For pre-processing, following [7], we subtract the mean of the training set from the images.

We run our experiments on a regular desktop machine with a NVIDIA GTX 770 GPU and a Core i5-4570 CPU with 3.20 GHz and implement our method in Theano [36]. For training parameters (learning rate, momentum, weight decay) we use the standard Caffe learning parameters for the CIFAR-10 Quick archi-tecture. As network architecture we use a larger version of the CIFAR-10 Quick architecture, which is proposed by Cogswell et al. [7]. The architecture is C-$64 \times 5 \times 5$, MP-$3 \times 3(2 \times 2)$, C-$64 \times 5 \times 5$, AP-$3 \times 3(2 \times 2)$, C-$128 \times 5 \times 5$, AP-$3 \times 3(2 \times 2)$, FC-128, FC-128, FC-D, where C-$F \times S \times S$, denotes a convolu-tion layer with F filters of size $S \times S$, MP-$N \times N(S \times S)$ denotes a max pooling layer of size $N \times N$ with stride $S \times S$, AP-$N \times N(S \times S)$ denotes an average pooling layer of size $N \times N$ with stride $S \times S$, and FC-N denotes a fully connected layer of size N. Each layer except the last two fully connected layers are followed by a ReLU nonlinearity. The last fully connected layer is our output layer, which has an output dimensionality of $D = 10$ in our CIFAR-10 experiments and a dimensionality of $D = 100$ in our CIFAR-100 experiments and uses a softmax nonlinearity. The last *hidden* layer uses no nonlinearity (i.e. is a linear layer).

4.1   Comparison with Negative Correlation Learning

In this section we compare maximizing cross-entropy loss between classifiers to minimizing the correlation, as proposed by negative correlation learning [32]. Hence, we directly penalize the correlation of the ensemble for a training sample with the following function:

$$\frac{1}{D} \sum_{j=1}^{D} \sum_{b=1}^{B} \sum_{b' \neq b} (\sigma(c_{bj}) - \sigma(o_j)) \cdot (\sigma(c_{b'j}) - \sigma(o_j)), \tag{16}$$

where D is the number of classes, c_{bj} are logits of the bth classifier for the jth class and o_j is the logit of the ensemble output. We also experimented by directly penalizing the logits as opposed to their softmax activations, but achieved best results with the above formulation. We compare negative correlation learning to the cross entropy loss and the absolute correlation, i.e.:

$$\frac{1}{D}\sum_{j=1}^{D}\sum_{b=1}^{B}\sum_{b'\neq b}|(\sigma(c_{bj}) - \sigma(o_j)) \cdot (\sigma(c_{b'j}) - \sigma(o_j))|. \qquad (17)$$

We run the experiments on CIFAR-10 and summarize the results in Table 1. We observe more accurate results for maximizing cross-entropy than for minimizing correlation. We hypothesize that this is because cross-entropy is a more natural fit to measure diversity for classifiers which output a probability distribution compared to correlation. Further, penalizing the absolute value of the correlation works better for classification problems, since it encourages classifers to be weakly correlated, as opposed to negatively correlated.

Table 1. Comparison to negative correlation learning on CIFAR-10.

Method	Test acc.
Cross entropy	**82.3**
Absolute correlation	81.28
Negative correlation	80.9
Baseline	80.86

4.2 Diversity

In this section we analyze the effect on the diversity of our ensemble and compare it to a network trained with Dropout. To measure diversity, we count the number of disagreements of all classifier pairs.

$$\frac{0.5}{B \cdot (B-1)}\sum_{b=1}^{B}\sum_{b'\neq b}\frac{1}{N}\sum_{i=1}^{N}f_b(x_i) \neq f_{b'}(x_i), \qquad (18)$$

where B is the number of classifiers and $f_b(x_i)$ is the bth classifier output for the ith sample (i.e. the label prediction). The higher this number, the more diverse the classifier outputs are.

Since Dropout is an approximate average of an exponential number of neural networks, we sub-sample 16 sub-networks and analyze their correlation on the validation set. We execute this experiment 10 times and compare this to a network trained with our method consisting of 16 sub-networks on CIFAR-10. We report the diversity of individual sub-networks, the average accuracy of these 16 individual sub-networks and the accuracy of the full ensemble in Table 2. Since for the Dropout experiments, we sub-sample 16 sub-networks and repeat the experiment 10 times, we report mean and standard deviation of the diversity and the average accuracy of the individual sub-networks.

Interestingly, in Table 2 we see that our method trains sub-networks which are more diverse on the validation set compared to Dropout. Further, individual sub-networks are less accurate compared to Dropout, but complement each other better, since we tie them together with a global loss.

Table 2. Diversity of a network trained with Dropout and our method on CIFAR-10.

Method	Diversity	Avg. sub-network acc.	Ensemble acc.
Dropout	0.071 ± 0.0025	$\mathbf{0.799} \pm 0.00073$	81.07
Ours	**0.240**	0.744	**82.3**

4.3 CIFAR-10

In this section, we evaluate our method on the CIFAR-10 dataset. To make a fair comparison, we use the same architecture as proposed in [7], i.e. we double the number of hidden units and convolution filters of the Caffe 10 Quick architecture and add an additional fully connected layer to our network. This architecture will be denoted "Baseline". We split the training set into 10,000 validation images and 40,000 training images to determine hyperparameters (i.e. our weighting parameter, dropout rates, DeCov [7] hyperparameters). Our method benefits from a large number of non-overlapping groups, as diversity can be easier maximized if no parameters are shared among groups. To enable a fair comparison to existing work, we fix the hidden layer size to 128. We divide the last hidden layer into non-overlapping groups with 8 hidden units, as sub-networks with a smaller number of hidden units fail in our experiments to learn anything meaningful. As in [7] our network takes 32×32 patches as input and we do not apply any kind of data augmentation. We shuffle the training dataset after each epoch and employ early stopping.

We apply our ensembling method on top of the output layer of the neural network and report our results in Table 3. We see that our method can significantly outperform Dropout [6] and achieves similar results to DeCov [7]. Additionally, our method can benefit from DeCov as well as Dropout. We hypothesize that DeCov helps a neural network to develop more decorrelated features, which help building more diverse classifiers. With Dropout the generalization performance of individual networks of our ensemble increases, hence the performance of the full ensemble improves. For a fair comparison, we also apply re-shuffling and early stopping to DeCov [7], which improves the overall accuracy by 0.38.

To show that our method works on auxiliary layers, we additionally apply our loss on the first fully connected layer. We observe a notable increase in accuracy from 82.3 to **83.44** for our method.

4.4 CIFAR-100

We re-use the same network architecture for the CIFAR-100 experiment. Since the number of hidden units (128) is quite small compared to the number of

Table 3. CIFAR-10 classification accuracy.

Method	Test acc.
DeCov [7]	81.68
DeCov + re-shuffling and early stopping	82.06
Baseline	80.86
Dropout	81.07
DivLoss	82.3
DivLoss + Dropout	82.52
DivLoss + Decov	**82.95**

classes, we perform weight sharing for our classifiers. We fix the number of hidden units of a single classifier to 64 and randomly group hidden units to a classifier. We use 16 classifiers in our experiments. Further, we split the dataset into 10,000 validation images and 40,000 training images and determine our hyperparameters (i.e. the weighting parameter, dropout rates, DeCov hyperparameter) on the validation set.

Our results are summarized in Table 4. All 3 regularization methods (DivLoss, Dropout, DeCov) achieve similar results on CIFAR-100 and can significantly improve over a baseline method which just uses weight decay as regularization. Further, we observe that we can combine DivLoss with Dropout and DeCov, to increase accuracy.

Table 4. CIFAR-100 classification accuracy.

Method	Test acc.
DeCov [7]	45.10
DeCov + re-shuffling and early stopping	49.61
Baseline	47.38
Dropout	49.44
DivLoss	49.42
DivLoss + Dropout	49.9
DivLoss + DeCov	**50.08**

When we additionally apply our loss function as auxiliary layer on the first fully connected layer, we observe an increase in accuracy from 49.42 to **50.32**.

5   Conclusion

We proposed an ensemble method which improves the generalization performance of neural networks by efficient model averaging. Motivated by learning theory, we propose to optimize a loss function to increase the diversity of the individual classifiers of the ensemble. Our method can be trained end-to-end with stochastic gradient descent and momentum. Further, we showed that our method outperforms or achieves competitive performance compared to Dropout and DeCov on the challenging CIFAR datasets. Since we setup our method so that it can be mapped back to a regular neural network, no additional runtime cost is incurred at test time. At training time we impose negligible additional runtime cost for computing the responses and the loss for our sub-networks. This overhead is, however, negligible, since most of the time during training is spent computing forward and backward passes of convolution layers.

Our experiments show that our method benefits especially from very wide networks where the number of hidden units is large compared to the number of classes. In such networks diversity of sub-networks can be better maximized as they have less shared parameters.

Compared to Dropout, which is an approximate ensemble of exponentially many classifiers sharing the same parameters, our method relies on a smaller number of classifiers. To enforce diversity, Dropout relies on randomly omitting neurons from the hidden layers. In contrast to that, our method employs a loss function to encourage diversity of individual classifiers. Due to backpropagation, the diversity also affects the hidden layers (i.e. the feature representation) of the network and, similar to Dropout, encourages a diverse feature representation.

Future work will analyze larger sub-networks consisting of several layers with separate (non-shared) weights.

Acknowledgement. This work was supported by the Austrian Research Promotion Agency (FFG) project Vision+.

References

1. Breiman, L.: Bagging predictors. Mach. Learn. (ML) **24**, 123–140 (1996)
2. Freund, Y., Schapire, R.E.: A decision-theoretic generalization of on-line learning and an application to boosting. J. Comput. Syst. Sci. (JCSS) **55**, 119–139 (1997)
3. Breiman, L.: Random forests. Mach. Learn. (ML) **45**, 5–32 (2001)
4. Krizhevsky, A., Sutskever, I., Hinton, G.E.: ImageNet classification with deep convolutional neural networks. In: Advances in Neural Information Processing Systems (NIPS) (2012)
5. Simonyan, K., Zisserman, A.: Very deep convolutional networks for large-scale image recognition. In: Proceedings of the International Conference on Learning Representations (ICLR) (2014)
6. Srivastava, N., Hinton, G., Krizhevsky, A., Sutskever, I., Salakhutdinov, R.: Dropout: a simple way to prevent neural networks from overfitting. J. Mach. Learn. Res. (JMLR) **15**, 1929–1958 (2014)
7. Cogswell, M., Ahmed, F., Girshick, R.B., Zitnick, L., Batra, D.: Reducing overfitting in deep networks by decorrelating representations. In: Proceedings of the International Conference on Learning Representations (ICLR) (2016)
8. Nair, V., Hinton, G.E.: Rectified linear units improve restricted boltzmann machines. In: Proceedings of the International Conference on Machine Learning (ICML) (2010)
9. He, K., Zhang, X., Ren, S., Sun, J.: Delving deep into rectifiers: surpassing human-level performance on imageNet classification. In: Proceedings of the IEEE International Conference on Computer Vision (ICCV) (2015)

10. Maas, A.L., Hannun, A.Y., Ng, A.Y.: Rectifier nonlinearities improve neural network acoustic models. In: Proceedings of the International Conference on Machine Learning (ICML) (2013)
11. Goodfellow, I., Warde-Farley, D., Mirza, M., Courville, A., Bengio, Y.: Maxout networks. In: Proceedings of the International Conference on Machine Learning (ICML) (2013)
12. Clevert, D., Unterthiner, T., Hochreiter, S.: Fast and accurate deep network learning by exponential linear units (ELUs). In: Proceedings of the International Conference on Learning Representations (ICLR) (2016)
13. Hochreiter, S., Bengio, Y., Frasconi, P., Schmidhuber, J.: Gradient flow in recurrent nets: the difficulty of learning long-term dependencies. In: Kremer, S.C., Kolen, J. (eds.) Field Guide to Dynamical Recurrent Neural Networks. IEEE Press (2001)
14. Srivastava, R.K., Greff, K., Schmidhuber, J.: Training very deep networks. In: Advances in Neural Information Processing Systems (NIPS) (2015)
15. He, K., Zhang, X., Ren, S., Sun, J.: Deep residual learning for image recognition. (2015). arXiv:1512.03385
16. Duchi, J., Hazan, E., Singer, Y.: Adaptive subgradient methods for online learning and stochastic optimization. J. Mach. Learn. Res. (JMLR) **12**, 2121–2159 (2011)
17. Tieleman, T., Hinton, G.: Lecture 6.5–RmsProp: divide the gradient by a running average of its recent magnitude. COURSERA: Neural Networks for Machine Learning (2012)
18. Zeiler, M.D.: ADADELTA: an adaptive learning rate method (2012). http://arxiv.org/abs/1212.5701
19. Kingma, D.P., Ba, J.: Adam: a method for stochastic optimization. In: Proceedings of the International Conference on Learning Representations (ICLR) (2015)
20. Ioffe, S., Szegedy, C.: Batch normalization: accelerating deep network training by reducing internal covariate shift. In: Proceedings of the International Conference on Machine Learning (ICML) (2015)
21. Mishkin, D., Matas, J.: All you need is a good in it. In: Proceedings of the International Conference on Learning Representations (ICLR) (2016)
22. Glorot, X., Bengio, Y.: Understanding the difficulty of training deep feedforward neural networks. In: Artificial Intelligence and Statistics Conference (AISTATS) (2010)
23. Krähenbühl, P., Doersch, C., Donahue, J., Darrell, T.: Data-dependent initializations of convolutional neural networks. In: Proceedings of the International Conference on Learning Representations (ICLR) (2016)
24. Szegedy, C., Liu, W., Jia, Y., Sermanet, P., Reed, S., Anguelov, D., Erhan, D., Vanhoucke, V., Rabinovich, A.: Going deeper with convolutions. In: Proceedings of the IEEE Conference on Computer Vision and Pattern Recognition (CVPR) (2015)
25. Lee, C.Y., Xie, S., Gallagher, P., Zhang, Z., Tu, Z.: Deeply-supervised nets. In: Artificial Intelligence and Statistics Conference (AISTATS) (2015)
26. Lin, M., Chen, Q., Yan, S.: Network in network. In: Proceedings of the International Conference on Learning Representations (ICLR) (2014)
27. Sermanet, P., LeCun, Y.: Traffic sign recognition with multi-scale convolutional networks. In: Proceedings of the International Joint Conference on Neural Networks (IJCNN) (2011)
28. Hinton, G., Vinyals, O., Dean, J.: Distilling the knowledge in a neural network (2015). http://arxiv.org/abs/1503.02531

29. Romero, A., Ballas, N., Kahou, S.E., Chassang, A., Gatta, C., Bengio, Y.: Fit-Nets: hints for thin deep nets. In: Proceedings of the International Conference on Learning Representations (ICLR) (2015)
30. Wan, L., Zeiler, M., Zhang, S., Cun, Y.L., Fergus, R.: Regularization of neural networks using dropConnect. In: Proceedings of the International Conference on Machine Learning (ICML), JMLR Workshop and Conference Proceedings (2013)
31. Zeiler, M.D., Fergus, R.: Stochastic pooling for regularization of deep convolutional neural networks. In: Proceedings of the International Conference on Learning Representations (ICLR) (2013)
32. Liu, Y., Yao, X.: Ensemble learning via negative correlation. Neural Netw. **12**, 1399–1404 (1999)
33. Duda, R.O., Hart, P.E., Stork, D.G.: Pattern Classification, 2nd edn. Wiley, New York (2001)
34. Ueda, N., Nakano, R.: Generalization error of ensemble estimators. In: Proceedings of the IEEE International Conference on Neural Networks (ICNN) (1996)
35. Krizhevsky, A.: Learning Multiple Layers of Features from Tiny Images (2009)
36. The Theano Development Team: Theano: a Python framework for fast computation of mathematical expressions (2016). arXiv.org/abs/1605.02688

Joint Training of Generic CNN-CRF Models with Stochastic Optimization

A. Kirillov[1], D. Schlesinger[1([⊠])], S. Zheng[2], B. Savchynskyy[1], P.H.S. Torr[2], and C. Rother[1]

[1] Dresden University of Technology, Dresden, Germany
dmytro.shlezinger@tu-dresden.de
[2] University of Oxford, Oxford, England

Abstract. We propose a new CNN-CRF end-to-end learning framework, which is based on joint stochastic optimization with respect to both Convolutional Neural Network (CNN) and Conditional Random Field (CRF) parameters. While stochastic gradient descent is a standard technique for CNN training, it was not used for joint models so far. We show that our learning method is (i) general, i.e. it applies to arbitrary CNN and CRF architectures and potential functions; (ii) scalable, i.e. it has a low memory footprint and straightforwardly parallelizes on GPUs; (iii) easy in implementation. Additionally, the unified CNN-CRF optimization approach simplifies a potential hardware implementation. We empirically evaluate our method on the task of semantic labeling of body parts in depth images and show that it compares favorably to competing techniques.

1 Introduction

Deep learning have tremendous success since a few years in many areas of computational science. In computer vision, Convolutional Neural Networks (CNNs) are successfully used in a wide range of applications – from low-level vision, like segmentation and optical flow, to high-level vision, like scene understanding and semantic segmentation. For instance in the VOC2012 object segmentation challenge[1] the use of CNNs has pushed the quality score by around 28% (from around 50% to currently around 78% [1]). The main contribution of CNNs is their ability to adaptively fine-tune millions of features to achieve best performance for the task at hand. However, CNNs have also their shortcomings. One limitation is that often a large corpus of labeled training images is necessary. Secondly, it is difficult to incorporate prior knowledge into the CNN architecture. In contrast, graphical models like Conditional Random Fields (CRFs) [2] overcome these two limitations. CRFs have been used to model geometric properties, such as object shape, spatial relationship between objects, global properties like object connectivity, and many others. Furthermore, CRFs designed based on e.g. physical properties are able to achieve good results even with few training images. For

---

[1] http://host.robots.ox.ac.uk:8080/leaderboard.

© Springer International Publishing AG 2017
S.-H. Lai et al. (Eds.): ACCV 2016, Part II, LNCS 10112, pp. 221–236, 2017.
DOI: 10.1007/978-3-319-54184-6_14

these reasons, a recent trend has been to explore the combination of these two modeling paradigms by using a CRF, whose factors are dependent on a CNN. By doing so, CRFs are able to use the incredible power of CNNs, to fine-tune model features. On the other hand, CNNs can more easily capture global properties such as object shape and contextual information. The study of this fruitful combination (sometimes called "deep structured models" [3]) is the main focus of our work. We propose a generic joint learning framework for the combined CNN-CRF models, based on a sampling technique and a stochastic gradient optimization.

Related Work. The idea of making CRF models more powerful by allowing factors to depend on many parameters has been explored extensively over the last decade. One example is the Decision Tree Field approach [4] where factors are dependent on Decision Trees. In this work, we are interested in making the factors dependent on CNNs. Note that one advantage of CNNs over Decision Trees is that CNNs learn the appropriate features for the task at hand, while Decision Trees, as many other classifiers, only combine and select from a pool of simple features, see e.g. [5,6] for a discussion on the relationship between CNNs and Decision Trees. We now describe the most relevant works that combine CNNs and CRFs in the context of semantic segmentation, as one of the largest application areas of this type of models. The framework we propose in this work is also evaluated in a similar scenario, although its theoretical basis is application-independent.

Since CNNs have been used for semantic segmentation, this field has made a big leap forward, see e.g. [7,8]. Recently, the advantages of additionally integrating a CRF model have given a further boost in performance, as demonstrated by many works. To the extent that the work [1] is currently leading the VOC2012 object segmentation challenge, as discussed below. In [9] a fully connected Gaussian CRF model [10] was used, where the respective unaries were supplied by a CNN. The CRF inference was done with a Mean Field approximation. This separate training procedure was recently improved in [11] with an end-to-end learning algorithm. To achieve this, they represent the Mean Field iterations as a Recurrent Neural Network. The same idea was published in [12]. In [10], the Mean Field iterations were made efficient by using a so-called permutohedral lattice approximation [13] for Gaussian filters. However, this approach allows for a special class of pairwise potentials only. Besides the approaches [11] and [12], there are many other works that consider the idea of backpropagation with a so-called unrolled CRF-inference scheme, such as [14–20]. These inference steps mostly correspond to message passing operations of e.g. Mean Field updates or Belief Propagation. However the number of inference iterations in such learning schemes remains their critical parameter: too few iterations lead to a quality deterioration, whereas more iterations slow down the whole learning procedure.

Likelihood maximization is NP-hard for CRFs, which implies that it is also NP-hard for joint CNN-CRF models. To avoid this problem, *piece-wise* learning [21] was used in [1]. Instead of likelihood maximization a surrogate loss is considered which can be minimized efficiently. However, there are no guarantees that minimization of the surrogate loss will lead to maximization of the true likelihood. On the positive side, the method shows good practical results and leads the VOC2012 object segmentation competition at the moment.

Another likelihood approximation, which is based on fractional entropy and a message passing based inference, was proposed in [3]. However, there is no clear evidence that the fractional entropy always leads to tight likelihood approximations. Another point relates to the memory footprint of the method. To avoid the time consuming, full inference, authors of [3] interleave gradient steps w.r.t. the CNN parameters and minimization over the dual variables of the LP-relaxation of the CRF. This allows to solve the issue with a small number of inference iterations comparing to the unrolled inference schemes. However, it requires to store current values of the dual variables for *each* element of a training set. The number of the dual variables is proportional to the number of labels in the used CRF as well as to the number of its pairwise factors. Therefore, the size of such a storage can significantly exceed the size required for the training set itself. We will discuss this point in more details in Sect. 4.

Contribution. Inspired by the contrastive divergence approach [22], we propose a *generic joint maximum likelihood learning framework* for the combined CNN-CRF models. In this context, *"generic"* means that (i) factors in our CRF are of a non-parametric form, in contrast to e.g. [11], where Gaussian pairwise potentials are considered; and (b) we maximize the likelihood itself instead of its approximations. Our framework is based on a sampling technique and stochastic gradient updates w.r.t. both CNN and CRF parameters. To avoid the time consuming, full inference we interleave sampling-based inference steps with CNN parameters updates. In terms of the memory overhead, our method stores only a single (current) labeling for each element of the training set during learning. This requires less memory than the training set itself. Our method is efficient, scalable and highly parallelizable with a low memory footprint, which makes it an ideal candidate for a GPU-based implementation.

We show the efficiency of our approach on the task of semantic labeling of body parts in depth images.

2 Preliminaries

Conditional Random Fields. Let $y = (y_1, \ldots, y_N)$ be a random *state* vector, where each coordinate is a random variable y_i that takes its values from a finite set $\mathcal{Y}_i = \{1, \ldots, |\mathcal{Y}_i|\}$. Therefore $y \in \mathcal{Y} := \prod_{i=1}^N \mathcal{Y}_i$, where $\prod$ stands for a Cartesian product. Let x be *an observation* vector, taking its values in some set $\mathcal{X}$. The *energy function* $E \colon \mathcal{Y} \times \mathcal{X} \times \mathbb{R}^m \to \mathbb{R}$ assigns a score $E(y, x, \theta)$ to a pair (y, x) of a state and an observation vector and is parametrized by a *parameter* vector $\theta \in \mathbb{R}^m$. An exponential posterior distribution related to the energy E reads

$$p(y|x, \theta) = \frac{1}{Z(x, \theta)} \exp(-E(y, x, \theta)). \tag{1}$$

Here $Z(\boldsymbol{x}, \boldsymbol{\theta})$ is a partition function, defined as

$$Z(\boldsymbol{x}, \boldsymbol{\theta}) = \sum_{y \in \mathcal{Y}} \exp(-E(\boldsymbol{y}, \boldsymbol{x}, \boldsymbol{\theta})). \tag{2}$$

Let $I = 1, ..., N$ be a set of *variable indexes* and 2^I denote its powerset. Let also $\mathcal{Y}_A$ stand for the set $\prod_{i \in A} \mathcal{Y}_i$ for any $A \subseteq I$. In CRFs, the energy function E can be represented as a sum of its components depending on the subsets of variables $\boldsymbol{y}_f \in \mathcal{Y}_f$, $f \subset I$:

$$E(\boldsymbol{y}, \boldsymbol{x}, \boldsymbol{\theta}) = \sum_{f \in \mathcal{F} \subset 2^I} \psi_f(\boldsymbol{y}_f, \boldsymbol{x}, \boldsymbol{\theta}). \tag{3}$$

The functions $\psi_f \colon \mathcal{Y}_f \times \mathcal{X} \times \mathbb{R}^m \to \mathbb{R}$ are usually called *potentials*. For example, in [9,11] only CRFs with *unary* and *pairwise* potentials are considered, i.e. $|f| \leq 2$ for any $f \in \mathcal{F}$.

In what follows, we will assume that each ψ_f is potentially a non-linear function of $\boldsymbol{\theta}$ and $\boldsymbol{x}$. It can be defined by e.g. a CNN with the input $\boldsymbol{x}$ and weights $\boldsymbol{\theta}$.

Inference is a process of estimating the state vector $\boldsymbol{y}$ for an observation $\boldsymbol{x}$. There are several inference criteria, see e.g. [23]. In this work we will stick to the so called *maximum posterior marginals*, or shortly *max-marginal* inference

$$y_i^* = \arg \max_{y_i \in \mathcal{Y}_i} p(y_i | \boldsymbol{x}, \boldsymbol{\theta}) := \arg \max_{y_i \in \mathcal{Y}_i} \sum_{(y' \in \mathcal{Y}\colon y_i' = y_i)} p(y' | \boldsymbol{x}, \boldsymbol{\theta}) \qquad \text{for all } i. \tag{4}$$

Though maximization in (4) can be done directly due to the typically small size of the sets $\mathcal{Y}_i$, computing the marginals $p(y_i | \boldsymbol{x}, \boldsymbol{\theta})$ is NP-hard in general. Summation in (4) can not be performed directly due to the exponential size of the set $\mathcal{Y}$. In our framework we approximate the marginals with Gibbs sampling [24]. The corresponding estimates converge to the true marginals in the limit. We detail this procedure in Sect. 3.

Learning. Given a training set $\{(\boldsymbol{x}^d, \boldsymbol{y}^d) \in (\mathcal{X} \times \mathcal{Y})\}_{d=1}^D$, we consider the maximum likelihood learning criterion for estimating $\boldsymbol{\theta}$:

$$\arg \max_{\boldsymbol{\theta} \in \mathbb{R}^m} \sum_{d=1}^D \log p(\boldsymbol{y}^d | \boldsymbol{x}^d, \boldsymbol{\theta}) = \arg \max_{\boldsymbol{\theta} \in \mathbb{R}^m} \sum_{d=1}^D \left[-E(\boldsymbol{y}^d, \boldsymbol{x}^d, \boldsymbol{\theta}) - \log Z(\boldsymbol{x}^d, \boldsymbol{\theta}) \right]. \tag{5}$$

Since a (stochastic) gradient descent is used for CNN training, we stick to it for estimating (5) as well. The gradient of the objective reads:

$$\frac{\partial \sum_{d=1}^{D} \log p(\boldsymbol{y}^d|\boldsymbol{x}^d,\boldsymbol{\theta})}{\partial \boldsymbol{\theta}} = \sum_{d=1}^{D} \left[-\frac{\partial E(\boldsymbol{y}^d,\boldsymbol{x}^d,\boldsymbol{\theta})}{\partial \boldsymbol{\theta}} - \frac{\partial \log Z(\boldsymbol{x}^d,\boldsymbol{\theta})}{\partial \boldsymbol{\theta}} \right]$$

$$= \sum_{d=1}^{D} \left[-\frac{\partial E(\boldsymbol{y}^d,\boldsymbol{x}^d,\boldsymbol{\theta})}{\partial \boldsymbol{\theta}} - \frac{1}{Z(\boldsymbol{x}^d,\boldsymbol{\theta})} \frac{\partial \sum_{y\in\mathcal{Y}} \exp(-E(\boldsymbol{y},\boldsymbol{x}^d,\boldsymbol{\theta}))}{\partial \boldsymbol{\theta}} \right]$$

$$= \sum_{d=1}^{D} \left[-\frac{\partial E(\boldsymbol{y}^d,\boldsymbol{x}^d,\boldsymbol{\theta})}{\partial \boldsymbol{\theta}} + \sum_{y\in\mathcal{Y}} \frac{\exp(-E(\boldsymbol{y},\boldsymbol{x}^d,\boldsymbol{\theta}))}{Z(\boldsymbol{x}^d,\boldsymbol{\theta})} \frac{\partial E(\boldsymbol{y},\boldsymbol{x}^d,\boldsymbol{\theta})}{\partial \boldsymbol{\theta}} \right]$$

$$= \sum_{d=1}^{D} \left[-\frac{\partial E(\boldsymbol{y}^d,\boldsymbol{x}^d,\boldsymbol{\theta})}{\partial \boldsymbol{\theta}} + \sum_{y\in\mathcal{Y}} p(\boldsymbol{y}|\boldsymbol{x}^d,\boldsymbol{\theta}) \frac{\partial E(\boldsymbol{y},\boldsymbol{x}^d,\boldsymbol{\theta})}{\partial \boldsymbol{\theta}} \right]$$

$$= \sum_{d=1}^{D} \left[-\frac{\partial E(\boldsymbol{y}^d,\boldsymbol{x}^d,\boldsymbol{\theta})}{\partial \boldsymbol{\theta}} + \mathbb{E}_{p(\boldsymbol{y}|\boldsymbol{x}^d,\boldsymbol{\theta})} \frac{\partial E(\boldsymbol{y},\boldsymbol{x}^d,\boldsymbol{\theta})}{\partial \boldsymbol{\theta}} \right]. \tag{6}$$

Direct computation of the gradient (6) is infeasible due to an exponential number of possible variable configurations $\boldsymbol{y}$, which must be considered to compute $\mathbb{E}_{p(\boldsymbol{y}|\boldsymbol{x}^d,\boldsymbol{\theta})} \frac{\partial E(\boldsymbol{y},\boldsymbol{x}^d,\boldsymbol{\theta})}{\partial \boldsymbol{\theta}}$. Inspired by [22], in our work we employ sampling based approximation of (6) instead, which we detail in Sect. 3.

Stochastic Approximation. The stochastic gradient approximation proposed in [25] is a common way to learn parameters of a CNN nowadays. It allows to perform parameter updates for a single randomly selected input observation, or a small subset of observations, instead of computing the update step for the whole training set at once, as the latter can be very costly. Assume that the gradient of some function $f(\theta)$ can be represented as follows:

$$\nabla_\theta f = \mathbb{E}_{p(y|\theta)} \nabla_\theta g(y,\theta). \tag{7}$$

Then under mild technical conditions the following procedure

$$\theta_{i+1} = \theta_i - \eta_i \nabla_\theta g(y',\theta_i), \text{ where } y' \sim p(y|\theta_i) \tag{8}$$

and η_i is a diminishing sequence of step-sizes, converges to a critical point of the function $f(\theta)$. We refer to [25,26] for details, for the cases of both convex and non-convex functions $f(\theta)$.

3   Stochastic Optimization Based Learning Framework

Stochastic Likelihood Maximization. Since the value $\frac{\partial E(\boldsymbol{y}^d,\boldsymbol{x}^d,\boldsymbol{\theta})}{\partial \boldsymbol{\theta}}$ does not depend on y, we can rewrite the gradient (6) as

$$\frac{\partial \sum_{d=1}^{D} \log p(\boldsymbol{y}^d|\boldsymbol{x}^d,\boldsymbol{\theta})}{\partial \boldsymbol{\theta}} = \sum_{d=1}^{D} \mathbb{E}_{p(\boldsymbol{y}|\boldsymbol{x}^d,\boldsymbol{\theta})} \left[-\frac{\partial E(\boldsymbol{y}^d,\boldsymbol{x}^d,\boldsymbol{\theta})}{\partial \boldsymbol{\theta}} + \frac{\partial E(\boldsymbol{y},\boldsymbol{x}^d,\boldsymbol{\theta})}{\partial \boldsymbol{\theta}} \right]. \tag{9}$$

The summation over samples from the training set can be seen as an expectation over a uniform distribution and therefore the index d can be seen as drawn from this uniform distribution. According to this observation we can rewrite (9) as

$$\frac{\partial \sum_{d=1}^{D} \log p(\boldsymbol{y}^d | \boldsymbol{x}^d, \boldsymbol{\theta})}{\partial \boldsymbol{\theta}} \doteq D \cdot \mathbb{E}_{p(\boldsymbol{y}, d | \boldsymbol{x}^d, \boldsymbol{\theta})} \left[-\frac{\partial E(\boldsymbol{y}^d, \boldsymbol{x}^d, \boldsymbol{\theta})}{\partial \boldsymbol{\theta}} + \frac{\partial E(\boldsymbol{y}, \boldsymbol{x}^d, \boldsymbol{\theta})}{\partial \boldsymbol{\theta}} \right], \quad (10)$$

where $p(\boldsymbol{y}, d | \boldsymbol{x}^d, \boldsymbol{\theta}) = p(d) p(\boldsymbol{y} | \boldsymbol{x}^d, \boldsymbol{\theta})$ and $p(d) = \frac{1}{D}$. Assume that we can obtain i.i.d. samples $\boldsymbol{y}'$ from $p(\boldsymbol{y} | \boldsymbol{x}^d, \boldsymbol{\theta})$. Then the following iterative procedure converges to a critical point of the likelihood (5) according to (7) and (8)

$$\boldsymbol{\theta}_{i+1} = \boldsymbol{\theta}_i - \eta_i \left[-\frac{\partial E(\boldsymbol{y}^d, \boldsymbol{x}^d, \boldsymbol{\theta}_i)}{\partial \boldsymbol{\theta}} + \frac{\partial E(\boldsymbol{y}', \boldsymbol{x}^d, \boldsymbol{\theta}_i)}{\partial \boldsymbol{\theta}} \right], \quad (11)$$

where d is uniformly sampled from $\{1, \ldots, D\}$ and $\boldsymbol{y}' \sim p(\boldsymbol{y} | \boldsymbol{x}^d, \boldsymbol{\theta}_i)$.

Now we turn to the computation of the stochastic gradient $-\frac{\partial E(\boldsymbol{y}^d, \boldsymbol{x}^d, \boldsymbol{\theta})}{\partial \boldsymbol{\theta}} + \frac{\partial E(\boldsymbol{y}', \boldsymbol{x}^d, \boldsymbol{\theta})}{\partial \boldsymbol{\theta}}$ itself, provided $\boldsymbol{y}^d, \boldsymbol{y}', \boldsymbol{x}^d$ and $\boldsymbol{\theta}$ are given. In the *overcomplete representation* [23] the energy (3) reads

$$E(\boldsymbol{y}, \boldsymbol{x}, \boldsymbol{\theta}) = \sum_{f \in \mathcal{F}} \sum_{\hat{\boldsymbol{y}}_f \in \mathcal{Y}_f} \psi_f(\hat{\boldsymbol{y}}_f, \boldsymbol{x}, \boldsymbol{\theta}) \cdot [\![\boldsymbol{y}_f = \hat{\boldsymbol{y}}_f]\!], \quad (12)$$

where expression $[\![A]\!]$ equals 1 if A is true and 0 otherwise. Therefore $\frac{\partial E(\boldsymbol{y}, \boldsymbol{x}, \boldsymbol{\theta})}{\partial \psi_f(\hat{\boldsymbol{y}}_f, \boldsymbol{x}, \boldsymbol{\theta})} = [\![\boldsymbol{y}_f = \hat{\boldsymbol{y}}_f]\!]$. If the potential $\psi_f(\hat{\boldsymbol{y}}_f, \boldsymbol{x}, \boldsymbol{\theta})$ is an output of a CNN, then the value $-\frac{\partial E(\boldsymbol{y}^d, \boldsymbol{x}^d, \boldsymbol{\theta})}{\partial \psi_f(\hat{\boldsymbol{y}}_f, \boldsymbol{x}^d, \boldsymbol{\theta})} + \frac{\partial E(\boldsymbol{y}', \boldsymbol{x}^d, \boldsymbol{\theta})}{\partial \psi_f(\hat{\boldsymbol{y}}_f, \boldsymbol{x}^d, \boldsymbol{\theta})} = -[\![\boldsymbol{y}_f^d = \hat{\boldsymbol{y}}_f]\!] + [\![\boldsymbol{y}'_f = \hat{\boldsymbol{y}}_f]\!]$ is the error to propagate to the CNN. During the back-propagation of this error all parameters $\boldsymbol{\theta}$ of the CNN are updated. The overall stochastic maximization procedure for the likelihood (5) is summarized in Algorithm 1. The algorithm is fully defined up to sampling from the distribution $p(\boldsymbol{y} | \boldsymbol{x}^d, \boldsymbol{\theta})$ in Step 5. We discuss different approaches in the next subsection.

Algorithm 1. Sampling-based maximization of the likelihood (5)

1: Initialize parameters $\boldsymbol{\theta}_0$ of the CNN-CRF model.
2: **for** $i = 1$ to M (*max. number of iterations*) **do**
3:     Uniformly sample d from $\{1, \ldots, D\}$
4:     Perform forward pass of the CNN to get $\psi_f(\hat{\boldsymbol{y}}, \boldsymbol{x}^d, \boldsymbol{\theta}_{i-1})$ for each $f \in \mathcal{F}$ and $\hat{\boldsymbol{y}}_f \in \mathcal{Y}_f$
5:     Sample $\boldsymbol{y}'$ from the distribution $p(\boldsymbol{y} | \boldsymbol{x}^d, \boldsymbol{\theta}_{i-1})$ defined by (1)
6:     Compute the error $-[\![\boldsymbol{y}_f^d = \hat{\boldsymbol{y}}_f]\!] + [\![\boldsymbol{y}'_f = \hat{\boldsymbol{y}}_f]\!]$ for each $f \in \mathcal{F}$ and $\hat{\boldsymbol{y}}_f \in \mathcal{Y}_f$
7:     Back propagate the error through CNN to obtain a gradient ∇_θ
8:     Update the parameters $\boldsymbol{\theta}_i := \boldsymbol{\theta}_{i-1} - \eta_i \nabla_\theta$
9: **return** $\boldsymbol{\theta}_M$

Sampling. Obtaining an exact sample from $p(\boldsymbol{y}|\boldsymbol{x}, \boldsymbol{\theta})$ is a difficult problem for a general CRF due to the exponential size of the set $\mathcal{Y} \ni \boldsymbol{y}$ of all possible configurations. There are, however, ways to mitigate it. The full Markov Chain Monte Carlo (MCMC) sampling method [27] starts from an arbitrary variable configuration $\boldsymbol{y} \in \mathcal{Y}$ and generates the next one $\boldsymbol{y}'$. In our case this generation can be done with e.g. Gibbs sampling [24], as presented in Algorithm 2. Algorithm 2 passes over all variables y_n and updates each of them according to the conditional distribution $p(y_n|\boldsymbol{y}_{\backslash n}, \boldsymbol{x}, \boldsymbol{\theta})$, where $\backslash n$ denotes all variable indexes except n. Let $\mathrm{nb}(n) = \{k \in I | \exists f \in \mathcal{F}: n, k \in f\}$ denote all neighbors of the variable n. Note, that due to the Markov property of CRFs [28], it holds

$$p(y_n|\boldsymbol{y}_{\backslash n}, \boldsymbol{x}, \boldsymbol{\theta}) = p(y_n|\boldsymbol{y}_{\mathrm{nb}(n)}, \boldsymbol{x}, \boldsymbol{\theta}) \propto \exp\left(-\sum_{f \in \mathcal{F}: n \in f} \psi_f(\boldsymbol{y}_f, \boldsymbol{x}, \boldsymbol{\theta})\right). \quad (13)$$

Therefore, sampling from this distribution can be done efficiently, since it requires evaluating only those potentials $\psi_f(\boldsymbol{y}_f, \boldsymbol{x}, \boldsymbol{\theta})$ which are dependent on the variable y_n, i.e. for $f \in \mathcal{F}$ such that $n \in f$. Algorithm 2 summarizes one iteration of the sampling procedure. Note that it is highly parallelizable [29] and allows for efficient GPU implementations. Under mild technical conditions the MCMC sampling process converges to a stationary distribution after a finite number of iterations [27]. This distribution coincides with $p(\boldsymbol{y}|\boldsymbol{x}, \boldsymbol{\theta})$. However, such a sampling is time-consuming, because convergence to the stationary distribution may require many iterations and must be performed after each update of the parameters $\boldsymbol{\theta}$.

To overcome this difficulty a contrastive-divergence (CD) method was proposed in [30] and theoretically justified in [31]. For a randomly generated index $d \in \{1, \dots, D\}$ of the training sample one performs a single step of the MCMC procedure starting from a ground-truth variable configuration, which in our case boils down to a single run of Algorithm 2 for $\boldsymbol{y} = \boldsymbol{y}^d$. Unfortunately, the sufficient conditions needed to justify this method according to [31] do not hold for CRFs in general. Nevertheless, we provide an experimental evaluation of this method in Sect. 5 along with a different technique described next.

Persistent contrastive divergence (PCD) [32] is a further development of contrastive divergence, where one step of the MCMC method is performed starting from the sample obtained on a previous learning iteration. It is based on the assumption that the distribution $p(\boldsymbol{y}, \boldsymbol{x}, \boldsymbol{\theta})$ changes slowly from iteration to iteration and a sample from $p(\boldsymbol{y}|\boldsymbol{x}^d, \boldsymbol{\theta}_{i-1})$ is close enough to a sample from $p(\boldsymbol{y}|\boldsymbol{x}^d, \boldsymbol{\theta}_i)$. Moreover, when getting closer to a critical point, the gradient becomes smaller and therefore $p(\boldsymbol{y}, \boldsymbol{x}, \boldsymbol{\theta}_i)$ deviates less from $p(\boldsymbol{y}, \boldsymbol{x}, \boldsymbol{\theta}_{i-1})$. Therefore, close to a critical point the generated samples can be seen as samples from the stationary distribution of the full MCMC method, which coincides with the desired one $p(\boldsymbol{y}|\boldsymbol{x}, \boldsymbol{\theta})$.

With the above description of the possible sampling procedures the whole joint CNN-CRF learning Algorithm 1 is well-defined.

Algorithm 2. Gibbs sampling

Require: A variable configuration $\boldsymbol{y} \in \mathcal{Y}$
1: **for** n = 1, ..., N **do**
2:    y'_n is sampled from $p(y_n | \boldsymbol{y}_{\setminus n}, \boldsymbol{x}, \boldsymbol{\theta})$
3:    $y_n \leftarrow y'_n$
4: **return** $\boldsymbol{y}$

---

4   Comparison to Alternative Approaches

Unrolled Inference. In contrast to the learning method with the unrolled inference proposed in [11,12], our approach is not limited to Gaussian pairwise potentials. In our training procedure the potentials $\phi_f(\boldsymbol{y}_f, \boldsymbol{x}, \boldsymbol{\theta})$ can have arbitrary form.

The piece-wise training method [1] is able to handle arbitrary potentials in CRFs. However, maximization of the likelihood (5) in that work is substituted with

$$(\text{arg}) \max_{\boldsymbol{\theta}} \sum_{d=1}^{D} \sum_{f \in \mathcal{F}} \left[-\psi_f(\boldsymbol{y}_f^d, \boldsymbol{x}^d, \boldsymbol{\theta}) - \log \sum_{\boldsymbol{y}_f \in \mathcal{Y}_f} \exp(-\psi_f(\boldsymbol{y}_f, \boldsymbol{x}^d, \boldsymbol{\theta})) \right], \quad (14)$$

which lacks a sound theoretical justification.

LP-relaxation and fractional entropy based approximation is employed in [3]. As mentioned above, there is no clear evidence that the fractional entropy always leads to tight likelihood approximations. Additionally, the method requires a lot of memory: to avoid the time consuming, full message passing based inference, the gradient steps w.r.t. the CNN parameters $\boldsymbol{\theta}$ and minimization over the dual variables of the LP-relaxation of the CRF are interleaved with each other. This requires to store current values of the dual variables for *each* training sample. The number of dual variables is proportional to the number of labels used in the CRF as well as to the number of its factors. So, for example in our experiments we use a dataset containing 2000 images of the approximate size 320×120. The corresponding CRF has 20 labels and around 10^6 pairwise factors (see Sect. 5 for details). The dual variables stored by the method [3] would require around 200MB per image and 0.4 TB for the whole dataset. Note that our approach requires to store only the current variable configuration $\boldsymbol{y}$ for each of the D training samples, when used with the PCD sampling. Therefore, it requires only 78 MB of working storage for the whole dataset. The difference between our method and the method proposed in [3] gets even more pronounced for larger problems and datasets, such as the augmented Pascal VOC dataset [33,34] containing 10000 images with 500×300 pixels each.

5   Experiments

In the experimental evaluation we consider the problem of semantic body-parts segmentation from a single depth image [35]. We specify a CRF, which has unary potentials dependent on a CNN. We test different sampling options in Algorithm 1 and compare our approach with another CRF-CNN learning framework proposed in [11]. Additionally, we analyze the trained model, in order to understand whether it can capture an object shape and contextual information.

Dataset and Evaluation. We apply our approach to the challenging task of predicting human body parts from a depth image. To the best of our knowledge, there is no publicly available dataset for this task that contains real depth images. For this reason, in [35], a set of synthetically rendered depth images along with the corresponding ground truth labelings were introduced (see examples in Fig. 1 (left column)). In total there are 19 different body part labels, and one additional label for the background. The dataset is split into 2000 images for training and 500 images for testing. As a quality measure, the authors use the averaged per-pixel accuracy for body parts labeling, excluding the background. This makes sense since the background can be easily identified from the depth map.

Our Model. Following [4], in our experiments, we use a pixel-level CRF that is able to capture geometrical layout and context. The state vector y defines a per pixel labeling. Therefore the number N of coordinates in y is equal to the number of pixels in a depth image, which has dimensions varying around 130×330. For all $n \in \{1, \ldots, N\}$ the label space is $\mathcal{Y}_n = \{1, \ldots, 20\}$. The observation x represents a depth image. Our CRF has the following energy function $E(y, x, \theta)$:

$$E(y, x, \theta) = \sum_{n=1}^{N} \psi_n(y_n, x, \theta) + \sum_{c \in \mathcal{C}} \sum_{(i,j) \in E_c} \psi_c(y_i, y_j, \theta), \qquad (15)$$

where $\psi_n(y_n, x, \theta)$ are unary potentials that depend on a CNN. Our CRF has $|\mathcal{C}|$ classes of pairwise potentials. All potentials of one class are represented by a learned value table, which they share. The neighborhood structure of the CRF is visualized in Fig. 2b. All pixels are connected to 64 neighbors, apart from those close to the image border.

The local distribution (13) used by the sampling Algorithm 2 takes the form:

$$p_i(y_i{=}l|x, y_{R \setminus i}; \theta) \propto \exp\Big[-\psi_i(l) - \sum_c \big(\psi_c(l, y_{j'}) + \psi_c(y_{j''}, l)\big)\Big]. \qquad (16)$$

Note that according to our CRF architecture there are exactly two edges (apart from the nodes close to the image border) in each edge class c that are incident to a given node i. The corresponding neighboring nodes are denoted by j' and j'' in (16).

| | | 86.53% | 90.88% | 90.38% | 93.14% |

| | | 83.93% | 88.20% | 87.74% | 90.27% |

| | | 90.15% | 92.62% | 91.98% | 93.58% |

Fig. 1. Results. (From left to right). The input depth image. The corresponding ground truth labeling for all body parts. The result of a trained CNN model. The result of [11] using an end-to-end training procedure. Our results with separate learning and joint learning, respectively. Below each result we give the averaged pixel-wise accuracy for all body parts.

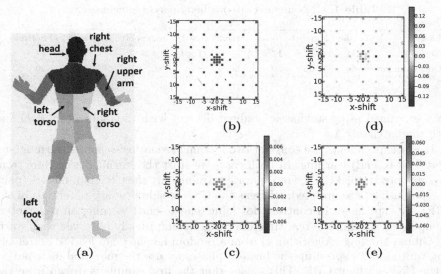

Fig. 2. Model Insights. (a) Illustrating the 19 body parts of a human. (c–e) Weights of pairwise factors for different pairs of labels, see details below. (b) Neighborhood structure for pairwise factors. The center pixel (red) is connected via pairwise factors to all green pixels. Note that "opposite" edges share same weights, e.g. the edge with x, y-shift $(5, 10)$ has the same weights as the edge with x, y-shift $(-5, -10)$. (c) Weights for pairwise potentials that connect the label "head" with the label "foot". Red means a high energy value, i.e. a discouraged configuration, while blue means the opposite. Since there is no sample in the training dataset where a foot is close to a head, all edges are positive or close to 0. Note that the zero weights can occur even for very unlikely configurations. The reason is that during training these unlikely configurations did not occur. (d) Weights for pairwise potentials that connect the label "left torso" with the label "right torso". The potentials enforce a straight, vertical border between the two labels, i.e. there is a large penalty for "left torso" on top (or below) of "right torso" (x-shift 0, y-shift arbitrary). Also, it is encouraged that "right torso" is to the right of the "left torso" (Positive x-shift and y-shift 0). (e) Weights for pairwise potentials that connect the label "right chest" with the label "right upper arm". It is discouraged that the "right upper arm" appears close to "right chest", but this configuration can occur at a certain distance. Since the training images have no preferred arm-chest configurations, all directions have similar weights. (Color figure online)

As mentioned above, the unary terms of our CRF model depend on the image via a CNN. Since most existing pre-trained CNNs [7,36,37] use RGB images as input, for the depth input we use our own fully convolutional architecture and train it from scratch. Moreover, since some body parts, such as hands, are relatively small, we use the architecture that does not reduce the resolution in intermediate layers. This allows us to capture fine details. All intermediate layers have 50 output channels and a stride of one. The final layer has 20 output channels that correspond to the output labels. The architecture of our CNN is summarized in Table 1. During training, we optimize the cross-entropy loss. The

Table 1. CNN architecture for body parts segmentation.

Layer	conv1	relu1	conv2	relu2	conv3	maxpool1	relu3	conv4	Softmax
Kernel size	41×41	-	17×17	-	11×11	3×3	-	5×5	-
Output channels	50	50	50	50	50	50	50	20	20

CNN is trained using stochastic gradient descent with the momentum 0.99 and with the batch size 1.[2]

In our experiments, we consider two learning scenarios: *separate* learning and *joint* (end-to-end) learning. In both cases we start the learning procedure from the same pre-trained CNN. For separate learning only the CRF parameters (pairwise potentials) are updated, whereas the CNN weights (unary potentials of the CRF) are kept fixed. In contrast, for joint (end-to-end) learning all parameters are updated. During the test-time inference we empirically observed that starting Gibbs sampling (Algorithm 2) from a random labeling can lead to extremely long runtimes. To speed-up the burn-in-phase, we use the marginal distribution of the CNN without CRF. This means that the first sample is drawn from the marginal distribution of the pre-trained CNN.

We also experiment with different sampling strategies during the training phase: we considered (i) the contrastive-divergence with K sampling iterations, denoted as CD-K for K equal to 1, 2, 5 and (ii) the persistent contrastive-divergence PCD.

Baselines. We compare our approach to the method of [35], which introduced this dataset. Their approach is based on a random forest model. Unfortunately, we were not able to compare to the recent work [38], which extends [35], and is also based on random forests. The reason is that in the work [38] its own evaluation measure is used, meaning that the accuracy of only a small subset of pixels is evaluated. This subset is chosen in such a way that each of the 20 classes is represented by the same number of pixels. We are concerned, however, that such small pixel subsets may introduce a bias. Furthermore, we did not have this subset at our disposal. Since our main aim is to evaluate CNN-based CRF models, we compare to the approach [11]. As described above, they incorporate a densely connected Gaussian CRF model into the CNN as a Recurrent Neuronal Network of the corresponding Mean Field inference steps. This approach has recently been the state-of-the-art in the VOC2012 object segmentation challenge.

Results. Qualitative and quantitative results are shown in Fig. 1 and Table 2 respectively. Our method with joint learning is performing best. In particular, the persistent contrastive-divergence version shows the best results, which conforms to the observations made in other works [32]. The CNN-CRF approach

---

[2] We use the commonly adopted terminology from the CNN literature for technical details, to allow reproducibility of our results.

Table 2. Average per-pixel accuracy for all foreground parts. *Separate* learning means that weights of the respective CNN were trained prior to CRF parameters. In contrast, *joint* training means that all weights were learned jointly, starting with a pre-trained CNN. We obverse that joint training is superior to separate training, and furthermore that the model of [11], which is based on a dense Gaussian CRF, is inferior to our generic CRF model.

Method	Learning	Accuracy
Online random forest [35]	-	$\approx 79.0\%$
CNN	-	84.47
CNN + CRF [11]	separate	86.55%
CNN + CRF [11]	Joint	88.17%
CNN + CRF (ours) PCD	Separate	87.62%
CNN + CRF (ours) CD-1	Joint	88.17%
CNN + CRF (ours) CD-2	Joint	88.15%
CNN + CRF (ours) CD-5	Joint	88.23%
CNN + CRF (ours) PCD	Joint	**89.01%**

of [11] is inferior to ours. Note that the accuracy difference of 1% can mean that e.g. a complete hand is incorrectly labeled. We attribute this to the fact that for this task the spatial layout of body parts is of particular importance. The underlying dense Gaussian CRF model of [11] is rotational invariant and cannot capture contextual information such as "the head has to be above the torso". Our approach is able to capture this, which we explain in detail in Figs. 2 and 3. We expect that even higher levels of accuracy can be achieved by

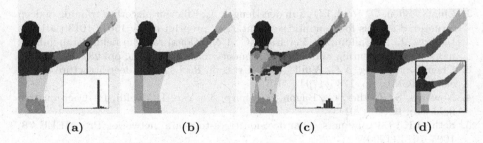

(a)   (b)   (c)   (d)

Fig. 3. Model Insights. (a) The most likely labeling for a separately trained CNN. For the circled pixel, the local marginal distribution is shown. (b) Max. marginal labeling of a separately trained CRF, which uses the CNN unaries from (a), i.e. our approach with separate learning. We observe that unaries are spatially smoothed-out. (c) Most likely labeling of a CNN that was jointly trained with the CRF. The labeling looks worse than (a). However, the main observation is that the pixel-wise marginal distributions are more ambiguous than in (a), see the circled pixel. (d) The final, max-marginal labeling of the jointly trained CRF model, which is considerably better than the result in (b). The reason is that due to the ambiguity in the local unary marginals, the CRF has more power to find the correct body part configuration. The inlet shows the ground truth labeling.

exploring different network designs and learning strategies, which we leave for future work.

6   Discussion and Future Work

We have presented a generic CRF model where a CNN models unary factors. We have introduced an efficient and scalable maximum likelihood learning procedure to train all model parameters jointly. By doing so, we were able to train and test on large-size factor graphs. We have demonstrated a performance gain over competing techniques for semantic labeling of body parts. We have observed that our generic CRF model can capture the shape and context information of relating body parts.

There are many exciting avenues for future research. We plan to apply our method to other application scenarios, such as semantic segmentation of RGB images. In this context, it would be interesting to combine the dense CRF model of [11] with our generic CRF model. Note that a dense CRF is implicitly modeling the property that objects have a compact color distribution, see [39], which is a complementary modeling power to our generic CRF model.

Acknowledgements. This work was supported by: European Research Council (ERC) under the European Union's Horizon 2020 research and innovation programme (grant agreement No. 647769); German Federal Ministry of Education and Research (BMBF, 01IS14014A-D); EPSRC EP/I001107/2; ERC grant ERC-2012-AdG 321162-HELIOS. The computations were performed on an HPC Cluster at the Center for Information Services and High Performance Computing (ZIH) at TU Dresden.

References

1. Lin, G., Shen, C., Reid, I.D., van den Hengel, A.: Efficient piecewise training of deep structured models for semantic segmentation. preprint arXiv:1504.01013 (2015)
2. Lafferty, J., McCallum, A., Pereira, F.C.: Conditional random fields: probabilistic models for segmenting and labeling sequence data. In: ICML, pp. 282–289 (2001)
3. Chen, L., Schwing, A.G., Yuille, A.L., Urtasun, R.: Learning deep structured models. In: ICML, pp. 1785–1794 (2015)
4. Nowozin, S., Rother, C., Bagon, S., Sharp, T., Yao, B., Kohli, P.: Decision tree fields. In: ICCV (2011)
5. Sethi, I.K.: Entropy nets: from decision trees to neural networks. Proc. IEEE **78**, 1605–1613 (1990)
6. Richmond, D.L., Kainmueller, D., Yang, M.Y., Myers, E.W., Rother, C.: Relating cascaded random forests to deep convolutional neural networks for semantic segmentation. preprint arXiv:1507.07583 (2015)
7. Long, J., Shelhamer, E., Darrell, T.: Fully convolutional networks for semantic segmentation. preprint arXiv:1411.4038 (2014)
8. Farabet, C., Couprie, C., Najman, L., LeCun, Y.: Learning hierarchical features for scene labeling. TPAMI **35**, 1915–1929 (2013)
9. Chen, L.C., Papandreou, G., Kokkinos, I., Murphy, K., Yuille, A.L.: Semantic image segmentation with deep convolutional nets and fully connected CRFs. preprint arXiv:1412.7062 (2014)

10. Krähenbühl, P., Koltun, V.: Efficient inference in fully connected CRFs with gaussian edge potentials. In: NIPS (2011)
11. Zheng, S., Jayasumana, S., Romera-Paredes, B., Vineet, V., Su, Z., Du, D., Huang, C., Torr, P.H.S.: Conditional random fields as recurrent neural networks. In: Proceedings of ICCV (2015)
12. Schwing, A.G., Urtasun, R.: Fully connected deep structured networks. preprint arXiv:1503.02351 (2015)
13. Adams, A., Baek, J., Davis, M.A.: Fast high-dimensional filtering using the permutohedral lattice. In: Computer Graphics Forum, vol. 29. Wiley Online Library (2010)
14. Domke, J.: Learning graphical model parameters with approximate marginal inference. TPAMI **35**, 2454–2467 (2013)
15. Kiefel, M., Gehler, P.V.: Human pose estimation with fields of parts. In: Fleet, D., Pajdla, T., Schiele, B., Tuytelaars, T. (eds.) ECCV 2014. LNCS, vol. 8693, pp. 331–346. Springer, Heidelberg (2014). doi:10.1007/978-3-319-10602-1_22
16. Barbu, A.: Training an active random field for real-time image denoising. IEEE Trans. Image Process. **18**, 2451–2462 (2009)
17. Ross, S., Munoz, D., Hebert, M., Bagnell, J.A.: Learning message-passing inference machines for structured prediction. In: Proceedings of CVPR (2011)
18. Stoyanov, V., Ropson, A., Eisner, J.: Empirical risk minimization of graphical model parameters given approximate inference, decoding, and model structure. In: Proceedings of AISTATS (2011)
19. Tompson, J.J., Jain, A., LeCun, Y., Bregler, C.: Joint training of a convolutional network and a graphical model for human pose estimation. In: Proceedings of NIPS (2014)
20. Liu, Z., Li, X., Luo, P., Loy, C.C., Tang, X.: Semantic image segmentation via deep parsing network. In: Proceedings of ICCV (2015)
21. Sutton, C., McCallum, A.: Piecewise training of undirected models. In: Conference on Uncertainty in Artificial Intelligence (UAI) (2005)
22. He, X., Zemel, R.S., Carreira-perpiñán, M.Á.: Multiscale conditional random fields for image labeling. In: CVPR. Citeseer (2004)
23. Wainwright, M.J., Jordan, M.I.: Graphical models, exponential families, and variational inference. Found. Trends® Mach. Learn. **1**, 1–305 (2008)
24. Geman, S., Geman, D.: Stochastic relaxation, gibbs distributions, and the Bayesian restoration of images. TPAMI **6**, 721–741 (1984)
25. Robbins, H., Monro, S.: A stochastic approximation method. Ann. Math. Stat. 400–407 (1951)
26. Spall, J.C.: Introduction to Stochastic Search and Optimization: Estimation, Simulation, and Control, vol. 65. Wiley, Hoboken (2005)
27. Geyer, C.J.: Practical Markov chain Monte Carlo. Stat. Sci. 473–483 (1992)
28. Lauritzen, S.L.: Graphical Models. Oxford University Press, Oxford (1996)
29. Gonzalez, J., Low, Y., Gretton, A., Guestrin, C.: Parallel Gibbs sampling: from colored fields to thin junction trees. In: International Conference on Artificial Intelligence and Statistics. pp. 324–332 (2011)
30. Hinton, G.: Training products of experts by minimizing contrastive divergence. Neural Comput. **14**, 1771–1800 (2002)
31. Yuille, A.L.: The convergence of contrastive divergences. In: NIPS (2004)
32. Tieleman, T.: Training restricted Boltzmann machines using approximations to the likelihood gradient. In: ICML. ACM, New York (2008)
33. Everingham, M., Van Gool, L., Williams, C.K.I., Winn, J., Zisserman, A.: The PASCAL Visual Object Classes Challenge 2012 (VOC 2012) Results

34. Hariharan, B., Arbelaez, P., Bourdev, L., Maji, S., Malik, J.: Semantic contours from inverse detectors. In: International Conference on Computer Vision (ICCV) (2011)
35. Denil, M., Matheson, D., de Freitas, N.: Consistency of online random forests. In: ICML (2013)
36. Krizhevsky, A., Sutskever, I., Hinton, G.E.: Imagenet classification with deep convolutional neural networks. In: Pereira, F., Burges, C.J.C., Bottou, L., Weinberger, K.Q. (eds.) Advances in Neural Information Processing Systems, vol. 25, pp. 1097–1105. Curran Associates Inc., New York (2012)
37. Simonyan, K., Zisserman, A.: Very deep convolutional networks for large-scale image recognition. CoRR abs/1409.1556 (2014)
38. Ren, S., Cao, X., Wei, Y., Sun, J.: Global refinement of random forest. In: CVPR (2015)
39. Cheng, M.M., Prisacariu, V.A., Zheng, S., Torr, P.H.S., Rother, C.: Densecut: densely connected CRFs for realtime Grabcut. Comput. Graph. Forum **34**, 193–201 (2015)

Object-Aware Dictionary Learning with Deep Features

Yurui Xie[1,2,3]([✉]), Fatih Porikli[1,2], and Xuming He[1,2]

[1] Australian National University, Canberra, Australia
gloriousxyr@163.com
[2] Data61/CSIRO, Canberra, Australia
[3] University of Electronic Science and Technology of China,
Chengdu, China

Abstract. Visual dictionary learning has the capacity to determine sparse representations of input images in a data-driven manner using over-complete bases. Sparsity allows robustness to distractors and resistance against over-fitting, two valuable attributes of a competent classification solution. Its data-driven nature is comparable to deep convolutional neural networks, which elegantly blend global and local information through progressively more specific filter layers with increasingly extending receptive fields. One shortcoming of dictionary learning is that it does not explicitly select and focus on important regions, instead it generates responses on uniform grid of patches or entire image. To address this, we present an object-aware dictionary learning framework that systematically incorporates region proposals and deep features in order to improve the discriminative power of the combined classifier. Rather than extracting a dictionary from all fixed sized image windows, our methods concentrates on a small set of object candidates, which enables consolidation of semantic information. We formulate this as an optimization problem on a new objective function and propose an iterative solver. Our results on benchmark datasets demonstrate the effectiveness of our method, which is shown to be superior to the state-of-the-art dictionary learning and deep learning based image classification approaches.

1 Introduction

Dictionary learning (DL) has attracted considerable amount of attentions in the past few years. The goal of DL is to learn an over-complete collection of atoms for representation in a data-driven manner. The main property of learned dictionary is that it is capable of approximating an input signal as a linear combination of a small number of atoms. Recently, dictionary learning approaches have widely applied to various problems of computer vision area, such as image denoising [1,2], image restoration [3], image synthesis [4,5], visual tracking [6] and image classification [7–9].

The original intention of DL methods is to reconstruct the input data faithfully by a learned over-complete dictionary. Therefore, they are not appropriate

© Springer International Publishing AG 2017
S.-H. Lai et al. (Eds.): ACCV 2016, Part II, LNCS 10112, pp. 237–253, 2017.
DOI: 10.1007/978-3-319-54184-6_15

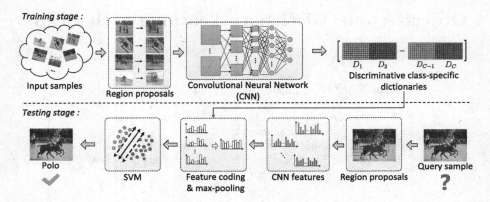

Fig. 1. Overall framework of our object-aware dictionary learning method.

for the visual recognition task. To overcome this problem, recent work [7–19] attempt to learn discriminative dictionaries in order to achieve better performance in classification problems. However, one main drawback of existing visual dictionary learning methods is that they are unable to select and focus on important image regions explicitly. Instead, these methods only generate responses on the regular patches or the entire image. For the visual recognition problem, especially in complex scenes, localized semantic information of image often provides crucial visual cues for improving the discriminative power of feature representation. Nevertheless, few efforts have been made to explore useful discriminative semantic information within local regions in the dictionary learning for image classification.

Recently, Convolutional Neural Networks (CNN) [21] have been shown to be successful in numerous visual recognition problems. One main advantage of the CNN is that it allows integrating the global context and local cues through multiple filter layers with increasingly extending receptive fields thanks to the pooling operations. Inspired by this property of the CNN, we propose an Object-Aware Dictionary Learning (OADL) framework to address the above shortcoming of dictionary learning. To this end, we incorporate a region proposal mechanism into the deep features extraction by the CNN to discover underlying local semantic information in the image. We design a new object category aware objective for dictionary learning and then feed the deep features of region proposals to extract multiple discriminative class-specific dictionaries.

Unlike conventional dictionary learning approaches that extract a dictionary from all the fixed sized image patches or entire image, our method concentrates on a small set of object candidates. We first extract the region proposals, which encode the local semantic information of image and provide important visual cues for recognition task. This facilitates the feature representation to consolidate the semantic information and suppress the distraction due to the background. We then learn the class-aware dictionary of the object candidate features. To minimize our dictionary learning objective, we derive an efficient iterative optimization procedure that alternatively solves several simpler subproblems. In the recognition stage, the learned discriminative dictionaries are used to encode the deep features of all object candidates within image and generates a global image representation by max-pooling. Figure 1 shows the framework of this new OADL method.

The remainder of this paper is organized as follows: The related works are briefly reviewed in Sect. 2. Section 3 presents the proposed object-aware dictionary learning framework that integrates with the region proposals and deep features systematically to improve the discriminative power of feature representation. The optimization algorithm of the OADL is also described in this section. Experimental results are given in Sect. 4.

2   Related Work

In order to enhance the representation power of image feature, many works aim to learn a discriminative dictionary for different visual recognition tasks. Existing dictionary learning approaches can be grouped as unsupervised and supervised methods. The goal of unsupervised dictionary learning is to compose an overcomplete dictionary by minimizing the reconstruction error. A typical example for unsupervised dictionary learning would be the KSVD algorithm [15], which iteratively applies SVD to fit the atoms of a single dictionary to reconstruction error. To reduce the time complexity, Lee et al. [20] cast the standard sparse representation to the least squares problem.

To obtain a feature representation with more discriminative power, supervised dictionary learning incorporates additional classification objective into reconstruction loss using the labeled data. The existing supervised dictionary learning can be further grouped into two categories. Methods in the first category aim to make the representation coefficients discriminative by learning a single dictionary across all classes. The common characteristic of recent methods [11–13,16,17] is to combine a classification loss term with the standard sparse dictionary learning formulation. Similarly, Jiang et al. [9] incorporate both a label consistent constraint and a linear classification cost into the KSVD objective for improving the discriminative power of feature encoding.

Methods in the second category [7,8,10,14,18,19] learn a set of class-specific sub-dictionaries, and then these multiple sub-dictionaries are concatenated together to form a structured dictionary for feature representation. Specifically, Mairal et al. [18] integrate a softmax discriminative function with the KSVD model. Ramirez et al. [14] impose an incoherent constrain in the standard dictionary learning model, which encourages the learned class-specific dictionaries to be as independent as possible. Yang et al. [7] incorporate the Fisher Discriminant criterion into the dictionary learning for further improving the discriminative capability of class-specific dictionaries. Zhou et al. [8] propose to learn multiple class-specific dictionaries and a shared dictionary for the groups of classes that have the visually similar patterns. Gao et al. [19] also propose to train the class-specific dictionaries and a shared dictionary for addressing the fine-grained recognition problem. In addition, Gu et al. [10] learn a structured synthesis dictionary and a structured analysis dictionary simultaneously for enhancing the representation power of feature. However, the existing dictionary leaning schemes are not capable of selecting and focusing on the important regions within image. Instead, these methods extract a visual dictionary from the fixed sized image patches or the entire image. For this reason, the underlying image regions with useful semantic cues and background clutters cannot be distinguished in the dictionary learning procedure.

For the feature generation task, the convolutional neural networks [21–23] provide powerful solutions. One advantage of CNN based methods is that they allow fusing the global and local information through gradually more specific filter layers with increasing receptive fields.

Recently, the region proposals approaches [24–27] provide an effective option to generate the object candidates from image. These methods utilize objectness measures derived from different visual cues. Compared with the traditional image interest points and sliding windows, region proposals are capable of detecting regions with high-level semantic meanings.

3   The Proposed Method

We first introduce our object-aware dictionary learning framework, which consists of object proposal generation, class-aware dictionary learning, and image-level feature representation. In Sect. 3.4, we then present the optimization algorithm for solving all the variables in our OADL objective function.

3.1   Region Proposal Generation

In order to explore the local semantic information, we propose to adopt convolutional neural network (CNN) features and further integrate deep features with the region proposals in our method. Compared with the fixed size of image patch, each region proposal is a mid-level element of image, which allows us to extract important region information in image for the recognition task.

We use the EdgeBox [27] algorithm to generate a set of initial region proposals within image. Then the non-maximum suppression (NMS) is adopted to refine these region windows, where the overlap rate of NMS is set to 0.8 IoU. Afterward, the deep feature is utilized to describe each region proposal in image. Finally, all the CNN features of region proposals from the training samples are fed into our OADL model to obtain a set of discriminative class-specific dictionaries.

3.2   Object-Aware Dictionary Learning (OADL)

Let $X = [X_1, X_2, \ldots, X_C]$ be a set of training data with C classes, where $X_i \in \mathbf{R}^{d \times N_i}, i = 1, 2, \ldots, C$ denotes the training samples corresponding to class i, d is the dimension of the feature and N_i denotes the number of samples from class i. Each column in X_i is the CNN feature of one region proposal. The goal of the OADL objective is to learn a structured dictionary $D = [D_1, D_2, \ldots, D_C, D_{C+1}] \in \mathbf{R}^{d \times K}$, which is used to transform the CNN features into a discriminative feature space. $K = \sum_{i=1}^{C+1} K_i$ is the number of visual atoms in dictionary D, where K_i denotes the number of visual atoms in class-specific dictionary D_i. Since different object classes may have the similar visual patterns, we further incorporate a shared dictionary D_{C+1} to represent the common visual patterns in the OADL model. We formulate the OADL model for C classes as follows:

$$
\min_{A_i, Z_i, D_i, D_{C+1}, w_i, b_i} \sum_{i=1}^{C} \{ \cdot \sum_{n=1}^{N_i} [\|X_i^n - D_{\in i} Z_i^n\|_F^2 + \|X_i^n - D A_i^n\|_F^2
$$
$$
+ \ \alpha \ \|A_i^n - Z_i^n\|_F^2 + J(w_i, b_i, A_i^n, y_i) + \ \lambda_1 \ \|A_i^n\|_1 \quad (1)
$$
$$
+ \ \lambda_2 \ \|Z_i^n\|_1] + \ \beta \sum_{j=1, j \neq i}^{C+1} \|D_i^T D_j\|_F^2 \}
$$

where X_i^n, $n = 1, 2, \ldots, N_i$ denotes all the feature data of n-th image from class i, dictionary $D_{\in i}$ is a sub-dictionary associated with class i, defined as $[O_{d \times \Sigma_{q=1}^{i-1} K_q}, D_i, O_{d \times \Sigma_{q=i+1}^{C+1} K_q}] \in \mathbf{R}^{d \times K}$, and O is the zero matrix. Z_i^n is the class-specific representation coefficients of X_i^n on dictionary $D_{\in i}$. We define $D = [D_1, D_2, \ldots, D_C, D_{C+1}] \in \mathbf{R}^{d \times K}$ as the structured dictionary that concatenates all the class-specific dictionaries $D_i, i = 1, 2, \ldots, C$ and the additional background dictionary D_{C+1} together. A_i^n is the representation coefficients of X_i^n on the structured dictionary D. $J(\cdot)$ corresponds to a hinge loss for classification (see below for details). $\|A_i^n\|_1$, $\|Z_i^n\|_1$ are the sparsity constrains imposed on the representation coefficients A_i^n and Z_i^n. $\alpha, \beta, \lambda_1, \lambda_2$ are the weighting parameters to balance the different terms in the objective function. We now describe each term in detail in the following.

Discriminative Reconstruction Terms: The first two terms of Eq. (1) are the reconstruction residual terms. These two terms ensure the input data from class i not only to be represented using the class-specific dictionary $D_{\in i}$, but also be reconstructed by the structured dictionary D. Due to this property, the learned class-specific dictionaries have both the data generation and discriminative properties.

Coefficients Consistency Constraint Term: The third term $\|A_i^n - Z_i^n\|_F^2$ aims to make the indexes of non-zero entries in representation coefficients A_i^n and Z_i^n to be the same as far as possible. In this energy term, A_i^n is the representation coefficients of X_i^n using the structured dictionary D, and Z_i^n denotes the representation coefficients of X_i^n on the class-specific dictionary $D_{\in i}$. This penalty term encourages the consistency between the representation coefficients A_i^n and Z_i^n. Therefore, the non-zero entries of A_i^n only appear on the indexes of visual atoms associated with class-specific dictionary D_i. In other words, it indicates that the structured dictionary D tends to represent the samples X_i^n of class i by choosing the visual atoms in dictionary D_i. Due to this consistency property, the discriminative power of feature representation can be strengthened.

Classification Error Term: The fourth term $J(w_i, b_i, A_i^n, y_i)$ of Eq. (1) is a loss function to measure the classification error. In our method, we incorporate a SVM hinge loss into our objective, which is defined as $J(w_i, b_i, A_i^n, y_i) = \| w_i \|^2 + R(w_i, b_i, A_i^n, y_i)$, where $R(w_i, b_i, A_i^n, y_i) = \eta \sum_{j=1}^{P_i^n} [\max(0, y_i \cdot w_i^T A_i^{n,j} + b_i - 1)]^2$ is the quadratic hinge loss due to the differentiable property [28]. w_i, b_i are the parameters of SVM classifier and η is a constant. P_i^n denotes the number of region proposals in the n-th image from class i, y_i is the label of sample corresponding to class i, and $A_i^{n,j}$ denotes the representation feature of j-th region proposal within the n-th image from class i. The A_i, Z_i denote the representation coefficients of X_i on dictionaries D and $D_{\in i}$, respectively. The minimization of this term is to guide the dictionary learning process, which is beneficial to obtain a discriminative feature representation.

Dictionary Incoherent Constraint: The coefficient consistency constraint focuses on the representation coefficients to promote the discriminative power of A_i. We further incorporate a dictionary incoherent constraint [14] in the last term to ensure the multiple class-specific dictionaries to be as independent as possible. It is another way to improve the discrimination of A_i. Besides, we impose the incoherent constraint between all the class-specific dictionaries $D_i, i = 1, 2, \ldots, C$ and the additional background dictionary D_{C+1}, which is used to separate the shared visual patterns and the class-specific visual patterns for all classes.

3.3   Construction of Image-Level Feature

We first describe our proposed feature representation strategy for each region proposal using a group sparsity constraint. With the region features, we then introduce the construction of image-level feature for final recognition task.

Given the learned discriminative structured dictionary $D = [D_1, D_2, \ldots, D_C, D_{C+1}]$, we propose to encode the deep feature of an object proposal with the l_1/l_2-norm group sparsity constraint. Mathematically, the feature coding step is solved by the following l_1/l_2-norm regularized least squares problem.

$$\min_{B_i^n} \| X_i^n - D_{/C+1} B_i^n \|_2^2 + \rho \sum_{m=1}^{C} \| B_{i,m}^n \|_2 \tag{2}$$

where the dictionary $D_{/C+1}$ denotes the structured dictionary D when removing the visual atoms associated with the background class $C+1$. Instead of using the overall dictionary D for feature representation, the shared visual patterns of all classes corresponding to potential background/common information can be separated by the dictionary $D_{/C+1}$. B_i^n is the representation coefficients of X_i^n on dictionary $D_{/C+1}$. In the feature coding step, we divide the representation coefficients into C non-overlapping groups, where $B_{i,m}^n$, $m = 1, 2, \ldots C$ denotes the m-th group of representation coefficients B_i^n. The entry indexes of $B_{i,m}^n$ is associated to the class-specific dictionary D_m. This feature representation strategy with l_1/l_2-norm sparsity encourages the dictionary $D_{/C+1}$ to represent feature sample by selecting the groups of visual atoms corresponding to the class-specific

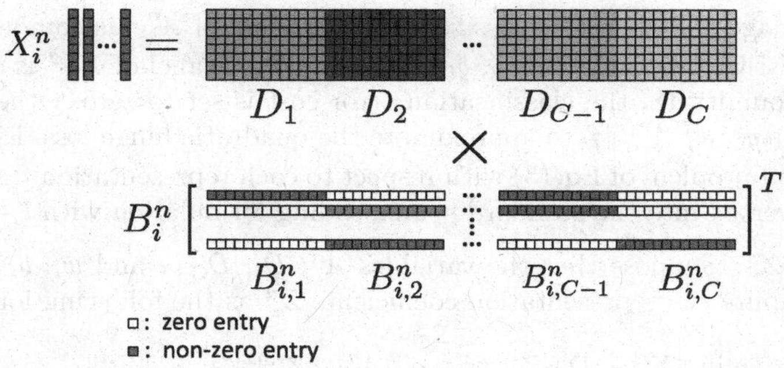

X_i^n

$D_1 \quad D_2 \qquad D_{C-1} \quad D_C$

$\times$

$B_i^n \begin{bmatrix} & & & \\ & & & \\ & & & \end{bmatrix}^T$

$B_{i,1}^n \quad B_{i,2}^n \qquad B_{i,C-1}^n \quad B_{i,C}^n$

□: zero entry
■: non-zero entry

Fig. 2. Visual interpretation of the proposed feature coding strategy.

dictionaries. Therefore, the discriminative power of feature representation can be promoted effectively. ρ is a weighting parameter to balance the reconstruction term and the sparsity constrain in the objective function.

Figure 2 depicts the proposed feature representation property. In our method, we use the SLEP tool [29] to solve the minimization problem of Eq. (2). Once all the feature representations of region proposals within an image are computed, we then use them to construct the image-level feature by max-pooling all the object proposal features in an image for the recognition task.

3.4 Optimization Algorithm for Dictionary Learning

To solve the OADL problem, we derive an iterative algorithm to optimize the objective with respect to the variables in Eq. (1) alternatively. The detailed optimization procedures can be divided into the following five sub-problems: (1) updating variable A_i with fixed variables Z_i, D_i, D_{C+1} and w_i, b_i; (2) computing Z_i by fixing A_i, D_i, D_{C+1} and w_i, b_i; (3) updating dictionary D_i when fixing A_i, Z_i, D_{C+1} and w_i, b_i; (4) updating dictionary D_{C+1} with fixed A_i, Z_i, D_i, w_i, b_i. (5) updating w_i, b_i while fixing variables A_i, Z_i, D_i, D_{C+1}. We now describe each subproblem in detail.

Updating A_i^n: With fixing the representation coefficients Z_i^n, dictionaries D_i, D_{C+1} and classifier parameters w_i, b_i, we can reduce the objective function of Eq. (1) with respect to A_i^n into the following optimization task:

$$\min_{A_i^n} \|X_i^n - DA_i^n\|_F^2 + \alpha \|A_i^n - Z_i^n\|_F^2 + R(w_i, b_i, A_i^n, y_i) + \lambda_1 \|A_i^n\|_1. \qquad (3)$$

We note that the objective can be optimized by considering each region proposal separately as they are decoupled. The representation coefficients A_i^n can be rewritten as $A_i^n = [A_i^{n,1}, A_i^{n,2}, \ldots, A_i^{n,P_i^n}] \in \mathbf{R}^{K \times P_i^n}$, where $A_i^{n,j} \in \mathbf{R}^{K \times 1}$, $j = 1, 2, \ldots, P_i^n$ denotes the representation feature of j-th region proposal in the n-th image from class i, P_i^n is the number of region proposals in the n-th image from class i. We first assign the image label to these region proposals associated with this image and then the classification error cost of $A_i^{n,j}$ is computed using a linear SVM with parameters w_i, b_i. If the predicted label of $A_i^{n,j}$ is consistent with the groundtruth, the classification error cost is set to zero. Otherwise, we use $\|b_i - 1 + y_i \cdot w_i^T A_i^{n,j}\|_F^2$ to approximate the quadratic hinge loss. Finally, the minimization problem of Eq. (3) with respect to each representation feature $A_i^{n,j}$ can be converted into the standard sparse coding formulation with l_1-norm.

Updating Z_i^n: Suppose that the variables A_i^n, D_i, D_{C+1} and w_i, b_i are fixed, we can compute the representation coefficients Z_i^n as the following form:

$$\min_{Z_i^n} \|X_i^n - D_{\in i} Z_i^n\|_F^2 + \alpha \|A_i^n - Z_i^n\|_F^2 + \lambda_2 \|Z_i^n\|_1 . \tag{4}$$

We can rewrite the above equation into a standard form,

$$\min_{Z_i^n} \left\| \begin{pmatrix} X_i^n \\ \sqrt{\alpha} A_i^n \end{pmatrix} - \begin{pmatrix} D_{\in i} \\ \sqrt{\alpha} I \end{pmatrix} Z_i^n \right\|_F^2 + \lambda_2 \|Z_i^n\|_1 \tag{5}$$

where $I \in \mathbf{R}^{K \times K}$ denotes an identity matrix, K is the number of visual atoms in structured dictionary D. Let $\widetilde{X_i^n} = (X_i^n, \sqrt{\alpha} A_i^n)^T$, $\widetilde{D_{\in i}} = (D_{\in i}, \sqrt{\alpha} I)^T$, the minimization formulation of Eq. (5) is converted to a sparse coding problem. In our method, we use SPAMS solver [30] to find the optimal solution.

Updating D_i: When the representation coefficients A_i, Z_i, background dictionary D_{C+1} and classifier parameters w_i, b_i are fixed, each class-specific dictionary D_i can be updated separately. We compute the dictionary D_i by removing the terms that are independent of class-specific dictionary D_i, and the optimization objective function (1) with respect to D_i is reduced to the following form:

$$\min_{D_i} \|X_i - D_{\in i} Z_i\|_F^2 + \|X_i - DA_i\|_F^2 + \beta \sum_{j=1, j \neq i}^{C+1} \|D_i^T D_j\|_F^2 . \tag{6}$$

To find the optimal dictionary, we propose to compute each visual atom of dictionary $D_i = [d_i^1, d_i^2, \ldots, d_i^{K_i}] \in \mathbf{R}^{d \times K_i}$ one by one. Specifically, when we compute the t-th visual atom d_i^t, the other visual atoms of D_i are fixed. We rewrite the representation coefficients Z_i and A_i as $Z_i = [z_i^1; z_i^2; \ldots; z_i^K] \in \mathbf{R}^{K \times N_i}$, $A_i = [a_i^1; a_i^2; \ldots; a_i^K] \in \mathbf{R}^{K \times N_i}$, where $z_i^t \in \mathbf{R}^{1 \times N_i}$, $a_i^t \in \mathbf{R}^{1 \times N_i}$, $t = 1, 2, \ldots, K$ denote the t-th row vector of Z_i and A_i, respectively. To update visual atom d_i^t, we let the first derivative of d_i^t equal to zero. Therefore, the t-th visual atom in dictionary D_i is computed as the closed-form

$$d_i^t = (\|z_i^t\|_2^2 I + \|a_i^t\|_2^2 I + \beta H_1 H_1^T)^{-1} \cdot (Y_1 \cdot z_i^{t^T} + Y_2 \cdot a_i^{t^T}) \tag{7}$$

where $Y_1 = X_i - \sum_{u=1, u \neq t}^{K_i} d_i^u z_i^u$, $Y_2 = X_i - \sum_{u=1, u \neq t}^{K_i} d_i^u a_i^u - \sum_{h=1, h \neq i}^{C+1} D_h A_i^h$ and $H_1 = [D_1, D_2, \ldots, D_{i-1}, \mathbf{O}_{d \times K_i}, D_{i+1} \ldots, D_C, D_{C+1}]$. A_i^h denotes the submatrix of representation coefficients A_i corresponding to the indexes of h-th class, $\mathbf{O}$ is a zero matrix associated with the indexes of class-specific dictionary D_i.

As an visual atom in dictionary, the atom d_i^t is further normalized by the l_2-norm, i.e. $\hat{d}_i^t = d_i^t / \|d_i^t\|_2$. Similarly, we can compute all the visual atoms of class-specific dictionary D_i.

Updating D_{C+1}: In order to compute the background dictionary D_{C+1}, the other variables A_i, Z_i, D_i, w_i, b_i are fixed. When removing the independent terms with respect to D_{C+1}, the minimization formulation of Eq. (1) is converted into the following optimization problem.

$$\min_{D_{C+1}} \|X_i - DA_i\|_F^2 + \beta \sum_{j=1, j \neq C+1}^{C+1} \|D_{C+1}^T D_j\|_F^2 \tag{8}$$

In our method, we update each visual atom of background dictionary D_{C+1} one by one. When the t-th visual atom is updated, the rest of visual atoms in D_{C+1} are kept fixed, and we can compute the t-th visual atom d_{c+1}^t by a closed-form solution,

$$d_{c+1}^t = (\|a_i^t\|_2^2 I + \beta H_2 H_2^T)^{-1} \cdot (Y_3 \cdot a_i^{t^T}) \tag{9}$$

where $Y_3 = X_i - \sum_{u=1, u \neq t}^{K_{C+1}} d_{c+1}^u a_i^u - \sum_{h=1, h \neq C+1}^{C+1} D_h A_i^h$ and $H_2 = [D_1, D_2, \ldots, D_C, \mathbf{O}_{d \times K_{C+1}}]$. Here a_i^t, $t = 1, 2, \ldots K$ denote the t-th row vector of representation coefficients A_i, and A_i^h is the sub-matrix of A_i corresponding to the h-th class, $\mathbf{O}$ denotes a zero matrix associated with background dictionary D_{C+1}. The updated dictionary atom is then normalized by the l_2-norm. Once all the visual atoms in D_{C+1} is computed, the background dictionary D_{C+1} is updated.

Updating w_i, b_i: To update the classifier parameters w_i, b_i, the other variables are fixed. In our method, we cast the SVM classifier learning problem with C classes into the C one-vs-all SVM sub-problems. More specifically, we first assign the image label to the region proposals in that image. Then all the feature representations of region proposals across all classes are used to train multiple SVM classifiers. In our OADL, a linear SVM solver [28] is adopted to learn the parameters of SVM classifiers.

Initialization: To start the iterative procedure, we need to initialize the variables $\{D_i, Z_i, A_i, i = 1, 2, \ldots, C\}$ and D_{C+1}. For class-specific dictionary D_i, it is initialized by the K-SVD [15] algorithm using all the region proposals of images from class i. We also adopt the K-SVD algorithm to initialize the background dictionary D_{C+1} using the region proposals of training samples across all classes. The representation coefficients Z_i and A_i are initialized by solving the sparse coding problem with $l_{2,1}$-norm: $\min_{Z_i} \|X_i - D_{\in i} Z_i\|_F^2 + \rho_1 \|Z_i\|_{2,1}$, $\min_{A_i} \|X_i - DA_i\|_F^2 + \rho_2 \|A_i\|_{2,1}$, respectively. ρ_1, ρ_2 are the scale parameters to balance the different energy terms. The proposed optimization procedures of the OADL objective are summarized in Algorithm 1.

Algorithm 1. Object-Aware Dictionary Learning

Input: training samples $X = [X_1, X_2, \ldots, X_C]$, the number of visual atoms K_i, $i = 1, 2, \ldots, C, C + 1$ for each class-specific dictionary and an additional background dictionary, parameters α, β, λ_1 and λ_2.

Output: class-specific dictionary D_i, $i = 1, 2, \ldots, C$, background dictionary D_{C+1}

1: **initialize** D_i, Z_i, A_i, $i = 1, 2, \ldots, C$, and D_{C+1}
2: **while** not convergence and the maximum number of iterations is not reached **do**
3:    **for** $i = 1 \longrightarrow C$ **do**
4:        update representation coefficients A_i^n by solving Eq. (3);
5:        update representation coefficients Z_i^n using Eq. (5);
6:        **for** $t = 1 \longrightarrow K_i$ **do**
7:            update the t-th visual atom d_i^t of class-specific dictionary D_i by Eq. (7);
8:        **end for**
9:        **for** $t = 1 \longrightarrow K_{C+1}$ **do**
10:            update the t-th visual atom d_{c+1}^t in background dictionary D_{C+1} by Eq. (9);
11:        **end for**
12:        compute the parameters w_i, b_i of SVM classifier ;
13:    **end for**
14: **end while**

---

4 Experiments

In this section, we verify the effectiveness of our method with other competing methods on the UIUC8 Sport [31] and Graz-02 [32] public datasets. The goal of our OADL method is to learn a feature subspace for improving the discriminative power of feature representation. In the experiments, we adopt two convolutional neural networks: VGG-F [22] and VGG-VeryDeep19 [23] models to generate the CNN features for evaluating our method. Specifically, the deep features generated by the last fully-connected layer in CNN models [22, 23] before the 1000-way softmax operation are used to learn our dictionary representations. The OADL(F) and OADL(VD19) denote our method integrated with the different deep features for brevity. For the OADL method, the dimension of deep feature is reduced to 1000 by PCA in the experiments. The weighting parameters $\alpha, \beta, \lambda_1, \lambda_2$ of the OADL model are empirically set as 0.01, constant η is set to 0.2. In the recognition stage, the scale parameter ρ for the regularization term of group sparsity is set to 0.5. Finally, the obtained global image representation is fed into a linear SVM classifier for predicting the label of image.

4.1 UIUC8 Sport Dataset

We first evaluate our method and several competing approaches on the UIUC8 Sport [31] event recognition dataset. This dataset have eight sport classes and 1792 images in total, including rowing, badminton, polo, bocce, snowboarding, croquet, sailing and rockclimbing. The number of images from each class varies from 137 to 250. Several example images of this dataset are shown in Fig. 3.

Fig. 3. Sample images from different classes on the UIUC8 Sport dataset.

Following the common experimental setting on this dataset [33], we randomly choose 70 images from each class as the training samples and randomly select 60 images from the rest images as the testing samples in the experiment. As for our OADL model, we learn the class-specific dictionary with 200 visual atoms for each class. The number of visual atoms in background dictionary is also set to 200. We show the confusion matrices of our method on the UIUC8 Sport dataset in Fig. 4. More specifically, the confusion matrix of OADL incorporating with the deep feature by VGG-F model is demonstrated in Fig. 4(a). Figure 4(b) shows the obtained confusion matrix by our OADL with the deep feature of VGG-VeryDeep19 model on the UIUC8 Sport dataset. Moreover, we evaluate our method with several competing approaches on this dataset, such as KSPM [34], ScSPM [28], LLC [35], KSVD [15], SPMSM [36], LRSC [37], VLAD [38], VC+VQ [39], OB [40], ISPR [41], RSP [42], LPR [43], LSC [33], LScSPM [44], Fusion [45], DSFL+DdCAF [46] and the two deep features by VGG-F [22] and VGG-VD19 [23] models. The recognition results of different methods are summarized in Table 1. It is noticed that our OADL model outperforms the state-of-the-art methods, including the recent dictionary learning and two powerful deep learning based image classification approaches. Finally, our method achieves the highest performance on the UIUC8 Sport dataset. In addition, we can observe that the OADL(VD19) gains the better recognition accuracy than the OADL(F). It indicates the discrimination of feature representation by our OADL method can be further enhanced with the increase of depth in convolutional network.

Furthermore, we evaluate the effects of different components in the proposed scheme. In details, we first test the sensitivity of region proposal in the overall framework. Then we verify the impact of dictionary learning component of our scheme. Table 2 summarizes the recognition results on the UIUC8 Sport dataset. Specifically, *Proposed (overlap = 0.5)*, *Proposed (overlap = 0.8)* denote our approach with different overlap rate of NMS for generating the region proposals by the EdgeBox [27] objectness method. The *Proposed (EdgeBox, dim = 300)* and *Proposed (Selective Search, dim = 300)* denote the proposed approach integrated with different objectness methods [27,47] when the dimensionality of deep feature is reduced to 300 by PCA, respectively. *Baseline (NoDL)* is a designed baseline method that purely uses the extracted 4096-dimensional CNN features

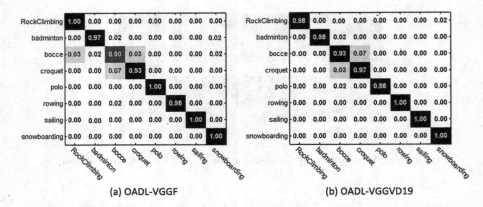

Fig. 4. Confusion matrices on the UIUC8 Sport dataset for our method. (a) Confusion matrix with the deep feature of VGG-F model. (b) Confusion matrix with the deep feature of VGG-VeryDeep19 model.

Table 1. Performance comparisons between our method and the state-of-the-art approaches on the UIUC8 Sport dataset.

Method	Accuracy (%)	Method	Accuracy (%)
KSPM [34]	79.98	RSP [42]	79.6
ScSPM [28]	82.74	LPR [43]	86.25
LLC [35]	81.77	LSC [33]	82.79
K-SVD [15]	82.21	LScSPM [44]	85.31
SPMSM [36]	83.0	Fusion [45]	94.8
LRSC [37]	88.17	DSFL+DdCAF [46]	96.78
VLAD [38]	79.16	VGG-F [22]	94.5
VC+VQ [39]	88.4	VGG-VD19 [23]	95.45
OB [40]	77.88	OADL(F)	96.9
ISPR [41]	89.5	OADL(VD19)	**98.09**

of region proposals [27] within image to form the global image feature by max-pooling. Table 2 shows that the component of region proposal has only slight effect on the recognition accuracy. Besides, it is noticed that our dictionary learning objective has the significant impact to the performance of overall scheme.

4.2   Graz-02 Dataset

The Graz-02 dataset contains 1096 images with three classes, including bike, car and people. It is also a challenge object recognition dataset because the objects from each class have the large intra-class differences in location, scale and viewpoint, as shown in Fig. 5. The effectiveness of our method is also tested on this dataset following the standard evaluation setting [32]. In detail, the

Table 2. The effects of different components in the proposed approach.

Variants of proposed method	Accuracy (%)
Proposed (overlap = 0.5)	97.4
Proposed (overlap = 0.8)	**98.09**
Proposed (EdgeBox [27], dim = 300)	93.11
Proposed (Selective Search [47], dim = 300)	93.81
Baseline (NoDL)	93.08

Fig. 5. Sample images of different classes from the Graz-02 dataset.

class-specific dictionary with 400 visual atoms is learned for each class. For the background dictionary in OADL model, the number of visual atoms is also set to 400 in the experiment. Moreover, we compare the OADL method with several competing approaches [48–50] and the two CNN features by VGG-F [22] and VeryDeep-19 [23] models on this dataset. The recognition results of different methods are summarized in Table 3. As can been seen, our OADL method is superior to the deep features and other competing approaches on the Graz-02 dataset. It is also observed that the discriminative power of feature generated by our OADL method can be promoted effectively with the increasing depth of convolutional network.

In addition, we give the computation time of our method on the Graz-02 dataset in Fig. 6. Specifically, the number of training samples across all classes is first fixed to 8874 in the experiments, then we vary the number of visual atoms

Table 3. Recognition results of different methods on the Graz-02 dataset.

Method	Bike	Car	People	Total
[48]	89.5	80.2	85.2	84.9
[49]	91.2	87.5	85.3	88.0
[50]	-	-	-	82.2
VGG-F [22]	94.44	96.05	85.71	92.48
VGG-VD19 [23]	96.91	**97.74**	89.29	94.98
OADL(F)	98.15	97.18	88.57	94.99
OADL(VD19)	**98.77**	**97.74**	**91.43**	**96.24**

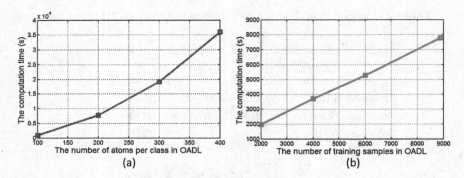

Fig. 6. Computation time analysis on the Graz-02 dataset. (a) Training time of OADL as a function of the number of visual atoms per class. (b) Running time of OADL with the increase of training samples across all classes.

per class with $[100, 200, 300, 400]$. The running time of OADL during each iteration as a function of the number of visual atoms is shown in Fig. 6(a). We can see that the computation time of OADL increases with the number of visual atoms per class gradually. With fixed the number of visual atoms per class to 200, we change the number of training samples from all classes in the range $[8874, 6000, 4000, 2000]$. Figure 6(b) demonstrates that the runtime of OADL increases with the growth of training samples. All experiments are performed using a single CPU Intel Core at 3.0 GHz.

5   Conclusion

Visual dictionary learning provides a data-driven manner to represent image data as a linear combination of a few atoms from an over-complete dictionary. However, a critical problem of existing dictionary learning approaches is that they do not focus on the important image regions explicitly. Thus, discriminative semantic information within image regions cannot be selected effectively for the recognition task during dictionary learning procedure. Currently, the convolutional neural network (CNN) has the capacity to combine the global and local information within image by means of designed specific filter layers with the increasingly receptive fields. Motivated by the advantage of deep feature, we proposed an object-aware dictionary learning framework that integrates the deep features and region proposals to overcome this problem. Instead of extracting a dictionary from all the fixed size of image patches or entire image, our method focuses on the small amounts of object candidates, which ensure the local semantic information can be encoded into the feature representation of image. We treat this as a unified optimization problem and derive an iterative algorithm to solve it. The experimental results on two public benchmark datasets demonstrate that our method outperforms the state-of-the-art dictionary learning and several deep learning based image classification approaches.

References

1. Elad, M., Aharon, M.: Image denoising via learned dictionaries and sparse representation. In: CVPR (2006)
2. Fu, Y., Lam, A., Sato, I., Sato, Y.: Adaptive spatial-spectral dictionary learning for hyperspectral image denoising. In: ICCV (2015)
3. Bao, C., Cai, J.F., Ji, H.: Fast sparsity-based orthogonal dictionary learning for image restoration. In: ICCV (2013)
4. Wang, S., Zhang, L., Liang, Y., Pan, Q.: Semi-coupled dictionary learning with applications to image super-resolution and photo-sketch synthesis. In: CVPR (2012)
5. Huang, D.A., Wang, Y.C.F.: Coupled dictionary and feature space learning with applications to cross-domain image synthesis and recognition. In: ICCV (2013)
6. Wang, N., Wang, J., Yeung, D.: Online robust non-negative dictionary learning for visual tracking. In: ICCV (2013)
7. Yang, M., Zhang, D., Feng, X., Zhang, D.: Fisher discrimination dictionary learning for sparse representation. In: ICCV (2011)
8. Zhou, N., Shen, Y., Peng, J., Fan, J.: Learning inter-related visual dictionary for object recognition. In: CVPR (2012)
9. Jiang, Z., Lin, Z., Davis, L.: Label consistent K-SVD: learning a discriminative dictionary for recognition. IEEE Trans. Pattern Anal. Mach. Intell. **35**, 2651–2664 (2013)
10. Gu, S., Zhang, L., Zuo, W., Feng, X.: Projective dictionary pair learning for pattern classification. In: NIPS (2014)
11. Cai, S., Zuo, W., Zhang, L., Feng, X., Wang, P.: Support vector guided dictionary learning. In: Fleet, D., Pajdla, T., Schiele, B., Tuytelaars, T. (eds.) ECCV 2014. LNCS, vol. 8692, pp. 624–639. Springer, Heidelberg (2014). doi:10.1007/978-3-319-10593-2_41
12. Mairal, J., Ponce, J., Sapiro, G., Zisserman, A., Bach, F.R.: Supervised dictionary learning. In: Koller, D., Schuurmans, D., Bengio, Y., Bottou, L. (eds.) NIPS (2009)
13. Yang, J., Yu, K., Huang, T.: Supervised translation-invariant sparse coding. In: CVPR (2010)
14. Ramirez, I., Sprechmann, P., Sapiro, G.: Classification and clustering via dictionary learning with structured incoherence and shared features. In: CVPR (2010)
15. Aharon, M., Elad, M., Bruckstein, A.: K-SVD: an algorithm for designing overcomplete dictionaries for sparse representation. IEEE Trans. Signal Process. **54**, 4311–4322 (2006)
16. Zhang, Q., Li, B.: Discriminative K-SVD for dictionary learning in face recognition. In: CVPR (2010)
17. Yang, L., Jin, R., Sukthankar, R., Jurie, F.: Unifying discriminative visual codebook generation with classifier training for object category recognition. In: CVPR (2008)
18. Mairal, J., Bach, F., Ponce, J., Sapiro, G., Zisserman, A.: Discriminative learned dictionaries for local image analysis. In: CVPR (2008)
19. Gao, S., Tsang, I.H., Ma, Y.: Learning category-specific dictionary and shared dictionary for fine-grained image categorization. IEEE Trans. Image Process. **23**, 623–634 (2014)
20. Lee, H., Battle, A., Raina, R., Ng, A.Y.: Efficient sparse coding algorithms. In: NIPS (2007)

21. Krizhevsky, A., Sutskever, I., Hinton, G.E.: ImageNet classification with deep convolutional neural networks. In: NIPS (2012)
22. Chatfield, K., Simonyan, K., Vedaldi, A., Zisserman, A.: Return of the devil in the details: delving deep into convolutional nets. In: BMVC (2014)
23. Simonyan, K., Zisserman, A.: Very deep convolutional networks for large-scale image recognition. In: ICLR (2014)
24. Alexe, B., Deselaers, T., Ferrari, V.: What is an object? In: CVPR (2010)
25. Manen, S., Guillaumin, M., Gool, L.V.: Prime object proposals with randomized prim's algorithm. In: ICCV (2013)
26. Cheng, M.M., Zhang, Z., Lin, W.Y., Torr, P.H.S.: BING: Binarized normed gradients for objectness estimation at 300 fps. In: CVPR (2014)
27. Zitnick, C.L., Dollár, P.: Edge boxes: locating object proposals from edges. In: Fleet, D., Pajdla, T., Schiele, B., Tuytelaars, T. (eds.) ECCV 2014. LNCS, vol. 8693, pp. 391–405. Springer, Heidelberg (2014). doi:10.1007/978-3-319-10602-1_26
28. Yang, J., Yu, K., Gong, Y., Huang, T.: Linear spatial pyramid matching using sparse coding for image classification. In: CVPR (2009)
29. Liu, J., Ji, S., Ye, J.: SLEP: sparse learning with efficient projections. Arizona State University (2009)
30. Mairal, J., Bach, F., Ponce, J., Sapiro, G.: Online learning for matrix factorization and sparse coding. J. Mach. Learn. Res. 11, 19–60 (2010)
31. Li, L.J., Fei-Fei, L.: What, where and who? Classifying events by scene and object recognition. In: ICCV (2007)
32. Marszałek, M., Schmid, C.: Accurate object localization with shape masks. In: CVPR (2007)
33. Liu, L., Wang, L., Liu, X.: In defense of soft-assignment coding. In: ICCV (2011)
34. Lazebnik, S., Schmid, C., Ponce, J.: Beyond bags of features: spatial pyramid matching for recognizing natural scene categories. In: CVPR (2006)
35. Wang, J., Yang, J., Yu, K., Lv, F., Huang, T., Gong, Y.: Locality-constrained linear coding for image classification. In: CVPR (2010)
36. Kwitt, R., Vasconcelos, N., Rasiwasia, N.: Scene recognition on the semantic manifold. In: Fitzgibbon, A., Lazebnik, S., Perona, P., Sato, Y., Schmid, C. (eds.) ECCV 2012. LNCS, vol. 7575, pp. 359–372. Springer, Heidelberg (2012). doi:10.1007/978-3-642-33765-9_26
37. Zhang, T., Ghanem, B., Liu, S., Xu, C., Ahuja, N.: Low-rank sparse coding for image classification. In: ICCV (2013)
38. Jégou, H., Douze, M., Schmid, C., Pérez, P.: Aggregating local descriptors into a compact image representation. In: CVPR (2010)
39. Li, Q., Wu, J., Tu, Z.: Harvesting mid-level visual concepts from large-scale internet images. In: CVPR, pp. 851–858 (2013)
40. Li, L.-J., Su, H., Lim, Y., Fei-Fei, L.: Objects as attributes for scene classification. In: Kutulakos, K.N. (ed.) ECCV 2010. LNCS, vol. 6553, pp. 57–69. Springer, Heidelberg (2012). doi:10.1007/978-3-642-35749-7_5
41. Lin, D., Lu, C., Liao, R., Jia, J.: Learning important spatial pooling regions for scene classification. In: CVPR (2014)
42. Jiang, Y., Yuan, J., Yu, G.: Randomized spatial partition for scene recognition. In: Fitzgibbon, A., Lazebnik, S., Perona, P., Sato, Y., Schmid, C. (eds.) ECCV 2012. LNCS, vol. 7573, pp. 730–743. Springer, Heidelberg (2012). doi:10.1007/978-3-642-33709-3_52

43. Sadeghi, F., Tappen, M.F.: Latent pyramidal regions for recognizing scenes. In: Fitzgibbon, A., Lazebnik, S., Perona, P., Sato, Y., Schmid, C. (eds.) ECCV 2012. LNCS, vol. 7576, pp. 228–241. Springer, Heidelberg (2012). doi:10.1007/978-3-642-33715-4_17
44. Gao, S., Tsang, I.W.H., Chia, L.T., Zhao, P.: Local features are not lonely: Laplacian sparse coding for image classification. In: CVPR (2010)
45. Koskela, M., Laaksonen, J.: Convolutional network features for scene recognition. In: ACMMM (2014)
46. Zuo, Z., Wang, G., Shuai, B., Zhao, L., Yang, Q., Jiang, X.: Learning discriminative and shareable features for scene classification. In: Fleet, D., Pajdla, T., Schiele, B., Tuytelaars, T. (eds.) ECCV 2014. LNCS, vol. 8689, pp. 552–568. Springer, Heidelberg (2014). doi:10.1007/978-3-319-10590-1_36
47. Van de Sande, K.E.A., Uijlings, J.R.R., Gevers, T., Smeulders, A.W.M.: Segmentation as selective search for object recognition. In: ICCV (2011)
48. Tuytelaars, T.: Vector quantizing feature space with a regular lattice. In: ICCV (2007)
49. Krapac, J., Verbeek, J., Jurie, F.: Learning tree-structured descriptor quantizers for image categorization. In: BMVC (2011)
50. Hong, Y., Li, Q., Jiang, J., Tu, Z.: Learning a mixture of sparse distance metrics for classification and dimensionality reduction. In: ICCV (2011)

People Tracking and Action Recognition

Gait Energy Response Function for Clothing-Invariant Gait Recognition

Xiang Li[1,2], Yasushi Makihara[2](✉), Chi Xu[1,2], Daigo Muramatsu[2],
Yasushi Yagi[2], and Mingwu Ren[1]

[1] School of Computer Science and Engineering, Nanjing University of Science and Technology, Nanjing, China
lixiangmzlx@gmail.com, xuchisherry@gmail.com, renmingwu@mail.njust.edu.cn
[2] Institute of Scientific and Industrial Research, Osaka University, Osaka, Japan
{makihara,muramatsu,yagi}@am.sanken.osaka-u.ac.jp

Abstract. This paper describes a method of clothing-invariant gait recognition by modifying intensity response function of a silhouette-based gait feature. While a silhouette-based representation such as gait energy image (GEI) has been popular in gait recognition community due to its simple yet effective property, it is also well known that such a representation is susceptible to clothes variations since it significantly changes silhouettes (e.g., down jacket, long skirt). We therefore propose a gait energy response function (GERF) which transforms an original gait energy into another one in a nonlinear way, which increases discrimination capability under clothes variation. More specifically, the GERF is represented as a vector of components of a lookup table from an original gait energy to another one and its optimization process is formulated as a generalized eigenvalue problem considering discrimination capability as well as regularization on the GERF. In addition, we apply Gabor filters to the GEI transformed by the GERF and further apply a spatial metric learning method for better performance. In experiments, the OU-ISIR Treadmill dataset B with the largest clothing variation was used to measure the performance both in verification and identification scenarios. The experimental results show that the proposed method achieved state-of-the-art performance in verification scenarios and competitive performance in identification scenarios.

1 Introduction

Gait recognition [1] is one of behavioral biometrics and advantageous over the other biometrics (*e.g.*, face, iris, finger vein) because it can be used even at a distance from a camera since it does not require a high image resolution. In addition, gait is usually captured as an unconscious behavior, and hence it does not require subject cooperation in general. Due to these characteristics, it can be applied to many areas (*e.g.*, surveillance, forensics, criminal investigation [2–4]).

Gait recognition approaches can be divided into two main groups: model-based approaches [5–10] and model-free (appearance-based) approaches [11–16]. The model-based approaches have greater invariant properties and are better at

© Springer International Publishing AG 2017
S.-H. Lai et al. (Eds.): ACCV 2016, Part II, LNCS 10112, pp. 257–272, 2017.
DOI: 10.1007/978-3-319-54184-6_16

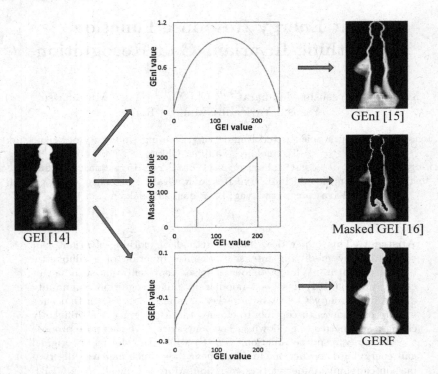

Fig. 1. Concept of the proposed GERF compared with existing intensity transformation-based approaches.

handling occlusion, noise, scale, and rotation. These approaches, however, require higher resolution images for model fitting and have relatively high computational cost.

The appearance-based approaches directly use input or silhouette images in a holistic way to extract gait features without modeling, and hence they generally work well even for relatively low-resolution images. In particular, silhouette-based representations such as gait energy image (GEI) [14], frequency-domain feature [17], chrono-gait image [18], Gabor GEI [19], are dominant in gait recognition community due to its simple yet effective property. The appearance-based approaches, however, often suffer from many covariates (*e.g.*, clothing, view, speed, and carrying status) since the appearance-based features of individuals are significantly affected by them, which causes a rapid decline in recognition rate. Among these covariates, clothing is one of the most challenge covariates [11,20–23].

There are two major categories to address the clothing-invariant problem in appearance based approaches: (1) spatial metric learning-based approaches and (2) intensity transformation-based approaches. In addition, the spatial metric learning-approaches further fall into two families: whole-based approaches [14–16,24] and part-based approaches [21–23,25,26].

While the whole-based approaches usually apply discriminative projections to the holistic appearance-based features (*e.g.*, linear discriminant analysis (LDA) [14,18,27] in conjunction with principal component analysis (PCA), discriminant analysis with tensor representation (DATER) [28,29], random subspace method (RSM) [23,26] to gain robustness to the clothes variation, the part-based approaches firstly divide the whole body into multiple body parts and then exploit the body parts which are not so much affected by the clothes variation by adaptively assigning weights for individual body parts [21] or finding the effective body parts [22], which mitigates the effect of clothes variation.

Whereas the above mentioned approaches mainly focus on the metric learning aspect, the intensity transformation-based approaches more focus on gait representation aspect. Since the clothes variation affects more on static parts (*e.g.*, torso and limb shapes) than on dynamic parts (*e.g.*, leg and arm motion), gait entropy image (GEnI) [15] extracts the dynamic parts from GEI by computing its Shannon entropy, where gait energy for each pixel is regarded as a foreground probability. For example, the pixels with large and small gray values (*e.g.*, 255 and 0) in GEI become small in GEnI, while the pixels with middle values (*e.g.*, 127) become large (see Fig. 1, the top row). The static parts (*i.e.*, complete foreground and background), however, still have discrimination capability to some extent even under clothes variation, and hence GEnI discards such useful information. Moreover, GEnI treats two different gait energies which are symmetric with just the middle value (*i.e.*, 127.5), as the same value, and hence it loses discrimination capability (*e.g.*, gait energies $64(= 127.5 - 63.5)$ and $191(= 127.5 + 63.5)$ returns the same value in GEnI).

In order to solve the latter problem, masked GEI [16] is proposed, where gait energies whose corresponding gait entropy is smaller than a certain threshold (*i.e.*, more static parts) are masked out and are set to zero, while the other gait energies are kept as their original values (see Fig. 1, the middle row). Masked GEI is, however, dependent on choice of the threshold to mask out and also still discards useful static information.

Because both GEnI and masked GEI are generated from GEI, we can regard this as a sort of gait energy transformation process via a gait energy response function (GERF). While both GEnI and masked GEI employ hand-crafted GERFs to focus on the dynamic parts, we may generate more discriminative features under clothes variation by designing the GERF in a more general and data-driven way.

We therefore propose to introduce the GERF to transform GEI into more discriminative feature and show its effectiveness on gait recognition under clothes variation. The contributions of this work are three-fold.

1. **A data-driven approach to intensity transformation**
   While the existing intensity transformation-based methods such as GEnI and masked GEI are designed in a handcrafted way, the proposed method learn the GERF in a data driven way. More specifically, we train the GERF so as to maximizing the discrimination capability using the training set including clothes variation. This enables us to realize a good tradeoff between static and dynamic parts, unlike the existing method discard the static parts.

2. **A closed-form solution to optimize the GERF**

We train the GERF so as to maximize dissimilarity for different subjects' pairs while to minimize dissimilarities for the same subjects' pairs, and consequently formulate its optimization process as a generalized eigenvalue problem. We therefore obtain an analytic solution in a closed form without any iterations and hence avoids troublesome convergence problems which is inseparable from a nonlinear optimization framework.

2. **State-of-the-art performance on clothing-invariant gait recognition**

We achieved the state-of-the-art performance on clothing-invariant gait recognition using publicly available gait database containing the largest clothes variations up to 32 types, in conjunction with Gabor filtering and spatial metric learning.

2   Gait Recognition Using GERF

2.1   Representation of GERF

In this section, we introduce the GERF for the most widely used gait feature, *i.e.*, GEI. For this purpose, we briefly describe the GEI at first. The GEI [14] a.k.a. averaged silhouette [30] is a size-normalized and registered silhouette averaged over one gait period (cycle) T defined as

$$I(x,y) = \frac{1}{T} \sum_{t=1}^{T} B(x,y,t), \tag{1}$$

where $B(x,y,t)$ is a size-normalized and registered binary silhouette value (0 and I_{max}[1] for background and foreground, respectively) at the position (x,y) at the n-th frame, and $I(x,y)$ is a gait energy (averaged silhouette) at the position (x,y). While the domain of the gait energy is real number, *i.e.*, $I(x,y) \in \mathbb{R}$, we approximate it as an integer number, *i.e.*, $I(x,y) \in \{0, 1, \ldots, I_{max}\}$ for simplicity.

A transformation from an original gait energy $I(x,y)$ to another one $I'(x,y)$ is then defined via the GERF f as

$$I'(x,y) = f(I(x,y)) \ \forall (x,y). \tag{2}$$

Since the original gait energy takes one of $(I_{max} + 1)$ integer numbers from 0 to I_{max}, the GERF is represented as a lookup table $\boldsymbol{f} = [f_0, ..., f_{I_{max}}]^T \in \mathbb{R}^{I_{max}+1}$, where f_i represent a transformed gait energy from an original gait energy i.

Next, we consider a dissimilarity measure between a pair of GEIs transformed from original GEIs I_1 and I_2. We simply adopt Euclidean distance between them and define its squared distance d_{I_1,I_2}^2 and further formulate it in a quadratic form of $\boldsymbol{f}$ as

$$d_{I_1,I_2}^2 = \sum_{x,y} (f_{I_1(x,y)} - f_{I_2(x,y)})^2 = \boldsymbol{f}^T A_{I_1,I_2} \boldsymbol{f}, \tag{3}$$

where $A_{I_1,I_2} \in \mathbb{R}^{(I_{max}+1) \times (I_{max}+1)}$ is a coefficient matrix for quadratic-form representation and its (l,m) component is obtained using the Kronecker delta $\delta_{i,j}$ as

---

[1] I_{max} is usually 255 for 8-bit depth.

$$(A_{I_1,I_2})_{l,m} = \sum_{x,y} (\delta_{I_1(x,y),l} \delta_{I_1(x,y),m} + \delta_{I_2(x,y),l} \delta_{I_2(x,y),m}$$

$$- \delta_{I_1(x,y),l} \delta_{I_2(x,y),m} - \delta_{I_2(x,y),l} \delta_{I_1(x,y),m}). \tag{4}$$

2.2 Training of GERF

In order to make the transformed GEI discriminative under clothes variation, we optimize the GERF using a training set including the clothes variation. The whole training set is composed of two subsets $\mathcal{S}$ and $\mathcal{D}$, where the subset $\mathcal{S}$ is a set of GEI pairs of the same subject, while the subset $\mathcal{D}$ is a set of GEI pairs of different subjects. For better discrimination, it is preferable to make it larger the sum of squared distances $D_\mathcal{S}$ for the same subject pairs $\mathcal{S}$ while make it smaller the sum of squared distances $D_\mathcal{D}$ for the different subject pairs. Here, $D_\mathcal{S}$ and $D_\mathcal{D}$ are calculated as

$$D_\mathcal{S} = \sum_{(I_1,I_2) \in \mathcal{S}} d_{I_1,I_2}^2 = \boldsymbol{f}^T S_\mathcal{S} \boldsymbol{f}$$

$$D_\mathcal{D} = \sum_{(I_1,I_2) \in \mathcal{D}} d_{I_1,l_2}^2 = \boldsymbol{f}^T S_\mathcal{D} \boldsymbol{f}, \tag{5}$$

where $S_\mathcal{S} \in \mathbb{R}^{(I_{max}+1) \times (I_{max}+1)}$ and $S_\mathcal{D} \in \mathbb{R}^{(I_{max}+1) \times (I_{max}+1)}$ are computed as $S_\mathcal{S} = \sum_{(I_1,I_2) \in \mathcal{S}} A_{I_1,I_2}$ and $S_\mathcal{D} = \sum_{(I_1,I_2) \in \mathcal{D}} A_{I_1,I_2}$, respectively.

Moreover, in order to make the GERF smoother, we also introduce a regularizer D_R, which is defined as

$$D_R = w_1 \sum_{i=1}^{I_{max}} (f_i - f_{i-1})^2 + w_2 \sum_{i=1}^{I_{max}-1} (f_{i+1} - 2f_i + f_{i-1})^2$$

$$= \boldsymbol{f}^T (w_1 S_{R_1} + w_2 S_{R_2}) \boldsymbol{f}$$

$$= \boldsymbol{f}^T S_R \boldsymbol{f}, \tag{6}$$

where w_1 and w_2 are weighting parameters for the first-order and second-order smoothness, and $S_{R_1} \in \mathbb{R}^{(I_{max}+1) \times (I_{max}+1)}$ and $S_{R_2} \in \mathbb{R}^{(I_{max}+1) \times (I_{max}+1)}$ are coefficients matrices for the first-order and the second-order smoothness, which are defined as

$$S_{R_1} = \begin{bmatrix} 1 & -1 & 0 & \cdots & 0 \\ -1 & 2 & -1 & \ddots & \vdots \\ 0 & \ddots & \ddots & \ddots & 0 \\ \vdots & \ddots & -1 & 2 & -1 \\ 0 & \cdots & 0 & -1 & 0 \end{bmatrix}, \quad S_{R_2} = \begin{bmatrix} 1 & -2 & 0 & \cdots\cdots\cdots & 0 \\ -2 & 5 & -4 & \ddots & & \vdots \\ 0 & -4 & 6 & -4 & \ddots & \vdots \\ \vdots & \ddots & \ddots & \ddots & \ddots & \vdots \\ \vdots & & \ddots & -4 & 6 & -4 & 0 \\ \vdots & & & \ddots & -4 & 5 & -2 \\ 0 & \cdots\cdots\cdots & & 0 & -2 & 1 \end{bmatrix}. \tag{7}$$

Finally, the GERF is optimized so as to maximize the ratio between the sum of squared distances $D_\mathcal{D}$ for the different subject pairs and those $D_\mathcal{S}$ for the same subject pairs plus the regularizer D_R under an L_2 norm constraint on $\boldsymbol{f}$ as

$$\boldsymbol{f}^* = \arg\max_{\boldsymbol{f}} \frac{\boldsymbol{f}^T S_\mathcal{D} \boldsymbol{f}}{\boldsymbol{f}^T (S_\mathcal{S} + S_R) \boldsymbol{f}} \quad \text{s.t.} \quad \|\boldsymbol{f}\| = 1. \tag{8}$$

In an analogous fashion to well-known LDA formulation, we can formulate this optimization problem as the following generalized eigenvalue problem

$$S_\mathcal{D} \boldsymbol{f} = \lambda (S_\mathcal{S} + S_R) \boldsymbol{f} \quad \text{s.t.} \quad \|\boldsymbol{f}\| = 1, \tag{9}$$

where λ is an eigenvalue, and $\boldsymbol{f}$ is regarded as a corresponding eigenvector. We therefore analytically obtain the optimal GERF $\boldsymbol{f}^*$ in a closed-form solution by assigning the eigenvector corresponding to the largest eigenvalue.

2.3   Gabor Filtering

In order to further improve the performance, we introduce two sequential processes after obtaining a GEI transformed with the optimal GERF (call it GEI-GERF later), since the proposed GERF can be jointly used with other filtering and spatial metric learning techniques.

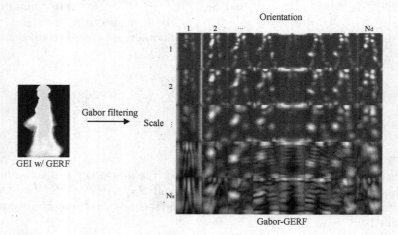

Fig. 2. An example of Gabor-GERF. The rows show different scales and the columns show different orientations. In this figure, $N_s = 5$ and $N_d = 8$.

The first one is Gabor filtering, which has been successfully employed in gait recognition because of its effectiveness [19,31]. In a similar way to [19], we will briefly describe Gabor functions. The Gabor function are defined by multiplying an elliptical Gaussian envelope function with a complex oscillation, defined as

$$\psi_{s,d}(\boldsymbol{p}) = \frac{|k_{s,d}|^2}{\delta^2} \exp\left\{-\frac{|k_{s,d}|^2 \|\boldsymbol{p}\|^2}{2\delta^2}\right\} \left[\exp\left(\boldsymbol{k}(jk_{s,d}) \cdot \boldsymbol{p}\right) - \exp\left(-\frac{\delta^2}{2}\right)\right], \tag{10}$$

where a vector $p = [x, y]^T$ is the spatial location in Gabor kernel window, a complex number $k_{s,d} = \theta_s e^{j\phi_d}$ determines the scale ($s = 0, \ldots, N_s - 1$) and orientation ($d = 0, \ldots, N_d - 1$) of the Gabor kernel function, j is an imaginary unit, and $k(\cdot)$ is a function to convert a complex number to a two-dimensional real vector. Specifically, $\theta_s = 2^{-s}(\pi/2)$ is the scale item, and $\phi_d = \pi d/N_d$ is the direction item. Since we have N_s scales and N_d orientations, we then get a total of $N_s N_d$ Gabor functions.

Given a GEI-GERF whose width and height are W and H, respectively, it is convolved with all the Gabor kernel functions and further down-sampled into half size (*i.e.*, W/2 by H/2) for computational efficiency in the same way as [31]. We then concatenate all $N_s N_d$ downsampled Gabor-filtered images into a single image, where scale and orientation components are concatenated along row and column directions, respectively. As a result, we obtain a concatenated image whose width and height are $W' = N_d W/2$ and $H' = N_s H/2$, respectively. In this paper, we call it Gabor-GERF later and show an example of the Gabor-GERF in Fig. 2.

2.4 Spatial Metric Learning

Once we obtain the Gabor-GERF, we introduce a spatial metric learning, *i.e.*, two-dimensional LDA (2DLDA) in conjunction with preceding dimension reduction by two-dimensional PCA (2DPCA) [32]. Unlike PCA and LDA handle one-dimensional vector unfolded from an image matrix, a covariance matrix for 2DPCA and within-class and between-class matrices for 2DLDA are directly constructed using the original image matrices and result in smaller size of covariance/within-class/between-class matrices, which ensures lower time complexity and less singularity than PCA and LDA, respectively. We therefore adopt a combination of 2DPCA and 2DLDA (call it 2DPCA+2DLDA later) for spatial metric learning.

Suppose that we have M samples of Gabor-GERFs $\{X_i \in \mathbb{R}^{H' \times W'}\}$($i = 1, \ldots, M$) in the training set, and its mean is denoted by $\bar{X}$. The covariance matrix $S_T \in \mathbb{R}^{W' \times H'}$ for 2DPCA (projection for column direction) is

$$S_T = \frac{1}{M} \sum_{i=1}^{M} (X_i - \bar{X})^T (X_i - \bar{X}). \tag{11}$$

We then obtain a projection matrix $P \in \mathbb{R}^{W' \times W''}$ composed of a set of W'' eigenvectors of the covariance matrix S_T. In this paper, we set the reduced dimension W'' so as to keep more than 99% variance (*i.e.*, less than 1% information loss).

After applying 2DPCA to the Gabor-GERF and obtaining projected matrices $Y_i = (X_i - \bar{X})P$, ($i = 1, \ldots, M$), we subsequently calculate the within-class scatter matrix $S_w \in \mathbb{R}^{H' \times H'}$ and between-class scatter matrix $S_b \in \mathbb{R}^{H' \times H'}$ as

$$S_w = \sum_{i=1}^{M} (Y_i - \bar{Y}_{l_i})(Y_i - \bar{Y}_{l_i})^T \tag{12}$$

$$S_b = \sum_{c=1}^{N_c} M_c(\bar{Y}_c - \bar{Y})(\bar{Y}_c - \bar{Y})^T, \tag{13}$$

where l_i is the class label (subject ID) for the i-th sample, $\bar{Y}_c$ is a mean for the c-th class, $\bar{Y}$ is a total mean, N_c is the number of classes, and M_c is the number of samples for the c-th class. Finally, the optimal projection $\boldsymbol{w}^*$ for 2DLDA is obtained as

$$\boldsymbol{w}^* = \arg\max_{\boldsymbol{w}} \frac{\boldsymbol{w}^T S_b \boldsymbol{w}}{\boldsymbol{w}^T S_w \boldsymbol{w}}. \tag{14}$$

We then reformulate Eq. (14) as a generalized eigenvalue problem and obtain a projection matrix $R \in \mathbb{R}^{H' \times H''}$ composed of a set of eigenvectors corresponding to the H'' largest eigenvalues.

Once we obtain the projection matrices P and R, we project the Gabor-GERF X_i into dimension reduced matrix Z_i in the 2DPCA+2DLDA space as

$$Z_i = R^T(X_i - \bar{X})P. \tag{15}$$

Finally, matching for a pair of Gabor-GERFs is done based on Euclidean distance in the 2DPCA+2DLDA space.

3   Experiments

3.1   Data Set

We used the OU-ISIR Gait Database, Treadmill Dataset B [33] for our experiments, since it has the largest clothing variations. It includes 68 subjects with at most 32 combinations of different clothing. The whole dataset is divided into three subsets: training set, gallery set, and probe set. In the training set, there are 446 sequences of 20 subjects with the range of 15 to 28 different combinations of clothing. The gallery set and probe set form the testing set composed of 48 subjects, which were disjoint from the 20 subjects in the training set. The gallery contains only standard clothing type (*i.e.*, regular pant and full shirt), while the probe set includes 856 sequences of other remaining clothing types.

3.2   Parameter Setting

There are two main hyper parameters of our GERF in the training stage: weighting parameters of the regularizer w_1 and w_2. We experimentally set $w_1 = w_2 = 5000$. About parameters in Gabor filtering, we set Gabor kernel window size to 41×41 and set parameter δ (in Eq. (10)) to 2π. The number of scales N_s and orientations N_d are set to 5 and 8, respectively. Since the silhouette image resolution provided in the database is 128×88, and hence the resolution of the Gabor-GERF is 320×352.

3.3 Comparison with Intensity Transformation-Based Methods

To investigate the effectiveness of the proposed GERF module, we firstly conducted comparison experiments with a family of intensity transformed-based methods, *i.e.*, GEnI and Masked GEI as well as GEI as a baseline.

We show examples for the four gait features, *i.e.*, GEI, GEnI, Masked GEI, and GEI w/GERF, as well as cropped original images in Fig. 3. Note that the trained GERF is depicted as a red curve at the bottom row of Fig. 1. The profile of the GERF for smaller gait energy (*e.g.*, gait energy from 0 to 127) is similar to the profile of GEnI, which suggests to emphasize the difference from background to middle-level gray value. On the other hand, the profile of the GERF for larger gait energy is approximately flat and hence the complete background and foreground is still differentiated, unlike GEnI or Masked GEI confuse it. In this way, the proposed GERF can highlight differences in dynamic parts on one hand, and it keeps static information on the other hand.

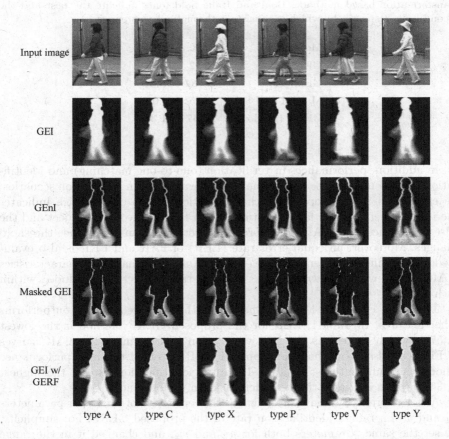

Fig. 3. Examples of extracted features for intensity transformation-based methods.

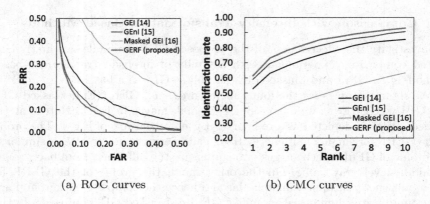

(a) ROC curves   (b) CMC curves

Fig. 4. ROC and CMC curves for intensity transformation-based methods.

Table 1. EER [%] and rank-1 identification rate (denoted as Rank-1) [%] for intensity transformation-based methods. Bold and Italic bold fonts indicate the best and the second best, respectively, which is consistent throughout this paper.

Method	EER	Rank-1
GEI [14]	16.12	52.80
GEnI [15]	*12.81*	*59.00*
Masked GEI [16]	28.15	28.04
GERF (proposed)	**11.57**	**61.33**

In addition, performances in verification (one-to-one matching) and identification (one-to-many matching) scenarios are evaluated. In verification scenarios, we employ an receiver operating characteristics (ROC) curve which indicates the tradeoff between the false rejection rate (FRR) of the same subject and the false acceptance rate (FAR) of different subjects when an acceptance threshold changes. Moreover, an equal error rate (EER) of FAR and FRR is also evaluated. In identification scenarios, we employ cumulative matching characteristics (CMC) curve which shows the rates that the true subjects are included within each of rank.

The ROC curves in Fig. 4(a) show that the proposed GERF outperforms other features. In Table 1, EER for the proposed GERF method is the lowest 11.57%, which represents the best verification performance. The CMC curves in Fig. 4(b) also show that the proposed GERF yielded the best performance among the four features. In Table 1, rank-1 identification rate is the highest 61.33%, which indicates the best identification performance.

As for reference, we have investigated the sensitivity of the hyper parameters w_1 and w_2 on rank-1 identification rate of the proposed GERF. For simplicity, we set the same parameters both for w_1 and w_2, and changed it in the range from 1 to 10,000 as shown in Fig. 5. As a result, rank-1 identification rate is not

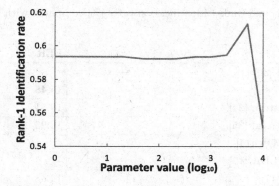

Fig. 5. Sensitivity analysis of the hyper parameters on rank-1 identification rates. The horizontal axis is shown by log-scale.

so much degraded for smaller range of the hyper parameters (less than 5,000) and is still better than the second best method, *i.e.*, GEnI with 59.0% rank-1 identification rate. It is therefore turned out that the proposed method is not so insensitive to the setting of parameter w_1 and w_2 as long as we use less than 5,000.

3.4 Comparison with the State-of-the-Arts Methods

In verification scenarios, we compare the proposed method with the frequency-domain feature (denoted as whole-based) [17], part-based method with adaptive weight control (denoted as part-based) [21], GEI with LDA (denoted as LDA) [34], SVB frieze pattern [24] and gait components-based method (denoted as components-based) [35] to confirm its effectiveness. The performance is evaluated by ROC curves in Fig. 6(a). As a result, the proposed method gets the state-of-the-art performance in contrast to the other methods.

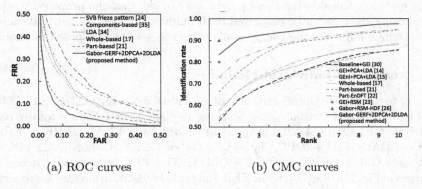

(a) ROC curves   (b) CMC curves

Fig. 6. ROC and CMC curves compared with the state-of-the-arts.

Table 2. EERs [%] and rank-1 identification rates [%] compared with the state-of-the-arts.

Method	EER	Rank-1
Baseline+GEI [30]	-	52.8
LDA [34]	15.48	-
SVB frieze pattern [24]	19.81	-
Components-based [35]	18.25	-
GEI+PCA+LDA [14]	-	54.3
GEnI+PCA+LDA [15]	-	57.4
Whole-based [17]	14.88	58.1
Part-based [21]	*10.26*	66.3
Part-EnDFT [22]	-	72.8
GEI+RSM [23]	Not applicable	80.4
Gabor+RSM-HDF [26]	Not applicable	**90.7**
Gabor-GERF+2DPCA+2DLDA (proposed method)	**6.19**	*83.4*

In identification scenarios, we compare the proposed method with the averaged silhouette (denoted as baseline+GEI) [30], GEI+PCA+LDA [14], GEnI+PCA+LDA [15], whole-based [17], part-based [21], part-EnDFT [22], GEI+RSM [23], Gabor+RSM-HDF [26] to confirm its effectiveness. Note that this different list of benchmarks in identification scenario from that in verification scenario comes from the difference in the availabilities of reported results in each paper. The performance is evaluated by CMC curve s in Fig. 6(b). In addition, the Rank-1 identification rate is shown in Table 2. The proposed method gets the second best performance, lower than the Gabor+RSM-HDF [26]. However, we need to point out that the RSM framework cannot guarantee a stable accuracy because of its randomness. Moreover, the RSM framework is only applicable to identification scenarios since it relies on a framework of majority voting to all of the galleries. Considering these points, we can say that the proposed method is promising since it can be employed both in identification and verification scenarios, which indicates the widely application range of the proposed method.

3.5   Analysis of Individual Modules

In order to investigate the effectiveness of individual modules (GEI/GEnI/GERF, Gabor filtering and spatial metric learning methods), we compare totally eight methods: GEI+2DPCA+2DLDA, GEnI+2DPCA+2DLDA, GERF, GERF+ PCA +LDA, GERF+2DPCA+2DLDA, Gabor-GERF, Gabor-GERF+PCA+ LDA and Gabor-GERF+2DPCA+2DLDA. The ROC and CMC curves are reported in Fig. 7(a) and (b), while EER and rank-1 identification rate are reported in Table 3, respectively. If we exclude Gabor filtering and 2DPCA+2DLDA from the full proposed method (Gabor-GERF+2DPCA+2DLDA), rank-1 identification rates drops by approximately 10%, and EER increases by approximately 2%, and hence we confirmed that Gabor filtering and spatial metric learning success-fully enhance the proposed GERF framework.

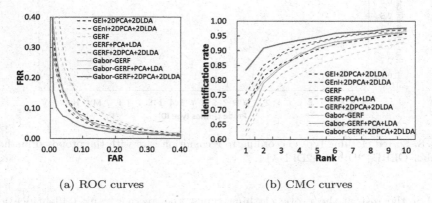

(a) ROC curves   (b) CMC curves

Fig. 7. ROC and CMC curves of GERF and Gabor-GERF w/metric learning.

Table 3. EER [%] and rank-1 identification rate (denoted as Rank-1) [%] of GERF and Gabor-GERF w/metric learning.

Method	EER	Rank-1
GEI+2DPCA+2DLDA	8.91	70.68
GEnI+2DPCA+2DLDA	*7.48*	*75.47*
GERF	11.57	61.33
GERF+PCA+LDA	10.98	62.85
GERF+2DPCA+2DLDA	7.94	71.50
Gabor-GERF	8.41	71.50
Gabor-GERF+PCA+LDA	8.53	65.00
Gabor-GERF+2DPCA+2DLDA	**6.19**	**83.41**

3.6 Analysis of Difficulty Levels by Clothing Type

To evaluate the difficulty levels of clothes variation for the proposed method, we compute the rank-1 identification rates for all probe clothing types and list them in descending order as shown in Fig. 8. For the first 21 probe clothes types, the proposed method achieves over 85% rank-1 identification rate, which include more complex clothing type, such as clothing type B (*i.e.*, regular pants + down jacket), clothing type 6 (*i.e.*, regular pants + long coat + muffler). It is therefore validated that the proposed method effectively gains the discriminative features under a certain clothes types.

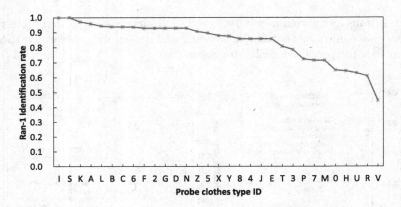

Fig. 8. Sorted clothing types according to recognition rate with the proposed method (Gabor-GERF+2DPCA+2DLDA).

For the rest of the probe clothing types, the average rank-1 identification rate drops to approximately 67%. Specifically, the lowest two clothing types are type V (*i.e.*, skirt + down jacket) with 44% rank-1 identification rate and type R (*i.e.*, raincoat) with 61% rank-1 identification rate. In fact, the clothing type V has quite different appearance from the others as shown in Fig. 3, and even the proposed GERF suffers from large intra-subject variations. Use of a single common GERF for all the clothes type may cause this performance degradation, one of future research avenues is a clothes type-adaptive selection of a suitable GERF from multiple GERFs in future.

4   Conclusion

The paper described a data-driven framework to learn GERF for clothes-invariant gait recognition. The GERF transforms an original gait energy into another one so as to make it more discriminative under clothes variation. The GERF is represented as a look-up table vector and is optimized through efficient generalized eigenvalue problem, which enables us to obtain analytical solution in a closed form without any iterations. In addition, in order to boost the GERF performance, Gabor filtering and 2DPCA+2DLDA are employed. Through comprehensive experiments, the proposed method shows the state-of-the-art performance in verification scenarios and competitive performance in identification scenarios.

Since, we only use the eigenvector corresponding to the largest eigenvalue as the GERF, the use of multiple eigenvectors will be investigated in the future. Moreover, since we use a common GREF regardless clothes type and spatial positions, further performance improvement is expected by introducing adaptive selection of GREF in future.

Acknowledgement. This work was supported by JSPS Grants-in-Aid for Scientific Research (A) JP15H01693, the JST CREST "Behavior Understanding based on Intention-Gait Model" project and Nanjing University of Science and Technology.

References

1. Nixon, M.S., Tan, T.N., Chellappa, R.: Human Identification Based on Gait. International Series on Biometrics. Springer, Heidelberg (2005)
2. Bouchrika, I., Goffredo, M., Carter, J., Nixon, M.: On using gait in forensic biometrics. J. Forensic Sci. **56**, 882–889 (2011)
3. Iwama, H., Muramatsu, D., Makihara, Y., Yagi, Y.: Gait verification system for criminal investigation. IPSJ Trans. Comput. Vis. Appl. **5**, 163–175 (2013)
4. Lynnerup, N., Larsen, P.: Gait as evidence. IET Biometrics **3**, 47–54 (2014)
5. Bobick, A., Johnson, A.: Gait recognition using static activity-specific parameters. In: Proceedings of the 14th IEEE Conference on Computer Vision and Pattern Recognition, vol. 1, pp. 423–430 (2001)
6. Cunado, D., Nixon, M., Carter, J.: Automatic extraction and description of human gait models for recognition purposes. Comput. Vis. Image Underst. **90**, 1–41 (2003)
7. Urtasun, R., Fua, P.: 3D tracking for gait characterization and recognition. In: Proceedings of the 6th IEEE International Conference on Automatic Face and Gesture Recognition, pp. 17–22 (2004)
8. Wagg, D., Nixon, M.: On automated model-based extraction and analysis of gait. In: Proceedings of the 6th IEEE International Conference on Automatic Face and Gesture Recognition, pp. 11–16 (2004)
9. Yam, C., Nixon, M., Carter, J.: Automated person recognition by walking and running via model-based approaches. Pattern Recogn. **37**, 1057–1072 (2004)
10. Zhao, G., Liu, G., Li, H., Pietikainen, M.: 3D gait recognition using multiple cameras. In: 7th International Conference on Automatic Face and Gesture Recognition (FGR 2006), pp. 529–534 (2006)
11. Sarkar, S., Phillips, P.J., Liu, Z., Vega, I.R., Grother, P., Bowyer, K.W.: The humanid gait challenge problem: data sets, performance, and analysis. IEEE Trans. Pattern Anal. Mach. Intell. **27**, 162–177 (2005)
12. Murase, H., Sakai, R.: Moving object recognition in eigenspace representation: gait analysis and lip reading. Pattern Recogn. Lett. **17**, 155–162 (1996)
13. Wang, L., Tan, T., Ning, H., Hu, W.: Silhouette analysis-based gait recognition for human identification. IEEE Trans. Pattern Anal. Mach. Intell. **25**, 1505–1518 (2003)
14. Han, J., Bhanu, B.: Individual recognition using gait energy image. IEEE Trans. Pattern Anal. Mach. Intell. **28**, 316–322 (2006)
15. Bashir, K., Xiang, T., Gong, S.: Gait recognition using gait entropy image. In: Proceedings of the 3rd International Conference on Imaging for Crime Detection and Prevention, pp. 1–6 (2009)
16. Bashir, K., Xiang, T., Gong, S.: Gait recognition without subject cooperation. Pattern Recogn. Lett. **31**, 2052–2060 (2010)
17. Makihara, Y., Sagawa, R., Mukaigawa, Y., Echigo, T., Yagi, Y.: Gait recognition using a view transformation model in the frequency domain. In: Leonardis, A., Bischof, H., Pinz, A. (eds.) ECCV 2006. LNCS, vol. 3953, pp. 151–163. Springer, Heidelberg (2006). doi:10.1007/11744078_12

18. Wang, C., Zhang, J., Wang, L., Pu, J., Yuan, X.: Human identification using temporal information preserving gait template. IEEE Trans. Pattern Anal. Mach. Intell. **34**, 2164–2176 (2012)

19. Tao, D., Li, X., Wu, X., Maybank, S.J.: General tensor discriminant analysis and gabor features for gait recognition. IEEE Trans. Pattern Anal. Mach. Intell. **29**, 1700–1715 (2007)

20. Matovski, D., Nixon, M., Mahmoodi, S., Carter, J.: The effect of time on gait recognition performance. IEEE Trans. Inf. Forensics Secur. **7**, 543–552 (2012)

21. Hossain, M.A., Makihara, Y., Wang, J., Yagi, Y.: Clothing-invariant gait identification using part-based clothing categorization and adaptive weight control. Pattern Recogn. **43**, 2281–2291 (2010)

22. Rokanujjaman, M., Islam, M., Hossain, M., Islam, M., Makihara, Y., Yagi, Y.: Effective part-based gait identification using frequency-domain gait entropy features. Multimedia Tools Appl. **74**, 3099–3120 (2015)

23. Guan, Y., Li, C.T., Hu, Y.: Robust clothing-invariant gait recognition. In: 2012 Eighth International Conference on Intelligent Information Hiding and Multimedia Signal Processing (IIH-MSP), pp. 321–324 (2012)

24. Lee, S., Liu, Y., Collins, R.: Shape variation-based frieze pattern for robust Gait recognition. In: Proceedings of the 2007 IEEE Computer Society Conference on Computer Vision and Pattern Recognition, Minneapolis, USA, pp. 1–8 (2007)

25. Boulgouris, N., Chi, Z.: Human gait recognition based on matching of body components. Pattern Recogn. **40**, 1763–1770 (2007)

26. Guan, Y., Li, C.T., Roli, F.: On reducing the effect of covariate factors in gait recognition: a classifier ensemble method. IEEE Trans. Pattern Anal. Mach. Intell. **37**, 1521–1528 (2015)

27. Liu, Z., Sarkar, S.: Effect of silhouette quality on hard problems in gait recognition. Trans. Syst. Man Cybern. Part B Cybern. **35**, 170–183 (2005)

28. Yan, S., Xu, D., Yang, Q., Zhang, L., Tang, X., Zhang, H.J.: Discriminant analysis with tensor representation. In: Proceedings of the IEEE Computer Society Conference Computer Vision and Pattern Recognition, pp. 526–532 (2005)

29. Xu, D., Yan, S., Tao, D., Zhang, L., Li, X., Zhang, H.J.: Human gait recognition with matrix representation. IEEE Trans. Circ. Syst. Video Technol. **16**, 896–903 (2006)

30. Liu, Z., Sarkar, S.: Simplest representation yet for gait recognition: averaged silhouette. In: Proceedings of the 17th International Conference on Pattern Recognition, vol. 1, pp. 211–214 (2004)

31. Xu, D., Huang, Y., Zeng, Z., Xu, X.: Human gait recognition using patch distribution feature and locality-constrained group sparse representation. IEEE Trans. Image Process. **21**, 316–326 (2012)

32. Yang, J., Zhang, D., Frangi, A.F., Yang, J.Y.: Two-dimensional PCA: a new approach to appearance-based face representation and recognition. IEEE Trans. Pattern Anal. Mach. Intell. **26**, 131–137 (2004)

33. Makihara, Y., Mannami, H., Tsuji, A., Hossain, M., Sugiura, K., Mori, A., Yagi, Y.: The OU-ISIR gait database comprising the treadmill dataset. IPSJ Trans. Comput. Vis. Appl. **4**, 53–62 (2012)

34. Liu, Z., Sarkar, S.: Improved gait recognition by gait dynamics normalization. IEEE Trans. Pattern Anal. Mach. Intell. **28**, 863–876 (2006)

35. Li, X., Maybank, S., Yan, S., Tao, D., Xu, D.: Gait components and their application to gender recognition. Trans. Syst. Man Cybern. Part C **38**, 145–155 (2008)

Action Recognition Based on Optimal Joint Selection and Discriminative Depth Descriptor

Haomiao Ni, Hong Liu[✉], Xiangdong Wang, and Yueliang Qian

Beijing Key Laboratory of Mobile Computing and Pervasive Device,
Institute of Computing Technology, Chinese Academy of Sciences,
Beijing 100190, China
hliu@ict.ac.cn

Abstract. This paper proposes a novel human action recognition using the decision-level fusion of both skeleton and depth sequence. Firstly, a state-of-the-art descriptor RBPL, relative body part locations, is adopted to represent skeleton. But the original RBPL employs all the available joints, which may introduce redundancy or noise. This paper proposes an adaptive optimal joint selection model based on the distance traveled by joints before RBPL for each different action, which can reduce redundant joints. Then we use dynamic time warping to handle temporal misalignment and adopt KELM, kernel-based extreme learning machine, for action recognition. Secondly, an efficient feature descriptor DMM-disLBP, depth motion maps-based discriminative local binary patterns, is constructed to describe depth sequences, and KELM is also used for classification. Finally, we present an effective decision fusion for action recognition based on the maximum sum of decision values from skeleton and depth maps. Comparing with the baseline methods, we improve the performance using either skeleton or depth information, and achieve the state-of-the-art average recognition accuracy on the public dataset MSR Action3D using proposed fusing strategy.

1 Introduction

Human action recognition is an active area in computer vision and has various applications in surveillance, health care and video games. In the past decades, many researchers focused on recognizing action from RGB videos, which is a difficult task due to illumination changes and background clutter [1].

Recently, with the introduction of some cost-effective depth sensors, such as Microsoft Kinect, both depth maps and 3D skeleton are provided. Unlike traditional RGB images, depth maps are insensitive to light condition and cluttered background. And skeleton contains much action-relevant information because it describes human body naturally. These recent advances facilitate a lot of research on skeleton-based and depth maps-based methods for action recognition.

For skeleton-based human action recognition, Xia et al. [2] proposed HOJ3D (Histograms Of 3D Joint locations) to represent poses. They constructed this view-invariant feature by placing joint locations into 3D spatial bins, which is real time but depends on estimated skeleton root position [3]. Vemulapalli et al. [1]

© Springer International Publishing AG 2017
S.-H. Lai et al. (Eds.): ACCV 2016, Part II, LNCS 10112, pp. 273–287, 2017.
DOI: 10.1007/978-3-319-54184-6_17

proposed a skeletal representation RBPL (Relative Body Part Locations) that explicitly models the 3D geometric relationships between various body parts.[1] Using RBPL representation, they modeled human actions as curves in Lie group and then mapped them into Lie algebra, and finally used SVM for classification. However, this approach considers all body parts, which may involves confusion or redundancy. More recently, Du et al. [4] proposed an end-to-end hierarchical recurrent neural network (RNN) for skeleton-based action recognition. They divided skeleton into five parts according to human physical structure, and then five subnets were used for training and testing separately.

On the other hand, for depth maps-based action recognition, Li et al. [5] presented a bag of 3D points for action recognition. They used a small number of 3D points as a descriptor of the 3D shape of each pose. However, it is difficult to sample the interest points robustly due to the large intra-class variability [3]. Yang et al. [6] proposed a method based on DMM (Depth Motion Maps) to capture motion cues and use HOG (Histogram of Oriented Gradients) on DMM to extract features. Recently, Chen et al. [7] proposed another descriptor DMM-LBP, which extracted LBP (Local Binary Patterns) on DMM from three views to describe texture features. This method outperformed DMM-HOG on public dataset MSR Action3D [5].

The above methods only use single information for action recognition, however, each information has its own advantages and may complements each other. Skeleton joints have stronger representation power than depth maps while depth maps suffer less distortion than skeleton joints. Thus, some researchers [8–10] pay attention to fusing multiple information for action recognition. Althloothi et al. [8] employed MKL (Multiple Kernel Learning) to fuse two features extracted from depth images and skeleton. But they only used six specified body parts to describe actions, which may overlook other important ones for certain action. And they used spherical harmonics coefficients as shape feature to represent depth maps, which may be unreliable due to the huge intra-class variability of body silhouettes. This method only achieves average accuracy of 79.27% on MSR Action3D dataset. Recently, Liu et al. [10] proposed 3D-based deep convolutional neural network to directly learn spatio-temporal features from depth sequences and computed a so-called JointVector feature to describe skeleton sequences. Finally, the SVM classification results from high-level feature and JointVector are fused. They achieved average accuracy of 84.07%.

The above analysis shows that it is important to make sure the high accuracy using single information before fusion. This paper focuses on the improvement of feature description ability for skeleton-based and depth maps-based action recognition. Then we further present a decision fusion strategy to combine these two types of information to improve the final recognition performance. The overview of our method is illustrated in Fig. 1.

Firstly, for skeleton-based feature description, we model the human skeleton by RBPL [1]. However, the original RBPL may introduce redundancy due to

---

[1] The authors don't name their method. In order to facilitate the writing, we name it RBPL (Relative Body Part Locations).

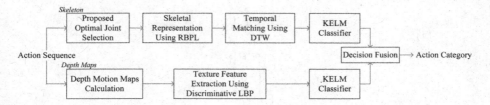

Fig. 1. Overview of our proposed method.

considering all the available joints. We propose an adaptive optimal joint selection based on the distance traveled by joints (OJSDTJ) before RBPL to reduce redundance. This optimal joint selection measures the importance of joint by the distance it travels firstly and a two-layered model is further used to enhance the robustness and discriminative power of selected joints for each action. Then we use DTW (Dynamic Time Warping) [11] to handle temporal misalignment and perform the final classification by KELM (Kernel-based Extreme Learning Machine) [12]. Compared with the original RBPL framework, proposed method OJSDTJ achieves an average accuracy of 89.66% on MSR Action3D, which is improved by 1.56%, while the feature dimensionality is reduced by 69.56%.

Secondly, for depth maps-based feature description, inspired by DMM-LBP descriptor [7], we propose DMM-disLBP (Depth Motion Maps-based discriminative Local Binary Patterns) descriptor to improve texture description ability for depth sequence. We first calculate DMMs from three views on depth sequences to capture motion information. Then disLBP operator [13] is applied to over-lapped blocks of the DMM. This operator is based on the optimal pattern subsets learned from a three-layered model, rather than using all predefined patterns like conventional LBP. We subsequently concatenate the disLBP histograms of the blocks for each DMM to form the feature vectors. And KELM is used for classification. Compared with DMM-LBP, our DMM-disLBP achieves an average accuracy of 94.28% on MSR Action3D, which is improved by 2.14%.

Thirdly, to combine above two methods, we present an effective decision fusion based on the maximum sum of decision values from multiple KELMs. Compared with the results using single information (89.66%, 94.28%), this fusion achieves average accuracy of 97.30%, which shows the reliability of this fusion.

The contributions of this paper can be summarized as follows. (1) An optimal joint selection method based on the distance traveled by joints is proposed to reduce redundant joints and improves recognition performance. (2) A discriminative descriptor DMM-disLBP to describe depth maps is proposed, which improves feature description ability. (3) Finally, we propose an effective decision fusion strategy based on the maximum sum of decision values from skeleton-based and depth maps-based action recognition. And the proposed fusing method achieves average recognition accuracy of 97.3% on public dataset MSR Action3D, which is the state-of-the-art result on this dataset.

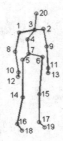

Fig. 2. Human skeleton captured by Kinect.

2   Skeleton-Based Action Recognition

Figure 2 shows a human skeleton with 20 joints and 19 body parts,[2] which is captured by Kinect. Most existing skeleton-based methods simply extract features from all the joints [1] or some predefined joints [8]. This paper proposes a novel adaptive optimal joint selection method for each different action as Fig. 3 shows. Firstly, we preprocess raw skeletal data to ensure skeleton scale-invariant and view-invariant, and then Savitzky-Golay filter [4] is employed to reduce noise. Secondly, we propose a two-layered model to select discriminative joints for each action. The histogram of this model represents the distance traveled by all joints of a skeleton in Fig. 3. Layer 1 aims to determine the dominant joint subset from the initial joint set of each training sequence. And Layer 2 is used to select the discriminative dominant joint set of each action class by taking intersection of the joint subsets from Layer 1. Thirdly, RBPL is employed to represent skeleton.[3] Finally, we use DTW to handle temporal misalignment and KELM to perform classification. The details of each step are as follows.

2.1   Basic Preprocessing

We adopt the methods in [1,4] to preprocess raw skeletal data. Generally speaking, actions are independent of the performer's location. Therefore, it is essential step to transform raw skeletal data from real world coordinates to human-centered coordinates. This can be completed by placing the origin of coordinate system on the hip center (joint 7 in Fig. 2). Besides, it is obvious that the lengths of body parts vary from person to person. To ensure skeleton scale-invariant, we take one skeleton as reference to normalize the lengths of body parts of all the other skeletons. To eliminate view-dependency, the vector from left hip (joint 6) to right hip (joint 5) is first projected to the ground plane. And skeleton is then rotated to where this projection is parallel to the global x-axis.

---

[2] http://research.microsoft.com/en-us/um/people/zliu/actionrecorsrc/
SkeletonModelMSRAction3D.jpg.
[3] The diagram of RBPL is quoted from [1].

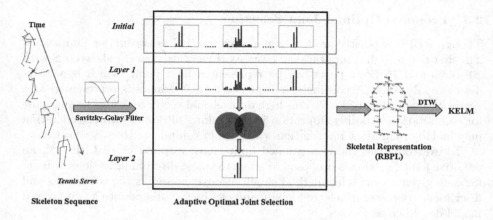

Fig. 3. The main steps of our skeleton-based method.

Next, to handle noise issues, a simple Savitzky-Golay smoothing filter is employed. The filter is designed as follows.

$$f_i = (-3x_{i-2} + 12x_{i-1} + 17x_i + 12x_{i+1} - 3x_{i+2})/35 \tag{1}$$

where x_i denotes the skeleton joint coordinates in the ith frame, and f_i denotes the filtering result.

2.2 Skeletal Representation

In order to make it easier to understand our method, here we first introduce a state-of-the-art skeletal representation in [1], relative body part locations (RBPL). The main idea of RBPL is described as follows. Given a pair of body parts, each one of them can be rotated and translated to the orientation and location of another part. Therefore, we can use the structure consisting of this rotation and translation to represent their relative geometric relationship. Mathematically, this structure is proven to be a member of the special Euclidean group $SE(3)$, which is a matrix Lie group. That is, we can represent the relative 3D geometry between a pair of body parts as a point in $SE(3)$. Accordingly, a skeleton S at frame T including various pairs of body parts can be represented as a point in $SE(3) \times \ldots \times SE(3)$, *i.e.* Lie group. Therefore, an action consisting of a skeleton sequence can be represent as a curve in Lie group. For the subsequent processing, each sequence should have an equal number of samples. Since we represent a sample as a point in Lie group, it is easy to use the interpolation algorithm in [1] to complete this task. Finally, we map the action curves from Lie group to its Lie algebra, a vector space, where standard classification is directly applicable.

2.3   Proposed Optimal Joint Selection

Though RBPL explicitly models human skeleton, it has two major limitations. The first one is high computational cost. As is mentioned in [1], skeleton S with 20 joints and 19 body parts can be represented as a point, which is a 2052-dimensional vector. Accordingly, a sequence with N frames is represented as a vector of dimension $2052 \times N$. This high dimensional vector may lead to huge cost for computation. Another limitation is that taking all joints into consideration may include redundant information which could complicate the classification.

To overcome the above-mentioned limitations, inspired by disLBP [13], an adaptive joint selection is proposed to select the most discriminative joint subsets for each given action. This method simultaneously considers the robustness and discriminative power of selected joints, and can be formulated into a two-layered model described as follows.

Layer 1 aims to select the most dominant and robust joints from each training skeleton sequence. The main idea of this layer comes from an observation that the importance of a joint is associated with the distance traveled by it. Specifically speaking, given an action, the joints which have longer distance traveled during this action process may play more important role in depicting it. In contrast, the joints which almost keep still may contribute little to recognizing this action. And sometimes they may involve noise, which harms the subsequent feature extraction. The *Algorithm 1* describes the details of Layer 1.

Algorithm 1 (Layer 1). Determine the dominant joint subset of each training sequence x_i from the training set T_{train}.

Input: at $t = 1, 2, \ldots, T$, the joint positions $JP_j(t)$ from the initial joint set of each sequence x_i, where $j = 1, 2, \ldots, m$ (m denotes the number of the initial joints). And the threshold parameter n to determine the proportions of dominant joints selected from each training sequence.

Output: Dominant joint set J_i with respect to each training sequence x_i.

1. FOR $j = 1$ TO m
2. $s_i[j] = s_{i,j} = \sum_{t=2}^{T} \| JP_j(t) - JP_j(t-1) \|_2$, where $\| \bullet \|$ denotes l_2 norm
3. END FOR
4. Initialize reference vectors V, where $V[j] = j$ $(j = 1, \ldots, m)$
5. Sort s_i in descending order to obtain $\hat{s}_i$. Change the configuration of V according to element order of $\hat{s}_i$, resulting in $\hat{V}$
6. FOR $k = 1$ TO m
7.   IF $\left(\sum_{l=1}^{k} \frac{\hat{s}_{i,l}}{\sum_{l=1}^{m} \hat{s}_{i,l}} \right) \geqslant n$
8.   BREAK
9.   END IF
10. END FOR
11. RETURN $J_i = \left\{ \hat{V}[1], \ldots, \hat{V}[k] \right\}$ as a dominant joint subset for sequence $x_i \in T_{train}$

In the second layer, the most discriminative joint subsets for each class are estimated by minimizing inter-class similarity based on joints selected from

Layer 1. Theoretically, different sequences belonging to the same action category should have the same dominant joint subset. However, due to distortion or noise, these sequences from the same class usually have different joint subsets. To minimize class ambiguity, we take the intersection of dominant joint subsets across all training sequences belonging to the same class. The *Algorithm 2* describes the details of Layer 2.

Algorithm 2 (Layer 2). Select the discriminative dominant joint set of class j

input: Dominant joint sets $J_1, J_2, \ldots, J_{n_J}$ of n_J sequences belonging to class j obtained from *Algorithm 1*.

output: Discriminative dominant joint set JC_j of class j.

1. Initialize $JC_j = J_1$
2. FOR each sequence $k = 2$ TO n_J belonging to class j
3. $JC_j = JC_j \cap J_k$
4. END FOR
5. RETURN JC_j

Since RBPL representation is applied to body parts rather than joints, we should transform our joint sets to body part sets. Considering that RBPL models the relative 3D geometry between various body parts, we first add joint *torso* (joint 4) to strengthen this relative relationship. And then joint sets are translated to body part sets.

Let $E = \{e_1, e_2, \ldots, e_{m-1}\}$ denotes the initial body part set, where m denotes the number of the initial joints, and $e_j = \{e_{j_1}, e_{j_2}\}$ denotes the body part e_j connecting joint e_{j_1} to joint e_{j_2}, the *Algorithm 3* explains the transform as follows.

Algorithm 3. Transform discriminative dominant joint set JC_j of class j to the corresponding discriminative dominant body part set EC_j

input: The discriminative dominant joint set $JC_j = \{J_1, J_2, \ldots, J_c\}$ for each class j obtained from *Algorihtm 2*, where $c = |JC_j|$, and the initial body part set E.

output: Discriminative dominant joint set EC_j of class j.

1. Initialize $JC_j = JC_j \cup J_{torso}$, $EC_j = \oslash$
2. FOR $i = 1$ TO c
3.   FOR $k = 1$ TO $m - 1$
4.   IF $J_i \in \{e_{k_1}, e_{k_2}\}$
5. $EC_j = EC_j \cup e_k$
6.   END IF
7.   END FOR
8. END FOR
9. RETURN EC_j

Through the selection of Layer 1 and Layer 2, we have obtained discriminative joint subset for each action. After applying the transform algorithm, we also get the discriminative body part subset for each action, which has removed the confusion and redundancy. Then we apply RBPL to our body parts set EC_j

of class j, we obtain the so-called OJSDTJ skeletal representation. Comparing with the original RBPL, it has several advantages of lower computational cost, higher robustness and stronger discriminative power.

2.4   Temporal Modeling and Classification

The above mentioned OJSDTJ algorithm is used to extract feature from each frame of a skeleton sequence. However, the action sequences may have different lengths due to the various motion for different persons and different repeats. Following [1], DTW is used to handle rate variations. We first apply DTW to compute the nominal curves across all training sequences belonging to the same action. And then all curves are warped into the nominal curves of each class. Since the warping process requires all curves to have equal length, we employ the interpolation algorithm [1] to re-sample the curves before mapping to vector space. Unlike the proposed method in [1], we directly perform classification rather than applying Fourier temporal pyramid (FTP) [14] before. The reason is that two major functions of FTP, temporal matching and denoising, can be replaced by DTW and Savitzky-Golay smoothing filter respectively. One-vs-all linear KELM is employed to the final classification. The experimental evaluation will be described in Subsect. 5.1.

3   Depth Maps-Based Action Recognition

Due to the distortion of skeleton in some sequences, it is insufficient to only use skeletal feature to robustly describe action. This paper further improves depth maps-based method based on recent DMM-LBP method [7], which has the highest average recognition rate among depth maps-based approaches on MSR Action3D dataset. The main steps of our proposed depth maps-based method DMM-disLBP are shown in Fig. 4. DMM is first used to capture motion cues from three views and then disLBP operator [13] is applied to DMM to extract texture feature. Finally, KELM is used for action classification. The details of each step are described as follows.

We first employ depth motion maps (DMM) to capture motion information. The main idea behind DMM is as follows. Given a depth sequence with N frames, we project all the frames onto three orthogonal Cartesian planes from three project views [front(f), side(s), top(t)]. This projection produces three 2D images, denoted by map_f, map_s, map_t. Then the following equation is applied to generate three DMMs [7].

$$DMM_{\{f,s,t\}} = \sum_{j=1}^{N-1} |map_{\{f,s,t\}}^{j+1} - map_{\{f,s,t\}}^{j}| \tag{2}$$

where j is the frame index.

As is shown in Fig. 4, DMMs contain abundant texture information, which reflects the temporal motion characteristics. Therefore, an effective texture

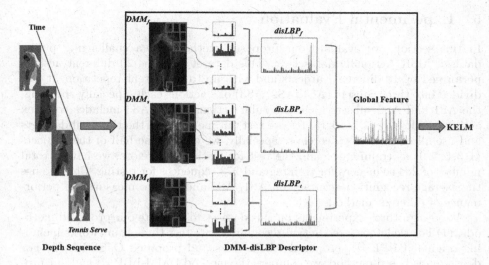

Fig. 4. The main steps of our depth maps-based method.

descriptor LBP (local binary patterns) is then used to extract the texture feature. Unlike using conventional LBP in [7], we employ a novel LBP named disLBP (discriminative LBP). This kind of LBP is based on the optimal pattern subset learned from a three-layered model in [13]. And thus, it performs better than conventional LBP. We apply disLBP operator to overlapped blocks of the DMMs and calculate disLBP histograms for each block. The resulted histograms of the blocks in a DMM are represented as a feature vector. And disLBP feature vector for each DMM is then concatenated as a global feature vector. KELM is used for final classification. The experimental evaluation will be described in Subsect. 5.2.

4   Decision-Level Fusion Based on the Maximum Sum of Decision Values

Based on above proposed skeleton-based method and depth maps-based method, we further introduce an efficient decision-level fusion to combine two complementary methods. It can be precisely described as follows.

Given a testing sequence x, let n denotes the total number of action category, we assume that the KELM in our skeleton-based method gives the decision values of each class i $(i = 1, \ldots, n)$ as: $v_1, \ldots, v_n$, and the KELM in our depth maps-based method gives the decision values of each class i as: $\hat{v}_1, \ldots, \hat{v}_n$. Then we can calculate the class label l of sequence x by the following equation:

$$l = \underset{i}{\mathrm{argmax}}\{v_1 + \hat{v}_1, \ldots, v_n + \hat{v}_n\} \tag{3}$$

This method is called decision fusion based on the max sum of decision values, or DFMSDV for short.

5   Experimental Evaluation

In this section, we evaluate our proposed methods on a challenging public dataset, MSR Action3D dataset [5]. This dataset contains 20 different actions performed by 10 different subjects and with up to 3 different repetition. It was divided into three subsets (AS1, AS2, AS3) of 8 actions each. Specially speaking, the AS1 and AS2 subsets include similar actions, while AS3 includes complex actions. Following [4,7–10,15–18], we test our methods on the three subdatasets and use all of the 557 video clips. Specially, we choose one half of the subjects (1, 3, 5, 7, 9) as training set and the rest for testing. Therefore, we have a total number of 284 sequences for training and 273 sequences for testing. Here we use the average recognition accuracy on AS1, AS2 and AS3 to measure the performance of different methods.

We design three experiments on this dataset to evaluate our proposed methods. (1) For skeleton-based action recognition, we test three frameworks including original RBPL [1], proposed RBPL-Ours and proposed OJSDTJ. (2) For depth maps-based method, we compare the method DMM-LBP [7] and our proposed DMM-disLBP with the same parameters setting. (3) To illustrate the effectiveness of fusing method, we evaluate recognition performance of our proposed fusing strategy DFMSDV and also compare our proposed methods with various state-of-the-art methods.

5.1   Skeleton-Based Approach Evaluation

As is mentioned in Sect. 2, we propose optimal joint selection model OJSDTJ to combine RBPL for skeletal representation.

Parameters Setting. For the data reliability, we remove the first and the last n frames of skeleton sequences, here we set $n = 1$. Our OJSDTJ bases on the distance traveled by joints and we find that the proportion of the distances traveled by hands, arms and elbows in most actions are more than 98% on training set. Therefore, only joints belonging to hands, arms and elbows are selected for these actions unless threshold parameter n is higher than 0.98. So we set $n = 0.99$. After optimal joint selection, we use RBPL to extract skeleton feature and employ KELM for classification as is mentioned in Sect. 2. For KELM, we choose linear kernel and set parameter $C = 1000$.

Table 1. Recognition accuracy (%) of three different frameworks on MSR Action3D.

Method	AS1	AS2	AS3	Avg.
RBPL (RBPL+DTW+FTP+SVM) [1]	86.67	83.04	94.59	88.10
RBPL-Ours (Filter+RBPL+DTW+KELM)	87.62	**84.82**	94.59	89.01
OJSDTJ (Filter+OJSDTJ+RBPL+DTW+KELM) (Ours)	**90.47**	83.92	94.59	**89.66**

Experimental Results. As Table 1 shows, we compare three methods: the original framework RBPL [1], our proposed framework RBPL-Ours, and our proposed framework OJSDTJ. The original framework RBPL employs a combination of RBPL, DTW and FTP for skeleton feature description, and SVM is used for final classification. By contract, our proposed framework RBPL-Ours employs a combination of Savitzky-Golay filter, RBPL and DTW for skeleton feature description, and KELM is used for final classification. Note that the accuracy of the original framework using RBPL is different from [1], this is because their experimental parameters setting is different. We obtained the results of original RBPL [1] within the same experimental setting as ours, such as subjects (1, 3, 5, 7, 9) as training set and the rest for testing. Compared with the original framework, our proposed framework RBPL-Ours achieves higher accuracy on AS1 and AS2, and the average accuracy is also improved by 0.91%. This result shows that our framework RBPL-Ours outperforms the original one with the same skeletal representation. The main reason may be that the combination of Savitzky-Golay filter and DTW performs better than FTP on denoising and temporal misalignment. Compared with RBPL-Ours, the framework using OJSDTJ achieved much higher accuracy on AS1, and the average accuracy is also 0.65% higher. However, this method performs worse slightly on AS2 when using OJSDTJ. Further analysis on AS2 (Table 2) shows us that only two action *high arm wave* and *draw circle* are performed worse. The reason may be that Layer 1 fails to select the key joints of these two actions for some noisy training sequences, which may lead to the missing of the key joints when we take the intersection set in Layer 2. Nevertheless, the average accuracy of our framework OJSDTJ still performs the best among these three frameworks for action recognition.

Moreover, OJSDTJ has lower computational cost than RBPL. Here we analyze the dimension reduction after using OJSDTJ. Since we have 20 actions and 20 joints, the initial total number of joints is $20 \times 20 = 400$ and each sequence is represented as a vector of 155,952 dimensions on MSR Action3D dataset for initial RBPL. For proposed OJSDTJ, we only have 139 joints in total and each sequence is represented as a vector of 47,469 dimensions on average, which is reduced by 69.56%. The above experimental results show that our proposed framework OJSDTJ has 1.56% improvement of average accuracy with better computational efficiency than original RBPL framework [1] as Table 1 shows.

Table 2. Accuracy (%) comparison of RBPL-Ours and OJSDTJ on AS2.

	High arm wave	Hand catch	Draw cross	Draw tick	Draw circle	Two-hand wave	Forward kick	Side-boxing
RBPL-Ours	91.67	33.33	76.92	93.33	73.33	100	100	100
OJSDTJ	**83.33**	33.33	84.62	100	**60.00**	100	100	100

Table 3. Recognition accuracy (%) of DMM-LBP and DMM-disLBP with $m = 4, r = 1$.

Method	AS1	AS2	AS3	Avg.
DMM-LBP [6]	95.24	88.39	92.79	92.14
DMM-disLBP (Ours)	**97.14**	**89.29**	**96.40**	**94.28**

5.2   Depth Maps-Based Approach Evaluation

As is mentioned in Sect. 3, we employ DMM-disLBP descriptor to characterize depth sequences and use KELM for classification.

Parameters Setting. To ensure data reliability, we remove the first and the last n frames of depth sequences, here we set $n = 1$.

Following [7], we calculate DMMs from three projection views [front(f), side(s), top(t)]. The sizes of DMM_f, DMM_s, and DMM_t are normalized to be 102×54, 102×75, and 75×54 respectively. The block sizes of the DMMs were considered to be 25×27, 25×25, and 25×27 corresponding to DMM_f, DMM_s, and DMM_t. Then we applied disLBP on DMMs to extract texture feature. Following [13], we set the threshold parameter $n = 0.90$ for disLBP. We also set the radius r and the number of the neighbors m for disLBP descriptor. We initially want to follow the setting of LBP in [7]. But the authors don't tell us the parameters setting when testing on three subsets. So we set $r = 1$ and $m = 4$, a default setting in the authors' code. Within this setting, DMM-LBP performs a little worse compared with the results listed in [7]. And we perform classification by KELM. Following [7], we choose RBF kernel and set $C = 1000$, $Gamma = 10.5$.

Experimental Results. As is shown in Table 3, we compare two approaches: DMM-LBP and DMM-disLBP with parameter $m = 4$, $r = 1$. Comparing with DMM-LBP, our proposed method performs better on AS1, AS2, and AS3. The average accuracy is 2.14% higher than the original method. The result shows that disLBP is more effective than conventional LBP to extract texture feature.

5.3   Results of Decision Fusion

To combine our skeleton-based and depth maps-based method, we propose a fusion strategy DFMSDV for action recognition as is mentioned in Sect. 4.

The parameters setting is as the same as the Subsects. 5.1 and 5.2.

Table 4. Recognition accuracy (%) of OJSDTJ, DMM-disLBP, and DFMSDV.

Method	AS1	AS2	AS3	Avg.
OJSDTJ	90.47	83.92	94.59	89.66
DMM-disLBP	97.14	89.29	96.40	94.28
DFMSDV	**99.05**	**93.75**	**99.10**	**97.30**

We compare our three proposed approaches in Table 4. As Table 4 shows, compared with OJSDTJ based on skeleton, fused method DFMSDV improves the accuracy on AS1, AS2, AS3 by 8.58%, 9.83%, 4.51% respectively and achieves 7.64% higher average accuracy. Compared with DMM-disLBP, DFMSDV improves the accuracy on AS1, AS2, AS3 by 1.91%, 4.46%, 2.7% respectively and achieves 3.02% higher average accuracy. These results shows that our decision fusion strategy combines these two complementary methods well. Also, we compare our proposed approaches with various state-of-the-art human action recognition approaches [1,4,7–10,15–18] on MSR Action3D dataset. The detailed results are shown in Table 5. We can see that our proposed method DFMSDV achieves the highest accuracy on AS1 (99.05%) and the highest average accuracy (97.30%). Note that we still list the best average accuracy of DMM-LBP, 94.9% rather than 92.14% mentioned in Subsect. 5.2. But our proposed method DFMSDV also performs better than this best result. The results shows that DFMSDV outperforms all the other method using only single information or combining multiple information, which is also shown in Fig. 5.

In our survey, we find that the average accuracy in paper [19] is 98.2% on MSR Action3D dataset, which uses subjects (2, 3, 5, 7, 9) as training set rather than (1, 3, 5, 7, 9) as Table 5 shows. We also add extra experiments using subjects (2, 3, 5, 7, 9) as training set and keep the same parameters setting as our above experiments. The final average accuracy is 97.62%, which is a little lower than [19]. Compared with the results in Table 5, the average accuracy of our fusion method has increased by 0.32%. The average accuracy of our skeleton-based method has increased by 3.38% while the average accuracy of our depth maps-based approach has reduced by 0.93%. This result shows that the parameters of our depth method should be tuned with the change of training set.

Table 5. Recognition accuracy (%) comparison of our proposed approaches and various state-of-the-art human action recognition on MSR Action3D dataset.

Method	Information	Year	AS1	AS2	AS3	Avg.
Althloothi et al. [8]	Skeleton+Depth	2014	74.3	76.8	86.7	79.27
Evangelidis et al. [15]	Skeleton	2014	88.39	86.61	94.59	89.86
Theodorakopoulos et al. [16]	Skeleton	2014	91.23	90.09	**99.5**	93.61
Vemulapalli et al. [1]	Skeleton	2014	86.67	83.04	94.59	88.10
Vieira et al. [17]	Depth	2014	91.7	72.2	98.6	87.5
Shen et al. [18]	Depth	2014	90.6	81.4	94.6	88.87
Du et al. [4]	Skeleton	2015	93.33	**94.64**	95.50	94.49
Chen et al. [7]	Depth	2015	98.1	92.0	94.6	94.9
Liu and Pei [9]	Skeleton+Depth	2015	91.55	84.67	93.06	89.76
Liu et al. [10]	Skeleton+Depth	2016	86.79	76.11	89.29	84.07
DMM-disLBP (Ours)	Depth	2016	97.14	89.29	96.40	94.28
OJSDTJ (Ours)	Skeleton	2016	90.47	83.92	94.59	89.66
DFMSDV (Ours)	Skeleton+Depth	2016	**99.05**	93.75	99.10	**97.30**

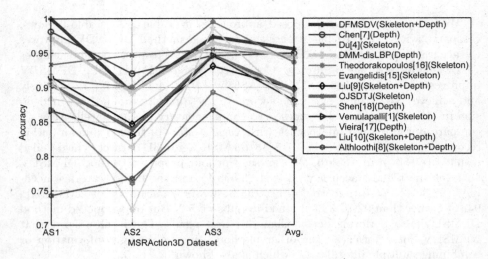

Fig. 5. Comparison of our proposed methods and various state-of-the-art approaches.

6   Conclusion

This paper proposes an adaptive optimal joint selection model based on distance traveled by joints to select discriminative joint subsets for each given action. Using this framework, we achieve higher average accuracy for action recognition while reducing feature dimensionality considerably. We also introduce an effective depth descriptor DMM-disLBP to improve the feature description ability of depth maps. Finally, we propose a decision fusion scheme based on the maximum sum of decision values to combine skeleton and depth maps information. Our final fusing results outperform most state-of-the-art action recognition approaches published recently on MSR Action3D dataset.

Acknowledgement. This work is supported in part by Beijing Natural Science Foundation: 4142051.

References

1. Vemulapalli, R., Arrate, F., Chellappa, R.: Human action recognition by representing 3D skeletons as points in a lie group. In: Proceedings of the IEEE Conference on Computer Vision and Pattern Recognition, pp. 588–595 (2014)
2. Xia, L., Chen, C.C., Aggarwal, J.: View invariant human action recognition using histograms of 3D joints. In: 2012 IEEE Computer Society Conference on Computer Vision and Pattern Recognition Workshops (CVPRW), pp. 20–27. IEEE (2012)
3. Ye, M., Zhang, Q., Wang, L., Zhu, J., Yang, R., Gall, J.: A survey on human motion analysis from depth data. In: Grzegorzek, M., Theobalt, C., Koch, R., Kolb, A. (eds.) Time-of-Flight and Depth Imaging. Sensors, Algorithms, and Applications. LNCS, vol. 8200, pp. 149–187. Springer, Heidelberg (2013). doi:10.1007/978-3-642-44964-2_8

4. Du, Y., Wang, W., Wang, L.: Hierarchical recurrent neural network for skeleton based action recognition. In: Proceedings of the IEEE Conference on Computer Vision and Pattern Recognition, pp. 1110–1118 (2015)
5. Li, W., Zhang, Z., Liu, Z.: Action recognition based on a bag of 3D points. In: 2010 IEEE Computer Society Conference on Computer Vision and Pattern Recognition Workshops (CVPRW), pp. 9–14. IEEE (2010)
6. Yang, X., Zhang, C., Tian, Y.: Recognizing actions using depth motion maps-based histograms of oriented gradients. In: Proceedings of the 20th ACM International Conference on Multimedia, pp. 1057–1060. ACM (2012)
7. Chen, C., Jafari, R., Kehtarnavaz, N.: Action recognition from depth sequences using depth motion maps-based local binary patterns. In: 2015 IEEE Winter Conference on Applications of Computer Vision (WACV), pp. 1092–1099. IEEE (2015)
8. Althloothi, S., Mahoor, M.H., Zhang, X., Voyles, R.M.: Human activity recognition using multi-features and multiple kernel learning. Pattern Recogn. **47**, 1800–1812 (2014)
9. Liu, T., Pei, M.: Fusion of skeletal and STIP-based features for action recognition with RGB-D devices. In: Zhang, Y.-J. (ed.) ICIG 2015. LNCS, vol. 9218, pp. 312–322. Springer, Heidelberg (2015). doi:10.1007/978-3-319-21963-9_29
10. Liu, Z., Zhang, C., Tian, Y.: 3D-based deep convolutional neural network for action recognition with depth sequences. Image Vis. Comput. **55**, 93–100 (2016)
11. Müller, M.: Information Retrieval for Music and Motion, vol. 2. Springer, Heidelberg (2007)
12. Huang, G.B., Zhou, H., Ding, X., Zhang, R.: Extreme learning machine for regression and multiclass classification. IEEE Trans. Syst. Man Cybern. Part B: Cybern. **42**, 513–529 (2012)
13. Guo, Y., Zhao, G., Pietikälnen, M.: Discriminative features for texture description. Pattern Recogn. **45**, 3834–3843 (2012)
14. Wang, J., Liu, Z., Wu, Y., Yuan, J.: Mining actionlet ensemble for action recognition with depth cameras. In: 2012 IEEE Conference on Computer Vision and Pattern Recognition (CVPR), pp. 1290–1297. IEEE (2012)
15. Evangelidis, G., Singh, G., Horaud, R.: Skeletal quads: human action recognition using joint quadruples. In: ICPR 2014-International Conference on Pattern Recognition (2014)
16. Theodorakopoulos, I., Kastaniotis, D., Economou, G., Fotopoulos, S.: Pose-based human action recognition via sparse representation in dissimilarity space. J. Vis. Commun. Image Represent. **25**, 12–23 (2014)
17. Vieira, A.W., Nascimento, E.R., Oliveira, G.L., Liu, Z., Campos, M.F.: On the improvement of human action recognition from depth map sequences using space-time occupancy patterns. Pattern Recogn. Lett. **36**, 221–227 (2014)
18. Shen, X., Zhang, H., Gao, Z., Xue, Y., Xu, G.: Human behavior recognition based on axonometric projections and phog feature. J. Comput. Inf. Syst. **10**, 3455–3463 (2014)
19. Zhu, Y., Chen, W., Guo, G.: Fusing multiple features for depth-based action recognition. ACM Trans. Intell. Syst. Technol. (TIST) **6**, 18 (2015)

Model-Free Multiple Object Tracking
with Shared Proposals

Gao Zhu[1(✉)], Fatih Porikli[1,2,3], and Hongdong Li[1,3]

[1] Australian National University, Canberra, Australia
gao.zhu@anu.edu.au
[2] Data61/CSIRO, Canberra, Australia
[3] ARC Centre of Excellence for Robotic Vision, Canberra, Australia

Abstract. Most previous methods for tracking of multiple objects follow the conventional "tracking by detection" scheme and focus on improving the performance of category-specific object detectors as well as the between-frame tracklet association. These methods are therefore heavily sensitive to the performance of the object detectors, leading to limited application scenarios. In this work, we overcome this issue by a novel model-free framework that incorporates generic category-independent object proposals without the need to pretrain any object detectors. In each frame, our method generates a small number of target object proposals that are shared by multiple objects regardless of their category. This significantly improves the search efficiency in comparison to the traditional dense sampling approach. To further increase the discriminative power of our tracker among targets, we treat all other object proposals as the negative samples, i.e. as "distractors", and update them in an online fashion. For a comprehensive evaluation, we test on the PETS benchmark datasets as well as a new MOOT benchmark dataset that contains more challenging videos. Results show that our method achieves superior performance in terms of both computational speed and tracking accuracy metrics.

1 Introduction

Single object tracking attained considerable success thanks to the advances in "tracking-by-detection" that demonstrated improved performance on standard benchmarks [1–3]. Compared to single-object tracking counterpart, multiple-object tracking is a more challenging task due to the frequent occlusions between the target objects [4] and typical similarities in their motion patterns as well as visual appearances. Moreover, the background scenes also tend to be more cluttered due to the presence of other moving objects [3,5].

In model-based tracking-by-detection of multiple objects, an offline trained category-specific object detector, e.g., DPM [6] or R-CNN [7], is applied at every frame to generate high quality object hypotheses, and then graph-based methods such as max-flow [8,9] are used to solve the subsequent multi-frame multi-target association problem. These multiple object tracking methods, however, depend heavily on the performance of category-specific object detectors, which often miss

S.-H. Lai et al. (Eds.): ACCV 2016, Part II, LNCS 10112, pp. 288–304, 2017.
DOI: 10.1007/978-3-319-54184-6_18

Fig. 1. Results obtained using our model-free multiple object tracking method. Bounding boxes of the same color denote the same tracked object. After initialization, our method tracks each object without any pretrained models. (Color figure online)

objects or generate false positives that are induced by the discrepancy between the training dataset and the test conditions of individual deployments [10].

Being constrained to a specific object class also limits the applicability of the tracker to a certain setting, for example, multiple vehicle tracking in traffic scenes. In practice, however, various applications demand tracking of different types of objects undergoing complex motions as shown in Fig. 1.

On the other end of the spectrum, "model-free" approaches aim to track arbitrary (category-independent) objects [11–15]. They initiate a single bounding box on the target in the first frame and then employ either a generative [16–19] or a discriminative [20–23] strategy to train their object models online. These methods are successfully applied for single-object tracking. However, extending "model-free" methods to multiple tracking task is not a straightforward problem due to two major reasons:

- Computational efficiency – Since each tracker searches around the previous location to localize the object, the time cost is proportional to the number of objects.
- Interactions – Objects contact or occlude each other. They often have similar appearances. Blindly and independently applying single-object trackers multiple times for different targets leads to ambiguities and tracking failures.

To overcome the above challenges, we propose a model-free multiple object tracking framework based on generic object proposals. We take advantage of the proposals in both online training and testing of the tracker.

In the testing stage, a small set of object candidates are generated based on simple objectness cues first. Notice, this set is shared by *all* trackers and it provides two benefits: (i) a significant reduction of the number of candidates, and

(ii) tracking accuracy improvement since many false positives can be eliminated at this stage. The proposals are then assigned to trackers based on the classifier confidence and temporal smoothness measures. The number of proposals can be as many as hundreds while the number of objects might be only a few. We use the Hungarian algorithm [23,24] with appropriate modifications to reduce the computational cost during the data association stage. Other association methods [1–3] can also be used, yet we observe that the computationally efficient Hungarian method works favorably when we build discriminative classifiers based on the generated proposals.

In the training stage, we collect the proposals as *hard* negative samples instead of manual selecting around positive samples. These proposals are expected to contain the other targets and object-like background clutter. Mining explicitly for such hard negative samples and employing hard negatives in the training of individual object models significantly improves the discriminative power of the object models. We update the classifiers at certain time intervals in an online fashion to compensate for object appearances changes over time and incorporate *new* distractors. A few local candidates sampled around the previous object locations are included in the negative set to further improve tracking precision.

We focus on a challenging scenario of multi-object tracking where each object may move **very fast** in an **irregular** fashion. To our knowledge, this challenge has not been widely researched and there are only a few benchmarks (e.g. PETS [5]) available for investigation. Therefore, we collected an extensive set of challenging video sequences from various sources and manually labeled the ground-truth object locations for a comprehensive experimental evaluation.

Our method is conceptually simple, easy to implement, and most importantly, achieves superior performance in comparison to several state-of-the-art techniques in terms of both tracking accuracy metrics and computational efficiency.

2   Related Work

Here we give a brief review to previous methods for multi-object tracking that are most related to this paper. For more comprehensive literature surveys the reader is referred to [3,11–13].

Multiple Target Tracking. As aforementioned in Sect. 1, most multiple object tracking methods focus on the data association problem, assuming sufficiently long and accurate tracklets are provided by using advanced object detectors [3]. For example, [25] considers motion dynamics as the major cue to distinguish different targets with similar appearance. It solves the problem as generalized linear assignment (GLA) of tracklets, which are incrementally joined forming longer trajectories based on their similar dynamics. The work in [1] observes that motion cues are not always reliable for this task, due to for example abrupt camera movement. As a remedy a structured motion constraint between objects is therefore proposed to address this issue.

Tracker in [2] proposes an online discriminative appearance learning approach to handle similar appearances of different objects in tracklet association. This method is similar to our method to be described in this paper; however, in their work those negative training samples are only collected around the tracklets, while ours pivots on the hard negative ones.

Model-Free Object Tracking. Model-free object tracking algorithms are proposed primarily for solving single object tracking applications [11, 12]. The work in [26] tries to improve the identification of a single target object by also tracking stable features in the background, thereby improving the location prior for the target object. [27] proposes a context-aware tracker which considers a set of auxiliary objects as the contextual information for the foreground. These auxiliary objects must satisfy conditions such as having persistent co-occurrence with the foreground and consistent motion correlation.

The tracker in [28] is probably the most closely related work to ours. However, they assume spatial relationship between objects. For instance, nearby objects tend to move along the same direction. The appearance models of all the objects and the structural constraints between these objects are jointly trained in an online structured support vector machine framework. Our framework has no such an assumption and can track arbitrarily moving objects.

Object Proposals for Visual Tracking. As reported in [29, 30], using object proposal improves the object detection benchmark along with the convolutional neural nets. Since, a subset of high-quality candidates are used for detection, object proposal methods boost not only the speed but also the accuracy by reducing false positives. The top performing detection methods [31, 32] for PASCAL VOC [33] use detection proposals. Among the existed proposal methods, the EdgeBox method [30] proposes object candidates based on the observation that the number of contours wholly enclosed by a bounding box is an indicator of the likelihood of the box containing an object. It is designed as a fast algorithm to balance between speed and proposal recall, comparing to BING [34] and region proposal network (RPN) [7].

Many work exist adopting the object proposals for the model-free single object tracking. A straightforward strategy based on linear combination of the original tracking confidence and an adaptive objectness score obtained by BING is employed in [35]. In [36], a detection proposal scheme is applied as a post-processing step, mainly to improve the tracker's adaptability to scale and aspect ratio changes. EBT [37] employs the EdgeBoxes method to globally track the object, disregarding potentially fast or drastic object motion. In contrast, our work utilizes the shared proposals for efficient handling of multiple trackers. [38] deals with generic object tracking for street scenes by generating multi-scale candidates from the point-density map. Tracking is performed using the pseudo-Boolean optimization (QPBO) method. In comparison, our method is applied to more generic object categories rather than street scenes. Besides, our object models is built taking advantage of the proposals, while [38] adopts a generative model using RGB feature distance.

3   Multiple Object Tracking with Proposals

As illustrated in Fig. 2, our framework starts with a few manually initialized bounding boxes on the target objects to be tracked in the first frame of the video. This is similar to the single object online visual tracking task [11–13]. Given these initial bounding boxes, denoted as $\{B^i_{t=1}\}$, $i = 1, \ldots, N_o$, where N_o is the total number of objects, the multiple object tracking problem then aims to find the locations and bounding boxes of the multiple objects in the remainder of the video while maintaining the correct identity of each object.

Following the tracking-by-detection framework, we train the object appearance models for each object. We have an option to use either the generative or discriminative learning strategy. Recent literature on object tracking resort to the discriminative learning to maximize the inter-class separability between the object and background regions and report improved performance as the discriminative learning is more robust to distractions from the background. This property is especially important in multiple object tracking [2,39] where the objects exhibit similar appearance and interact frequently, as depicted in Fig. 2.

As explained in Sect. 1, we do not independently initialize N_o classifiers by collecting locally and densely sampled negative patches as training samples, a scheme that conventional online single object trackers typically employ.

Instead, we incorporate object proposals [29,30] to generate a small number of shared object candidates. Notice that, we are not simply using the original

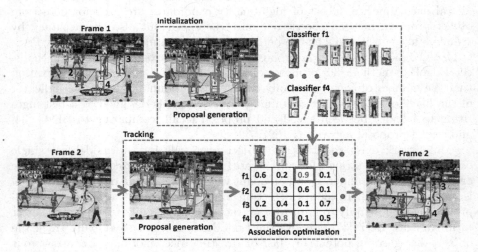

Fig. 2. The structure of our model-free multiple object tracker. The only input is the bounding boxes at the first frame. Our method then initializes multiple classifiers for each object taking advantage of a small set of object proposals generated from the frame. In the next frame, these classifiers are used to assign confidence scores for the candidate proposals. The final trajectories are obtained after solving the optimal association problem. Note that, we also apply the proposals to online update the classifiers to make them more robust to distractors.

object proposals either, since the sizes of the objects usually change during the tracking. We impose the proposal bounding box sizes to be within a certain range of the object sizes. More details about this can be found in Sect. 3.1.

Suppose the object proposal bounding boxes are $\{\hat{B}_{t=1}^j\}$, $j = 1, \ldots, N_p^{t=1}$, where $N_p^{t=1}$ is the total number of proposals in the first frame. We train the classifiers with the corresponding positive samples $B_{t=1}^i$ that are not in the common negative set $\{\hat{B}_{t=1}^j\}$. The initialized classifiers are denoted as

$$f_{t=1}^i(B), \quad i = 1, \ldots, N_o, \tag{1}$$

We additionally select a small set of local candidates sampled around the object to further improve the discriminative power, thus the localization precision, of the classifier as [37].

In the consecutive frame, we generate a set of proposals $\{\hat{B}_{t=2}^j\}$, $j = 1, \ldots,$ $N_p^{t=2}$, to be shared and tested by all classifiers $\{f_{t=1}^i(B)\}$. Considering the temporal smoothness between the object $B_{t=1}^i$ and the proposal $\hat{B}_{t=2}^j$, (spatial distance between them), we build an association matrix that will be efficiently optimized by a modified Hungarian algorithm [23,24]. The new object locations are then determined as the optimal solution of this association problem. More details about it can be found in Sect. 3.2.

To adapt the object appearance changes as well as to increase the discriminative power of the classifiers against newly appeared distractors, we incrementally update the classifiers by treating the estimated bounding box in current frame as the positive sample and object proposals as the negative samples as we did in the first frame. More information is in Sect. 3.3.

3.1   Object Proposal Generation

As mentioned in Sect. 2, various object proposal algorithms exist. We employ EdgeBox [30] as it strikes a good balance between recall and speed. In our experimental analysis, we also test other proposal methods such as BING [34] and region proposal network (RPN) [7].

Two important factors should be noticed here. The first one is the about the sizes of the generated object proposals, termed as size adaption ratio and denoted as $\alpha \in [0, 1]$. We allow the size of the proposals maximally differ the target with a bounding box intersection-over-union (IoU) [33] of ratio α. To be specific, we consider $\hat{B}_t^j$ only when

$$\max_i(\text{IoU}(\hat{B}_t^j, B_{t-1}^i,)) > \alpha, \quad i \in [1, \ldots, N_o] \tag{2}$$

This setting significantly reduces the number of proposals while permitting the object window to adapt the target size changes at the same time. We use $\alpha = 0.8$ and test other values in the experimental part.

The other factor is the maximal number of object proposals generated. Edge-Box does not output a fixed number of proposals. The number of proposals could be any depending on the threshold of the "objectness" score (set as 0.01 as recommended). An appropriate maximal number of proposals needs to be used as its lower values may result in missing the object window in the proposal set while its higher values would cause an extensive number of distractors. We set this number at 500 for all experiments. We also run test other values of the maximal number of proposals in Sect. 4.2.

Similar to [37], we generate a fixed number of bounding boxes, $\{\tilde{B}_t^k\}_{t-1}^i$, $k = 1, \ldots, N_s$, by sampling only around the previous object location B_{t-1}^i for each object (as in traditional methods). This set $\{\tilde{B}_t^k\}_{t-1}^i$ is only tested by the corresponding classifier $f_{t-1}^i(B)$ and they are useful to smoothen the trajectory as the object proposal component works independently at each frame, which may result in temporally inconsistent proposals. Thus, a combined set of $\{\hat{B}_t^j\} \cup \{\tilde{B}_t^k\}_{t-1}^i$ is used during the test stage for the classifier $f_{t-1}^i(B)$. However, we only update the classifier when the estimated one comes from the proposal set $\{\hat{B}_t^j\}$ to attain resistance to potential corruptions. We sample $N_s = 80$ patches uniformly within a 30-pixels radius. More details are in Sect. 3.3.

3.2   Optimal Target Association

Given N_o targets and $(N_p^t + N_s \times N_o)$ candidates, the target association stage therefore aims to find the optimal non-repetitive N_o candidates for the N_o targets, such that the overall *gain* is maximized. Note that, the candidates $\{\tilde{B}_t^k\}_{t-1}^i$ are only allowed to link with target i, thus we set the *gain values* of linking them to other targets to zero.

The gain value $P(B_t, i)$ of linking a candidate B_t to target i is designed base on both classifier confidence score and temporal smoothness,

$$P(B_t, i) = f_{t-1}^i(B_t) + s(B_t, B_{t-1}^i). \tag{3}$$

$s(B_t, B_{t-1}^i)$ is a term representing the temporal smoothness between the previous target bounding box B_{t-1}^i and the candidate box B_t. We use a simple function in this paper: $s(B_t, B_{t-1}^i) = \exp(-\frac{1}{2\sigma^2}\|c(B_t) - c(B_{t-1}^i)\|^2)$, where $c(B_t)$ is the center of bounding box B_t and σ is a value controlling the impact of the temporal smoothness term. We set $\sigma = R_i$, where R_i is half of the diagonal length of the initialized bounding box B_1^i. We also test other values as in Sect. 4.2.

Once the gain values are set, the standard Hungarian algorithm [23,24] can be modified to optimally solve the association problem. As $(N_p^t + N_s \times N_o)$ is usually much larger than N_o (a few hundreds vs. a few), available fast implementation [40] is too slow to be applied directly. We thus firstly find top N_o candidates for each target i locally and separately. As the global optimal assignment for that target i must be one of them, we then combine those found local candidates into a small matrix in which the optimal solution is exact the same global optimal solution to the original association problem. Notice that, the standard Hungarian algorithm solves the minimization problem, thus a simple modification is required before feeding the small matrix to it.

3.3 Online Updating with Proposals

To update the classifier, $f_{t-1}^i \rightarrow f_t^i$, we also generate a few local samples, $\{\tilde{B}_t^k\}_t^i$, $k = 1, \ldots, N_s$, around the estimated object location B_t^i. They are helpful to increase the discriminative power of the classifier, as the object proposals alone represent other good "object-like" regions and training with them increases the discriminative power among "objects-like" candidates, while the negative sample space contains a lot more other negative samples, thus more negative samples help. The updating procedure is applied every 5 frames to balance computational time and minimize potential drift.

As mentioned in the last paragraph of Sect. 3.1, we treat the estimated result B_t^i as an indication for model updating. This is to say, when $B_t^i \in \{\tilde{B}_t^k\}_{t-1}^i$, we assume that there is no good object proposal and the current estimation is a compromise for trajectory smoothness, thus skipping the model updating. If $B_t^i \in \{\hat{B}_t^j\}$, then it suggests a good estimation which has both desirable classifier response and high "objectness", then we update the object model $f_{t-1}^i(B)$ immediately.

3.4 Proposed Tracker: PMOT

Various object models can be integrated into our framework. We choose a popular structured support vector machine (SSVM) method [41], as it shows good performance on several benchmarks [11,12]. The tracker is denoted as PMOT to reflect the concepts of shared proposals and multiple object tracking.

Denote the support vector set trained in the SSVM as $\mathcal{V}_{t-1}$, the classification function can then be expressed as a weighted sum of affinities between the candidate bounding box and the support vectors [41,42]:

$$f_{t-1}^i(B_t) = \sum_{\bar{B}^m \in \mathcal{V}_{t-1}} w^m k(\bar{B}^m, B_t), \quad m = 1, \ldots, |\mathcal{V}_{t-1}| \tag{4}$$

where w^m is a scalar weight associated with the support vector $\bar{B}^m$. Kernel function $k(\bar{B}^m, B_t)$ calculates the affinity between two feature vectors extracted from $\bar{B}^m$ and B_t respectively. The classifier is updated in an online fashion using [43,44] with a budget [45]. Intersection kernel is used and other parameters are set same as [41]. We use histogram features obtained by concatenating 16-bin intensity histograms from a spatial pyramid of 5 levels and RGB channels separately. At each level L, the patch is divided into $L \times L$ cells, resulting in a 2640-D feature vector.

4 Experiments

4.1 Full Benchmark Evaluations

To evaluate the performance of the proposed multiple object tracking method, we collect 10 videos from various sources, including TB50 [15], OTB [11] and

VOT2015 [13]. We denote this dataset as MOOT (Multiple Object Online Tracking) and a few samples can be seen in Fig. 5. The number of targets in these videos ranges from 2 to 5. This dataset contains extremely challenging scenarios, including repetitive mutual occlusion (videos "liquor" and "skating2") and similar appearance among the targets (videos "bolt1", "bolt2", "football" and "basketball").

We also evaluate the proposed method on the video sequences from Performance Evaluation of Tracking and Surveillance (PETS) 2015 [5]. These videos are from surveillance cameras and all targets are humans. We list the details of the four sequences in Table 1 with corresponding challenges featured. As we can see, all sequences contain challenging aspects, while video "A1_ARENA-15_06_TRK_RGB_2" (row 2 in Fig. 3) is the most difficult one containing both deformation and occlusion challenges.

Compared Trackers and Evaluation Metrics. Our method (PMOT) is compared with several state-of-the-art methods. Specifically, we compare our method

Table 1. Attributes of the four video sequences from the PETS dataset.

Video	#humans	#frames	Challenge
N1_ARENA-01_02_TRK_RGB_2	3	115	Size change
W1_ARENA-11_03_ENV_RGB_3	2	107	Body deformation
W1_ARENA-11_03_TRK_RGB_1	2	101	Body deformation
A1_ARENA-15_06_TRK_RGB_2	3	121	Occlusion and body deformation

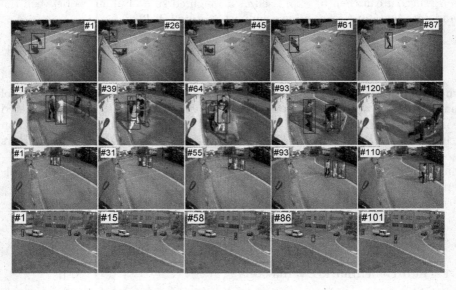

Fig. 3. Sample sequences from the PETS benchmark dataset [5] with ground truth object windows (blue). (Color figure online)

with SPOT [28] which addresses a similar task as ours and it deploys a structure preserving model. We also compare with several single online object trackers to corroborate the point that by sharing and building discriminative classifiers based on proposals, our method is more robust to drifting. MEEM [20], KCF [22] and Struck [41] are three top-ranked trackers in recent large benchmarks [11,12,15,46] for single online object tracking. For all the trackers, we use their default settings and separately initialize on each object for each video. We also modify the PMOT for the single object case, denoted as PMOTsingle. This allows us to precisely analyze the improvement of adopting the proposal sharing scheme, in term of both the tracking metrics and computational efficiency.

We use the single online object tracking metrics to measure the tracking performance, similar to [28]. Evaluation metrics and code are provided by the benchmark [11,15]. We employ the one-pass evaluation (OPE) and use two metrics: *precision plot* and *success plot*. The former one calculates the percentage (*precision score*, PS) of frames whose center location is within a certain threshold distance with the ground truth. A commonly used threshold is 20 pixels. The latter one calculates a same percentage but based on bounding box overlap threshold. We utilize the area under curve (AUC) as an indicative measurement for it.

Experimental Setting. Our tracker is implemented using C++ and MATLAB, on an i7-2600 3.40 GHz desktop with a 8 GB RAM. For the EdgeBox proposal method and SSVM applied, we use the default setting recommended by the authors, except those specified otherwise. We further discuss some parameters in Sect. 4.2.

Benchmark Results. The results are summarized in Fig. 4 and Table 2. We can see that the SPOT tracker achieves undesirable results, significantly lagging behind other compared methods. In term of the PS metric, it is 27.3% worse than Struck, the second worst tracker. It is not particularly surprising though, as can be seen in Fig. 5, where we draw the visual comparison between the proposed PMOT and SPOT. It clearly demonstrates that the SPOT tracker presumes a strong spatial structure exhibited among the objects, while it does not always hold. As shown in the video "bolt1" (row 1 in Fig. 5), the four dash-line windows (SPOT) still maintain the relative positions while drifting away the true objects. In contrast, our method robustly and consistently tracks the objects even they are not moving coherently.

When comparing to the single object online tracking methods, the improvement is clearly shown. On the challenging MOOT dataset, our PMOT tracker outperforms the second best tracker by a large margin, with 9% and 14.7% in term of AUC and PS respectively. We can also see the clear advantage of applying the proposal based approach. Even the single object tracking variant, PMOTsingle, outperforms the best non-proposal tracker, MEEM, by 7.8% and 3.8% in AUC and PS respectively. This is partly contributed by the online updating strategy of collecting the proposals as hard negative samples to improve the

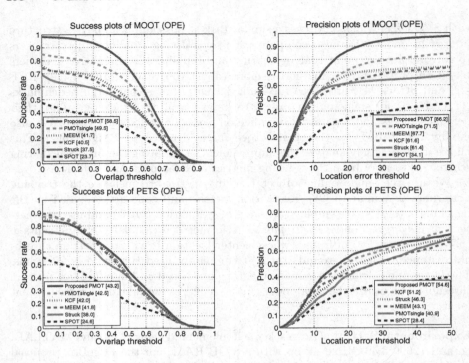

Fig. 4. *Success plot* and *precison plot* on two datasets: MOOT and PETS. Algorithms are ranked by the area under the curve (AUC) and the *precision score* (20 pixels threshold, PS). Our method achieves consistently superior performance, especially on the more challenging MOOT dataset.

Table 2. Area Under Curve (AUC) of *success plot* and *precision score* (PS) with 20 pixels threshold on the MOOT dataset for the one-pass evaluation (OPE). Cell values: AUC/PS

MOOT	PMOT	PMOTsingle	SPOT [28]	MEEM [20]	KCF [22]	Struck [41]
ball1	**66.2**/99.0	66.0/**99.3**	30.6/67.4	51.3/74.5	48.5/83.1	52.7/86.0
basketball	**61.5/84.0**	60.2/81.7	11.6/8.6	46.2/70.9	51.3/59.8	38.5/50.3
bolt1	**47.4/93.8**	36.6/71.6	0.5/0.5	23.5/50.6	34.3/70.6	33.9/73.8
bolt2	50.8/89.0	38.6/69.9	0.6/0.8	47.3/90.4	50.9/93.6	**57.4/97.7**
football	**62.0**/94.6	57.8/88.9	23.4/41.5	60.7/**97.0**	49.5/69.1	57.5/79.7
human4	60.7/93.5	34.5/48.5	**61.5/99.5**	57.4/91.2	50.2/75.7	62.7/94.7
jogging	**67.4/97.6**	63.8/89.7	12.3/13.5	60.6/88.4	15.5/19.9	15.0/19.7
liquor	**61.0/79.8**	41.6/51.0	32.8/38.2	10.6/16.8	18.8/24.6	7.2/8.9
skating1	56.5/71.2	46.5/55.4	55.5/78.4	62.2/**92.3**	**62.8**/89.6	35.9/50.0
skating2	**50.8/44.9**	48.1/43.7	34.6/25.8	35.9/28.4	33.7/37.1	26.7/18.2
Mean	**58.5/86.2**	49.5/71.5	23.7/34.1	41.7/67.7	40.5/61.6	37.5/61.4

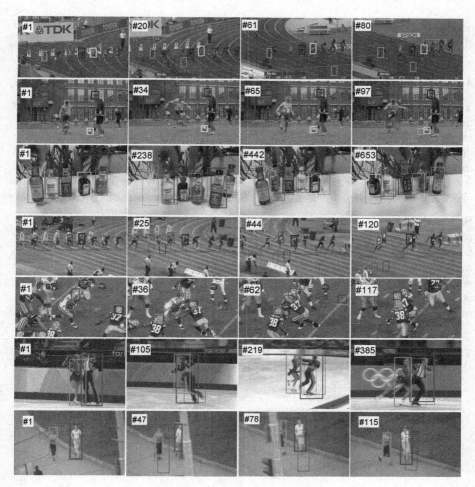

Fig. 5. Qualitative comparisons with the proposed PMOT tracker (solid lines) against the SPOT tracker (dash lines) on videos "bolt1", "ball1", "liquor", "bolt2", "football", "skating2" and "jogging" from MOOT dataset (from top to bottom). Our method exhibits robustness in challenging scenarios such as repetitive mutual occlusions and similar target appearances.

discriminative power of the classifier, hence is robust to the distractions from other objects as well as potential distractors in the background.

For the PETS dataset, we can see that the improvement of PMOT is not great, outperforming the second best tracker, by 3.4% and 0.7% in the PS and AUC metrics, respectively. This is partly due to the fact that there is no significant interactions presented among the objects on PETS, except the video "A1_ARENA-15_06_TRK_RGB_2". Therefore, our proposed multiple object tracking system is unable to take a strong advantage of the proposal sharing benefit.

4.2   Further Remarks

Temporal Smoothness. The smoothness term $s(B_t, B_{t-1}^i)$ (3) discussed in Sect. 3.2 controls the temporal consistency of the trajectory. This is especially important in our formulation as the object proposals are generated independently in each frame, which results temporal inconsistencies inevitably. We test different σ values and include the results in Table 3. We observe that a small σ leads to a strong smoothness constraint, which harms the performance when objects are occluded, while a large σ tends to result in unstable trajectories.

Size Adaption Ratio. The size adaption ratio α in (2) allows the target window to adapt the object size changes naturally once set properly. A smaller α leads to a larger set of object proposals with a more significant size variance, which harms both the computational efficiency and trajectory stability. We validate it with different values and results is in Table 3. It corroborates that a larger value is preferable, but the performance drops when $\alpha = 0.9$, as it constrains the sizes of object proposals too tight that it fails to adapt the object size changes.

Table 3. Area Under Curve (AUC) of *success plot* and *precision score* (20 pixels threshold) results of PMOT with different temporal smoothness constraints and size adaption ratios.

	Temporal smoothness			Size adaption ratio		
	$\sigma = 0.5R_i$	$\sigma = R_i$	$\sigma = 2R_i$	$\alpha = 0.7$	$\alpha = 0.8$	$\alpha = 0.9$
AUC	51.0	58.5	56.2	49.5	58.5	57.9
PS	72.3	86.2	84.1	70.5	86.2	84.9

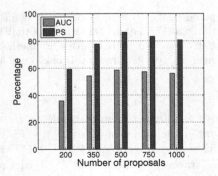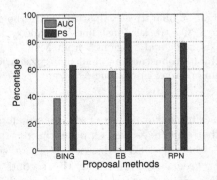

Fig. 6. Area Under Curve (AUC) of *success plot* and *precision score* (20 pixels threshold) results of PMOT with different maximal numbers of proposals and various proposal methods.

Table 4. Processing times (frames per second, fps) of PMOT on videos containing different number of objects.

	PMOT				PMOTsingle
# of targets N_o	2	3	4	5	1
fps	4.1	3.3	2.6	1.9	5.3

Maximal Number of Object Proposals. We test 5 variants with the maximal object proposal number set at 200, 350, 500, 750 and 1000, respectively. The results are reported in term of AUC/PS metrics as included in Fig. 6. As discussed in Sect. 3.1, using insufficient number of proposal leads to a bad coverage of the false positives as well as the object, while using a large number of proposals attracts spurious candidates.

Alternative Object Proposal Methods. We evaluate using other two popular object proposal methods, BING [34] and region proposal network (RPN) [7], instead of EdgeBox for proposals. Results are in Fig. 6. Both performances are worse than the EdgeBox method. This is expected. As shown [29,30], BING results in a relatively low recall of the objects, while RPN performs undesirably for small-size objects.

Failure Mode. Our method may not find every single object in every frame since we use object proposals as object candidates. Thus it may miss the object under, for example, extreme conditions (severe blur, distortion). Such miss detections, however, do not occur all the time. A temporary failure does not harm the overall performance since the model is incrementally and selectively updated.

Computational Efficiency. Since the object proposals are shared among the classifiers of multiple targets, we reduce the computational load by not repeating the proposal generation and feature extraction for each target. Table 4 shows the processing times (frames per second, fps) for different number of targets. We categorize the test videos according to the number of targets in them. For PMOTsingle, the number of targets is always 1. As we can see, our system is computationally efficient.

5   Conclusion

We proposed a computationally efficient and accurate model-free multiple object tracking method. It takes the advantage of the object proposals and generates a small and shared set of object hypotheses in the frame. Then it initializes multiple classifiers for each target using the shared set. In consecutive frames, the application and update of the classifiers are also achieved by using the detected proposals. We evaluated our method on both PETS and a newly introduced dataset. The results show superior performance against the state-of-the-art.

Acknowledgement. This work was supported under the Australian Research Council's Discovery Projects funding scheme (project DP150104645, DP120103896), Linkage Projects funding scheme (LP100100588), ARC Centre of Excellence on Robotic Vision (CE140100016).

References

1. Yoon, J.H., Lee, C.R., Yang, M.H., Yoon, K.: Online multi-object tracking via structural constraint event aggregation. In: CVPR (2016)
2. Bae, S.H., Yoon, K.J.: Robust online multi-object tracking based on tracklet confidence and online discriminative appearance learning. In: CVPR (2014)
3. Milan, A., Leal-Taixé, L., Reid, I.D., Roth, S., Schindler, K.: MOT16: a benchmark for multi-object tracking. CoRR (2016)
4. Zhang, X., Hu, W., Qu, W., Maybank, S.: Multiple object tracking via species-based particle swarm optimization. IEEE Trans. Circ. Syst. Video Technol. **20**, 1590–1602 (2010)
5. Li, L., Nawaz, T., Ferryman, J.: PETS 2015: datasets and challenge. In: AVSS (2015)
6. Felzenszwalb, P.F., Girshick, R.B., McAllester, D., Ramanan, D.: Object detection with discriminatively trained part-based models. TPAMI **32**(9), 1627–1645 (2010)
7. Ren, S., He, K., Girshick, R., Sun, J.: Faster R-CNN: towards real-time object detection with region proposal networks. In: NIPS (2015)
8. Leal-Taixe, L., Fenzi, M., Kuznetsova, A., Rosenhahn, B., Savarese, S.: Learning an image-based motion context for multiple people tracking. In: CVPR (2014)
9. Milan, A., Roth, S., Schindler, K.: Continuous energy minimization for multitarget tracking. TPAMI **36**(1), 58–72 (2014)
10. Torralba, A., Efros, A.A.: Unbiased look at dataset bias. In: CVPR (2011)
11. Wu, Y., Lim, J., Yang, M.H.: Online object tracking: a benchmark. In: CVPR (2013)
12. Smeulders, A.W.M., Chu, D.M., Cucchiara, R., Calderara, S., Dehghan, A., Shah, M.: Visual tracking: an experimental survey. TPMAI **36**(7), 1442–1468 (2014)
13. Kristan, M., Matas, J., Leonardis, A., Felsberg, M., Cehovin, L., Fernandez, G., Vojir, T., Hager, G., Nebehay, G., Pflugfelder, R.: The visual object tracking VOT2015 challenge results. In: ICCVW (2015)
14. Felsberg, M., Berg, A., Hager, G., Ahlberg, J., Kristan, M., Matas, J., Leonardis, A., Cehovin, L., Fernandez, G., Vojir, T., et al.: The thermal infrared visual object tracking VOT-TIR2015 challenge results. In: ICCVW (2015)
15. Wu, Y., Lim, J., Yang, M.: Object tracking benchmark. TPAMI **37**(9), 1834–1848 (2015)
16. Ross, D.A., Lim, J., Lin, R.S., Yang, M.H.: Incremental learning for robust visual tracking. IJCV **77**(1–3), 125–141 (2008)
17. Mei, X., Ling, H.: Robust visual tracking using L1 minimization. In: ICCV (2009)
18. Li, H., Shen, C., Shi, Q.: Real-time visual tracking using compressive sensing. In: CVPR (2011)
19. Jia, X., Lu, H., Yang, M.H.: Visual tracking via adaptive structural local sparse appearance model. In: CVPR (2012)

20. Zhang, J., Ma, S., Sclaroff, S.: MEEM: robust tracking via multiple experts using entropy minimization. In: Fleet, D., Pajdla, T., Schiele, B., Tuytelaars, T. (eds.) ECCV 2014. LNCS, vol. 8694, pp. 188–203. Springer, Heidelberg (2014). doi:10.1007/978-3-319-10599-4_13

21. Zhu, G., Porikli, F., Ming, Y., Li, H.: Lie-struck: affine tracking on Lie groups using structured SVM. In: WACV (2015)

22. Henriques, J.F., Caseiro, R., Martins, P., Batista, J.: High-speed tracking with kernelized correlation filters. TPAMI 37(3), 583–596 (2015)

23. Babenko, B., Yang, M.H., Belongie, S.: Visual tracking with online multiple instance learning. TPAMI 33(8), 1619–1632 (2011)

24. Munkres, J.: Algorithms for the assignment and transportation problems. J. Soc. Ind. Appl. Math. 5(1), 32–38 (1957)

25. Dicle, C., Camps, O.I., Sznaier, M.: The way they move: tracking multiple targets with similar appearance. In: ICCV (2013)

26. Duan, G., Ai, H., Cao, S., Lao, S.: Group tracking: exploring mutual relations for multiple object tracking. In: Fitzgibbon, A., Lazebnik, S., Perona, P., Sato, Y., Schmid, C. (eds.) ECCV 2012. LNCS, vol. 7574, pp. 129–143. Springer, Heidelberg (2012). doi:10.1007/978-3-642-33712-3_10

27. Yang, M., Wu, Y., Hua, G.: Context-aware visual tracking. TPAMI 31(7), 1195–1209 (2009)

28. Zhang, L., van der Maaten, L.: Preserving structure in model-free tracking. TPAMI 36(4), 756–769 (2014)

29. Hosang, J., Benenson, R., Schiele, B.: How good are detection proposals, really? In: BMVC (2014)

30. Zitnick, C.L., Dollár, P.: Edge boxes: locating object proposals from edges. In: Fleet, D., Pajdla, T., Schiele, B., Tuytelaars, T. (eds.) ECCV 2014. LNCS, vol. 8693, pp. 391–405. Springer, Heidelberg (2014). doi:10.1007/978-3-319-10602-1_26

31. Girshick, R., Donahue, J., Darrell, T., Malik, J.: Rich feature hierarchies for accurate object detection and semantic segmentation. In: CVPR (2014)

32. Wang, X., Yang, M., Zhu, S., Lin, Y.: Regionlets for generic object detection. In: ICCV (2013)

33. Everingham, M., Eslami, S.M.A., Gool, L.V., Williams, C.K.I., Winn, J.M., Zisserman, A.: The pascal visual object classes challenge: a retrospective. IJCV 111(1), 98–136 (2015)

34. Cheng, M., Zhang, Z., Lin, W., Torr, P.H.S.: BING: binarized normed gradients for objectness estimation at 300fps. In: CVPR (2014)

35. Liang, P., Liao, C., Mei, X., Ling, H.: Adaptive objectness for object tracking. CoRR (2015)

36. Huang, D., Luo, L., Wen, M., Chen, Z., Zhang, C.: Enable scale and aspect ratio adaptability in visual tracking with detection proposals. In: BMVC (2015)

37. Zhu, G., Porikli, F., Li, H.: Beyond local search: tracking objects everywhere with instance-specific proposals. In: CVPR (2016)

38. Ošep, A., Hermans, A., Engelmann, F., Klostermann, D., Mathias, M., Leibe, B.: Multi-scale object candidates for generic object tracking in street scenes. In: ICRA (2016)

39. Possegger, H., Mauthner, T., Bischof, H.: In defense of color-based model-free tracking. In: CVPR (2015)

40. Cao, Y.: Hungarian algorithm for linear assignment problems (V2.3) (2008). http://www.mathworks.com/

41. Hare, S., Saffari, A., Torr, P.H.S.: Struck: structured output tracking with kernels. In: ICCV (2011)

42. Blaschko, M.B., Lampert, C.H.: Learning to localize objects with structured output regression. In: Forsyth, D., Torr, P., Zisserman, A. (eds.) ECCV 2008. LNCS, vol. 5302, pp. 2–15. Springer, Heidelberg (2008). doi:10.1007/978-3-540-88682-2_2
43. Bordes, A., Bottou, L., Gallinari, P., Weston, J.: Solving multiclass support vector machines with LaRank. In: ICML (2007)
44. Bordes, A., Usunier, N., Bottou, L.: Sequence labelling SVMs trained in one pass. In: Daelemans, W., Goethals, B., Morik, K. (eds.) ECML PKDD 2008. LNCS (LNAI), vol. 5211, pp. 146–161. Springer, Heidelberg (2008). doi:10.1007/978-3-540-87479-9_28
45. Wang, Z., Crammer, K., Vucetic, S.: Multi-class pegasos on a budget. In: ICML (2010)
46. Kristan, M., et al.: The visual object tracking VOT2014 challenge results. In: Agapito, L., Bronstein, M.M., Rother, C. (eds.) ECCV 2014. LNCS, vol. 8926, pp. 191–217. Springer, Heidelberg (2015). doi:10.1007/978-3-319-16181-5_14

Learning to Integrate Occlusion-Specific Detectors for Heavily Occluded Pedestrian Detection

Chunluan Zhou[✉] and Junsong Yuan

School of EEE, Nanyang Technological University, Singapore, Singapore
czhou002@e.ntu.edu.sg

Abstract. It is a challenging problem to detect partially occluded pedestrians due to the diversity of occlusion patterns. Although training occlusion-specific detectors can help handle various partial occlusions, it is a non-trivial problem to integrate these detectors properly. A direct combination of all occlusion-specific detectors can be affected by unreliable detectors and usually does not favor heavily occluded pedestrian examples, which can only be recognized by few detectors. Instead of combining all occlusion-specific detectors into a generic detector for all occlusions, we categorize occlusions based on how pedestrian examples are occluded into K groups. Each occlusion group selects its own occlusion-specific detectors and fuses them linearly to obtain a classifer. An L1-norm linear support vector machine (SVM) is adopted to select and fuse occlusion-specific detectors for the K classifiers simultaneously. Thanks to the L1-norm linear SVM, unreliable and irrelevant detectors are removed for each group. Experiments on the Caltech dataset show promising performance of our approach for detecting heavily occluded pedestrians.

1 Introduction

Pedestrian detection has many applications such as video surveillance and autonomous driving and much work has been done to improve its performance in recent years [1–6]. Despite recent progress, partial occlusions are still a great challenge for pedestrian detection. Some state-of-the-art approaches achieve promising results on pedestrian detection benchmarks like Caltech [7] when pedestrians to be detected are not occluded or only slightly occluded. When heavy occlusions are present, their performances decrease drastically. For example, the log-average miss rate of Checkerboard detector [4] on the Caltech benchmark is 18.5% when only unoccluded or partially occluded pedestrians are considered. Its performance drops to 77.5% when pedestrians to be detected are heavily occluded. More efforts are still required to improve the performance of detecting heavily occluded pedestrians.

A simple yet effective approach to occlusion handling for pedestrian detection is to train specific detectors for various occlusion patterns [5,8]. Occlusion

© Springer International Publishing AG 2017
S.-H. Lai et al. (Eds.): ACCV 2016, Part II, LNCS 10112, pp. 305–320, 2017.
DOI: 10.1007/978-3-319-54184-6_19

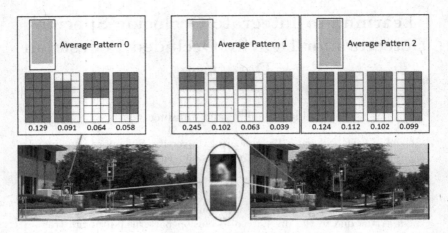

Fig. 1. Different integrations of occlusion-specific detectors. Left: The average pattern 0 is obtained by averaging visible parts of pedestrian examples in the Caltech dataset [7] and the top 4 occlusion patterns whose detectors have the highest weights are shown blow. Blue regions in the occlusion patterns indicate visible parts of the human body. The pedestrian which is heavily occluded in the image below is ranked at 4th among top 5 regions. Right: Our approach clusters the pedestrian examples into 2 groups each of which selects its own compatible occlusion-specific detectors. The detectors associated with the average pattern 1 rank the pedestrian in the image below at 1st. (Color figure online)

patterns can be manually designed according to prior knowledge [8] or automatically selected from a large pool of candidates [5]. After occlusion patterns are obtained, a specific detector is trained independently for each occlusion pattern. To make the approach work well, it is important to integrate the occlusion-specific detectors properly. Specifically, for a given image region, a detection score should be assigned to it based on the outputs of these detectors. A simple method is to assign the maximum among these outputs to the image region as adopted in [8]. This method needs a sophisticated score calibration to make the outputs of the independently trained detectors comparable. In [5], a weight vector is learned by training a linear support vector machine (SVM) to combine the outputs of occlusion-specific detectors. The important detectors would have large combination weights thus score calibration is not required. However, as this method only learns a single weight vector for all pedestrian examples, it may not be able to well separate these pedestrian examples from background especially when some of these examples are heavily occluded. A heavily occluded pedestrian usually receives low scores from the detectors of some occlusion patterns whose visible portion is incompatible with the visible part of the pedestrian. For example, the detector of the occlusion pattern in which only the lower body is visible would probably assign a low score to a pedestrian whose lower body is occluded. Therefore, with the single weight vector, heavily occluded pedestrians tend to have a relatively low detection score compared to fully visible pedestrians and are not easy to be distinguished from the background as illustrated in the left of Fig. 1. In addition, some occlusion detectors, especially those unreliable ones, tend to produce noisy outputs and including them may not help improve and even decrease the detection performance.

Following the framework in [5], we propose a new approach to integrate occlusion-specific detectors for heavily occluded pedestrian detection. Instead of only learning a single weight vector to fuse all occlusion-specific detectors, we cluster pedestrian examples into several groups according to their occlusion patterns and learn a weight vector for each group such that a pedestrian example in the group is scored by the linear combination of the outputs of the occlusion-specific detectors specified by the weight vector. To remove irrelevant and unreliable occlusion-specific detectors, we impose sparsity on the weight vector of each group. The weight vectors are learned simultaneously by training an L1-norm linear SVM [9]. Our integration approach selects highly compatible occlusion-specific detectors for each group and can better distinguish heavily occluded pedestrians from the background as illustrated in the right of Fig. 1. The effectiveness of our approach is demonstrated on the Caltech dataset [7].

2   Related Work

Occlusion handling is a difficult problem for pedestrian detection and some approaches have been proposed for this purpose. In [10], a pedestrian template is divided into several blocks and the occlusion is inferred by estimating the visibility statuses of these blocks. An implicit shape model (ISM) [11] is adopted in [12] to generate pedestrian hypotheses in an image with associated supporting regions which are further verified using local and global cues. As each pixel can only be assigned to the supporting region of one hypothesis, this approach is able to separate overlapping pedestrians. Several approaches [5,8,13,14] represent occlusions by a set of patterns for each of which a specific detector is learned. The approaches in [13,14] discover frequently occurring occlusion patterns from annotated training data and train a deformable part model (DPM) [15] for each occlusion pattern. In [5,8], occlusion patterns are manually defined and boosted detectors are adopted. Different from the occlusion patterns defined in [5,8] which only model a single pedestrian, the occlusion patterns used in [13,14] can model a single pedestrian or two overlapping pedestrians. For detection, these occlusion-specific detectors are integrated in a winner-take-all fashion [8,13,14] or using linear combination [5]. Our approach adopts the same framework as [5,8] with a new integration approach in which a sparse subset of occlusion-specific detectors are selected. Group sparsity has been exploited in [16] to select components for learning a mixture model, but we use it to simultaneously select and fuse detectors. In [17], multi-pedestrian detectors which are learned for detecting overlapping pedestrians are exploited to refine detections obtained by single-pedestrian detectors in a probabilistic framework. Some other approaches [18–24] learn a set of part detectors which are properly integrated for occlusion handling. In [18–20], a human body is divided into several parts and outputs from part detectors are integrated using a set of rules [18], by linear combination with weights learned from intensity, depth and motion cues [19] or in a Bayesian framework [20]. Different from

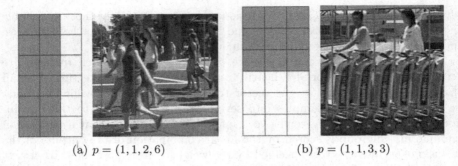

(a) $p = (1, 1, 2, 6)$ (b) $p = (1, 1, 3, 3)$

Fig. 2. Examples of occlusion patterns. (a) The right portion of the human body is occluded. Two instances marked by red rectangles are shown on the right: one is occluded by another pedestrian and the other is truncated by the image boundary. (b) The lower body is occluded. Two pedestrians on the right are occluded by carts. (Color figure online)

the approaches in [18–20], a human body is represented by a set of overlapping parts which are organized in a hierarchy structure in [21–24]. In [24], a set of rules are defined specifically for the hierarchy structure for inferring occlusion based on part visibility statuses. A discriminative deep model is used to learn correlations among part visibility statuses [21, 22] and the mutual visibility relationship among pedestrians [23] respectively. Then, pedestrian classification is done in a probabilistic framework with detection scores from part detectors as input and part statuses as hidden variables.

3   Training Occlusion-Specific Detectors

Occlusions may occur at different parts of a pedestrian and have various patterns. For example, the lower body of a pedestrian may be occluded by a car, and the left or right half body may be occluded by a pole or another pedestrian. As in [5], we construct a pool of occlusion patterns in which different parts of a human body are occluded. The human body is represented by an $R \times C$ grid, where R and C are the numbers of cells in one row and one column respectively. We sample all possible rectangular subregions of size $r \times c$ in the grid with $R_{min} \le r \le C$ and $C_{min} \le c \le C$. Each subregion together with the grid form one occlusion pattern in which the subregion represents the visible part of the human body. To avoid sampling subregions that are too small, we restrict the minimum height and width of a subregion. For each size $r \times c$, we sample subregions from top-left to bottom-right in the grid with a step length of one cell in both horizontal and vertical directions. Mathematically, an occlusion pattern can be represented by a 4-tuple $p = (x, y, r, c)$, where (x, y) are the top-left coordinates of the occlusion pattern in the grid and (r, c) specifies its width and height. We set $R_{min} = 2$, $C_{min} = 2$, $R = 6$ and $C = 3$ in our experiments, resulting in a pool of 45 occlusion patterns. The occlusion pattern pool can be

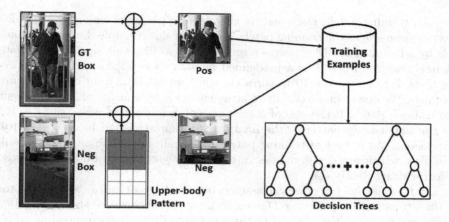

Fig. 3. Process of training a detector for the upper-body pattern.

expressed as $P = \{p_i | 1 \leq i \leq M\}$, where $p_i = (x_i, y_i, r_i, c_i)$ and $M = 45$. Figure 2 shows two examples of occlusion patterns.

For each occlusion pattern, we train a detector using locally decorrelated channel features (LDCF) [25] and RealBoost [26]. Figure 3 illustrates the training process for the upper-body pattern. LDCF belongs to a family of channel features [2–4,27] which are widely used for pedestrian detection. We consider the human body as a rigid object and model it with a template of $H \times W$ pixels, where H and W are the height and width, respectively. As the ground-truth bounding boxes of positive examples usually have different sizes and aspect ratios, we perform bounding box standardization as in [7]: for each ground-truth bounding box, we adjust its width such that the aspect ratio after adjustment is equal to $\frac{H}{W}$ while keeping its height and center coordinates unchanged. An example of bounding box standardization is given at the top-left of Fig. 3. The original bounding box is marked in red and the green rectangle is the new bounding box after standardization. It is pointed out in [1] that including a certain amount of background around a pedestrian as context can improve detection performance. To exploit the surrounding context, we add some padding to the template boundary. Denote the height and width of the padded template by H' and W' respectively. Accordingly, we expand the standardized bounding box of a positive example by including some background around it. The yellow rectangle at the top-left of Fig. 3 shows the expanded bounding box of the pedestrian example. Finally, we crop the region corresponding to the visible part of the occlusion pattern from the expanded bounding box as a positive training patch. Negative patches are sampled from background regions. For feature extraction, each training patch is scaled to the size of the visible part in the template. For example, the size of a training patch for the upper-body pattern is $\frac{H'}{2} \times W'$. We use the same channels as in [3] to compute features for training examples: normalized gradient magnitude (1 channel), LUV color channels (3 channels) and histograms of oriented gradients (6 channels), with a total of 10 feature channels for each training patch. Each feature channel is a per-pixel lookup table which has the same size as the training patch. We further downsample the feature channels by a factor of 2. LDCF learns a set of k $m \times m$ filters to locally decorrelate channel features where k is a predefined number (See [25] for details). Applying these k filters to the 10 features channels results in a total of $10k$ feature channels. We downsample the $10k$ feature channels by a factor of 2 to represent the training patch. If the size of a training patch is $s \times t$ ($s = \frac{H'}{2}$ and $t = W'$ for the upper-body pattern), the final feature representation has $\frac{s}{4} \times \frac{t}{4} \times 10k$ dimensions. With a set of training patches, we train and combine a set of weak classifiers which are decision trees in our implementation to obtain a boosted detector using RealBoost.

Let L be the number of weak classifiers of each detector and d_i be the detector of the i-th occlusion pattern p_i. Given an image patch x whose size is $H' \times W'$, we denote by $\psi_i(x)$ the channel features extracted from x corresponding to p_i. The detection score of x given by d_i is computed by

$$d_i(x) = \sum_{j=1}^{L} w_i^j f_i^j(\psi_i(x)), \tag{1}$$

where f_i^j is the j-th weak classifier of d_i and w_i^j is the corresponding weight. The set of detectors trained for the occlusion patterns in P can be expressed as $D = \{d_i | 1 \leq i \leq M\}$.

4 Integrating Occlusion-Specific Detectors

After obtaining the occlusion-specific detectors as described in Sect. 3, we need to integrate these detectors properly for pedestrian detection. Specifically, given an image patch x whose size is $H' \times W'$ pixels, we want to assign a score $g(x)$ indicating how likely the image patch x contains a pedestrian based on the detection scores $d_i(x)$ ($1 \leq i \leq M$) from the occlusion-specific detectors in D.

In [5], a linear SVM is trained to combine detection scores from different part detectors which are learned by deep neural networks. This method also applies to our detectors. Mathematically, the goal is to learn a weight vector $\mathbf{a} = [a_1, \ldots, a_M]$ to linearly combine the detection scores from the M detectors

$$g(x) = \sum_{i=1}^{M} a_i d_i(x) + b, \tag{2}$$

where b is a constant. Given a set of training examples (x_i, y_i) for $1 \leq i \leq N$ with x_i an image patch and $y_i \in \{1, -1\}$ the label of x_i, the weight vector $\mathbf{a}$ is obtained by solving the following optimization problem:

$$\min_{\mathbf{a}, b} ||\mathbf{a}||_2^2 + C \sum_{i=1}^{N} \max(0, 1 - y_i g(x_i)), \tag{3}$$

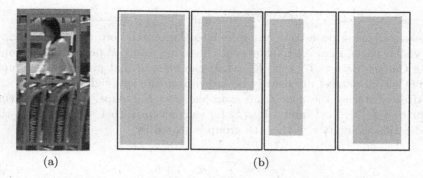

(a)   (b)

Fig. 4. Visible part clustering. (a) Red and green bounding boxes show the visible part and the full body of a pedestrian example respectively. (b) Four cluster centers obtained by K-means on pedestrian examples in the Caltech dataset. Cyan and white regions indicate visible and occluded parts respectively. (Color figure online)

where C is a parameter. By defining a score vector $\mathbf{s} = [d_1(x), \ldots, d_M(x)]$, we can write the score of x as $g(x) = \langle \mathbf{a} \cdot \mathbf{s} \rangle + b$, where $\langle \mathbf{a} \cdot \mathbf{s} \rangle$ is the inner product of $\mathbf{a}$ and $\mathbf{s}$.

Nevertheless, the above method has two limitations. First, it only learns a single weight vector for all pedestrian examples with various occlusion patterns. The decision boundary specified by $\mathbf{a}$ and b may not well separate these pedestrian examples from the background, especially when heavy occlusions are present. Actually, for a heavily occluded pedestrian, some occlusion-specific detectors are irrelevant and tend to give a low detection score. For example, an occlusion pattern in which the upper body is occluded does not have any overlap with a pedestrian whose lower body is occluded thus the pedestrian would probably receive a low detection score from the detector of that occlusion pattern. Therefore, heavily occluded examples usually have a relatively low score compared with fully visible pedestrians and are not easy to be distinguished from the background (See the left part of Fig. 1). Second, some occlusion-specific detectors are not reliable and may produce noisy outputs. Including these detectors may not help improve and even decrease the detection performance. To address the two limitations, we cluster pedestrian examples into several groups according to how they are occluded and learn a weight vector for each group such that only a subset of occlusion-specific detectors which are both relevant and reliable would contribute to the group.

Each pedestrian example is annotated with two bounding boxes which denote the visible part $B_{vis} = (x_{vis}^1, y_{vis}^1, x_{vis}^2, y_{vis}^2)$ and the full body $B_{full} = (x_{full}^1, y_{full}^1, x_{full}^2, y_{full}^2)$ respectively (See Fig. 4(a)). Let $w_{full} = x_{full}^2 - x_{full}^1 + 1$ and $h_{full} = y_{full}^2 - y_{full}^1 + 1$ be the width and height of B_{full} respectively. We normalize the visible part B_{vis} relative to the full body B_{full} to obtain a new bounding box $B_{norm} = (x_{norm}^1, y_{norm}^1, x_{norm}^2, y_{norm}^2)$ where $x_{norm}^1 = \frac{x_{vis}^1 - x_{full}^1}{w_{full}}$,

$y^1_{norm} = \frac{y^1_{vis}-y^1_{full}}{h_{full}}$, $x^2_{norm} = \frac{x^2_{vis}-x^1_{full}}{w_{full}}$ and $y^2_{norm} = \frac{y^2_{vis}-y^1_{full}}{h_{full}}$. Then, we use K-means to cluster pedestrian examples into K groups according to their normalized visible parts. Figure 4(b) shows the clustering result on pedestrian examples in the Caltech dataset [7] with $K = 4$. It can be seen that pedestrian examples in the Caltech dataset are usually occluded from the left, right or bottom.

After clustering the pedestrian examples into K groups, we learn a weight vector $\mathbf{c}_j = [c^1_j, \ldots, c^M_j]$ and a bias b_j for each group with $1 \leq j \leq K$ such that a pedestrian example x in the j-th group is scored by

$$g_j(x) = \sum_{i=1}^{M} c^i_j d_i(x) + b_j. \tag{4}$$

However, if we learn the weight vectors separately for each group, the final scores may not be comparable and a further calibration would be required. To solve this problem, we propose to learn $\mathbf{c}_j$ and b_j for $1 \leq j \leq K$ simultaneously by training a single L1-norm linear SVM [9]. The pedestrian examples comprise the positive training set and negative training examples are collected from background regions. Each training example x_i for $1 \leq i \leq N$ is labeled by $l_i = (y_i, g_i)$ where $y_i \in \{-1, 1\}$ indicates whether x_i is a pedestrian and $g_i \in \{1, \ldots, K\}$ is its group index. Let $\mathbf{s}_i = [d_1(x_i), \ldots, d_M(x_i)]$ be the detection scores of x_i from the M detectors. We represent x_i by a long feature vector $\mathbf{u}_i = [\mathbf{u}^1_i, \ldots, \mathbf{u}^K_i]$ where $\mathbf{u}^j_i$ ($1 \leq j \leq K$) is a $(M+1)$-dimensional feature vector corresponding to the j-th group and is defined by

$$\mathbf{u}^j_i = \begin{cases} [\mathbf{s}_i, B] & \text{if } g_i = j; \\ [\mathbf{0}, B] & \text{otherwise,} \end{cases} \tag{5}$$

where B is a constant feature. We learn a long weight vector $\mathbf{w} = [\mathbf{w}^1, \ldots, \mathbf{w}^K]$, where $\mathbf{w}^j = [\mathbf{w}^j(1), \ldots, \mathbf{w}^j(M+1)]$ ($1 \leq j \leq K$) is a $(M+1)$-dimensional weight vector corresponding to the j-th group, by solving the following optimization problem:

$$\min_{\mathbf{w}} ||\mathbf{w}||_1 + C \sum_{i=1}^{N} \max(0, 1 - y_i \langle \mathbf{w} \cdot \mathbf{u}_i \rangle). \tag{6}$$

The lasso penalty $||\mathbf{w}||_1$ is first proposed in [28] for regression problems and its L1 nature would cause some weights in $\mathbf{w}$ to be exactly zero when C is sufficiently small (See [9] for more discussions). For a specific group, irrelevant and unreliable occlusion-specific detectors are usually less important and tend to receive a weight of zero. After $\mathbf{w}$ is learned, we can obtain $\mathbf{c}_j$ and b_j by setting $\mathbf{c}_j = [\mathbf{w}^j(1), \ldots, \mathbf{w}^j(M)]$ and $b_j = B\mathbf{w}^j(M+1)$. We choose a large value for B to reduce the impact of its corresponding weights $\mathbf{w}^j(M+1)$ for $1 \leq j \leq K$ in the lasso penalty $||\mathbf{w}||_1$ ($B = 5000$ is used in our experiments). At testing stage, the score of an image patch x is determined by the largest score from the K groups: $g(x) = \max_{1 \leq j \leq K} g_j(x)$.

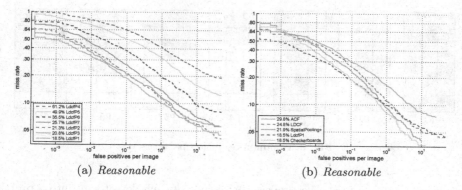

(a) *Reasonable*   (b) *Reasonable*

Fig. 5. (a) Results of occlusion-specific detectors. (b) Results of our LDCF implementation and some state-of-the-art approaches using channel features.

5   Experiments

We evaluate the proposed approach on the Caltech dataset [7]. Following the standard evaluation protocol, we use video sets S0–S5 for training and S6–S10 for testing. Detection performance is summarized by log average miss rate over 9 false positive per-image (FPPI) points ranging from 10^{-2} to 10^0. Three subsets from S6–S10 are used for testing: *Reasonable*, *Partial* and *Heavy*. In the *Reasonable* subset, only pedestrians with at least 50 pixels tall and under no or partial occlusion are used for evaluation. This subset is widely used for evaluating pedestrian detection approaches. In the *Partial* and *Heavy* subsets, pedestrians are at least 50 pixels tall and are partially occluded (1–35% occluded) and heavily occluded (35–80% occluded) respectively.

5.1   Implementation Details

For training detectors of occlusion patterns, we set the template size of the human body to 100×41 and the template size becomes 112×48 after an amount of padding is added. Four 5×5 filters are used for LDCF extraction, producing a total of 40 feature channels for an image patch. We sample training data from video sets S0–S5 at an interval of 2 frames. The maximum depth of a decision tree is set to 5 and a boosted detector of $L = 4096$ decision trees is trained for each occlusion pattern. We adopt five rounds of bootstrapping to learn 64, 512, 1024, 2048 and 4096 decision trees respectively.

For integrating occlusion-specific detectors, we sample pedestrians that are at least 50 pixels tall and are occluded no more than 60% as positive examples. Negative examples are collected by several rounds of hard mining. We use LIBLINEAR [29] and G-SVM [30] for solving linear SVMs and L1-norm linear SVMs respectively. Besides linear combination, we also implement a max integration approach in which the final score of an example x is defined as

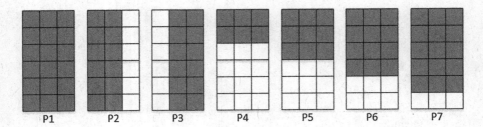

Fig. 6. Common occlusion patterns.

$g(x) = max_{1 \leq i \leq M} d_i(x)$. We convert the output of each detector d_i into a probability by logistic regression:

$$d_i'(x) = \frac{1}{1 + exp(-q_i d_i(x) - r_i)}, \tag{7}$$

where q_i and r_i are the parameters to be learned. Then, the final score becomes $g'(x) = max_{1 \leq i \leq M} d_i'(x)$.

5.2   Results of Occlusion-Specific Detectors

Figure 5(a) shows the results of occlusion-specific detectors of some common occlusion patterns given in Fig. 6 on the *Reasonable* subset. We can see that the log-average miss rates of these detectors vary largely from 18.5% to 61.2%. The detector of the full body pattern P1 performs best among the seven detectors. The detectors of P2 and P3 performs slightly worse than the detector of P1. From P4 to P7, the performance increases gradually when the amount of occlusion decreases. Generally, the detector of a heavily occluded pattern performs worse than the detector of a pattern that is only slightly or not occluded on the *Reasonable* subset. Figure 5(b) shows the comparison between our implementation of LDCF (LdcfP1) and some state-of-the-art approaches which are also based on channel features: ACF [3], LDCF [25], SpatialPooling+ [31] and Checkerboards [4]. LdcfP1 achieves a much better performance than the original LDCF and performs as well as Checkerboards.

5.3   Results of Different Integration Methods

We compare the proposed integration approach with three baselines: Max, Max-Cal and Linear. Linear is the approach proposed in [5] which learns a weight vector to linearly combine outputs of occlusion-specific detectors as described in Sect. 4. Max and MaxCal are two approaches using the max integration as described in Sect. 5.2 without and with score calibration respectively. For each baseline, we experiment with different numbers of occlusion-specific detectors. We set the number of occlusion-specific detectors m to 1, 5, 10, 15, 20, 25, 30, 35, 40 and 45 respectively. When $m = 1$, the detector of the full body pattern

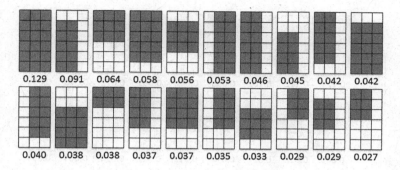

Fig. 7. Top 20 occlusion patterns of Linear with associated combination weights.

is used. For $m > 1$, we choose top m occlusion-specific detectors with the highest combination weights learned by Linear. Figure 7 shows the top 20 occlusion patterns of Linear. For our approach, we adjust the parameter C in Eq. (6) to control the number of occlusion-specific detectors to be selected.

Table 1 shows the results of the three baselines and our approach applied to different numbers of occlusion-specific detectors on the *Reasonable*, *Partial* and *Heavy* subsets. It can be seen that MaxCal outperforms Max for different values of m on all the three subsets. As our occlusion-specific detectors are trained independently, proper score calibration is necessary for the max integration approach to work well. On the *Reasonable* subset, integrating a number of occlusion-specific detectors ($m > 1$) using MaxCal does not help improve the performance compared to the detector of the full body pattern ($m = 1$). This is probably because the full-body detector already works reasonably well for detecting pedestrians which are slightly or not occluded, while for MaxCal, occlusion-specific detectors could introduce more false positives. When the amount of occlusion increases, the other detectors can complement the full-body detector for detecting partially and heavily occluded pedestrians. As shown in Table 1, the best performance of MaxCal is achieved when $m > 1$ on both *Partial* and *Heavy* subsets. Linear outperforms MaxCal for most values of m on the three subsets. In Linear, the outputs of occlusion-specific detectors are adjusted with combination weights and the complementarity among different detectors are exploited to better distinguish pedestrians from background. In most cases, integrating a number of occlusion-specific detectors using Linear outperforms a single detector as shown in Table 1. The best performances of Linear on the three subsets are achieved at $m = 20$, $m = 35$ and $m = 40$ respectively. Overall, our approach achieves better performance than Linear for most values of m on the three subsets. According to the results of Linear and Ours in Table 1, more detectors are needed with the increase of occlusion. Note that the best performance of Linear and Ours on each testing subset is not achieved at $m = 45$. This shows that including all the detectors does not necessarily have the best performance as some detectors can produce noisy outputs.

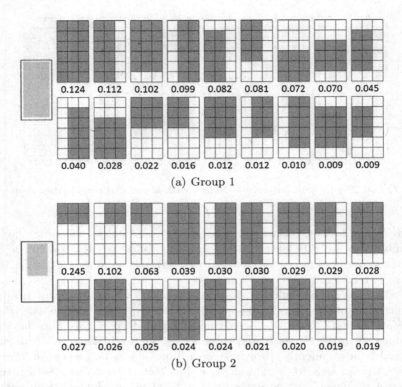

(a) Group 1

(b) Group 2

Fig. 8. Average occlusion patterns and top occlusion patterns of two groups of Ours-g2.

Table 1. Results of different integration methods. The bold number in each row indicates the best performance of the corresponding method.

	$m=1$	$m=5$	$m=10$	$m=15$	$m=20$	$m=25$	$m=30$	$m=35$	$m=40$	$m=45$
Max	**18.5**	25.5	26.0	24.8	23.5	26.6	28.6	29.1	29.1	29.2
MaxCal	**18.5**	19.0	20.2	20.2	18.8	18.8	18.6	19.1	18.9	**18.5**
Linear	18.5	17.4	21.2	16.4	**15.6**	17.3	16.8	16.0	16.3	16.2
Ours	18.5	15.9	15.8	15.4	**15.3**	**15.3**	**15.3**	**15.3**	15.5	16.0

(a) *Reasonable*

	$m=1$	$m=5$	$m=10$	$m=15$	$m=20$	$m=25$	$m=30$	$m=35$	$m=40$	$m=45$
Max	**39.3**	48.5	45.7	43.3	42.2	46.1	48.5	48.7	45.9	48.5
MaxCal	39.3	42.7	43.1	39.6	39.0	38.1	37.5	38.4	**36.9**	37.8
Linear	39.3	40.4	41.4	37.5	37.2	36.1	38.0	**35.0**	36.0	38.4
Ours	39.3	38.0	37.2	37.8	36.6	**36.1**	36.6	36.3	36.4	37.5

(b) *Partial*

	$m=1$	$m=5$	$m=10$	$m=15$	$m=20$	$m=25$	$m=30$	$m=35$	$m=40$	$m=45$
Max	75.7	**72.3**	78.2	78.3	77.9	76.8	73.8	73.4	73.1	73.4
MaxCal	75.7	**72.0**	74.7	74.6	74.5	74.0	72.6	73.1	72.9	72.4
Linear	75.7	77.0	71.9	74.6	74.2	73.0	79.2	75.3	**71.5**	75.0
Ours	75.7	75.0	74.6	74.4	74.4	73.8	72.8	72.7	**72.5**	74.0

(c) *Heavy*

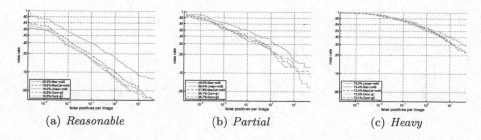

(a) *Reasonable*   (b) *Partial*   (c) *Heavy*

Fig. 9. Results of our integration approach and three baselines.

Figure 9 shows the results of the proposed integration approach and three baselines: Max-m45, MaxCal-m45 and Linear-m45. All the three baselines use $m = 45$ occlusion-specific detectors. Ours-gK is our approach with K groups. It can be seen that our approach performs best on all the three subsets. The difference between Linear-m45 and Ours-g1 is that the combination weights of Linear-m45 is learned by a linear SVM, while Ours-g1 adopts an L1-norm SVM to obtain its combination weight vector. With the sparsity constraint, Ours-g1 only selects 37 occlusion-specific detectors and discards the remaining 8 noisy ones. Ours-g1 outperforms Linear-m45 consistently on the *Reasonable*, *Partial* and *Heavy* subsets. Ours-g2 achieves a slightly worse performance than Ours-g1 on the *Reasonable* subset, but when the occlusion becomes more severe, Ours-g2 outperforms Ours-g1. Compared to Ours-g1, Ours-g2 achieves performance gains of 0.4% and 2.2% respectively on the *Partial* and *Heavy* subsets. The performance improvements of Ours-g2 over Linear-m45 on the three subsets are 0.9%, 2.7% and 4.9% respectively. The advantage of our approach over Linear-m45 becomes more obvious when heavy occlusions are present. Figure 10 shows the results of our approach with different numbers of groups ranging from 1 to 5. We can see that from $K = 2$, increasing the number of groups does not help much for the detection performance. Actually, most pedestrians in the Caltech dataset are occluded from the bottom according to the occlusion statistics in [7]. Figure 8 shows the average occlusion patterns and top occlusion patterns with the highest combination weights of two groups of Ours-g2. The first group corresponds to pedestrians which are slightly or not occluded and most pedestrians in the second group are occluded from the bottom. Ours-g2 gives high weights to the full-body pattern and partially occluded patterns in the first group (See the first 5 occlusion patterns), while occlusion patterns in which the human body is occluded from the bottom are more important in the second group, for example, the first three occlusion patterns.

Figure 11 shows the results of four integration approaches and some state-of-the-art approaches. Ours-g2 achieves the best performance among approaches using channel features on the three subsets. Compared with our own implementation of LDCF LdcfP1, Ous-g2 achieves performance gains of 3.0%, 3.6% and 5.6% respectively on the *Reasonable*, *Partial* and *Heavy* subsets. Linear-m45 also outperforms LdcfP1 consistently on the three subsets. Generally, properly

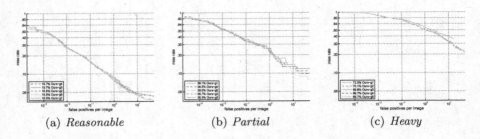

| (a) *Reasonable* | (b) *Partial* | (c) *Heavy* |

Fig. 10. Results of our integration approach with different numbers of groups.

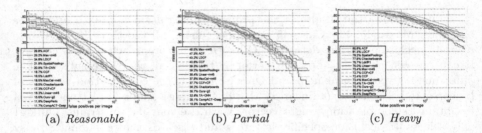

| (a) *Reasonable* | (b) *Partial* | (c) *Heavy* |

Fig. 11. Results of four integration approaches and some state-of-the-art approaches. Solid lines represent channel-feature based approaches and dashed lines represent deep-learning based approaches.

integrating a number of occlusion-specific detectors can achieve a better performance than the single full-body detector, especially when heavy occlusions are present. Currently, approaches using deep neural networks [5,32] or features learned by these networks [6,33] achieve better performance. Our detector integration approach still applies when the occlusion-specific detectors are replaced with detectors learned by these approaches.

6   Conclusions

In this paper, we propose a new approach to integrate occlusion-specific detectors for detecting heavily occluded pedestrians. Instead of linearly combining all occlusion-specific detectors into a generic detector for all occlusions, we categorize occlusions based on how pedestrian examples are occluded into K groups each of which selects its own occlusion-specific detectors and fuses them linearly to obtain a classifier. The K classifiers are learned simultaneously by an L1-norm linear SVM which can effectively remove irrelevant and unreliable occlusion-specific detectors for each group. The experiments on the Caltech dataset demonstrate the effectiveness of our approach.

Acknowledgement. This work is supported in part by Singapore Ministry of Education Academic Research Fund Tier 2 MOE2015-T2-2-114 and Tier 1 RG27/14.

References

1. Dalal, N., Triggs, B.: Histograms of oriented gradients for human detection. In: IEEE Conference on Computer Vision and Pattern Recognition (CVPR) (2005)
2. Dollar, P., Tu, Z., Perona, P., Belongie, S.: Integral channel features. In: British Machine Vision Conference (BMVC) (2009)
3. Dollar, P., Appel, R., Belongie, S., Perona, P.: Fast feature pyramids for object detection. IEEE Trans. Pattern Anal. Mach. Intell. (PAMI) **36**, 1532–1545 (2014)
4. Zhang, S., Benenson, R., Schiele, B.: Filtered channel features for pedestrian detection. In: IEEE Conference on Computer Vision and Pattern Recognition (CVPR) (2015)
5. Tian, Y., Luo, P., Wang, X., Tang, X.: Deep learning strong parts for pedestrian detection. In: International Conference on Computer Vision (ICCV) (2015)
6. Cai, Z., Saberian, M., Vasconcelos, N.: Learning complexity-aware cascades for deep pedestrian detection. In: International Conference on Computer Vision (ICCV) (2015)
7. Dollar, P., Wojek, C., Schiele, B., Perona, P.: Pedestrian detection: an evaluation of the state of the art. IEEE Trans. Pattern Anal. Mach. Intell. (PAMI) **34**, 743–761 (2012)
8. Mathias, M., Benenson, R., Timofte, R., Van Gool, L.: Handling occlusions with franken-classifiers. In: International Conference on Computer Vision (ICCV) (2013)
9. Zhu, J., Rosset, S., Hastie, T., Tibshirani, R.: 1-norm support vector machines. In: Advances in Neural Information Processing Systems (NIPS) (2004)
10. Wang, X., Han, T., Yan, S.: An HOG-LBP human detector with partial occlusion handling. In: International Conference on Computer Vision (ICCV) (2009)
11. Leibe, B., Leonardis, A., Schiele, B.: Combined object categorization and segmentation with an implicit shape model. In: ECCV Workshop on Statistical Learning in Computer Vision (2004)
12. Leibe, B., Seemann, E., Schiele, B.: Pedestrian detection in crowded scenes. In: IEEE Conference on Computer Vision and Pattern Recognition (CVPR) (2005)
13. Tang, S., Andriluka, M., Schiele, B.: Detection and tracking of occluded people. In: British Machine Vision Conference (BMVC) (2012)
14. Pepik, B., Stark, M., Gehler, P., Schiele, B.: Occlusion patterns for object class detection. In: IEEE Conference on Computer Vision and Pattern Recognition (CVPR) (2013)
15. Felzenszwalb, P., Girshick, R., McAllester, D., Ramanan, D.: Object detection with discriminatively trained part-based models. IEEE Trans. Pattern Anal. Mach. Intell. (PAMI) **32**, 1627–1645 (2010)
16. Chen, D., Batra, D., Freeman, W.: Group norm for learning structured SVMs with unstructured latent variables. In: International Conference on Computer Vision (ICCV) (2013)
17. Ouyang, W., Wang, X.: Single-pedestrian detection aided by multi-pedestrian detection. In: IEEE Conference on Computer Vision and Pattern Recognition (CVPR) (2013)
18. Shet, V., Neumann, J., Ramesh, V., Davis, L.: Bilattice-based logical reasoning for human detection. In: IEEE Conference on Computer Vision and Pattern Recognition (CVPR) (2007)
19. Enzweiler, M., Eigenstetter, A., Schiele, B., Gavrila, D.: Multi-cue pedestrian classification with partial occlusion handling. In: IEEE Conference on Computer Vision and Pattern Recognition (CVPR) (2010)

20. Wu, B., Nevatia, R.: Detection of multiple, partially occluded humans in a single image by Bayesian combination of edgelet part detectors. In: International Conference on Computer Vision (ICCV) (2005)

21. Ouyang, W., Wang, X.: A discriminative deep model for pedestrian detection with occlusion handling. In: IEEE Conference on Computer Vision and Pattern Recognition (CVPR) (2012)

22. Ouyang, W., Wang, X.: Joint deep learning for pedestrian detection. In: International Conference on Computer Vision (ICCV) (2013)

23. Ouyang, W., Zeng, X., Wang, X.: Modeling mutual visibility relationship in pedestrian detection. In: IEEE Conference on Computer Vision and Pattern Recognition (CVPR) (2013)

24. Duan, G., Ai, H., Lao, S.: A structural filter approach to human detection. In: Daniilidis, K., Maragos, P., Paragios, N. (eds.) ECCV 2010. LNCS, vol. 6316, pp. 238–251. Springer, Heidelberg (2010). doi:10.1007/978-3-642-15567-3_18

25. Nam, W., Dollar, P., Han, J.: Local decorrelation for improved pedestrian detection. In: Advances in Neural Information Processing Systems (NIPS) (2014)

26. Friedman, J., Hastie, T., Tibshirani, R.: Additive logistic regression: a statistical view of boosting. Ann. Stat. **28**, 337–407 (2000)

27. Benenson, R., Mathias, M., Tuytelaars, T., Van Gool, L.: Seeking the strongest rigid detector. In: IEEE Conference on Computer Vision and Pattern Recognition (CVPR) (2013)

28. Tibshirani, R.: Regression shrinkage and selection via the lasso. J. Roy. Stat. Soc. Ser. B **58**, 267–288 (1994)

29. Fan, R., Chang, K., Hsieh, C., Wang, X., Lin, C.: LIBLINEAR: a library for large linear classification. J. Mach. Learn. Res. **9**, 1871–1874 (2008)

30. Flamary, R., Jrad, N., Phlypo, R., Congedo, M., Rakotomamonjy, A.: Mixed-norm regularization for brain decoding. Comput. Math. Methods Med. (2014)

31. Paisitkriangkrai, S., Shen, C., Hengel, A.: Strengthening the effectiveness of pedestrian detection with spatially pooled features. In: Fleet, D., Pajdla, T., Schiele, B., Tuytelaars, T. (eds.) ECCV 2014. LNCS, vol. 8692, pp. 546–561. Springer, Heidelberg (2014). doi:10.1007/978-3-319-10593-2_36

32. Tian, Y., Luo, P., Wang, X., Tang, X.: Pedestrian detection aided by deep learning semantic tasks. In: IEEE Conference on Computer Vision and Pattern Recognition (CVPR) (2015)

33. Yang, B., Yan, J., Lei, Z., Li, S.: Convolutional channel features. In: International Conference on Computer Vision (ICCV) (2015)

Deep Second-Order Siamese Network for Pedestrian Re-identification

Xuesong Deng[1,2], Bingpeng Ma[1,2], Hong Chang[1(✉)], Shiguang Shan[1], and Xilin Chen[1]

[1] Key Lab of Intelligent Information Processing of Chinese Academy of Sciences (CAS),
Institute of Computing Technology, CAS, Beijing 100190, China
xuesong.deng@vipl.ict.ac.cn, bpma@ucas.ac.cn,
{changhong,sgshan,xlchen}@ict.ac.cn
[2] School of Computer and Control Engineering,
University of Chinese Academy of Sciences, Beijing 100049, China

Abstract. Typical pedestrian re-identification system consists of feature extraction and similarity learning modules. The learning methods involved in the two modules are usually designed separately, which makes them sub-optimal to each other, let alone to the re-identification target. In this paper, we propose a deep second-order siamese network for pedestrian re-identification which is composed of a deep convolutional neural network and a second-order similarity model. The deep convolutional network learns comprehensive features automatically from the data. The similarity model exploits second-order information, thus more suitable for re-identification setting than traditional metric learning methods. The two models are jointly trained over one unified large margin objective and the consistent convergence is guaranteed. Moreover, our deep model can be trained effectively with a small pedestrian re-identification dataset, through an irrelevant pre-training and relevant fine-tuning process. Experimental results on two public datasets illustrate the superior performance of our model over other state-of-the-art methods.

1 Introduction

Re-identifying a target pedestrian observed from a non-overlapping camera network is an important task in many real-world applications, such as threat detection, human retrieval and cross-camera tracking. During the past several years, it has drawn a lot of attentions from the field of computer vision and pattern recognition. Despite a lot of efforts spent on this task, pedestrian re-identification still remains largely unsolved due to low quality of images and complex variations in viewpoints, poses and illuminations. Some examples are shown in Fig. 1 to illustrate these difficulties.

Classical solutions for pedestrian re-identification mainly consist of two modules: feature extraction and similarity learning. Previous work usually focused on either feature extraction [1–10] or similarity learning [11–19].

© Springer International Publishing AG 2017
S.-H. Lai et al. (Eds.): ACCV 2016, Part II, LNCS 10112, pp. 321–337, 2017.
DOI: 10.1007/978-3-319-54184-6_20

Fig. 1. Examples of pedestrian images. The left five columns are from VIPeR dataset [1], and the right ones are from CUHK01 dataset [16]. Each column contains two images from same person in different cameras, indicated by red and blue boxes respectively. (Color figure online)

Early feature extraction methods mainly capture two clues: color and texture. They combine sophisticated low-level features such as HSV, Gabor, HOG to describe the appearance model of pedestrian images, from which the similarity between a pair of images can then be measured by some similarity metric learning methods. Farenzena et al. [2] proposed the Symmetry-Driven Accumulation of Local Features (SDALF). They exploited the symmetry structure of pedestrians to handle view variations. Ma et al. [4] combined Gabor filters and covariance descriptor to handle illumination changes. However, handcrafted features are extremely hard to design due to the complicated variations in the real world camera networks. People often fail to take all interference factors into consideration. Recently, a series of work by Zhao et al. [7–9] attempted to use the saliency parts of pedestrians to estimate whether they are the same person or not. However, the saliency parts are not always consistent under different camera views. A common situation is that the same part of a pedestrian may look quite different from two cameras due to complex lighting conditions and pose changes. Recently Zheng et al. [20] proposed a general feature fusion scheme for image search which can also be utilized in pedestrian re-identification. Given a query image, their method can automatically evaluate the effectiveness of a to-be-fused feature, and then make use of the good features and ignore the bad ones.

Recently, some researchers try to boost the performance of pedestrian re-identification in the metric learning context [13–15,21]. Zheng et al. [13] proposed the Probabilistic Relative Distance Comparison (PRDC) model to maximize the likelihood of genuine pairs (from the same person) having smaller distances than those of imposter pairs (from different persons). A simple though effective strategy to learn a distance metric, named KISSME, from equivalence constraints was proposed in [14]. Mignon and Jurie [15] proposed the Pairwise Constrained Component Analysis (PCCA) to learn a projection from raw input space into a latent space where a desired constraint is satisfied. Chen et al. [21] introduced a mixture of linear similarity functions that is able to discover different matching patterns in the polynomial kernel feature map. These typical metric learning (ML) methods generally aim to automatically learn similarity metrics from data under supervised or semi-supervised learning setting. To this end, the original data is usually transformed to another feature space where the distance measure is more ideal for the learning objective. However, metric learning for pedestrian re-identification is much more difficult than that for traditional learning tasks such as classification. As for pedestrian re-identification problems, we need to estimate whether a pair of images are from the same person or not. For this purpose, metric learning here should deal with the challenges of large number of classes (persons) and large within-class variance. Moreover, since the training and testing datasets contain totally different persons, metric learning for pedestrian re-identification is expected to generalize well to unseen categories. Therefore, common metric learning strategy of globally linear transformation in Euclidean or Cosine distance may not be an effective solution for this problem. Different from previous typical methods, Li et al. [17] proposed the Locally Adaptive Decision Function (LADF) model that exploits second-order information which is more effective to model the complex relations between pedestrian image pairs. This model achieved good performance on pedestrian re-identification task.

It is worth noting that all the above methods only focused on one aspect of the pedestrian re-identification task, either feature extractor or similarity learning model. However, designing features and learning similarity model separately cannot guarantee their optimality for each other, thus making the whole re-identification system sub-optimal. If the feature extraction part fails to capture consistent and comprehensive features, even a sophisticated similarity learning algorithm will behave poorly. On the other hand, when we have features containing rich information, we still cannot re-identify a pedestrian successfully with an ineffective similarity learning model.

Recently, deep networks [22–24] are proposed to tackle this problem. The end to end architecture can improve the optimality of their models. Specifically, [22] proposed a siamese network named the Deep Metric which is quite similar to an early work by Chopra et al. [11]. The main difference is that they use the Cosine norm to evaluate the similarity score while Euclidean norm is used in [11]. This model is similar to typical metric learning except that Deep Metric takes raw image pixels (instead of handcrafted features) and transforms them nonlinearly through a convolutional neural network (CNN), instead of traditional linear transformations. Therefore, the CNN in Deep Metric model actually works as both a feature extractor and a metric learner. This is too much duty for a single model. Even with nonlinear approximation ability, Deep Metric model may not perfectly fit for pedestrian re-identification task. In Deep Re-ID [23] model, several layers are designed to tackle different issues in pedestrian re-identification, such as photometric transforms, displacement and pose transforms. However, it is too ideal to expect each single layer to handle a complex variation. Ahmed et al. [24] proposed an improved deep architecture for pedestrian re-identification. Similar to Deep Re-ID, they formulated the problem of pedestrian re-identification as a binary classification problem. And they both defined a similar layer to encode the differences between features, which are extracted from previous convolutional layers. As they both train their models directly with a relatively small pedestrian re-identification dataset, the scale of their networks is limited. This limitation results in weaker hierarchical semantic abstraction ability of their model.

To address all the above problems for pedestrian re-identification task, we propose a deep second-order siamese network (DSSN) which is composed of a CNN model as feature extractor and a second-order similarity model as similarity learner. Deep CNN model succeeds in many computer vision applications as a feature extractor thanks to its highly nonlinearity and hierarchical semantic abstraction ability. However, it is quite hard to train a large scale CNN model with a relatively small pedestrian re-identification dataset. Inspired by RCNN [25], we design an irrelevant pre-training and relevant fine-tuning strategy to initialize the deep CNN. In the similarity learning part, we propose a model which encodes second-order information. The higher-order similarity layer can model more complex relation than typical ML methods, thus more suitable for pedestrian re-identification task. Both the deep CNN and the second-order similarity model are trained alternately with one unified energy based loss function to guarantee their optimality for each other. Moreover the energy based loss function leads to a large margin solution, which in turn enhances the generalization ability of the proposed method. Experimental results on benchmark pedestrian re-identification tasks verify the effectiveness of our proposed method. The main contributions in this paper can be summarized as follows:

- With irrelevant pre-training and relevant fine-tuning process, we succeed in training a large scale deep CNN with a small pedestrian re-identification dataset.
- A similarity model encoding second-order information is proposed to estimate the similarities between feature pairs, which is more effective than previous metric learning methods for pedestrian re-identification problem.
- Thanks to the alternate learning strategy, a reasonably optimal large margin solution is guaranteed.

2   Deep Second-Order Siamese Network

To estimate the similarity between two pedestrian images, we propose a method named Deep Second-order Siamese Network (DSSN). The architecture of our proposed method is shown in Fig. 2. It is composed of two identical convolutional neural networks (CNN) and a second-order similarity (SS) function. With a carefully designed energy based loss function, an ideal large margin solution for re-identification task is guaranteed. In the following subsections, we present the CNN model, the second-order similarity model and the energy based large margin solution in details.

2.1   Convolutional Neural Network (CNN)

The re-identification performance is affected by many factors such as low image resolution, illumination, viewpoint and pose variations. To overcome these interference factors, we need to extract comprehensive and robust features. Designing features with all interference factors taken into consideration is extremely hard. Nevertheless deep CNN model shows its superior power over handcrafted models thanks to its high nonlinearity and hierarchical semantic abstraction in many computer vision applications. And with a elaborately designed objective, deep CNN model can be learned to fit for a certain goal. Thus we believe features learned from a deep CNN are more powerful than handcrafted features or features learned via shallow networks.

To this end, we construct a deep convolutional neural network as feature extractor for pedestrian re-identification task. The network is same as that proposed in [26] except that we remove the softmax layer. The deep CNN contains five convolutional layers and two fully connected layers. The forward process of each layer is expressed in the following equations:

$$\mathbf{z}^{l+1} = \mathbf{W}^{l+1} * \mathbf{a}^l + \mathbf{b}^{l+1} \tag{1}$$

$$\mathbf{a}^{l+1} = \sigma(\mathbf{z}^{l+1}) \tag{2}$$

$$\sigma(x) = \max(0, x) \tag{3}$$

$\mathbf{W}^{l+1}$ and $\mathbf{a}^{l+1}$ are the parameter matrix and the activation of the $(l+1)^{th}$ layer respectively. $\sigma(\cdot)$ is the activation function and we use Rectified Linear Units (ReLU) in this paper. The details about the network can be found in [26]. The outputs of the last fully connected layer are taken as the learned deep features.

As illustrated in Fig. 2, two pedestrian images are split into several non-overlapping stripes. The two CNNs take each pair of the corresponding stripes as input and output the learned deep features for the following similarity model.

2.2   Second-Order Similarity Function

In standard siamese network [11,22], the distance between two deep features is usually calculated by a simple Euclidean or Cosine metric. Previous experimental results [22] of this type of siamese network on pedestrian re-identification task did not demonstrate superior power over handcrafted feature followed by metric learning methods [17]. Actually, extracting general sophisticated features with a deep network is not good enough for pedestrian re-identification task. Proper metric learning with respect to high-level re-identification objective is still necessary.

To handle the complex relation addressed above, a variety of metric learning (ML) methods [11–19] have been proposed. Despite of different objective functions, these metric learning models calculate the new distance similarly as:

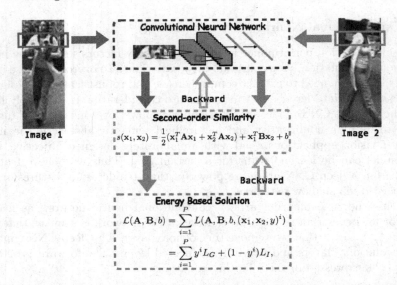

Fig. 2. An overview of the deep second-order siamese network (DSSN). In the training phase, the gradients of the energy based loss function with respect to all model parameters are back-propagated through the whole network which are indicated by the yellow arrows. (Color figure online)

$$d_{ML}(\mathbf{x}_1, \mathbf{x}_2) = (\mathbf{x}_1 - \mathbf{x}_2)^T \mathbf{L}^T \mathbf{L}(\mathbf{x}_1 - \mathbf{x}_2)$$
$$= \mathbf{x}_1^T \mathbf{M} \mathbf{x}_1 + \mathbf{x}_2^T \mathbf{M} \mathbf{x}_2 - 2\mathbf{x}_1^T \mathbf{M} \mathbf{x}_2 \tag{4}$$

$\mathbf{x}_1, \mathbf{x}_2 \in \mathbb{R}^d$ are features extracted from a pair of pedestrian images. $\mathbf{M} = \mathbf{L}^T \mathbf{L}$ is a real symmetric matrix. By learning a desired $\mathbf{L}$, samples from the same class get closer to each other while those from different classes farther in a projected latent space.

However, as pointed out in [17], there is an intrinsic mismatch between typical ML methods and re-identification task. The projection matrix $\mathbf{L}$ learned from training samples may not work well for testing samples from new categories. A desired model for pedestrian re-identification requires the ability of adapting locally rather than a simple global projection. Similar to [17], we define a *second-order similarity function* to capture local data structures as follows:

$$s(\mathbf{x}_1, \mathbf{x}_2) = \frac{1}{2}(\mathbf{x}_1^T \mathbf{A} \mathbf{x}_1 + \mathbf{x}_2^T \mathbf{A} \mathbf{x}_2) + \mathbf{x}_1^T \mathbf{B} \mathbf{x}_2 + b. \tag{5}$$

$s(\mathbf{x}_1, \mathbf{x}_2) > 0$ means $\mathbf{x}_1$ and $\mathbf{x}_2$ being a genuine pair, otherwise an imposter pair. $\{\mathbf{A}, \mathbf{B}, b\}$ are parameters for the similarity function. $\mathbf{A}$ and $\mathbf{B}$ are real symmetric matrices, and b is the bias term. And due to the symmetric constraint: $s(\mathbf{x}_1, \mathbf{x}_2) = s(\mathbf{x}_2, \mathbf{x}_1)$, we use $\mathbf{A}$ for both $\mathbf{x}_1$ and $\mathbf{x}_2$. Compared with typical ML methods, the second-order similarity metric can model much more complex relations due to three sets of parameters $\{\mathbf{A}, \mathbf{B}, b\}$.

2.3   Energy Based Large Margin Model

To learn improved deep features and an ideal second-order similarity function for pedestrian re-identification task, we adopt an energy based large margin model as it can generalize well to unseen examples. Suppose there are P pairs of labeled images, $(\mathbf{x}_1, \mathbf{x}_2, y)^i, i = 1, \ldots, P$, where $\mathbf{x}_1^i$ and $\mathbf{x}_2^i$ are the features extracted by the CNN model described before, and $y^i = 1$ indicates a genuine pair and $y^i = 0$ an imposter pair. We define the overall loss function as:

$$
\begin{aligned}
\mathcal{L}(\mathbf{\Theta}) &= \sum_i^P L(\mathbf{\Theta}, (\mathbf{x}_1, \mathbf{x}_2, y)^i) \\
&= \sum_i^P y^i L_G(\mathbf{\Theta}, E(\mathbf{x}_1^i, \mathbf{x}_2^i)) + (1 - y^i) L_I(\mathbf{\Theta}, E(\mathbf{x}_1^i, \mathbf{x}_2^i))
\end{aligned}
\tag{6}
$$

$\mathbf{\Theta}$ represents the whole set of parameters involved in the CNN and the similarity function. $L_G(\cdot)$ and $L_I(\cdot)$ are the loss functions for genuine and imposter pairs respectively. $E(\mathbf{x}_1^i, \mathbf{x}_2^i)$ is the energy function measuring the compatibility of a pair of the image features, which is defined as:

$$
E(\mathbf{x}_1^i, \mathbf{x}_2^i) = \frac{1}{Z} \exp\left(-\frac{s(\mathbf{x}_1^i, \mathbf{x}_2^i)}{\lambda_0}\right),
\tag{7}
$$

where

$$
Z = \sum_{i=1}^P \exp\left(-\frac{s(\mathbf{x}_1^i, \mathbf{x}_2^i)}{\lambda_0}\right)
\tag{8}
$$

is the partition function. Lower energy $E(\mathbf{x}_1^i, \mathbf{x}_2^i)$ indicates larger $s(\mathbf{x}_1^i, \mathbf{x}_2^i)$, which suggests $\mathbf{x}_1^i$ and $\mathbf{x}_2^i$ form a genuine pair. On the contrary, higher energy suggests the data being an imposter pair.

The two partial loss functions $L_G(\cdot)$ and $L_I(\cdot)$ take the forms of:

$$
L_G(\mathbf{x}_1^i, \mathbf{x}_2^i) = \alpha E(\mathbf{x}_1^i, \mathbf{x}_2^i)^2,
\tag{9}
$$

$$
L_I(\mathbf{x}_1^i, \mathbf{x}_2^i) = 2\exp\left(-\beta E(\mathbf{x}_1^i, \mathbf{x}_2^i)\right)
\tag{10}
$$

α and β are two constant parameters. The energy value $E > 0$ is guaranteed due to its definition. Clearly, $L_G(\cdot)$ is a monotonically increasing function and $L_I(\cdot)$ is a monotonically decreasing function, respect to E. It has been proved in [11] that with the partial losses of such monotonicity, minimizing the loss function $\mathcal{L}(\mathbf{\Theta})$ defined in Eq. (6) leads to a large margin solution which can generalize well to unseen persons.

To understand in a more loose way, minimizing $\mathcal{L}(\mathbf{\Theta})$ corresponds to decreasing the energy $E(\mathbf{x}_1^i, \mathbf{x}_2^i)$ for genuine pairs and increasing the energy for imposter pairs. Equivalently, the objective leads to larger similarity score $s(\mathbf{x}_1^i, \mathbf{x}_2^i)$ for genuine pairs and smaller similarity score for imposter pairs.

Researchers in [17] also proposed a method to approach a large margin solution for second-order metric model. They train their proposed model in SVM-like fashion. It should be noticed that the SVM-like solution is not quite straight-forward to combine with the back-propagation algorithm. That's why we choose the energy-based loss function.

2.4   Gradients

The parameters in both CNN and second-order similarity model are jointly opti-mized with the unified loss function 6. We use alternate training process to adjust each part of the deep second-order siamese network to get an optimal model. It should be pointed out that Li et al. [17] proposed a SVM-like objective function to approach the large margin solution. However, it is pretty hard for the SVM-like objective to optimize the CNN model. Instead, we adopt the energy based loss function, since it is quite straightforward to use back propagation (BP) algo-rithm to optimize both the CNN and similarity models under the energy based framework.

Optimizing the second-order similarity model is pretty straightforward. We can calculate the derivatives of the loss function (6) with respect to $\{\mathbf{A}, \mathbf{B}, b\}$ as follows:

$$\frac{\partial \mathcal{L}}{\partial \mathbf{A}} = \sum_{i=1}^{P} \nabla^i * \frac{\mathbf{x}_1^i {\mathbf{x}_1^i}^T + \mathbf{x}_2^i {\mathbf{x}_2^i}^T}{2} \tag{11}$$

$$\frac{\partial \mathcal{L}}{\partial \mathbf{B}} = \sum_{i=1}^{P} \nabla^i * \mathbf{x}_1^i {\mathbf{x}_2^i}^T \tag{12}$$

$$\frac{\partial \mathcal{L}}{\partial b} = \sum_{i=1}^{P} \nabla^i \tag{13}$$

with

$$\nabla^i = \frac{2y^i \alpha (E^3 - E^2) + 2(y^i - 1)\beta(E^2 - E)\exp(-\beta E)}{\lambda_0}, \tag{14}$$

where E is the abbreviation for $E(\mathbf{x}_1^i, \mathbf{x}_2^i)$. Then, we can use a gradient based optimization method like L-BFGS to obtain local optimal estimates for $\mathbf{A}, \mathbf{B}$ and b.

We use back-propagation (BP) algorithm to optimize the parameters of the convolutional neural network. Suppose the CNN has N layers and $\mathbf{W}^n$ is the parameter matrix of the n-th layer. The gradients of loss $\mathcal{L}$ with respect to $\mathbf{W}^n$ is calculated as:

$$\frac{\partial \mathcal{L}}{\partial \mathbf{W}^n} = \mathbf{a}^{n-1}\delta^n \tag{15}$$

$$\delta^N = \frac{\partial \mathcal{L}}{\partial \mathbf{a}^N} \odot \sigma'(\mathbf{z}^N) \tag{16}$$

$$\delta^n = (\mathbf{W}^{n+1^T}\delta^{n+1}) \odot \sigma'(\mathbf{z}^n) \tag{17}$$

$\mathbf{z}^n$ is the input of the activation function $\sigma(\cdot)$ in the n-th layer. $\mathbf{a}^n$ is the corre-sponding activation vector. δ^n is the *error term* defined in BP algorithm.

The inputs to the second-order metric model, $\mathbf{x}_1^i$ and $\mathbf{x}_2^i$ are also the outputs of the CNNs. Thus δ^N in our model can be calculated as:

$$\delta^N = \sum_i \frac{\partial \mathcal{L}}{\partial \mathbf{x}^i} \odot \sigma'(\mathbf{z}^N) \tag{18}$$

$$\frac{\partial \mathcal{L}}{\partial \mathbf{x}_1^i} = \nabla^i * (\mathbf{A}\mathbf{x}_1^i + \mathbf{B}\mathbf{x}_2^i) \tag{19}$$

$$\frac{\partial \mathcal{L}}{\partial \mathbf{x}_2^i} = \nabla^i * (\mathbf{A}\mathbf{x}_2^i + \mathbf{B}\mathbf{x}_1^i) \tag{20}$$

Once we get δ^N, the rest work can be done with typical BP algorithm.

As proved in [11], minimizing the loss function (6) with respect to the parameters of the CNN and the second-order similarity model will both lead to large margin solutions. Therefore, consistent convergence of our optimization method is guaranteed.

3   Learning Strategy

As large scale of parameters generally requires large scale of training samples, successful deep networks are usually trained on large datasets, such as ImageNet Large Scale Visual Recognition Challenge (ILSVRC) dataset [27] which contains millions of images. However, most of the public pedestrian re-identification datasets only contain several thousand images. It is unreasonable to directly train a deep model involving millions of parameters with such a small dataset.

In the work of Girshick et al. [25], researchers found a way to bridge the huge gap between small scale dataset and large scale model. They pre-train the model on ILSVRC dataset with supervision and then fine-tune the model with a domain-specific loss function on a small dataset. The reason behind the success is that lower level convolutional filters detecting edges, orientations, *etc.* can be shared across different sources of datasets, while higher level filters encoding rich semantic information must be task specific. Pre-training on a large dataset with supervision can generate effective low-level filters and a good initialization for high-level filters. After that fine-tuning process can further adjust the model to fit some specific task on a small dataset.

Inspired by this paradigm, our learning strategy consists of three stages. First, we pre-train the CNN model on the ILSVRC dataset. On the second stage, we fine-tune it with a pedestrian dataset with supervision. At last, we jointly minimize the loss function (6) with respect to both the CNN model and the second-order similarity model.

3.1   Supervised Pre-training and Fine-Tuning CNN

We pre-train the CNN model with the open source Caffe library [28] on ILSVRC dataset. The detailed training process is the same as [25]. After the pre-training

---

Algorithm 1. The main learning algorithm

Input: Dataset $\mathcal{D} = \{(\mathbf{x}_1, \mathbf{x}_2, y)^i\}_{i=1}^n$
    Random initialized $\mathbf{W}$ (CNN model)
    Random initialized $\{\mathbf{A}, \mathbf{B}, b\}$ (SS model)
Output: $\mathbf{W}_{opt}$ and $\{\mathbf{A}, \mathbf{B}, b\}_{opt}$
 1: Pre-train $\mathbf{W}$ on ILSVRC dataset
 2: Fine-tune $\mathbf{W}$ on $\mathcal{D}$ with Euclidean norm as similarity measure
 3: **while** until convergence **do**
 4:     Randomly select a batch of data samples from $\mathcal{D}$
 5:     Compute the value of Eq. 6 for this batch
 6:     Use back-propagation algorithm to calculate the gradients of $\mathbf{W}$ and $\{\mathbf{A}, \mathbf{B}, b\}$
 7:     Update $\mathbf{W}$ and $\{\mathbf{A}, \mathbf{B}, b\}$
 8: **end while**
 9: $\mathbf{W}_{opt} \leftarrow \mathbf{W}$
10: $\{\mathbf{A}, \mathbf{B}, b\}_{opt} \leftarrow \{\mathbf{A}, \mathbf{B}, b\}$
11: **return** $\mathbf{W}_{opt}$ and $\{\mathbf{A}, \mathbf{B}, b\}_{opt}$

---

step, the lower level convolutional filters are capable of extracting general instructive low-level image features.

The high-level filters in the pre-trained CNN model is suitable for image classification but may not fit for encoding significant information for pedestrian re-identification. Therefore, we fine-tune the CNN model using a set of pedestrian images. To this end, we construct genuine pairs and imposter pairs from these images, and minimize the loss function $\mathcal{L}(\mathbf{\Theta})$ on these sample pairs. In this stage, we replace the second-order similarity model in the DSSN model with a Euclidean norm. The effectiveness of the learned CNN model for pedestrian re-identification task is validated through experiments in Sect. 4.3.

3.2 Joint Optimization

After the fine-tuning process, we need to minimize the loss function (6) with respect to both the CNN and the second-order similarity model to guarantee their optimality for each other. To achieve this goal, we calculate the gradients of the second-order similarity model based on Eqs. 11–13. And the gradients of CNN in each layer can be computed through back-propagation algorithm. The error terms in the last full connected layer is in the form of Eqs. 18–20. Classical stochastic gradient descent (SGD) algorithm can be applied to optimize the CNN and second-order similarity model simultaneously. As discussed in Sect. 2.3, each step in the optimization process will lead to a large margin solution. Thus the DSSN model composed of the CNN model and second-order similarity model is guaranteed to converge to a local optimal state, which is fairly desirable for the pedestrian re-identification task.

To summarize, our overall learning algorithm is described in Algorithm 1.

4   Experiments and Analysis

We conduct our experiments on two benchmark datasets, *i.e.* the VIPeR [1] dataset and the CUHK01 [16] dataset. We compare our method to other deep models and state-of-the-art methods for pedestrian re-identification. The results validate the effectiveness of our proposed DSSN model.

4.1   Datasets

VIPeR dataset contains 632 people and 2 images for each pedestrian. Images are normalized to 48×128 for evaluations. A pair of images are captured from 2 different camera views (camera A and camera B). The viewpoint change is of $90°$ or more. Complex illumination conditions and huge pose variations make VIPeR dataset the most challenging pedestrian re-identification dataset.

CUHK01 dataset contains 971 people which are also captured from 2 camera views and are normalized to 60×160. And there are 2 images for each pedestrian in each camera view. Images in CUHK01 dataset have higher resolution. The illumination condition is more stable than VIPeR dataset. The better quality makes it possible for the DSSN to encode more information.

4.2   Evaluation Protocol

Our experiments on both datasets follow the evaluation protocol in [1]. The datasets are randomly partitioned into two even parts as training set and testing set. For VIPeR dataset 316 pedestrians are randomly picked up as training samples and 486 pedestrians for CUHK01 dataset. In the testing phase, the probe set is composed of images from camera A, and the gallery set is from camera B. We calculate the similarity scores between a target in the probe set and all candidates in the gallery set based on our proposed model. Then we can get the ranking of the candidates based on their similarity scores. The standard cumulative matching characteristic (CMC) curve is then reported to measure the performance over the whole probe set [29]. Generally higher CMC curve indicates better performance. To get stable statistics, all experiments are repeated 10 times with random training and testing partition on both datasets. And the average CMC curves over 10 trials are reported to evaluate the performance on both datasets.

4.3   Feature Learned via Deep CNN V.S. Handcrafted Feature

To evaluate the effectiveness of the deep CNN model, we compare the feature extracted from the fine-tuned deep CNN model (*Deep* feature) with several handcrafted features. They are HGR feature [30], eLDFV [3], eBiCov [4] and QALF [20]. Among them, HGR feature is used in [17] and achieved a good performance on pedestrian re-identification task. It is a hierarchical gaussianization representation based on simple patch color descriptors. It can be seen as the baseline of the handcrafted feature. eLDFV is a fusion of Weighted Color Histograms (wHSV), Maximally Stable Color Regions (MSCR) [2] and fisher vectors encoded local descriptors. eBiCov is a bio-inspired covariance descriptor fused with wHSV and MSCR. QALF is a self-adaptive feature fusion method that fuses several low-level features such as Color Histograms, Color Names, LBP, HOG. The results of different feature methods on VIPeR dataset are presented in Table 1. The best results at each rank are highlighted in bold face.

Table 1. Comparison with different features on VIPeR (Unit: %).

Method	Rank-1	Rank-5	Rank-10	Rank-20
HGR	9.46	22.50	31.80	40.41
eLDFV	22.34	46.92	60.04	71.81
eBiCov	20.66	42.62	56.11	67.67
QALF	30.17	51.60	62.44	73.81
Deep feature	**30.70**	**55.70**	**65.82**	**74.37**

From Table 1, we can tell that the *Deep* feature is much better than HGR, eLDFV and eBiCov features. The Rank-1 matching rate is around 20%, 8% and 10% higher than these three methods respectively. It is worth noting that eLDFV and eBiCov both fused several different features. And fusing features together generally can achieve better performance. QALF can even reward good features with higher weights and punish bad features. But the *Deep* feature alone is still slightly better than the QALF feature. These results demonstrate that due to the great semantic abstraction ability of the deep CNN model, the learned *Deep* feature is better than handcrafted ones for pedestrian re-identification task.

4.4   Comparison Between Similarity Methods

In our proposed model, the second-order similarity (SS) function is defined to model the complex relation between pair data. To validate the need for metric learning and the effectiveness of our SS model, we compare it with other similarity methods including Euclidean norm (Euc) and Mahanalobis based distance function (Eq. 4). Euc is compared as the baseline method. And Mahalanobis based distance is the similarity measure in most typical metric learning methods. We use Euc and Mahalanobis based distance as similarity function respectively to replace our SS model. And they are jointly trained with the CNN model under the same strategy described in Sect. 3. The results of different similarity methods on VIPeR and CUHK01 datasets are shown in Fig. 3.

It is clear showed in Fig. 3(a) and (b) that the performance of Mahalanobis based distance are better than Euc. The Rank-1 matching rate of Mahalanobis based distance are around 8% and 4% higher than Euc on VIPeR and CUHK01 respectively. While adopting the same CNN model and training strategy, the

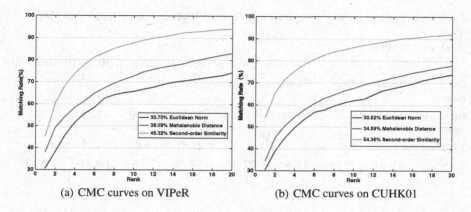

(a) CMC curves on VIPeR   (b) CMC curves on CUHK01

Fig. 3. Experimental results compared with different similarity methods.

only difference between these methods are the similarity measures. The better performance demonstrate that even processed by a strong CNN model, metric learning method is still a essential part in pedestrian re-identification.

We also notice that our proposed SS model achieves much better results than Mahalanobis based distance. The Rank-1 matching rate on two datasets are around 7% and 20% higher respectively. Mahalanobis based distance, like other typical ML method tries to find an ideal latent space through one single global projection while our SS model exploits the second-order information. Hence the better performance demonstrates the advantage of the second-order model over typical ML methods. This is all due to the ability of modeling more complex relation.

4.5   Comparison with Other Deep Models

As the proposed DSSN model is a deep model, we compare it with other deep models proposed for pedestrian re-identification task. These models include Deep Metric [22], FPNN [23] and Improved Deep [24]. Deep Metric is a siamese network but with similarity layer being a simple Cosine norm. FPNN model and Improved Deep model both formulate pedestrian re-identification task as a binary classification problem and train their models directly with relative small pedestrian re-identification datasets.

In Table 2, we present the experimental results of different deep models on VIPeR and CUHK01 datasets. Unavailable statistic data is denoted by −. And the best results at each rank are highlighted in bold face. From Table 2 it is obvious that our proposed DSSN achieves the best performance on both datasets. On VIPeR dataset, the Rank-1 matching rate of our DSSN model is 45.31% while the best of other methods is only 34.81%. This shows the superior power of our DSSN model over other deep models. Besides we notice that our DSSN model gets more improvement on VIPeR dataset than on CUHK01 dataset. On VIPeR dataset the Rank-1 matching rate is around 11% higher than the others while

Table 2. Comparison with deep models on VIPeR and CUHK01 (Unit: %).

Method	VIPeR				CUHK01			
	Rank-1	Rank-5	Rank-10	Rank-20	Rank-1	Rank-5	Rank-10	Rank-20
Deep metric	28.23	59.27	73.45	86.39	-	-	-	-
Improved deep	34.81	63.61	75.63	84.49	47.53	71.60	80.25	87.45
FPNN	-	-	-	-	27.87	59.64	73.53	87.34
DSSN	**45.31**	**78.00**	**87.81**	**92.37**	**54.36**	**78.53**	**86.19**	**92.11**

it is only around 7% on CUHK01 dataset. Considering the fact that VIPeR dataset suffers more from viewpoint changes and illumination variations than CUHK01 dataset, the larger improvement on VIPeR dataset indicates that our DSSN model is more robust to these interference factors than the other ones.

Compared with the Cosine norm of Deep Metric, our second-order similarity model can model more complex relation. And as for FPNN and Improved Deep they train their network directly while our DSSN adopts the pre-training and fine-tuning process. This makes our model deeper and better trained than theirs. Further more, the joint optimization leads our model to a overall optimal state. These differences result in the remarkable improvement over other deep models.

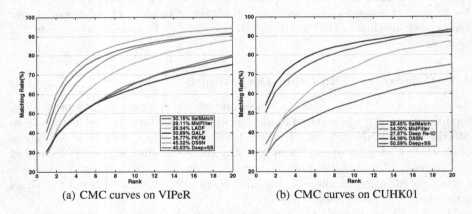

(a) CMC curves on VIPeR   (b) CMC curves on CUHK01

Fig. 4. Experimental results compared with state-of-the-art methods.

4.6   Comparison with State-of-the-Art Methods

After validating the effectiveness of each part in our DSSN model, we also compare our DSSN model with several state-of-the-art methods: SalMatch [8], Mid-Filter [9], LADF [17], QALF [20] and PKFM [21]. SalMatch and MidFilter define features based on the saliency parts in pedestrian images. LADF solved a SVM-like objective which leads its similarity model to a large margin solution. PKFM introduced a mixture of linear similarity functions to discover different matching patterns in the polynomial kernel feature maps.

CMC curves on VIPeR and CUHK01 dataset of different methods are shown in Fig. 4(a) and (b) respectively. We denote the methods of combining the *Deep* feature with our SS model as Deep+SS. The difference between Deep+SS and our full DSSN model is that the CNN model and the SS model of Deep+SS method are trained separately while our DSSN model are jointly optimized. From Fig. 4(a) we can see that on VIPeR dataset Deep+SS method achieves better result than other methods except for DSSN. The Rank-1 matching rate of PKFM is only 36.77% while Deep+SS method reaches at 40.83%. This result indicates that the fine-tuned CNN model and the SS model are quite effective. Simple combination achieves decent improvement over other methods. Furthermore, the performance of DSSN are better than Deep+SS on a remarkable scale. The Rank-1 matching rate of DSSN reaches at 45.31%. This indicates the proposed joint optimization algorithm further improves the CNN model and the second-order similarity model.

Similar results have been found on CUHK01 dataset in Fig. 4(b). Deep+SS outperforms others while DSSN achieves the best result. The Rank-1 matching rate of DSSN is 54.35%. These results further validate the effectiveness of the proposed DSSN model and the joint optimization algorithm.

5   Conclusion

In this paper, we propose a novel deep second-order siamese network for pedestrian re-identification which consists of feature extraction and similarity learning modules. The features learned via the deep CNN model encode effective information for pedestrian re-identification. And the second-order relation exploited in the similarity function makes the model more suitable for re-identification task. We propose an joint optimization process to train the model successfully. Therefore the feature learning and similarity learning modules are optimal for each other, which is rarely seen in previous related works. The experimental results validate the superiority of our model over other methods on two benchmark datasets.

Acknowledgment. This work was supported in part by National Basic Research Program of China (973 Program): 2015CB351802, and Natural Science Foundation of China (NSFC): 61272319, 61390515 and 61572465.

References

1. Gray, D., Tao, H.: Viewpoint invariant pedestrian recognition with an ensemble of localized features. In: Forsyth, D., Torr, P., Zisserman, A. (eds.) ECCV 2008. LNCS, vol. 5302, pp. 262–275. Springer, Heidelberg (2008). doi:10.1007/978-3-540-88682-2_21

2. Farenzena, M., Bazzani, L., Perina, A., Murino, V., Cristani, M.: Person re-identification by symmetry-driven accumulation of local features. In: 2010 IEEE Conference on Computer Vision and Pattern Recognition (CVPR), pp. 2360–2367. IEEE (2010)

3. Ma, B., Su, Y., Jurie, F.: Local descriptors encoded by fisher vectors for person re-identification. In: Fusiello, A., Murino, V., Cucchiara, R. (eds.) ECCV 2012. LNCS, vol. 7583, pp. 413–422. Springer, Heidelberg (2012). doi:10.1007/978-3-642-33863-2_41

4. Ma, B., Su, Y., Jurie, F.: BiCov: a novel image representation for person re-identification and face verification. In: British Machine Vision Conference, 11 p. (2012)

5. Li, W., Wang, X.: Locally aligned feature transforms across views. In: 2013 IEEE Conference on Computer Vision and Pattern Recognition (CVPR), pp. 3594–3601. IEEE (2013)

6. Kviatkovsky, I., Adam, A., Rivlin, E.: Color invariants for person reidentification. IEEE Trans. Pattern Anal. Mach. Intell. **35**, 1622–1634 (2013)

7. Zhao, R., Ouyang, W., Wang, X.: Unsupervised salience learning for person re-identification. In: 2013 IEEE Conference on Computer Vision and Pattern Recognition (CVPR), pp. 3586–3593. IEEE (2013)

8. Zhao, R., Ouyang, W., Wang, X.: Person re-identification by salience matching. In: 2013 IEEE International Conference on Computer Vision (ICCV), pp. 2528–2535. IEEE (2013)

9. Zhao, R., Ouyang, W., Wang, X.: Learning mid-level filters for person re-identification. In: 2014 IEEE Conference on Computer Vision and Pattern Recognition (CVPR), pp. 144–151. IEEE (2014)

10. Yang, Y., Yang, J., Yan, J., Liao, S., Yi, D., Li, S.Z.: Salient color names for person re-identification. In: Fleet, D., Pajdla, T., Schiele, B., Tuytelaars, T. (eds.) ECCV 2014. LNCS, vol. 8689, pp. 536–551. Springer, Heidelberg (2014). doi:10.1007/978-3-319-10590-1_35

11. Chopra, S., Hadsell, R., LeCun, Y.: Learning a similarity metric discriminatively, with application to face verification. In: IEEE Computer Society Conference on Computer Vision and Pattern Recognition, CVPR 2005, vol. 1, pp. 539–546. IEEE (2005)

12. Prosser, B., Zheng, W.S., Gong, S., Xiang, T., Mary, Q.: Person re-identification by support vector ranking. In: BMVC, vol. 2, p. 6 (2010)

13. Zheng, W.S., Gong, S., Xiang, T.: Person re-identification by probabilistic relative distance comparison. In: 2011 IEEE Conference on Computer Vision and Pattern Recognition (CVPR), pp. 649–656. IEEE (2011)

14. Kostinger, M., Hirzer, M., Wohlhart, P., Roth, P.M., Bischof, H.: Large scale metric learning from equivalence constraints. In: 2012 IEEE Conference on Computer Vision and Pattern Recognition (CVPR), pp. 2288–2295. IEEE (2012)

15. Mignon, A., Jurie, F.: PCCA: a new approach for distance learning from sparse pairwise constraints. In: 2012 IEEE Conference on Computer Vision and Pattern Recognition (CVPR), pp. 2666–2672. IEEE (2012)

16. Li, W., Zhao, R., Wang, X.: Human reidentification with transferred metric learning. In: Lee, K.M., Matsushita, Y., Rehg, J.M., Hu, Z. (eds.) ACCV 2012. LNCS, vol. 7724, pp. 31–44. Springer, Heidelberg (2013). doi:10.1007/978-3-642-37331-2_3

17. Li, Z., Chang, S., Liang, F., Huang, T.S., Cao, L., Smith, J.R.: Learning locally-adaptive decision functions for person verification. In: 2013 IEEE Conference on Computer Vision and Pattern Recognition (CVPR), pp. 3610–3617. IEEE (2013)

18. Zheng, W.S., Gong, S., Xiang, T.: Reidentification by relative distance comparison. IEEE Trans. Pattern Anal. Mach. Intell. **35**, 653–668 (2013)
19. Xiong, F., Gou, M., Camps, O., Sznaier, M.: Person re-identification using kernel-based metric learning methods. In: Fleet, D., Pajdla, T., Schiele, B., Tuytelaars, T. (eds.) ECCV 2014. LNCS, vol. 8695, pp. 1–16. Springer, Heidelberg (2014). doi:10.1007/978-3-319-10584-0_1
20. Zheng, L., Wang, S., Tian, L., He, F., Liu, Z., Tian, Q.: Query-adaptive late fusion for image search and person re-identification. In: IEEE Conference on Computer Vision and Pattern Recognition (CVPR) (2015)
21. Chen, D., Yuan, Z., Hua, G., Zheng, N., Wang, J.: Similarity learning on an explicit polynomial kernel feature map for person re-identification. In: Proceedings of the IEEE Conference on Computer Vision and Pattern Recognition, pp. 1565–1573 (2015)
22. Yi, D., Lei, Z., Liao, S., Li, S.Z.: Deep metric learning for person re-identification. In: 2014 22nd International Conference on Pattern Recognition (ICPR), pp. 34–39. IEEE (2014)
23. Li, W., Zhao, R., Xiao, T., Wang, X.: DeepReID: deep filter pairing neural network for person re-identification. In: 2014 IEEE Conference on Computer Vision and Pattern Recognition (CVPR), pp. 152–159. IEEE (2014)
24. Ahmed, E., Jones, M., Marks, T.K.: An improved deep learning architecture for person re-identification. Differences **5**, 25 (2015)
25. Girshick, R., Donahue, J., Darrell, T., Malik, J.: Rich feature hierarchies for accurate object detection and semantic segmentation. In: 2014 IEEE Conference on Computer Vision and Pattern Recognition (CVPR), pp. 580–587. IEEE (2014)
26. Krizhevsky, A., Sutskever, I., Hinton, G.E.: ImageNet classification with deep convolutional neural networks. In: Advances in Neural Information Processing Systems, pp. 1097–1105 (2012)
27. Deng, J., Dong, W., Socher, R., Li, L.J., Li, K., Fei-Fei, L.: ImageNet: a large-scale hierarchical image database. In: IEEE Conference on Computer Vision and Pattern Recognition, CVPR 2009, pp. 248–255. IEEE (2009)
28. Jia, Y., Shelhamer, E., Donahue, J., Karayev, S., Long, J., Girshick, R., Guadarrama, S., Darrell, T.: Caffe: convolutional architecture for fast feature embedding. In: Proceedings of the ACM International Conference on Multimedia, pp. 675–678. ACM (2014)
29. Gray, D., Brennan, S., Tao, H.: Evaluating appearance models for recognition, reacquisition, and tracking. In: Proceedings of IEEE International Workshop on Performance Evaluation for Tracking and Surveillance (PETS), vol. 3. Citeseer (2007)
30. Zhou, X., Cui, N., Li, Z., Liang, F., Huang, T.S.: Hierarchical gaussianization for image classification. In: 2009 IEEE 12th International Conference on Computer Vision, pp. 1971–1977. IEEE (2009)

Cost-Sensitive Two-Stage Depression Prediction Using Dynamic Visual Clues

Xingchen Ma[1], Di Huang[1(✉)], Yunhong Wang[1], and Yiding Wang[2]

[1] IRIP Lab, School of Computer Science and Engineering, Beihang University,
Beijing 100191, China
{chen.mxng,dhuang,yhwang}@buaa.edu.cn
[2] College of Information Engineering, North China University of Technology,
Beijing 100041, China
wangyd@ncut.edu.cn

Abstract. This paper presents a novel and effective approach to depression recognition in the visual modality of videos, which automatically predicts the depression level through two cost-sensitive stages. It delivers an improved solution in two ways compared with other vision based methods. On the one hand, current techniques regard depression recognition as either a classification or a regression problem, which tends to incur overfitting due to the high complexity of the model and the limited number of training samples. To handle such an issue, we propose a two-stage framework consisting of a coarse classifier and a fine regressor. The former makes use of a set of linear functions, corresponding to different depression intensities, to approximate the complex non-linear model, where a coarse range of the test sample is preliminarily located. The latter then predicts its precise depression level within the given range. On the other hand, depression recognition is different from the general classification and regression tasks, since its analysis is cost-sensitive as the diagnosis of heart diseases and cancers. However, this critical cue is not taken into account in the previous investigations, thus making their results problematic. To address this drawback, we embed the indicator of medical risk assessment into both the two stages by constraining the classifier using a weight matrix and loosening the regressor to an expanded range of depression level. The proposed method is evaluated on the Audio and Video Emotion Challenge (AVEC) 2013, and the performance is superior to the best one so far reported using the visual modality. Furthermore, it proves complementary to the audio based methods, and their joint use further ameliorates the accuracy. These facts clearly highlight the effectiveness of the proposed method on depression recognition.

1 Introduction

Major Depressive Disorder (MDD), often simply called depression, is a mental disorder characterized by a pervasive and persistent low mood, accompanied by low self-esteem and a loss of interest or pleasure in normally enjoyable activities. It adversely affects a person's family, work or school life, sleeping and eating

© Springer International Publishing AG 2017
S.-H. Lai et al. (Eds.): ACCV 2016, Part II, LNCS 10112, pp. 338–351, 2017.
DOI: 10.1007/978-3-319-54184-6_21

habits, and general health. When left untreated, depression can cause severe consequences, such as addiction, self-injury, reckless behavior and even suicide. Due to its harmfulness, in the recent years, MDD has received increasing attention within many related communities all over the world. Fortunately, according to some medical studies [1,2], it is treatable, and early detection of depression is extremely important, which has an immediate effect on easing the social and personal burden related to this illness.

Traditional methods of assessing psychopathology mostly depend on the verbal reports of patients, behaviors reported by friends and mental status experiences, such as the Scale for the Assessment of Negative Symptoms (SANS [3]), the Hamilton Rating Scale for Depression (HRSD [4]) and the Beck Depression Inventory (BDI-II [5]). They are all based on subjective ratings, and for lack of objective and quantitative measurements, the diagnosis results obtained for the same patient may be inconsistent at different time or various environment. In addition, they generally require extensive human expertise and are time-consuming. Therefore, Automatic Depression Detection (ADD) is very promising.

ADD is a young topic and it has not been discussed until 2009 [6]. To the best of our knowledge, the progress that has been made so far on affective computing mental disorder analysis is not so extensive. More recently, it has become one of most attention-getting topics, because the population increase of people with MDD, the technical development of artificial intelligence, and the public release of research data. In medical practice, it is not sufficiently informative to only decide whether the patients have depression or not, and it is expected to evaluate the severity of MDD. Taking BDI-II as an example, its depression value ranges from 0 to 63: 0–13 indicates the minimal state, 14–19 indicates the mild state, 20–28 indicates the moderate state, and 29–63 indicates the severe state. As a result, ADD should be formulated as a regression or a multi-classification problem.

Generally, there are two types of approaches towards this issue, *i.e.*, audio based and vision based. Although vision based methods have a variety of clues, such as facial behaviors and body gestures, their results are not as accurate as the audio based ones. On the one hand, there are more factors to consider in the visual channel, which makes the model more complex and the feature less robust [8,9], and non-linear classifiers or regressors are thus commonly employed for MDD analysis. On the other hand, MDD datasets are usually of small-scale for privacy protection demand as well as high acquisition cost. For instance, the size of training set in AVEC2013 is only 50, even less than the number of depression levels. Because of the limited training data and the complex visual model, the existing methods tend to be easily prone to overfit. As shown in Fig. 1, some visual features used in leading methods are depicted, which indicates that the separability of the samples is not so good as expected. Moreover, in serious illness analysis, such as the diagnosis of heart diseases and cancers disease, there is always a need for risk assessment [10], where the misclassification of patients for healthy ones leads to more side effects than the opposite situation. Additionally,

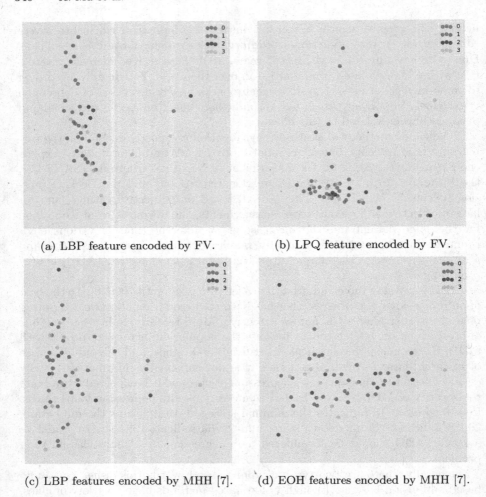

(a) LBP feature encoded by FV.   (b) LPQ feature encoded by FV.

(c) LBP features encoded by MHH [7].   (d) EOH features encoded by MHH [7].

Fig. 1. t-SNE visualization of some visual features used in state of the art work. It can be seen that in these feature spaces above, accurately classifying the samples into different group is a challenging task. Meanwhile, we can also see that if a single regression model is built to fit all these points, it tends to overfit.

in the context of using BDI-II as an indicator for depression level, misdiagnosing the depression state from minimal to severe brings in a higher cost than that from minimal to mild. However, the current investigations in ADD do not take such a factor into account, and treat the positive and negative errors equally, making their results not reliable enough.

This paper aims to contribute to this research area by addressing the two problems above, which proposes a novel approach for depression prediction in the visual modality of videos. Firstly, in contrast to the methods in the literature that make use of a single complex regression model to predict the depression state. It works in a two-stage manner consisting of a coarse classifier and a fine regressor. The former makes use of a set of linear functions, corresponding to the four depression levels defined by BDI-II (*i.e.*, minimal, mild, moderate, and severe), to approximate the complex non-linear model, where a coarse range of the test sample is preliminarily located. The latter then predicts its precise depression level within the given range. This framework alleviates the potential tendency to overfitting in training data. Secondly, we embed the indicator of medical risk assessment into both the two stages. In the first classification stage, to incorporate the requirements of cost sensitivity in medical diagnosis [11], we use a cost-sensitive classifier to make a optimal decision under asymmetric costs. In the second regression stage, to tolerate the possible classification error, we introduce a cost-sensitive loose which only approves to predict the depression state within to the coarse range as well as an adjacent more serious one.

To validate the effectiveness of our proposed approach, experiments are carried out on the depression prediction dataset, namely AVEC2013. The results achieved outperform the state of the art, and to the best of our knowledge, it ranks the first place in the visual modality. Meanwhile, the proposed method is well complementary to the audio based ones, evidenced by the improvement when they are joint used for multi-modal depression analysis.

2   Related Work

According to the information used, ADD methods can be categorized into audio based ones and vision based ones, and this study focuses on the latter. In this section, we give a brief introduction of recent ADD approaches.

Audio-based ADD methods can be found in [6, 12–14]. They analyze the audio features, such as spectrum, energy, Mel Frequency Cepstrum Coefficients (MFCC), *etc.*, which are supposed to be related to the depression emotion. Please refer to [15] for a more thorough review on ADD using the audio modality.

Vision based methods employ spatial and temporal information in the visual channel, where dynamic features are extracted from videos to capture depression related facial and body motions, and a standard classification and regression method is then used to predict the depression level.

As far as we know, the first published effort [16] by Wang *et al.* using visual clues towards schizophrenia, a kind of more serious mental disorder, dated back to 2008 from University of Pennsylvania. They proposed a computational framework that creates probabilistic expression profiles for video data and can potentially help to automatically quantify emotional differences between patients with neuropsychiatric disorders and healthy controls. They extracted geometric features based on facial landmarks and trained several probabilistic classifiers. To incorporate temporal information, they propagated classification results at each frame throughout the whole video. They pointed out that temporal dynamics are essential to capture subtle changes of facial expressions.

The following study [6] is the first one to address depression itself. They used manual FACS coding and Active Appearance Model (AAM) to represent facial expressions and adopted Support Vector Machine (SVM) and logistic regression for decision making respectively. They also attempted to fuse the contributions of audio and visual signals and claimed such combination improves the performance of depression detection. Their finding suggests the feasibility of ADD and has positive impacts on the clinical theory and practice.

More recently, Meng *et al.* [7] applied Motion History Histogram (MHH) to encode dynamic cues represented in the Local Binary Patterns (LBP) and Edge Orientation Histograms (EOH) feature spaces and used Partial Least Square (PLS) to predict depression levels, which ranked the first place in video-based methods in the AVEC2013 challenge. Cummins *et al.* [13] compared Space-Time Interest Points (STIPs) and Pyramid of Histogram of Gradients (PHOG) in their Support Vector Regression (SVR) based depression prediction system and found PHOG performed better in capturing the visual variations. Kächele *et al.* [17] presented a hierarchical classifier framework, which stacked a multilayer neural network over the SVR ensemble, to recognize the depression state and adopted the Kalman filter for the final audio-video decision fusion, which improved the prediction accuracy on the AVEC2013 dataset. Chao *et al.* [18], investigated the recent dominant deep learning models on this issue, and exploited Long Short Term Memory Recurrent Neural Network (LSTM-RNN) to describe dynamic temporal information. They used multi-task learning to boost the performance and reported very competitive results. Wen *et al.* [19] extracted Local Phase Quantization at Three Orthogonal Planes (LPQ-TOP) for representation of dynamic clues and utilized sparse coding and SVR for prediction.

In spite of the great improvement in ADD performance, these methods generally build the non-linear model in certain feature spaces and further apply a linear classifier or regressor for prediction. Figure 1 demonstrates that the features used for depression modeling are not competent enough as we expect. Furthermore, the high complexity of the features and the limited size of the training samples are prone to incur the problem of overfitting, dramatically degrading the generality. In addition, they treat ADD as a regular classification or regression task, and do not consider the medical analysis risk, which makes their result not so convincing in practice.

3   Cost-Sensitive Two-Stage Approach

To deal with the limitations of current vision based methods, we propose a cost-sensitive two-stage approach for ADD, which is composed of a coarse classifier and a fine regressor as illustrated in Fig. 2. The first stage employs a small set of linear functions, each of which corresponds to a depression level (in BDI-II), to approximate the complex non-linear model, and preliminarily localizes the coarse depression range of the test sample. The second stage then precisely predicts its level within the given range. Figure 3 shows such a motivation, where a single regressor is not sufficient to model the samples in certain visual feature space as in Fig. 3(a), but the piecewise approximation well addresses this problem as in Fig. 3(b). Furthermore, we take the medical risk into account in both the stages by constraining the classifier using a cost matrix and loosening the regressor

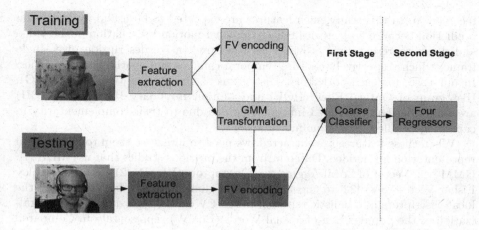

Fig. 2. Framework of the proposed two-stage ADD system.

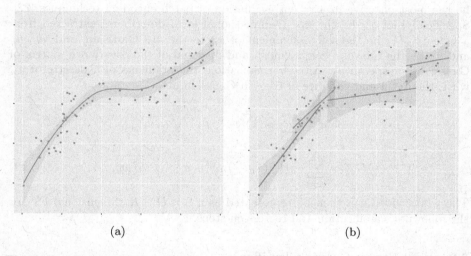

(a)   (b)

Fig. 3. Visualization of the advantage of the two stage framework. (a) Regression on the original data encoded by FV, where the samples are not linearly separable in the visual feature space, and a more complex model tends to overfit. (b) Regression on linear piecewise approximation of the non-linear model, where FVs are first classified into 4 groups and the regression is constrained to one of them for prediction.

to an expanded range of depression level. The details of each major steps are introduced in the subsequent.

3.1   Feature Encoding

For each video sample, the first step is to capture its motion variations. Dense trajectory features [20] have proved efficient for video representation in action recognition. In this study, such features are expected to be useful in depression prediction because trajectories reflect the local motion information of the video and dense representation provides very comprehensive description of a single frame, which respectively capture temporal changes and spatial characteristics. In our case, for each trajectory, Histogram of Orientation Gradients (HOG), Histogram of Optical Flow (HOF) and Motion Boundary Histogram (MBH) are computed and combined into a single one due to their complementarity in capturing detailed local dynamic variations.

When these features are extracted, we need to integrate them to obtain final representation for a video. Different from the previous studies that use MHH [7], GMM [12], Vector of Local Aggregated Descriptors (VLAD) [21], *etc.*, we employ Fisher Vector (FV) [22] to encode the visual clues in videos, by aggregating the local descriptors into holistic representation. FV can be regarded as the generalization of the popular Bag-of-Visual Words (BoVW) representation. Compared with bag of features, FV encodes both the first and second order statics between the video descriptors and a Gaussian Mixture Model (GMM). It is found to be the most effective in a recent evaluation of patch encoding techniques [22]. Specifically, let x_n be the n_{th} D-dimensional local descriptors extracted from a video, $\gamma_n(k)$ be the soft assignment of x_n to the kth Gaussian, and w_k, μ_k and σ_k be the mixture weight, mean and diagonal of the covariance matrix of Gaussian k respectively. After normalization, the gradients of a descriptor x_n w.r.t. the mean and variance of the kth Gaussian are:

$$G_{\mu_k} = \sum_{n=1}^{N} \gamma_n(k)(x_n - \mu_k)/\sqrt{\sigma_k w_k}, \tag{1}$$

$$G_{\sigma_k} = \sum_{n=1}^{N} \gamma_n(k)[(x_n - \mu_k)^2 - \sigma_k^2]/\sqrt{2\sigma_k^2 w_k}, \tag{2}$$

The whole video can then be represented as a $2 \times D \times K$ dimensional FV for the following classification and regression steps.

3.2   First Stage: Coarse Classifier

The first stage performs a typical classification process to provide a preliminary categorization for each FV, which represents a patient's interview video. According to BDI-II, we define 4 different groups, *i.e.* minimal, mild, moderate, and severe, which are used to build corresponding linear models for an approximation to the intrinsically nonlinear depression state. These groups of different depression levels have their own characteristics, and they should thus be treated differently in the following stage. In this paper, we explore two regular classification methods: logistic regression and Linear Discriminant Analysis (LDA), because they are computational efficient and less tendentious to overfit, and well match to FV [23].

Meanwhile, a regular classification problem treats all the types of misclassification errors equally. However, in depression analysis, the four classes denote increasing levels, and misclassifying a sample to the classes farther to the actual one has a higher cost than to the ones nearer to it. For example, misclassifying from the minimal depression state to the severe one should have a higher cost than to the mild one. Therefore, we consider this factor as a constraint when the classifier is trained, to make the results more convincing for medical practice. In this stage, we consider the misclassification caused by the whole class, *i.e.* class-dependent costs.

In this case, the cost associates to the level difference between the true and predicted label. The costs, denoted as $c(k,l)$ for predicting class k if the true label is l, are usually organized into a $K \times K$ matrix where K is the number of classes. Generally, it is assumed that the cost of predicting the correct class label y is minimal, *i.e.* $c(y,y) \leq c(k,y)$ for all $k = 1, \cdots, K$. In this work, we set the cost of correct classification to zero, *i.e.* $c(y,y) = 0$.

Table 1. Cost matrix for the coarse classifier.

True/pred	Minimal(0)	Mild(1)	Moderate(2)	Severe(3)
Minimal(0)	0^2	1^2	2^2	3^2
Mild(1)	1^2	0^2	1^2	2^2
Moderate(2)	2^2	1^2	0^2	1^2
Severe(3)	3^2	2^2	1^2	0^2

In multi-class cost-sensitive classification, we use the **rpart** package [24]. The cost matrix used in this paper is shown in Table 1, where the cost is the square of the difference between two labels, which penalizes misclassification.

3.3   Second Stage: Fine Regression

The second stage provides the final quantification of the depression state, which performs regression based on the results of the first stage. Here, we consider two simple regression methods: Ordinary Least Squares (OLS) and least absolute shrinkage and selection operator (LASSO). When making the final prediction, if the predicted response falls outside the range imposed by the coarse classifier, the value is cut off to its left or right boundary. For example, when regression is carried out on the minimal depression cluster, ranging from 0 to 13, if the predicted value is 20, it is assigned to 13 (Fig. 4).

However if the coarse classifier gives a wrong result, the fine regressor definitely fails. To make the method more robust, we propose a strategy which loosens the range to two adjacent clusters to tolerate possible errors incurred in the previous stage. On the other hand, in a normal regression problem, the confidence interval is two-sided. For example, in age estimation, it is the same if the age of 25 years old is predicted to 24 or 26. But in medical analysis, to control the medical risk, it is only allowed to predict the depression state to its actual level or a slightly more serious one rather than to a less serious one.

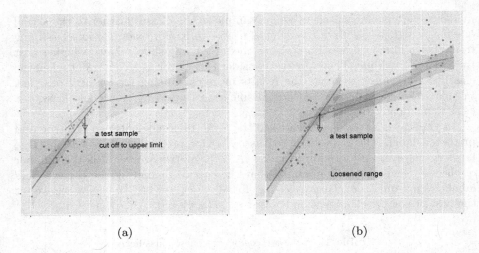

Fig. 4. Demonstration of the effectiveness of loosening the coarse range in the fine regression stage. (a) Regression without loosening coarse range and (b) regression with loosening coarse range.

Based on such consideration, in the second stage, we merge two consecutive clusters (the cluster given by the first stage and the adjacent one with a higher level), generating minimal and mild state, mild and moderate state, moderate and severe state and severe state. We then train regressors on them. In this way, when a sample is classified to the mild depression class, the regressor trained on the merged cluster of the mild and moderate states is used to predict its BDI-II score. This strategy, not only makes the entire system more reliable, but also better balances the medical diagnosis risk.

4   Experimental Evaluation

In this section, we describe the dataset, experimental settings, results, and the comparison with the state of the arts.

4.1   AVEC2013 Dataset

The proposed method is evaluated on the AVEC2013 dataset, which is a subset of the audio-visual depressive language corpus (AViD-Corpus). The whole dataset contains 340 video clips of 292 subjects performing a Human-Computer Interaction task while being recorded by a webcam and a microphone in a number of quiet settings. There is only one person per clip and all the participants are recorded from one to four times, with a period of two weeks between the measurements. 5 subjects appear in 4 recordings, 93 in 3, 66 in 2, and 128 in only one session. The length of these clips ranges from 20 to 50 min with the average of 25. The total duration of all clips lasts 240 h. The mean age of the subjects is 31.5 years old, in the interval of 18 to 63 with a standard deviation of 12.3. For the organization of the AVEC2013 challenge, the dataset is split into three parts: a training part, a development part, and a test part, each of which has 50 video clips.

4.2   Experimental Setting

When extracting dense trajectory features, we follow the parameter settings in [20]. The final dimensions of the descriptors are 30, 96, 108 and 192 for trajectory, HOG, HOF and MBH respectively. For feature encoding, we choose $K = 256$ as the number of Gaussian and randomly sample a subset of $1000 \times K$ descriptors from the training part of AVEC2013 to estimate the mixture model. After obtaining all the FVs, we apply power normalization and L2 normalization to them as in [23]. Principal Component Analysis (PCA) is adopted for FV dimensionality reduction. In the experiments, the AVEC2013 training subset is used for model learning and the test one is used for validation. The development subset is not used. The performance is measured in Mean Absolute Error (MAE) and Root Mean Square Error (RMSE) averaged over all the test samples as ruled by [25].

4.3   Results

The RMSE between the estimated BDI-II scores and the ground truth labels is 13.61 and the MAE is 10.88 [25], both of which are taken as the baseline results.

Evaluation of Two-Stage Model. In Table 2, we compare the performance of different combinations of classifiers (*i.e.* logistic regression and LDA) and regressors (*i.e.* OLS and LASSO) on the test part of AVEC2013. We can see that the combination of LDA and OLS achieves the best results in terms of both the measured errors, which are 7.26 and 8.91 for MAE and RMSE respectively. Moreover, it also displays the results of the traditional one-stage methods, and we find out that the proposed two-stage system performs much better than using a single regression model to predict the depression level. For example, single OLS achieves 7.63 and 9.51 for MAE and RMSE, which are obviously inferior to the ones reached by the proposed approach.

Comparison with State-of-the-Art. In Table 3, we compare our depression prediction results with the previous studies which also make use of visual clues on the test part in AVEC2013. The counterparts include the state of the art results so far reported since the AVEC2013 challenge was held. We can see that in the literature, the performance is slowly improved. While in this study, we reduce the best MAE and RMSE by 0.54 and 0.77 respectively. In particular, when compared with [19], the latest research on AVEC2013, our MAE and RMSE values are much lower, with the drop of 0.9 and 1.32 respectively.

Table 2. Comparison of different combinations of classifiers and regressors using the same FV features on the AVEC2013 test set.

Classifier	Regressor	MAE	RMSE
-	LASSO	7.74	9.67
Logistic regression	LASSO	7.69	9.59
LDA	LASSO	7.51	9.39
-	OLS	7.63	9.51
Logistic regression	OLS	7.53	9.32
LDA	OLS	**7.26**	**8.91**

Table 3. Comparison with previous vision-based studies in depression prediction on the AVEC2013 test set.

Methods	Modality	MAE	RMSE
Baseline [25] (2013)	Video	10.88	13.61
Meng *et al.* [7] (2013)	Video + Audio	8.72	10.96
Kächele *et al.* [17] (2014)	Video	8.97	10.82
Kaya *et al.* [14] (2014)	Video	7.86	9.72
Wen *et al.* [19] (2015)	Video	8.22	10.27
Ours	Video	**7.26**	**8.91**

Table 4. Multimodal fusion of vision and audio based methods at the score level.

Modality	MAE	RMSE
Audio	10.88	14.49
Video	7.26	8.91
Video + Audio	**6.75**	**8.29**

Multimodal Fusion. In the literature, it has been pointed out that visual and audio channels convey complementary information for depression recognition. We also evaluate contribution of the proposed method when it is combined with audio based ones. In our study, we make use of the top 20 MFCCs, the first and second order frame-to-frame difference coefficients. To achieve holistic representation for each video, we adopt BoVW, which has been successfully used for musical genre classification [26]. Next, we use the linear regressor, the linear opinion pool method is employed in the final multimodal fusion step due to its simplicity [7]. The fused depression score can be formulated as:

$$D_{final}(x) = \sum_{i=1}^{K} \alpha(i) D_i(x) \tag{3}$$

where x is a test sample, $D_i(x)$ is the decision value of ith modality seperately, and $\alpha(i)$ is the corresponding weight which satisfies $\sum_{i=1}^{K} \alpha(i) = 1$.

The results of multimodal fusion are shown in Table 4. We can see that although the audio based method used is a very simple one, whose MAE and RMSE are 10.88 and 14.49 respectively, when it is combined with the proposed vision based method, the performance is further improved. The MAE and RMSE values are decreased to 6.75 and 8.29, outperforming the ones of either of single modalities. It indicates that the our method presents complementary clues to the audio based method.

5 Conclusions

The contribution of this paper lies in two aspects. Firstly, we propose a novel two-stage framework for ADD in the visual modality, which consists of a coarse classier and a fine regressor. The classification step uses a set of linear functions to approximate the complex non-linear model and coarsely locates a range for the test sample, where the regression step further precisely predicts its depression level. It mitigates the potential tendency to overfitting to training data. Secondly, we present a new scheme in both the two stages, which takes the medical risk into account, making the results of depression prediction more convincing.

The proposed method is validated on the AVEC2013 test set, and to the best of knowledge, the result achieved is the best in the visual channel so far reported on this benchmark. Moreover, it is complementary to the audio based methods, and their combination further ameliorates the accuracy.

Acknowledgement. This work was supported in part by the National Key Research and Development Plan under Grant 2016YFC0801002, the Hong Kong, Macao, and Taiwan Science and Technology Cooperation Program of China under Grant L2015TGA9004, the National Natural Science Foundation of China under Grant 61540048, Grant 61673033, and Grant 61273263, and the Fundamental Research Funds for the Central Universities.

References

1. National Collaborating Centre for Mental Health (UK): Depression: the treatment and management of depression in adults (updated edition). National Institute for Health and Clinical Excellence: Guidance. British Psychological Society (2010)
2. Lejuez, C.W., Hopko, D.R., Hopko, S.D.: A brief behavioral activation treatment for depression treatment manual. Behav. Modif. **25**, 255–286 (2001)
3. Andreasen, N.C.: The Scale for the Assessment of Negative Symptoms (SANS): conceptual and theoretical foundations. Br. J. Psychiatry Suppl. **155**, 49–58 (1989)
4. Zimmerman, M., Chelminski, I., Posternak, M.: A review of studies of the Hamilton depression rating scale in healthy controls: implications for the definition of remission in treatment studies of depression. J. Nerv. Ment. Dis. **192**, 595–601 (2004)

5. Mcpherson, A., Martin, C.R.: A narrative review of the Beck Depression Inventory (BDI) and implications for its use in an alcohol-dependent population. J. Psychiatr. Ment. Health Nurs. **17**, 19–30 (2010)

6. Cohn, J.F., Kruez, T.S., Matthews, I., Yang, Y., Nguyen, M.H., Padilla, M.T., Zhou, F., La Torre, F.D.: Detecting depression from facial actions and vocal prosody. In: International Conference on Affective Computing and Intelligent Interaction and Workshops, pp. 1–7. IEEE (2009)

7. Meng, H., Huang, D., Wang, H., Yang, H., AI-Shuraifi, M., Wang, Y.: Depression recognition based on dynamic facial and vocal expression features using partial least square regression. In: International Workshop on Audio/Visual Emotion Challenge, pp. 21–30. ACM (2013)

8. Acharya, U.R., Sudarshan, V.K., Adeli, H., Santhosh, J., Koh, J.E., Puthankatti, S.D., Adeli, A.: A novel depression diagnosis index using nonlinear features in EEG signals. Eur. Neurol. **74**, 79–83 (2015)

9. Valenza, G., Garcia, R.G., Citi, L., Scilingo, E.P., Tomaz, C.A., Barbieri, R.: Nonlinear digital signal processing in mental health: characterization of major depression using instantaneous entropy measures of heartbeat dynamics. Front. Physiol. **6**, 74–81 (2015)

10. Alizadehsani, R., Hosseini, M.J., Sani, Z.A., Ghandeharioun, A., Boghrati, R.: Diagnosis of coronary artery disease using cost-sensitive algorithms. In: International Conference on Data Mining Workshops, pp. 9–16. IEEE (2012)

11. Vlahou, A., Schorge, J.O., Gregory, B.W., Coleman, R.L.: Diagnosis of ovarian cancer using decision tree classification of mass spectral data. BioMed Res. Int. **2003**, 308–314 (2003)

12. Williamson, J.R., Quatieri, T.F., Helfer, B.S., Horwitz, R., Yu, B., Mehta, D.D.: Vocal biomarkers of depression based on motor incoordination. In: International Workshop on Audio/Visual Emotion Challenge, AVEC 2013, pp. 41–48. ACM (2013)

13. Cummins, N., Joshi, J., Dhall, A., Sethu, V., Goecke, R., Epps, J.: Diagnosis of depression by behavioural signals: a multimodal approach. In: International Workshop on Audio/Visual Emotion Challenge, pp. 11–20. ACM (2013)

14. Kaya, H., Salah, A.A.: Eyes whisper depression: a CCA based multimodal approach. In: International Workshop on Audio/Visual Emotion Challenge, pp. 961–964. ACM (2014)

15. Cummins, N., Scherer, S., Krajewski, J., Schnieder, S., Epps, J., Quatieri, T.F.: A review of depression and suicide risk assessment using speech analysis. Speech Commun. **71**, 10–49 (2015)

16. Wang, P., Barrett, F., Martin, E., Milonova, M., Gur, R.E., Gur, R.C., Kohler, C., Verma, R.: Automated video-based facial expression analysis of neuropsychiatric disorders. J. Neurosci. Methods **168**, 224–238 (2008)

17. Kächele, M., Glodek, M., Zharkov, D., Meudt, S., Schwenker, F.: Fusion of audio-visual features using hierarchical classifier systems for the recognition of affective states and the state of depression. In: International Conference on Pattern Recognition Applications and Methods, pp. 671–678. ACM (2014)

18. Chao, L., Tao, J., Yang, M., Li, Y.: Multi task sequence learning for depression scale prediction from video. In: International Conference on Affective Computing and Intelligent Interaction, pp. 526–531. IEEE (2015)

19. Wen, L., Li, X., Guo, G., Zhu, Y.: Automated depression diagnosis based on facial dynamic analysis and sparse coding. IEEE Trans. Inf. Forensics Secur. **10**, 1432–1441 (2015)

20. Wang, H., Oneata, D., Verbeek, J., Schmid, C.: A robust and efficient video representation for action recognition. Int. J. Comput. Vis. **119**, 219–238 (2016)
21. He, L., Jiang, D., Sahli, H.: Multimodal depression recognition with dynamic visual and audio cues. In: International Conference on Affective Computing and Intelligent Interaction, pp. 260–266 (2015)
22. Chatfield, K., Lempitsky, V.S., Vedaldi, A., Zisserman, A.: The devil is in the details: an evaluation of recent feature encoding methods. In: BMVC, vol. 2, p. 8 (2011)
23. Sánchez, J., Perronnin, F., Mensink, T., Verbeek, J.: Image classification with the fisher vector: theory and practice. Int. J. Comput. Vis. **105**, 222–245 (2013)
24. Therneau, T., Atkinson, B., Ripley, B.: rpart: Recursive Partitioning and Regression Trees (2015). https://CRAN.R-project.org/package=rpart
25. Valstar, M., Schuller, B., Smith, K., Eyben, F., Jiang, B., Bilakhia, S., Schnieder, S., Cowie, R., Pantic, M.: AVEC 2013: the continuous audio/visual emotion and depression recognition challenge. In: International Workshop on Audio/Visual Emotion Challenge, pp. 3–10. ACM (2013)
26. Qin, Z., Liu, W., Wan, T.: A bag-of-tones model with MFCC features for musical genre classification. In: Motoda, H., Wu, Z., Cao, L., Zaiane, O., Yao, M., Wang, W. (eds.) ADMA 2013. LNCS (LNAI), vol. 8346, pp. 564–575. Springer, Heidelberg (2013). doi:10.1007/978-3-642-53914-5_48

Human Interaction Recognition by Mining Discriminative Patches on Key Frames

Dingyi Shan[1,2], Laiyun Qing[1,2(✉)], and Jun Miao[2]

[1] School of Computer and Control Engineering, University of Chinese Academy of Sciences, Beijing 100049, China
lyqing@ucas.ac.cn
[2] Key Lab of Intelligent Information Processing of Chinese Academy of Sciences (CAS), Institute of Computing Technology, CAS, Beijing 100190, China
dingyi.shan@vipl.ict.ac.cn, jmiao@ict.ac.cn

Abstract. In this paper, we propose a novel model for recognizing human interaction in videos via discriminative patches. Each frame is represented as a set of mid-level discriminative patches, which are extracted automatically by association rule mining on convolutional neural networks (CNN) activations. We further refine these patches based on the observation that discriminative patches usually occur in climax period of an interaction. The climax of an interaction in the paper is defined as the continuous frames which have more firing patches. The patches are further purified by a reward-punishment rule, which ensures that the discriminative patches emerge in climax period or key frames frequently and seldom occur in non-key frames. Finally, the label of an interaction video clip is determined by votes of each patch detected in it. The experimental results on UT-Interaction Set #1, Set #2 and BIT-Interaction Dataset show that the proposed discriminative patches obtain encouraging performances.

1 Introduction

Recently, interaction recognition has become a popular research topic due to its great scientific importance and practical applicability. Lots of applications have taken the advantage of interaction recognition, such as video analysis, surveillance and smart human-robot or human-computer interaction.

In close human interactions (e.g. shake-hands and hug), motion ambiguity increases significantly which leads that commonly used features such as interest points and trajectories are difficult to be uniquely assigned to a particular person. Therefore, recognizing human interactions become even more challenging. Mid-level discriminative patches proposed by Singh et al. [1] are clusters of image patches with rich semantic meanings discovered from a dataset where only image labels are available. A variety of state-of-the-art image classification methods [1–4] have validated that utilizing image patches which capture important aspects of objects is efficient in recognition field. Similar to recognizing objects from images, image patches can also be applied to human interaction recognition when the videos are treated as numerous image frames. However, a lot

© Springer International Publishing AG 2017
S.-H. Lai et al. (Eds.): ACCV 2016, Part II, LNCS 10112, pp. 352–367, 2017.
DOI: 10.1007/978-3-319-54184-6_22

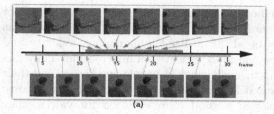 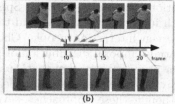

(a)   (b)

Fig. 1. An example of an interaction-sensitive patch and an non-interaction-sensitive patch. The interaction-sensitive patches are more discriminative and occur only within a particular period of time (the top of the pictures), while the non-interaction-sensitive ones span over the whole video (the bottom ones). (a) shake-hands; (b) kick.

of challenges will emerge while exploiting the image patches directly. The first reason is *frames explosion*. Since there will be thousands of frames in a short video clip, we need the extremely large memory during training. Efficient mid-level discriminative patches discovering method which can handle "big data" is emergency. The other one is existing *non-interaction-sensitive patches*. Unlike discovery objects from images, some non-interaction-sensitive patches may exist in a video all along. An example is illustrated in Fig. 1. From the figures, we can observe that in both video clips of shake-hands and kick, the most discriminative interaction-sensitive patches appear within a period of time (the patches on the top), while the non-interaction-sensitive patches span over the whole video (the bottom ones). It is essential to preserve interaction-sensitive patches and withdraw non-interaction-sensitive ones.

In this paper, we propose a human interaction recognition method based on key frames and discriminative patches in them (KFDP). Inspired by work of Poselets in object recognition [5] and mid-level discriminative patches [1], we represent videos in terms of discriminative patches rather than semantic parts or global feature vectors. The patches can be body parts or interaction parts, but are not restricted to them. The number of parts in each frame is also not limited. To avoid tedious key-point annotation in learning Poselets [5] and overcome the plague of frames explosion, we adopt mid-level deep pattern mining (MDPM) algorithm [6] to discover discriminative patches. Based on the observation that discriminative patches often emerge in the climax part rather than the whole process, we discover key frames and propose a reward-punishment rule to purify mid-level discriminative patches.

The remainder of the paper is organized as follows. In Sect. 2, related work is covered. The details of the proposed model are introduced in Sect. 3. Section 4 reports the experimental results and Sect. 5 concludes the paper.

2   Related Works

Human interaction recognition has been receiving much attention in computer vision. The key to interaction problem is how to represent the interactive information between people. A popular solution used in [7–12] is to learn joint motion state for interactions. Specifically, [8] utilized body part tracker to extract each individual in videos, and then described spatial and temporal connections between interacting people. Choi et al. [9] utilized human pose, velocity and spatial-temporal distribution of people to express action information. They further proposed a system that simultaneously tracked people and recognized their activities. Methods in [10,13] learned key-frames to represent complex actions. Although these approaches developed a good representation for videos, they heavily rely on the success of object detection and trackers algorithms.

To alleviate the dependence on trackers and detection, Brenderl et al. [14] over-segmented the whole video into some tubes first, and then adopted spatiotemporal graphs to learn the relationship among the parts. Raptis et al. [15] grouped the trajectories into clusters, each of which could be treated as an action part. Lan et al. [7] represented crowd context at both feature level and action context level. Kong et al. [16,17] proposed a method to describe complex interactions with rich semantic descriptions. Approaches in [18,19] regarded interacting people as a group and recognized their interactions based on group motion patterns. Sum-product networks [19] divided a video clip into multiple spatiotemporal volumes. Although these part-related methods achieved high performance, optimization algorithms in their models are complex and huge predicting parameter space overwhelmed the learning model.

Mid-level discriminative patch named by Singh et al. [1] are clusters of discriminative patches exacted through a cross-validation training strategy. Owing to the great success of mid-level discriminative patches, many works has been put forward based on that. Doersch et al. [2] formulated mid-level visual element discovery from the perspective of the well-known mean-shift algorithm. Bossard [20] discovered representative and discriminative superpixels using a Random Forest framework. Xu et al. [21] proposed an activity auto-completion (AAC) model, which explores discriminative patches for video representation and constructs prefix-candidate pairs for auto-completion.

In this paper, we adopt mid-level deep pattern mining (MDPM) [6] to discover candidate discriminative patches for human interaction recognition. The mid-level deep pattern mining (MDPM) algorithm adopts association rules on the powerful convolutional neural networks (CNN) features to mine mid-level visual element, which is suitable for "big data" in video analysis. We further make use of the climax stages of an interaction to discover key frames and propose a reward-punishment approach to refine the candidate patches. Finally, the label of an interaction is determined by a voting scheme based on these patches.

3   The Proposed KFDP Model

In this section, we first introduce how to mine candidate discriminative patches with mid-level deep pattern mining (MDPM) [6] in Sect. 3.1. Then the details of how to locate key frames or the climax stage and refine the candidate patches

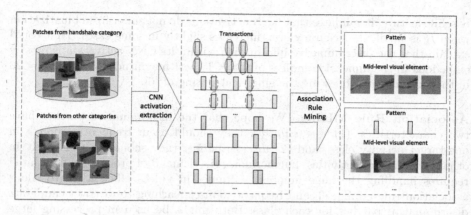

Fig. 2. Details of mining candidate patches. Given image patches sampled from both the target action (e.g. shake-hands) and the other categories, we extract every patch's CNN activation and regard it as a transaction of itemsets. Patterns are then obtained through association rule mining. Discriminative patch clusters are gained by searching image patches with the same patterns.

by the proposed reward-punishment rules are discussed in Sects. 3.2 and 3.3, respectively. In Sect. 3.4, human interaction recognition is showed.

3.1 Mining Candidate Patches

An overview of mining candidate patches is illustrated in Fig. 2. The approach is divided into two procedures: CNN activation extraction and association rule mining.

CNN Activation Extraction. We first sample $128 * 128$ patches with a stride of 32 pixels from each image. Then, for each image patch, we extract the 4096-dimensional output of the first fully-connected layer of *BVLC Reference CaffeNet* [22] or the 19-layer *VGG-VD* model [23]. To generate the final feature vector for each image, we consider two steps as follows.

(1) **Sparsified CNN.** Given the input feature as the first fully-connected layer of a CNN model, the CNN activation of an image patch is obtained by retaining only the K largest values of the vector and setting the remaining values as zero. That is, we force the features to be sparse.
(2) **Binarized CNN.** For each 4096-dimensional CNN activation of an image patch, we set the K largest values of the vector to one and the remaining elements to zero.

These two steps of processing CNN activation are critical to the appropriacy of such features to form the basis of a transaction-based approach. Work [6] has compared the "CNN-Sparsified" and "CNN-Binarized" counterpart with the baseline feature, it validates that CNN activations do not suffer from binarization when K is small, the accuracy even increases slightly in some cases. That could explain that the discriminative information within its CNN activation is mostly embedded in the dimension indices of the K largest magnitudes. We set $K = 20$ in the experiments, following the empirical setting of [6].

Association Rule Mining. We hope that the discriminative patches have two characteristics: (1) *repesentative:* they should occur frequently enough in one target activity class; and (2) *discriminative:* they should appear rarely in other categories of activities. Furthermore, the demand of processing "big data" requires handling the huge number of frames in videos efficiently. We adopt association rule mining [6] and frequent itemset learning to discover candidate discriminative patches for each class. Both might be used in processing large numbers of customer transactions to reveal information about their shopping behaviors.

More formally, let $A = \{a_1, a_2, ..., a_M\}$ denote a set of M items. A transaction T is a subset of A (i.e., $T \subseteq A$) which contains only a small number of items (i.e., $|T| \ll M$). We also define a transaction database $\mathcal{D}$ containing R (typically millions, or more) transactions (i.e., $\mathcal{D} = \{T_1, T_2, ..., T_R\}$). The frequent itemset and association rule are defined as follows. Note that the definitions are based on market analysis.

Definition 1 (Frequent Itemset). *Suppose that an itemset I is a subset of global itemset A. We are interested in the fraction of transactions $T \in \mathcal{D}$ which contain I. The support of I reflects this expression:*

$$\text{supp}(I) = \frac{|\{T | T \in \mathcal{D}, I \subseteq T\}|}{N} \in [0, 1] \tag{1}$$

*where $|\,.\,|$ means the cardinality. I is called a **frequent itemset** when $\text{supp}(I)$ is larger than a expected threshold.*

Definition 2 (Association Rule). *The confidence of an association rule* $\text{conf}(I \rightarrow 1)$ *can be taken to reflects this expression:*

$$\begin{aligned}
\text{conf}(I \rightarrow 1) &= \frac{\text{supp}(I \bigcup \{1\})}{\text{supp}(I)} \\
&= \frac{|\{T | T \in \mathcal{D}, (I \bigcup \{1\}) \subseteq T\}|}{T | T \in \mathcal{D}, I \subseteq T\}|} \in [0, 1]
\end{aligned} \tag{2}$$

In a traditional pattern mining model this might be taken to imply that customers who bought items in I are also likely to buy item 1. In practice, we are interested in "good" rules, meaning that the confidence of these rules should be reasonably high.

Given the transaction database $\mathcal{D}$, we use the Apriori algorithm to discover a set of patterns $\mathcal{P}$ through association rule mining. Each pattern $P \in \mathcal{P}$ must satisfy the following two criteria:

$$\text{supp}(P) > \text{supp}_{min}, \tag{3}$$

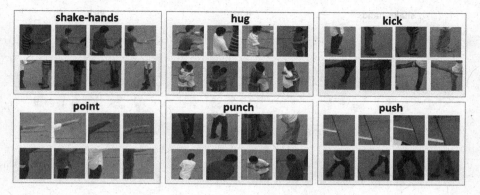

Fig. 3. Some examples of candidate patches discovered on the UT-Interaction Set #1, for each interaction, we present the top-1 patches (upper line) and top-2 patches (lower line).

$$\text{conf}(P \to pos) > \text{conf}_{min}, \tag{4}$$

where supp_{min} and conf_{min} are thresholds for the support value and confidence.

We now show how association rule mining implicitly satisfies the two requirements of candidate patches discovery, i.e., representativeness and discriminativeness. Specifically, based on Eqs. (3) and (4), we are able to rewrite Eq. (2), thus

$$\text{supp}(P \cup \{pos\}) = \text{supp}(P) \times \text{conf}(P \to pos)$$
$$> \text{supp}_{min} \times \text{conf}_{min} \tag{5}$$

where $\text{supp}(P \cup \{pos\})$ measures the frequent pattern P discovered in transactions of the target action among all the transactions. Hence, values of $\text{supp}(P)$ and $\text{conf}(P \to \{pos\})$ larger than their thresholds ensure that pattern P is found frequently in the target category, akin to the representativeness requirement (Eq. (5)). A high value of conf_{min} (Eq. (4)) also ensures that pattern P is more likely to be found in the target category rather than all other classes, reflecting the discriminativeness requirement.

Lastly, we merge similar patterns in an iterative procedure while training linear discriminant analysis (LDA) detectors, because some patches belonging to different patterns may overlap or describe the same visual concept. Figure 3 shows some candidate patch clusters.

3.2 Locating Key Frames

Usually an interaction has its start, climax and end periods, and the key elements to recognize human interaction is located in the climax part. We investigate the temporal distribution of the firing candidate patches and get some interesting observations. Figure 4 shows exemplars of the temporal distribution of the six interactions of the first people on UT Set #1. We can see that more candidate

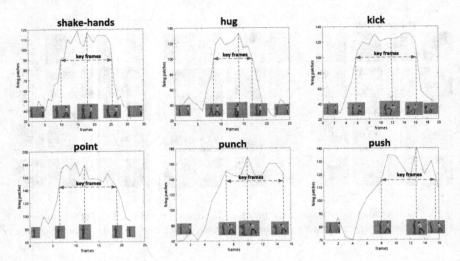

Fig. 4. Distribution of candidate patches in the interaction video exemplars. The key frames are in the time interval where more candidate patches response.

patches emerge in the middle part of the video clip whereas fewer are detected in the other parts. Hence, we develop a simple approach, using a threshold for judging whether a frame is a key frame or not, based on Eq. (6) as

$$\zeta^j = n_{min}^j + \frac{2}{3} \times (n_{max}^j - n_{min}^j), \tag{6}$$

where ζ^j is the threshold number of detected patches on a frame to become a key frame for video j. n_{max}^j and n_{min}^j are the maximum number and minimum number of firing patches on a frame of video j, respectively.

According to temporal distribution of the number of patches on each video and ζ^j defined by Eq. (6), we then define frames of video j where the firing candidate patches are no less than ζ^j as the discriminative frames or key frames.

3.3   Refining Discriminative Patches

As shown in Fig. 1 and the top-1 patch of "push" in Fig. 3, due to the effects of non-interaction-sensitive patches and backgrounds, patches mined in a class do not necessarily satisfy the aforementioned two characteristics: representativeness and discriminativeness. Therefore, we propose a reward-punishment approach to further refine the candidate patches.

Intuitively, if a patch cluster turns up in key frames frequently and seldom emerges in non-key frames, it will be highly discriminative. To refine the candidate patches, we proposed a *Reward-Punishment Rules*: reward the frequent patches in key frames and punish the patches occurring in non-key frames. The details of reward-punishment rules are as follows.

Fig. 5. Some examples of final discriminative patches discovered on the UT-Interaction Set #1, for each interaction, we present the top-1 patches (upper line) and top-2 patches (lower line).

(1) **Frequency Score:** Frequency score measures how often a patch occur during key frames. An ideal patch cluster is expected to occur in key frames frame by frame. Then if a patch appears in a key frame, this cluster will be rewarded, otherwise will be punished. The frequency score is summed by the scores during the period of key frames, formulated as

$$\phi_i = \sum_{t=T_{ks}}^{T_{ke}} frequency(\varphi_i(t)), \tag{7}$$

where φ_i is the ith cluster and $frequency(\varphi_i(t))$ means whether some patches of φ_i occur in the tth key frame, respectively. $frequency(\varphi_i(t))$ is 1 if φ_i exists, otherwise -1. T_{ks} and T_{ke} indicate the beginning and the ending timestamp of the key frames.

(2) **Discriminativeness Score:** Discrimination constraint requires patch clusters seldom occur in a non-key frame, otherwise their discrimination decreased and get punished. The discriminativeness score is summarized during the non-key frames, formulated in Eq. (8)

$$\psi_i = \sum_{t=T_s}^{T_{ks}} discrimination(\varphi_i(t)) + \sum_{t=T_{ke}}^{T_e} discrimination(\varphi_i(t)), \tag{8}$$

where φ_i is the ith cluster and $discrimination(\varphi_i(t))$ means whether some patches of φ_i occur in the tth non-key frame. $discrimination(\varphi_i(t))$ is -1 if φ_i exists, otherwise 0. T_s and T_e indicate the beginning and the ending timestamp of the video, respectively.

The final score of a patch cluster is obtained by merging its frequency score and discriminativeness score. We introduce λ to balance these two aspects, as shown in Eq. (9). We rank the scores of all clusters and select clusters which have top-N for each interaction.

$$\delta_i = \lambda \cdot \phi_i + (1 - \lambda) \cdot \psi_i, \tag{9}$$

where δ_i is the summation of ϕ_i and ψ_i with a certain balance coefficient λ. The impact of balance coefficient λ will be investigated in Sect. 4.2. Some examples of the refined discriminative patches are depicted in Fig. 5.

3.4   Interaction Recognition

We regard the top N clusters of each category as the final discriminative patches of each interaction and then learn N discriminative detectors for each interaction by SVM classifiers: *LIBSVM* [24]. The patches belonging to a particular cluster are used as positive samples and negative samples are from other categories. The feature of a patch is the first fully-connected layer of the 19-layer *VGG-VD* model [23], which has been extracted at the stage of mining candidate patches.

Given an unknown interaction video, we detect discriminative patches by sliding window with SVM detectors. The predicted label of the video clip is obtained by accumulating the votes of each detected patch.

4   Experiments

we conduct extensive experiments to evaluate the proposed interaction recognition method on UT-Interaction Set #1, Set #2 [25] and BIT-Interaction Dataset [16].

The UT-Interaction Set #1 and UT-Interaction Set #2 [25] are created for high-level human interaction analysis, and both of them consist of six different types of human interaction activities: shake-hands, hug, kick, point, punch and push, with 10 videos per activity class. Figure 6(a) shows some example snapshots of the six activities from two datasets. Backgrounds in Set #2 are more complex than those in Set #1 (e.g. tree moves, camera jitters). Following the experiment settings in [25], 10-fold leave-one-sequence-out cross validation setting is used.

The BIT-Interaction Dataset [16] consists of 8 classes of human interaction activities: bow, boxing, handshake, high-five, hug, kick, pat and push. Each class contains 50 videos, to provide a total of 400 videos. some example snapshots of the eight activities are shown in Fig. 6(b). In experiments on BIT dataset, 272 videos are randomly chosen for training and the remaining videos are used for testing, as following [26].

4.1   Experimental Parameters

We sample a frame at 5 frames interval for each video. For each image, we resize its dimension to $320 * 240$, then sample $128 * 128$ patches with a stride of 32 pixels, and calculate the CNN features using the 19-layer *VGG-VD* model [23] in caffe[1]. Because the number of patches sampled varies in different datasets,

---

[1] http://caffe.berkeleyvision.org/.

shake-hands point   kick   hug   push   punch

(a) Example snapshots of six different interactions (shake-hands, point, kick, hug, push and punch) from UT Set #1 (top) and UT Set #2 (bottom). Set #2 is more complicated than UT Set #1, since there are more tree moves, camera jitters, etc..

bow   boxing   handshake   high-five   hug   kick   pat   push

(b) Example snapshots of eight different interactions (bow, boxing, handshake, high-five, hug, kick, pat, push) from BIT-Interaction Dataset.

Fig. 6. Example snapshots of different datasets

two parameters $supp_{min}$ and $conf_{min}$ in the association rule mining in Sect. 3.1 are set according to each dataset so that at least 100 patterns are discovered for each category.

Number of Clusters. We first investigate the parameter of number of discriminative clusters for each action category. It is expected that recognition performance is not always getting better with the increasing number of final clusters and certain numbers of discriminative patches (e.g., 50) are enough to discriminate different actions. Table 1(a) illustrate the accuracy of the proposed model on UT Set #1 with different number of clusters in each class. It can be seen that ten to fifty patches get satisfying results. The best accuracy is gotten with top 20 discriminative patches and we fix this parameter as 20 in the following experiments.

Parameter of λ. The parameter λ in Eq. (9) controls the balance between frequency and discriminativeness in reward-punishment rules. Table 1(b) shows recognition accuracies of the proposed model on UT Set #s1 with different λ values. It can be seen that the best performance is gotten when $\lambda = 0.25$, which

Table 1. Impact of different factors on UT Set #1

(a) Impact of clusters

Clusters per class	Accuracy(%)
1	78.33%
5	88.33%
10	93.33%
20	**96.67%**
50	95.00%

(b) Impact of λ

Value of λ	Accuracy(%)
0	95.00%
0.25	**96.67%**
0.5	93.33%
0.75	91.67%
1	90.00%

(c) Impact of key frames

Methods	**Key frame**	non-Key frame
Accuracy(%)	**96.67%**	90.00%

means that discrimination factor plays a little more important role than the frequency factor. Hence, we fix $\lambda = 0.25$ in the following experiments. Please note that, when $\lambda = 0$, we only consider the discrinativeness. That is, we only concern whether the patch seldom occurs on non-key frames, without consideration about its appearance on key frames. However, remember that the candidate patches are mined by frequent rules so that they must appear frequently in the training samples. Thus it is reasonable that it's performance is acceptable though it not the best one. On the other hand, only the frequency is taken into consideration when $\lambda = 1$, which means that we do not concern the patch appear on non key frames or not. Therefore it may not be discriminative enough and the performance is not satisfying.

Key Frames. We also check the effect of key frames and the refinement of patches. The action recognition results on UT Set #1 are shown in Table 1(c). It can be seen that the patches refined by key frames is more effective than those extracted by only associate rules, which demonstrates the benefit brought by using key frames to purify the discriminative patches. It is interesting to observe that the accuracy without key frames and patch refinement (Non-key frame) is 90.00%, same as the result of $\lambda = 1$ in Table 1(b). Actually, in the case of $\lambda = 1$, the role of non-key frame is absent, which is similar with the case that all the frames are key frames, i.e., without refinement.

4.2 Experimental Results

In this section, we present the experimental results of our proposed KFDP model and compared with other methods on UT-Interaction Set #1, UT-Interaction Set #2 [25] and BIT-Interaction Dataset [16].

Result on UT-Interaction Sets. In the first experiment, we test the proposed method on UT-Interaction datasets. The confusion matrix is shown in Fig. 7. Our

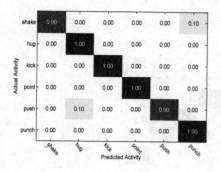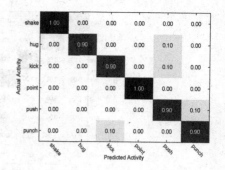

(a) Confusion Matrices on UT Set #1   (b) Confusion Matrices on UT Set #2

Fig. 7. Confusion matrices on UT sets

Table 2. Recognition accuracy on UT sets

(a) Accuracy on UT set #1	
Methods	Accuracy(%)
Ryoo & Aggarwal [18]	70.80%
Lan *et al.* [7]	78.33%
MSSC [27]	83.33%
Ryoo [28]	85.00%
Kong *et al.* [16]	88.33%
Vahdat *et al.* [10]	93.33%
Raptis *et al.* [13]	93.33%
Zhang *et al.* [29]	95.00%
Fu *et al.* [17]	91.67%
Ours	**96.67%**
(b) Accuracy on UT set #2	
Methods	Accuracy(%)
MSSC [27]	81.67%
HM [30]	83.33%
MTSSVM [31]	86.67%
Vahdat *et al.* [10]	90.00%
Zhang *et al.* [29]	90.00%
Kong *et al.* [26]	91.67%
Ours	**93.33%**

proposed method achieves 96.67% and 93.33% accuracy on UT-Interaction Set #1 and UT-Interaction Set #2, respectively. Table 2(a) and (b) compare the proposed method with some baseline methods [7, 10, 13, 16–18, 27–29] on the UT Set

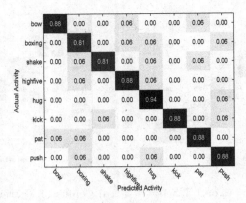

Fig. 8. Confusion matrix on BIT dataset.

Table 3. Recognition accuracy on BIT dataset.

Methods	Accuracy(%)
Bag-of-word	70.31%
Lan *et al.* [7]	83.33%
Kong *et al.* [16]	82.03%
Kong *et al.* [26]	85.38%
Ours	**86.72%**

#1 and #2, respectively. Compared with these methods, the performance gain achieved by our method is significant due to the use of mid-level discriminative knowledge of human interaction.

Most of confusions are arisen by visually similar movements and the influence of moving objects in the background. Our method can recognize human interactions in some challenging situations, e.g. partially occlusion and background clutter. On one hand, since there are 20 different kinds of classifiers for each category, even if some classifiers is out of work, the others would help to predict the label. On the other hand, because the final clusters refined on the key frames by reward-punishment rules, they can overcome background clutter clustering phenomenon to some extent.

Table 2(a) indicates that our method outperforms all the baseline methods [7,10,13,16–18,27–29] on the UT Set #1. Compared with all the methods, the performance gain achieved by our method is significant due to the use of mid-level discriminative knowledge of human interaction. Table 2(b) compare the results of our method with other leading approaches, our proposal utilizes discriminative patches to better represent complex human interactions and thus achieves the best results on the UT Set #2. Analysing the results, this may be because that with the interdependencies of both discriminative patches and key frames, our method can recognize some challenging interaction videos and thus achieves higher performance.

Result on BIT-Interaction Dataset. We test our method on BIT-Interaction Dataset and show the confusion matrix in Fig. 8. Our method achieves 86.72% recognition accuracy. Table 3 shows the comparative results with previous methods [7,16,26], which demonstrate the superiority of our proposed method. Compared with the baseline bag-of-words method, the result gain achieved by our method is significant due to the use of mid-level discriminative knowledge of human interaction. Our method also significantly outperforms Lan *etal.* and Kong *etal.*'s methods, which validates the effectiveness of the key frames in promoting the discrimination of interactive patches.

5 Conclusions

In this paper, we have proposed a novel model (KFDP) for recognizing human interaction in videos with discriminative patches. The candidate discriminative patches are first mined from videos of different interactions by applying association rules on convolutional neural networks (CNN) activations. Based on the observation that there are more discriminative patches occurring in the climax period of a human interaction, we define frames in such intervals as key frames and the final discriminative patches are obtained by further refining the candidate patches using reward-punishment rules, which ensure that a discriminative patch occur frequently during key frames and seldom emerge during non-key frames. The experimental results on UT-Interaction Sets and BIT-Interaction Dataset are comparable with the state-of-art, which indicates that the benefits of using discriminative patches in human interaction recognition.

Acknowledgments. This research is partially sponsored by Natural Science Foundation of China (Nos. 61472387, 61272320, and 61572004) and Beijing Natural Science Foundation (Nos. 4152005 and 4162058).

References

1. Singh, S., Gupta, A., Efros, A.A.: Unsupervised discovery of mid-level discriminative patches. In: Fitzgibbon, A., Lazebnik, S., Perona, P., Sato, Y., Schmid, C. (eds.) Computer Vision – ECCV 2012. LNCS, vol. 7573, pp. 73–86. Springer, Heidelberg (2012). doi:10.1007/978-3-642-33709-3_6
2. Doersch, C., Gupta, A., Efros, A.A.: Mid-level visual element discovery as discriminative mode seeking. In: Advances in Neural Information Processing Systems, pp. 494–502 (2013)
3. Juneja, M., Vedaldi, A., Jawahar, C., Zisserman, A.: Blocks that shout: distinctive parts for scene classification. In: Proceedings of the IEEE Conference on Computer Vision and Pattern Recognition, pp. 923–930 (2013)
4. Wang, X., Wang, B., Bai, X., Liu, W., Tu, Z.: Max-margin multiple-instance dictionary learning. In: Proceedings of the 30th International Conference on Machine Learning, pp. 846–854 (2013)
5. Bourdev, L., Malik, J.: Poselets: body part detectors trained using 3D human pose annotations. In: 2009 IEEE 12th International Conference on Computer Vision, pp. 1365–1372. IEEE (2009)

6. Li, Y., Liu, L., Shen, C., van den Hengel, A.: Mid-level deep pattern mining. In: 2015 IEEE Conference on Computer Vision and Pattern Recognition (CVPR), pp. 971–980. IEEE (2015)
7. Lan, T., Wang, Y., Yang, W., Robinovitch, S.N., Mori, G.: Discriminative latent models for recognizing contextual group activities. IEEE Trans. Pattern Anal. Mach. Intell. **34**, 1549–1562 (2012)
8. Ryoo, M.S., Aggarwal, J.K.: Recognition of composite human activities through context-free grammar based representation. In: 2006 IEEE Computer Society Conference on Computer Vision and Pattern Recognition, vol. 2, pp. 1709–1718. IEEE (2006)
9. Choi, W., Shahid, K., Savarese, S.: Learning context for collective activity recognition. In: 2011 IEEE Conference on Computer Vision and Pattern Recognition (CVPR), pp. 3273–3280. IEEE (2011)
10. Vahdat, A., Gao, B., Ranjbar, M., Mori, G.: A discriminative key pose sequence model for recognizing human interactions. In: 2011 IEEE International Conference on Computer Vision Workshops (ICCV Workshops), pp. 1729–1736. IEEE (2011)
11. Su, B., Ding, X.: Linear sequence discriminant analysis: a model-based dimensionality reduction method for vector sequences. In: ICCV, pp. 889–896 (2013)
12. Su, B., Zhou, J., Ding, X., Wang, H., Wu, Y.: Hierarchical dynamic parsing and encoding for action recognition. In: Leibe, B., Matas, J., Sebe, N., Welling, M. (eds.) ECCV 2016. LNCS, vol. 9908, pp. 202–217. Springer, Heidelberg (2016). doi:10.1007/978-3-319-46493-0_13
13. Raptis, M., Sigal, L.: Poselet key-framing: a model for human activity recognition. In: Proceedings of the IEEE Conference on Computer Vision and Pattern Recognition, pp. 2650–2657 (2013)
14. Brendel, W., Todorovic, S.: Learning spatiotemporal graphs of human activities. In: 2011 IEEE International Conference on Computer Vision (ICCV), pp. 778–785. IEEE (2011)
15. Raptis, M., Kokkinos, I., Soatto, S.: Discovering discriminative action parts from mid-level video representations. In: 2012 IEEE Conference on Computer Vision and Pattern Recognition (CVPR), pp. 1242–1249. IEEE (2012)
16. Kong, Y., Jia, Y., Fu, Y.: Learning human interaction by interactive phrases. In: Fitzgibbon, A., Lazebnik, S., Perona, P., Sato, Y., Schmid, C. (eds.) ECCV 2012. LNCS, vol. 7572, pp. 300–313. Springer, Heidelberg (2012). doi:10.1007/978-3-642-33718-5_22
17. Kong, Y., Jia, Y., Fu, Y.: Interactive phrases: semantic descriptions for human interaction recognition. IEEE Trans. Pattern Anal. Mach. Intell. **36**, 1775–1788 (2014)
18. Ryoo, M.S., Aggarwal, J.K.: Spatio-temporal relationship match: video structure comparison for recognition of complex human activities. In: 2009 IEEE 12th International Conference on Computer vision, pp. 1593–1600. IEEE (2009)
19. Amer, M.R., Todorovic, S.: Sum-product networks for modeling activities with stochastic structure. In: 2012 IEEE Conference on Computer Vision and Pattern Recognition (CVPR), pp. 1314–1321. IEEE (2012)
20. Bossard, L., Guillaumin, M., Gool, L.: Food-101 – mining discriminative components with random forests. In: Fleet, D., Pajdla, T., Schiele, B., Tuytelaars, T. (eds.) ECCV 2014. LNCS, vol. 8694, pp. 446–461. Springer, Heidelberg (2014). doi:10.1007/978-3-319-10599-4_29
21. Xu, Z., Qing, L., Miao, J.: Activity auto-completion: predicting human activities from partial videos. In: ICCV, pp. 3191–3199 (2015)

22. Jia, Y., Shelhamer, E., Donahue, J., Karayev, S., Long, J., Girshick, R., Guadarrama, S., Darrell, T.: Caffe: convolutional architecture for fast feature embedding. In: Proceedings of the ACM International Conference on Multimedia, pp. 675–678. ACM (2014)
23. Simonyan, K., Zisserman, A.: Very deep convolutional networks for large-scale image recognition. arXiv preprint arXiv:1409.1556 (2014)
24. Chang, C.C., Lin, C.J.: LIBSVM: a library for support vector machines. ACM Trans. Intell. Syst. Technol. **2**, 27:1–27:27 (2011). http://www.csie.ntu.edu.tw/cjlin/libsvm
25. Ryoo, M.S., Aggarwal, J.: Ut-interaction dataset, ICPR contest on semantic description of human activities (SDHA). In: IEEE International Conference on Pattern Recognition Workshops, vol. 2, p. 4 (2010)
26. Kong, Y., Fu, Y.: Close human interaction recognition using patch-aware models. IEEE Trans. Image Process. **25**, 167–178 (2016)
27. Lan, T., Chen, T.-C., Savarese, S.: A hierarchical representation for future action prediction. In: Fleet, D., Pajdla, T., Schiele, B., Tuytelaars, T. (eds.) ECCV 2014. LNCS, vol. 8691, pp. 689–704. Springer, Heidelberg (2014). doi:10.1007/978-3-319-10578-9_45
28. Ryoo, M.: Human activity prediction: early recognition of ongoing activities from streaming videos. In: 2011 IEEE International Conference on Computer Vision (ICCV), pp. 1036–1043. IEEE (2011)
29. Zhang, Y., Liu, X., Chang, M.-C., Ge, W., Chen, T.: Spatio-temporal phrases for activity recognition. In: Fitzgibbon, A., Lazebnik, S., Perona, P., Sato, Y., Schmid, C. (eds.) ECCV 2012. LNCS, vol. 7574, pp. 707–721. Springer, Heidelberg (2012). doi:10.1007/978-3-642-33712-3_51
30. Cao, Y., Barrett, D., Barbu, A., Narayanaswamy, S., Yu, H., Michaux, A., Lin, Y., Dickinson, S., Siskind, J., Wang, S.: Recognize human activities from partially observed videos. In: CVPR, pp. 2658–2665 (2013)
31. Kong, Y., Kit, D., Fu, Y.: A discriminative model with multiple temporal scales for action prediction. In: Fleet, D., Pajdla, T., Schiele, B., Tuytelaars, T. (eds.) ECCV 2014. LNCS, vol. 8693, pp. 596–611. Springer, Heidelberg (2014). doi:10.1007/978-3-319-10602-1_39

Stacked Overcomplete Independent Component Analysis for Action Recognition

Zhikang Liu[1], Ye Tian[2], and Zilei Wang[1(✉)]

[1] Department of Automation, University of Science and Technology of China,
Hefei 230027, China
lzk@mail.ustc.edu.cn, zlwang@ustc.edu.cn
[2] Institute for Computational and Mathematical Engineering, Stanford University,
Stanford, USA
yetian1@stanford.edu

Abstract. Generating the discriminative representations of video clips is of vital importance for human action recognition, especially for complex action scenarios. In this paper, we particularly introduce Overcomplete Independent Component Analysis (OICA) to directly learn structural spatio-temporal features from the raw video data. OICA as an unsupervised learning method can fully exploit the unlabeled videos, which is crucial for action recognition since labeling huge volume of video data is too effort-consumed in practice. In addition, features learned by OICA can more accurately describe the complex actions with enough details owing to the overcompleteness and independence constraints to the component bases. Furthermore, inspired by the layered structure of deep neural network, we also propose to stack OICA to form a two-layer network for abstracting robust high-level features. Such stacking is practically proved effective for boosting the recognition accuracy. We evaluate the proposed stacked OICA network on four benchmark datasets: Hollywood2, YouTube, UCF Sports and KTH, which cover the simple and complex action scenarios. The experimental results show that our method always outperforms the baselines, and achieves the state-of-the-art performance.

1 Introduction

Action recognition is one of the most active topics in video understanding, which can be applied to various visual systems, *e.g.*, intelligent video surveillance [23], and sport video analysis [21]. For the complex scenes and actions, however, the existing approaches still perform unsatisfactorily if the real-world videos are involved. Thus exploring more effective models of action recognition is always one of its core missions.

Action recognition as a specific recognition task generally follows the common pipeline consisting of the feature extraction and classification. Particularly, the feature extraction methods play a critical role due to its primarily determining the final performance [38]. Thus most of the previous works always put their emphasis on how to produce discriminative and robust features from video clips.

© Springer International Publishing AG 2017
S.-H. Lai et al. (Eds.): ACCV 2016, Part II, LNCS 10112, pp. 368–383, 2017.
DOI: 10.1007/978-3-319-54184-6_23

The traditional approach is to manually design some appropriate features, *e.g.*, transforming the well-performed 2D features into the corresponding 3D versions [13,33]. These hand-crafted features can provide good performance, but are difficult to adapt specific visual tasks as they are mainly constructed according to given prior knowledge. Consequently, the recognition accuracy may be greatly decreased when they are applied to some complex video scenarios, *e.g.*, web videos, movies, and TV shows.

To tackle this issue, some data-driven methods are proposed that can adaptively learn a proper feature extraction model using provided samples. For example, Convolutional Neural Network (CNN) as an end-to-end feature extractor is able to generate one complete feature for each video clip by straightforwardly processing the raw pixels [14]. However, CNN, which is usually used as one of the supervised learning methods, requires lots of labeled training data, and in practice supplying sufficient labels is very difficult for action recognition due to the massiveness and complication of video data. On the contrary, the unsupervised learning methods can fully utilize the unlabeled data to boost the performance [2], which is exactly the way we would investigate in this paper.

Unsupervised learning tends to find hidden structure by following the neurobiological organization of brains [7]. Recently, there can see growing interests in development of unsupervised feature learning methods; these include Deep Belief Nets (DBN) [18], Slow Feature Analysis (SFA) [34], Sparse Coding [17], and Independent Component Analysis (ICA) [8]. In particular, ICA and its variation Independent Subspace Analysis (ISA) have demonstrated impressive performance [20,26]. Moreover, in analogy to the mechanism of visual cortex in brain, the components learned by ICA are similar to the receptive fields of the V_1 area for static images and the MT area for sequences of images [36]. Therefore, it is believed that ICA has enormous potentiality in achieving better performance for action recognition.

The original ICA (ISA as well) has two major drawbacks in recognizing complex actions. First, the number of learned components cannot exceed the dimensionality of input data [19]. Consequently, one action is only described and reconstructed by limited components, which actually are not enough to completely represent complex actions. Meanwhile, it has been shown that the overcomplete feature learning would produce better performance, *i.e.*, the number of latent components is quite significant in unsupervised learning models [2]. Second, ICA is sensitive to the process of data whitening, especially for high-dimensional data [19], which, however, are unavoidable for robustly representing complex actions. Hence the inherent characteristics of ICA hinder further boosting the performance of action recognition. In this paper, we propose to introduce Overcomplete Independent Component Analysis (OICA) in automatically extracting the local features of video clips. OICA adopts the overcomplete independent bases to reconstruct the action elements (blocks of image sequences). Naturally, OICA would certainly produce more robust action representations *w.r.t*, image variations, owing to the overcompleteness of learned components.

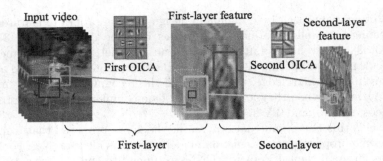

Fig. 1. The two-layer OICA model for action recognition. Samples from the input video are represented by the first-layer OICA bases. Then the combined responses are feed into the second-layer OICA to get a hierarchical representation.

For modern feature learning, the layered models are usually able to achieve more promising performance, *e.g.*, deep neural networks [14]. In philosophy, such layer-by-layer structure can hierarchically exploit the semantic features of input data, and the features generated in higher layers may be more robust in describing complex structural actions due to their high-level abstraction. Inspired by such a principle, we propose to stack OICA to form a two-layer OICA network, as illustrated in Fig. 1. Samples from the input video are represented by the first-layer OICA bases. Then the combined responses are feed into the second-layer OICA network to get high-level features. In practice, it is observed that the two-layer OICA certainly performs better than the one-layer version with involving a significant accuracy increase.

We evaluate the proposed stacked OICA network model on four well-known benchmark datasets: KTH [32], Hollywood2 [24], UCF Sports [29], and YouTube [23]. The datasets actually cover the simple and complex action scenarios. The experimental results show that our method always outperforms the baselines, especially for the complicated datasets, and achieves the state-of-the-art performance on these datasets.

2   Related Works

The key to action recognition is to generate the discriminative and robust video representations. For most of the modern approaches, such a process is generally completed by two successive steps: feature extraction and feature integration [38]. Here the former is to map the spatio-temporal pixels into many local features, and the latter is to integrate these features into video representations fed the final classifier. Traditionally, the two steps are conducted separately. Particularly, some sophisticated feature integration methods can be directly used by referring image classification, *e.g.*, Bag of Words (BoW) [27], LDC [39], and Fisher Vector [30]. Thus current works in action recognition mainly focus on how to extract better features from videos.

In the previous literatures, the hand-crafted features are first proposed, which are mostly constructed by transforming the corresponding 2D features, *e.g.*, HOG3D [13], 3D-SIFT [33], and Extended SURF [41]. These features can achieve noticeable performance on simple datasets (*e.g.*, Weizmann [6], and KTH [32]), but the accuracy would greatly decrease for complex action scenarios from the realistic videos. On the other hand, the feature detector also affects the recognition performance seriously as it controls the locations and number of extracting features [38]. Roughly, the detectors can be divided into two types: sparse and dense. The sparse way is to detect the spatio-temporal interest points according to specified criteria, such as Harris3D [16], Hessian detector [41], and temporal Gabor filter [5]. Particularly, the newly proposed Dense Trajectories [37] and Improved Trajectories [40] can also be viewed to extend this route by considering trajectories. On the contrary, the dense sampling strategy [9] is to extract features over multi-scale regular grids, which are usually predefined according to the requirements of a specific task. Wang *et al.* [38] empirically evaluated the combinations of different features and detectors. It is shown that the dense sampling consistently outperforms all of the sparse interest point detectors, and no hand-crafted feature always performs well for all of action datasets, *i.e.*, any feature has its own limitation.

Recently, There has been a growing interest in applying unsupervised learning methods to visual features for their ability of exploiting unlabeled data. For example, Kanan and Cottrell [12] show that ICA can be used to generate robust visual features for recognition. In [26], TICA, a extension of ICA, was proposed for image recognition and achieves state-of-the-art performance. Le *et al.* [20] propose ISA, another extension of ICA, to learn invariant spatio-temporal features. However, the features learned by these ICA based methods possess a crucial limitation, *i.e.*, the number of features cannot exceed the dimensionality of input data. Indeed, Coates *et al.* [2] have shown that the classification accuracy would be consistently increased if the latent components are overcomplete in unsupervised learning models, *e.g.*, RBM [18,35], and sparse autoencoder [1]. Hence such a drawback of ICA hinders its further application to recognise complex action scenarios.

Another interesting unsupervised learning work is the deeply-learned slow feature analysis (SFA) [34]. The model is particularly designed to learn the invariant and slowly varying features from noisy and quickly varying signals to represent motion patterns. The work in [34] similarly adopts stacking architecture of feature learning. However, SFA may suffer from the curse of dimensionality, since the dimensionality of the expanded function space increases very fast with the number of input signals [42]. Thus, SFA may be not suitable for visual tasks due to the natural high dimensionality of video data. Differently, Overcomplete Independent Component Analysis (OICA) we would adopt in this paper is inherently designed to deal with complex data [8].

3   Methology

In this section, we elaborate on the details of our proposed approach. We first give a brief introduction of the OCIA algorithm. Secondly, we apply the OICA algorithm to video data and build the proposed stacked OICA network. Then we illustrate the properties of the learned OICA features and explain how to compose local features. Finally, we describe the model we used to classify the local features.

3.1   OICA for Image Data

OICA is one of the unsupervised learning models, which has an inherent advantage of having capacity to learn features from unlabeled data. Given a set a unlabeled image patches $\{\mathbf{x}_1, \mathbf{x}_2, \cdots, \mathbf{x}_m\}$ where $\mathbf{x}_i \in \mathbb{R}^D$ denotes the grey-scale values in a patch. Both the OICA model and the traditional ICA model express $\mathbf{x}_i$ as a linear transformation of latent variables, $i.e.$, the independent components, which are required to be non-Gaussian and mutually independent:

$$\mathbf{x}_i = \mathbf{A}\mathbf{s} = \sum_{j=1}^{n} \mathbf{a}_j s_j. \tag{1}$$

where $\mathbf{A}$ is a full column rank matrix called the basis matrix and stay the same for all patched. Each column of $\mathbf{A}$, $i.e.$, $\mathbf{a}_j$ is called as a basis function. $\mathbf{s} = (s_1, s_2, \cdots, s_n)^T$ are the independent components (or source signals), different from patch to patch. Here n is the number of independent components.

In the traditional ICA [8], the optimization problem to solve the bases is defined as

$$\min_{\mathbf{W}} \sum_{i=1}^{m} \sum_{j=1}^{k} g(\mathbf{W}^{(j)}\mathbf{x}_i), \quad s.t. \quad \mathbf{W}\mathbf{W}^T = \mathbf{I}. \tag{2}$$

where $g(\cdot)$ is the measure function that can be any sufficiently regular, odd, nonlinear function, $e.g.$, $g(\cdot) = \tanh(\cdot)$. $\mathbf{W} \in \mathbb{R}^{k \times n}$ is the transformation matrix, also called as filter matrix. $\mathbf{W}^{(j)}$, which denoted the jth row of the transformation matrix, is a linear transformation. Here m is the number of the image patch, and k is the number of the linear transformations. The orthogonality constraint represented by $\mathbf{W}\mathbf{W}^T = \mathbf{I}$ prevents the linear transformations in $\mathbf{W}$ from becoming degenerate [8]. To speed up the optimization above, the unlabeled image patches are required to be whitened to have zero mean, $\sum_{i=1}^{m} \mathbf{x}_i = \mathbf{0}$, and unit covariance, $\frac{1}{m} \sum_{i=1}^{m} \mathbf{x}_i \mathbf{x}_i^T = \mathbf{I}$. The basis matrix equals the inverse of the transformation matrix, $e.g.$, $\mathbf{A} = \mathbf{W}^{-1}$. Note that $\mathbf{W}$ is orthonormal, therefore, $\mathbf{A} = \mathbf{W}^{-1} = \mathbf{W}^T$.

The traditional ICA requires that the dimension of input samples strictly equals the number of independent components, $i.e.$, $k \equiv n$. However, it has been shown that the overcomplete feature learning possesses more powerful ability to represent data [2]. The overcompleteness of ICA means $k > n$, and consequently the basis matrix $\mathbf{A}$ is not invertible any more, and the orthonormality constraint cannot be held completely. Instead, an approximate orthogonality constraint is needed. In particular, to compute overcomplete ICA representation, we adopt the method proposed by Le et $al.$ [19] to replaces the hard orthonormal constraint in Eq. (2) due to its low computation complexity. Specifically, let the basis matrix be $\mathbf{A} = \mathbf{W}^T$. Then the reconstruction cost of representing the input data $\mathbf{x}_i$ with the basis functions in $\mathbf{A}$ can be denoted as $\|\mathbf{W}^T\mathbf{W}\mathbf{x}_i - \mathbf{x}_i\|_2^2$. Consequently, the following unconstrained optimization problem is produced as

$$\min_W \frac{\lambda}{m} \sum_{i=1}^{m} \|\mathbf{W}^T\mathbf{W}\mathbf{x}_i - \mathbf{x}_i\|_2^2 + \sum_{i=1}^{m}\sum_{j=1}^{k} g(\mathbf{W}^{(j)}\mathbf{x}_i). \tag{3}$$

where λ is the parameter that controls the penalty of the reconstruction error. This optimization problem can be efficiently solved by the unconstrained optimizers such as L-BFGS [22] and CG [31]. Moreover, the introduced reconstruction penalty works well even when the input training samples are not completely whiten.

Once the transformation matrix for OICA model is learned, an image patch can be represented by the corresponding components:

$$\mathbf{s} = \mathbf{W}\mathbf{x}. \tag{4}$$

where $\mathbf{s}$ is exactly the feature representation of the image patch $\mathbf{x}$.

For the complex scenes in high-resolution videos, a single globe OICA kernel is difficult to represent large video blocks accurately and robustly. Next, we apply the OICA algorithm to video data and introduce a stacked model to learn hierarchical features.

3.2 OICA on Video Blocks

Figure 2 illustrates how we apply the OICA to the video data. First, we extract 3D video blocks instead of small image patches. Specifically, we first take a sequence of image patches and flatten them into a vector. Then, we whiten these vectors by removing the DC component. After that, PCA is employed to reduce

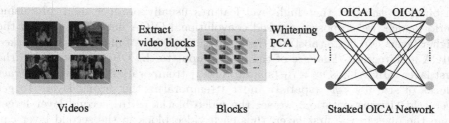

Fig. 2. Graphical depiction of applying the OICA algorithm to video data. We extract small video blocks from videos. Then the blocks are sent to stacked OICA network with whitening and PCA as preprocessing steps.

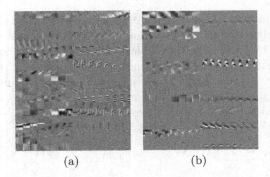

(a)   (b)

Fig. 3. Comparison of the video blocks extracted after (a) and before (b) applying the frame difference. Each row of (a) contains the illustrations of two video blocks applying the frame difference, and (b) is similar for random sampling. The columns of (a) and (b) indicate the patterns of the video blocks.

the dimension of the inputs. Finally, the OICA learns the transformation matrix $\mathbf{W}$ from the resulting vectors with the method detailed in Sect. 3.1.

Unsupervised algorithms are often trained on blocks which are randomly sampled from videos. Randomly sampling 3D video blocks may results that a large number of blocks only contain background information. However, the foreground information is more important in disambiguating similar actions. In order to enhance the foreground information, we design a video block extraction method based on frame difference. Specifically, for each training video, we randomly sample M blocks and compute differences between adjacent frames to detect moving pixels. Then the sampled blocks are sorted by their energies, which is the sum of intensities of all the pixels in that block. Finally, only the N highest energy blocks are kept. In our experiments, M and N are set to 400 and 200 respectively. Figure 3 compares the video blocks extracted after and before applying frame difference. The experimental comparison between our sampling strategy and random sampling is detailed in the Sect. 4.4.

3.3   Stacked Convolutional OICA Network

Inspired by the fact that high-level features usually produce more promising performance, we propose a stacked convolutional OICA network, in which the higher layers can better abstract the input video. Figure 4 illustrates the stacked OICA architecture. The size of the blocks input to the OICA algorithm in the first layer (OICA1), is $w_1 \times h_1$ (spatial) and t_1 (temporal). Similarly, we extract blocks of size $w_2 \times h_2$ (spatial) and t_2 (temporal)in the second layer. To get hierarchical representation, we set the video blocks in the second layer larger than the ones in the first layer, thus each video block in the second layer can be regarded as a collection of overlapping video blocks in the first layer. The top of Fig. 4 illustrate such construction wherein the biggest cube represents a second layer block. The inner cubes, such as the red, green ones, represent video

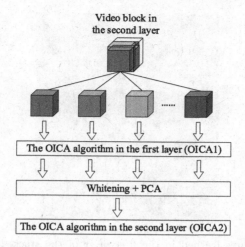

Fig. 4. An illustration of the stacked convolutional OICA network. The biggest cube on the top represents a block in the second layer. The inner cubes are input to the first layer and convolved with OICA1. The combined first-layer responses are processed by whitening and PCA, and then input to the second-layer OICA algorithm (OICA2).

blocks in the first layer. Then, the OICA1 is convolved with each inner block. The responses are combined and input to the OICA algorithm in the second layer (OICA2).

To speed up the computation, whitening and PCA are applied to reduce the dimensions of the inputs. Furthermore, our stacked network is trained greedily layerwise in the same manner as other methods proposed in the deep learning literature [1,18] to reduce the training time. More specifically, we train the first layer until convergence before training the second layer.

3.4 Properties of the Extracted Features from OICA Network

Figure 5 shows typical features learned by the OICA1 trained on video blocks of size $14 \times 14 \times 8$. It is observed that the OICA algorithm is able to learn Gabor features (edge filters) with diverse orientation and frequencies. The OICA model is especially intriguing for image modeling due to its characteristics closely related to overcomplete wavelet bases [28]. The large number of basis make OICA algorithm be good candidates for representing complex video data in comparison to other algorithms such as ICA, ISA.

3.5 Pooling and Local Features

The stacked OICA network detects simple features in the first layer and robust features in the second layer. As a second layer blocks contains several inner blocks, the number of simple features is much larger than the robust features. Meanwhile, the simple features are not as significant as the robust ones, so we

Fig. 5. Typical features learned in the first layer of the stacked OICA network by training the OICA1 on $14 \times 14 \times 8$ video blocks. The results are shown in the original space, *i.e.*, the inverse of the preprocessing (whitening) was performed.

try to reduce the contribution of the simple features by mean spatial-temporal pooling. Specially, the responses of the inner blocks in the same big block are processed with mean pooling. Then PCA is applied on the results to further reduce the dimension. Finally, the resulting simple feature and the second layer feature are concatenated to construct local features. In our experiments, the PCA layers keep 200 features in the first layer and 200 features in the second layers, so each local feature is a 400-dimensional vector.

3.6   Classification Model for Human Actions

We use the standard bag-of-feature (BOF) model to evaluate the performance of our local features. In particular, the stacked OICA network learns features from videos on a dense grid in which the second layer blocks overlap 50% in w, h and t dimensions. The codebook is obtained with k-means clustering of the learned features from training videos. Then, a video can be represented as a frequency histogram over the visual words.

We apply a non-linear SVM with $\mathcal{X}^2$ kernel to classify the human actions. For multi-class classification, we apply the "one versus all others" approach and select the class with the highest score. We achieve the best performance when the number of visual words equals 4500.

4   Experiments

In this section, we numerically compare our algorithm against the state-of-the-art action recognition algorithms. For fair comparison, we follow the pipeline in [38], which consists of the extraction of local features, Hard Vector Quantization with

k-means, and classification. The mean average precision (AP) over all classes is adopted as the metric to measure the performance of the action recognition methods.

In the experiment, the inputs to the first-layer OICA network are of size 14×14 (spatial) and 8 (temporal). The OICA1 learns 200 features, *i.e.*, there are 200 blue nodes in Fig. 2. The OICA2 is trained on blocks of size $18 \times 18 \times 12$. The convolution step is performed in the OICA2 with a stride of 4. The OICA2 learns 800 features, *i.e.*, there are 800 yellow nodes in Fig. 2, and then PCA to 200 features.

4.1 Datasets

We conduct the experimental evaluation on four standard human action datasets, *i.e.*, Hollywood2 [24], YouTube [23], UCF Sports [29] and KTH [32], see Fig. 6. These action datasets are very diverse. The Hollywood2 dataset is made up of real movies with significant background clutter, whereas the KTH views human actions in front of a uniform background. The YouTube dataset are low quality, whereas the UCF sports videos are high resolution.

Hollywood2 Action Dataset [24] is collected from 69 different Hollywood movies which are divided into 33 training movies and 36 test movies. The dataset provides 12 classes of human actions, *i.e.*, answering phone, driving car, eating, fighting person, getting out of car, hand shaking, hugging person, kissing, running, sitting down, sitting up, and standing up. The dataset is especially challenging since the scenes in the movies contain various complex context and background.

Fig. 6. Sample frames from video sequences of Hollywood2 (first row), YouTube (second row), UCF sports (third row) and KTH (last row) action datasets.

YouTube Dataset [23] is collected from YouTube video website. It contains 11 action categories, *i.e.*, basketball shooting, biking, tennis swinging, diving, soccer juggling, golf swinging, horseback riding, volleyball spiking, swinging, trampoline jumping, and walking with a dog. A total of 1600 video clips are available. We follow the original experimental in [23].

UCF Sports Action Dataset [29] contains 10 types of human actions: swinging (on the pommel horse and on the floor), diving, kicking weight-lifting, horse riding, running, skateboarding, swinging (at the high bar), golf swinging, and walking. The dataset consists 150 video samples and shows large intra-class variability. Here we follow the standard protocol in [38], *i.e.*, adding a horizontally flipped version of each sequence.

KTH Action Dataset [32] is a traditional dataset that contains 6 different human actions, *i.e.*, hand waving, hand clapping, jogging, walking, boxing, and running. Each action is performed multiple times by 25 subjects in four different conditions (outdoors, outdoors with scale variation, outdoors with different clothes, and indoors). The background in this dataset is static and homogeneous.

4.2   Classification Results

For the datasets of *Hollywood2*, *YouTube*, and *UCF Sports*, Tables 1, 2 and 3 provide the resulting classification performance, respectively. For each dataset, the performance of typical methods is also provided for comparison, which include the start-of-the-art performance in the previous works. From the results, it can be seen that our proposed method outperforms all the hand-crafted features on these datasets. In particular, for the complex datasets of *Hollywood2* and *YouTube*, our method involve a larger accuracy increase, which implies that the

Table 1. Average accuracy on the Hollywood2 dataset.

Algorithm	Mean AP
Hessian [41] + ESURF [15]	38.2%
Harris3D [25] + HOG/HOF [15] (from [38])	45.2%
Dense + HOG3D [13]	45.3%
Hessian [41] + HOG/HOF [15]	46.0%
Cuboid [5] + HOG/HOF [15]	46.2%
Convolutional GRBM [35]	46.6%
Dense + HOG/HOF [15]	47.7%
DL-SFA [34]	48.1%
Hierarchical ISA [20]	53.3%
One-layer OICA	**51.6%**
Random sampled OICA	**54.5%**
Our method	**55.1%**

Table 2. Average accuracy on the YouTube dataset.

Algorithm	Accuracy

Feature combining and pruning [23]
-Static features:
  HAR+HES+MSER+SIFT
-Motion features:
  Harris3D+Gradients+Heurristics
	71.2%

Hierarchical ISA [20]	75.8%
One-layer OICA	**74.9%**
Random sampled OICA	**78.5%**
Our method	**78.9%**

Table 3. Average accuracy on the UCF sports dataset.

Algorithm	Accuracy
Hessian [41] + ESURF [15]	77.3%
Harris3D [25] + HOG/HOF [15] (from [38])	78.1%
Hessian [41] + HOG/HOF [15]	79.3%
Dense + HOF [15]	82.6%
Cuboid [5] + HOG3D [13]	82.9%
Dense + HOG3D [13]	85.6%
Hierarchical ISA [20]	86.5%
DL-SFA [34]	86.6%
One-layer OICA	**82.9%**
Random sampled OICA	**86.5%**
Our method	**86.8%**

proposed OICA model is especially effective for the complex action scenarios. Let us focus on the unsupervised methods, DL-SFA [34] show the best performance among all baseline methods. But our method defeats it with big gaps, e.g., 55.1% vs. 48.1% on *Hollywood2*.

We report the performance on the *KTH* dataset in Table 4. Particularly, we compare the accuracy achieved by our method against the best results ever published. For this dataset, the background does not convey any meaningful information [38]. Therefore, we direct apply a Region of Interest (ROI) detector to this dataset. Naturally, the human detectors based on sliding windows (e.g., [4]) and trackers (e.g., [3]) can provide good bounding boxes. Finally, our method achieves the accuracy of 93.3%.

Table 4. Average accuracy on the KTH dataset. The symbol (∗) indicates that the method uses an interest point detector or other interest region detect technique.

Algorithm	Accuracy
(∗) Hessian [41] + ESURF [15]	81.3%
Dense + HOF [15]	88.0%
(∗) Cuboid [5] + HOG3D [13]	90.0%
Convolutional GRBM [35]	90.0%
3D CNN [11]	90.2%
(∗) Hierarchical ISA [20]	91.4%
HMAX [10]	91.7%
(∗) Harris3D [25] + HOG/HOF [15] (from [38])	91.8%
(∗) Harris3D [25] + HOF [15] (from [38])	92.1%
DL-SFA [34]	93.1%
(∗) One-layer OICA	**91.4%**
(∗) Random sampled OICA	**93.1%**
(∗) Our method	**93.3%**

4.3   Benefits of the Second Layer

Now we experimentally investigate the effect of the stacked OCIA on the classification performance compared to one-layer OICA. To this end, we further provide the classification accuracies of the one-layer OICA model. Here, the one-layer OICA is implemented by directly discarding the second-layer OICA and executing the following stages using the first-layer features. From the results, it can be seen that the classification performance is consistently decreased if the second-layer OICA is removed, and the gaps in four datasets are 3.5%, 4.0%, 3.9%, 1.9%, respectively. The results show that the features in the higher layer are indeed helpful to action recognition.

4.4   Benefits of Our Sampling Strategy

Our sampling strategy is detailed in Sect. 3.2. Here, we experimentally compare our sampling strategy with random sampling. To make this a fair comparison, we just replace our sampling strategy with random sampling, *e.g.*, we randomly sample 200 video blocks for each train video. The experimental results are presented in the tables for four datasets. All the results in the tables show that our sampling strategy is better than random sampling. The performance gaps in four datasets are 0.6%, 0.4%, 0.3%, 0.2%, respectively.

5   Conclusion

In this paper, we proposed a stacked Overcomplete Independent Component Analysis (OICA) model to extract the discriminative features from spatio-temporal videos for human action recognition. Firstly, we analyzed the properties of OICA as one of unsupervised learning models, and then introduced it to extracting local features of video clips for action recognition. Furthermore, we proposed to stack OICA to form a two-layer OICA network. Such layered structure provides an ability to produce more robust and discriminative high-layer features. Finally, we experimentally verified the effectiveness of the proposed method by the performance comparison on four benchmark datasets, especially for complex action scenarios. In the future, we plan to construct a more powerful OICA network by better incorporating different components.

Acknowledgement. This work is supported partially by the National Natural Science Foundation of China under Grant 61673362 and 61233003, and the Fundamental Research Funds for the Central Universities.

References

1. Bengio, Y., Lamblin, P., Popovici, D., Larochelle, H., et al.: Greedy layer-wise training of deep networks. In: NIPS (2007)
2. Coates, A., Ng, A.Y., Lee, H.: An analysis of single-layer networks in unsupervised feature learning. In: AISTATS (2011)
3. Comaniciu, D., Ramesh, V., Meer, P.: Kernel-based object tracking. IEEE Trans. PAMI **25**, 564–577 (2003)
4. Dalal, N., Triggs, B.: Histograms of oriented gradients for human detection. In: CVPR (2005)
5. Dollár, P., Rabaud, V., Cottrell, G., Belongie, S.: Behavior recognition via sparse spatio-temporal features. In: VS-PETS (2005)
6. Gorelick, L., Blank, M., Shechtman, E., Irani, M., Basri, R.: Actions as space-time shapes. IEEE Trans. PAMI **29**, 2247–2253 (2007)
7. Hastie, T., Tibshirani, R., Friedman, J.: Unsupervised learning. In: Liu, L., Özsu, M.T. (eds.) Encyclopedia of Database Systems. Springer, Heidelberg (2009)
8. Hyvärinen, A., Hurri, J., Hoyer, P.O.: Independent component analysis. In: Hyvärinen, A., Hurri, J., Hoyer, P.O. (eds.) Natural Image Statistics, vol. 39, pp. 151–175. Springer, Heidelberg (2009)
9. Jain, M., Jégou, H., Bouthemy, P.: Better exploiting motion for better action recognition. In: CVPR (2013)
10. Jhuang, H., Serre, T., Wolf, L., Poggio, T.: A biologically inspired system for action recognition. In: ICCV (2007)
11. Ji, S., Xu, W., Yang, M., Yu, K.: 3D convolutional neural networks for human action recognition. IEEE Trans. PAMI **35**, 221–231 (2013)
12. Kanan, C., Cottrell, G.: Robust classification of objects, faces, and flowers using natural image statistics. In: CVPR (2010)
13. Klaser, A., Marsza lek, M., Schmid, C.: A spatio-temporal descriptor based on 3D-gradients. In: BMVC (2008)
14. Krizhevsky, A., Sutskever, I., Hinton, G.E.: Imagenet classification with deep convolutional neural networks. In: NIPS (2012)
15. Laptev, I., Marsza lek, M., Schmid, C., Rozenfeld, B.: Learning realistic human actions from movies. In: CVPR (2008)
16. Laptev, I.: On space-time interest points. Int. J. Comput. Vis. **64**, 107–123 (2005)

17. Lee, H., Battle, A., Raina, R., Ng, A.Y.: Efficient sparse coding algorithms. In: NIPS (2006)
18. Lee, H., Grosse, R., Ranganath, R., Ng, A.Y.: Convolutional deep belief networks for scalable unsupervised learning of hierarchical representations. In: ICML (2009)
19. Le, Q.V., Karpenko, A., Ngiam, J., Ng, A.Y.: ICA with reconstruction cost for efficient overcomplete feature learning. In: NIPS (2011)
20. Le, Q.V., Zou, W.Y., Yeung, S.Y., Ng, A.Y.: Learning hierarchical invariant spatio-temporal features for action recognition with independent subspace analysis. In: CVPR (2011)
21. Li, H., Tang, J., Wu, S., Zhang, Y., Lin, S.: Automatic detection and analysis of player action in moving background sports video sequences. IEEE Trans. Circ. Syst. Video Technol. **20**, 351–364 (2010)
22. Liu, D.C., Nocedal, J.: On the limited memory BFGS method for large scale optimization. Math. Program. **45**, 503–528 (1989)
23. Liu, J., Luo, J., Shah, M.: Recognizing realistic actions from videos in the wild. In: CVPR (2009)
24. Marszalek, M., Laptev, I., Schmid, C.: Actions in context. In: CVPR (2009)
25. Mikolajczyk, K., Schmid, C.: An affine invariant interest point detector. In: Heyden, A., Sparr, G., Nielsen, M., Johansen, P. (eds.) ECCV 2002. LNCS, vol. 2350, pp. 128–142. Springer, Heidelberg (2002). doi:10.1007/3-540-47969-4_9
26. Ngiam, J., Chen, Z., Chia, D., Koh, P.W., Le, Q.V., Ng, A.Y.: Tiled convolutional neural networks. In: NIPS (2010)
27. Nowak, E., Jurie, F., Triggs, B.: Sampling strategies for bag-of-features image classification. In: Leonardis, A., Bischof, H., Pinz, A. (eds.) ECCV 2006. LNCS, vol. 3954, pp. 490–503. Springer, Heidelberg (2006). doi:10.1007/11744085_38
28. Olshausen, B.A., Field, D.J.: Sparse coding with an overcomplete basis set: a strategy employed by V1? Vis. Res. **37**, 3311–3325 (1997)
29. Rodriguez, M.D., Ahmed, J., Shah, M.: Action mach a spatio-temporal maximum average correlation height filter for action recognition. In: CVPR (2008)
30. Sánchez, J., Perronnin, F., Mensink, T., Verbeek, J.: Image classification with the fisher vector: theory and practice. Int. J. Comput. Vis. **105**, 222–245 (2013)
31. Schmidt, M.: Minfunc. Technical report (2005)
32. Schuldt, C., Laptev, I., Caputo, B.: Recognizing human actions: a local SVM approach. In: ICPR (2004)
33. Scovanner, P., Ali, S., Shah, M.: A 3-dimensional sift descriptor and its application to action recognition. In: ICME (2007)
34. Sun, L., Jia, K., Chan, T.H., Fang, Y., Wang, G., Yan, S.: DL-SFA: deeply-learned slow feature analysis for action recognition. In: CVPR (2014)
35. Taylor, G.W., Fergus, R., LeCun, Y., Bregler, C.: Convolutional learning of spatio-temporal features. In: Daniilidis, K., Maragos, P., Paragios, N. (eds.) ECCV 2010. LNCS, vol. 6316, pp. 140–153. Springer, Heidelberg (2010). doi:10.1007/978-3-642-15567-3_11
36. van Hateren, J.H., Ruderman, D.L.: Independent component analysis of natural image sequences yields spatio-temporal filters similar to simple cells in primary visual cortex. Proc. R. Soc. London B: Biol. Sci. **265**, 2315–2320 (1998)
37. Wang, H., Klaser, A., Schmid, C., Liu, C.L.: Action recognition by dense trajectories. In: CVPR (2011)
38. Wang, H., Ullah, M.M., Klaser, A., Laptev, I., Schmid, C.: Evaluation of local spatio-temporal features for action recognition. In: BMVC (2009)
39. Wang, Z., Feng, J., Yan, S., Xi, H.: Linear distance coding for image classification. IEEE Trans. Image Process. **22**, 537–548 (2013)

40. Wang, H., Oneata, D., Verbeek, J., Schmid, C.: A robust and efficient video representation for action recognition. Int. J. Comput. Vis. **119**, 1–20 (2015)
41. Willems, G., Tuytelaars, T., Gool, L.: An efficient dense and scale-invariant spatio-temporal interest point detector. In: Forsyth, D., Torr, P., Zisserman, A. (eds.) ECCV 2008. LNCS, vol. 5303, pp. 650–663. Springer, Heidelberg (2008). doi:10.1007/978-3-540-88688-4_48
42. Wiskott, L., Berkes, P., Franzius, M., Sprekeler, H., Wilbert, N.: Slow feature analysis. Scholarpedia **6**, 52–82 (2011)

Searching Action Proposals via Spatial Actionness Estimation and Temporal Path Inference and Tracking

Nannan Li[1], Dan Xu[2], Zhenqiang Ying[1], Zhihao Li[1], and Ge Li[1(✉)]

[1] Peking University Shenzhen Graduate School, Shenzhen,
People's Republic of China
gli@pkusz.edu.cn
[2] DISI, University of Trento, Trento, Italy

Abstract. In this paper, we address the problem of searching action proposals in unconstrained video clips. Our approach starts from actionness estimation on frame-level bounding boxes, and then aggregates the bounding boxes belonging to the same actor across frames via linking, associating, tracking to generate spatial-temporal continuous action paths. To achieve the target, a novel actionness estimation method is firstly proposed by utilizing both human appearance and motion cues. Then, the association of the action paths is formulated as a maximum set coverage problem with the results of actionness estimation as a priori. To further promote the performance, we design an improved optimization objective for the problem and provide a greedy search algorithm to solve it. Finally, a tracking-by-detection scheme is designed to further refine the searched action paths. Extensive experiments on two challenging datasets, UCF-Sports and UCF-101, show that the proposed approach advances state-of-the-art proposal generation performance in terms of both accuracy and proposal quantity.

1 Introduction

Video action analysis is an important research topic for human activity understanding, and has gained a wide attention in recent years. A common task of video action analysis is action recognition, which aims to identify which type of action is occurring in a video volume [1–3]. Compared to action recognition, action detection is a more difficult task, as it requires not only determining the action class, but also localizing the action in the video. Similar to the object detection task, in which reliable object proposals play a crucial role in the detection performance [4], action proposal is also a fundamental problem in action detection.

This paper focuses on generating high-quality action proposals in both spatial compactness and temporal continuity. Existing works in the literature have made different efforts to address the problem, including segmentation-and-merging strategy [5–8], dense motion features [9–11], human-centric models [12,13], and object proposals based approaches [14,15]. Despite promising results achieved

© Springer International Publishing AG 2017
S.-H. Lai et al. (Eds.): ACCV 2016, Part II, LNCS 10112, pp. 384–399, 2017.
DOI: 10.1007/978-3-319-54184-6_24

in these works, video action proposal generation is still a challenging problem due to the complex spatio-temporal relationship modeling involved in the task. The problem can be considered as a task with two essential steps, namely spatial (*i.e.* frame-level) actionness estimation and temporal (*i.e.* video-level) action path generation. For one aspect, because of the large diversity and variation of human actions, it is difficult to generate robust frame-level actionness proposals which contain meaningful action motion patterns and are clearly discriminative from the background in unconstrained videos. For the other aspect, as a fact that the whole number of potential actionness regions on each frame usually has an exponential growth of the video duration [16], it is extremely impractical to calculate on all possible connections of the regions for generating the action paths and guaranteeing each of them associated with the same actor(s).

To tackle the above mentioned issues in the generation of action proposals, we propose a novel framework based on spatial actionness estimation from multiple cues and temporal action path extraction from a fast inference and tracking. Firstly, unlike previous works using selective search [4] or edge boxes [17] for generating object proposals for actionness estimation, we employ more action related cues including both human and motion. Secondly, a deep Faster-RCNN [18] network is trained and fine-tuned on augmented action detection datasets for obtaining accurate human proposals. Then action motion patterns with Gaussian Mixture Models are modeled for motion estimation of each human proposal. Both human and motion estimations are feed into a proposed forward and backward search algorithm for video-level action path generation. Finally, we use a tracking-by-detection approach to refine the action path by supplement actionness proposals missing in frames.

The key contributions of this paper include three folds: (i) we construct an action detector at frame-level by taking both appearance and motion clues into account, which can handle the problem of detecting human with uncommon poses and discriminate actionness proposals containing meaningful motion patterns from the backgrounds; (ii) We formulate the action path generation as a maximum convergence problem [19]. We propose an improved optimization objective for the problem and provide a greedy search algorithm to solve it. (iii) Extensive experiments on UCF-Sports, UCF-101 datasets show that the proposed method achieves the state-of-the-art performance compared with other existing approaches.

2   Related Work

Traditionally, action localization or detection is performed by sliding window based approaches [20–23]. For instance, Siva and Xiang [20] proposed a supervised model based on multiple-instance-learning to slide over subvolumes both spatially and temporally for action detection. Instead of performing an exhaustive search through sliding over the whole video volumes, Oneata *et al.* [6] put forward a branch-and-bound search approach to achieve the time-efficiency. The main limitation of these sliding-window based approaches is that the detection results are confined by a video subvolume, and thus can not accurately capture the varying shape of the motion.

Some research works address the problem by employing segmentation-and-merging strategy. Generally, these methods include three steps: (i) segment the video; (ii) merge the segments to generate tube proposals; (iii) represent tubes with dense motion features and construct action classifier for recognition. For instance, in [8] action tubes are generated by hierarchically merging super-voxels. However, accurate video segmentation is a difficult problem especially under unconstrained environments. To alleviate the difficulty encountered with segmentation, some other methods use a figure-centric based model. In [12] the human and objects are detected first and then their interactions are described. Kläser et al. [13] detect human on each frame and track the detection results across frames using optical flow. Our approach also utilizes tracking, via a more robust tracking-by-detection approach [24,25] based on a combined feature representation of color and shape.

Recently, some methods built upon generation of action proposals are presented. Gkioxari and Malik [15] proposed to utilize Selective Search method for proposing actions on each frame, then scored those proposals by using features extracted by a two-streams Convolutional Neural Networks (CNN), and finally, linked them to formation tubes. Philippe et al. [9] adopted the same feature extraction procedure, then utilized a tracking-by-detection approach to link frame-level detections, in combination with a class-specific detector. Our method replaces object proposal method and two-stream CNN with the Faster R-CNN model for calculation efficiency. The most related work to ours is that presented in [14], in which actionness score is calculated for each action path and then a greedy search method is used to generate proposals. Our work differentiates from theirs in the following three aspects: (i) we train a Faster R-CNN model for human estimation, which has a stronger ability to differentiate human from backgrounds; (ii) compared with the optimization objective they proposed, our improved optimization objective simultaneously maximizing actionness score and member similarity in a path set, thus can effectively cluster the paths from the same actor into a group; (iii) we utilize a tracking-by-detection approach to supplement the missing detections.

3   The Proposed Approach

The proposed approach takes video clip as input and generates action proposal results. The framework of our approach is illustrated in Fig. 1. The main procedure consists of two stages: spatial actionness estimation and temporal action path extraction. Firstly, bounding boxes at frame-level that may contain meaningful motion are extracted by simultaneously considering appearance and motion cues; then action paths corresponding to the same actor at video-level are generated and linked to obtain action proposals. The details of our method will be elaborated in the following sections.

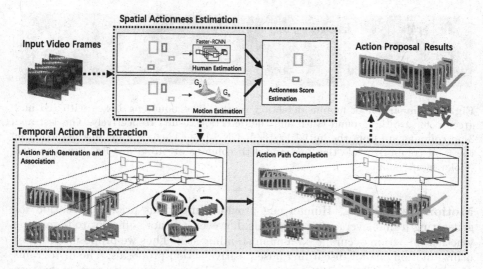

Fig. 1. The framework of the proposed action proposal generation approach.

3.1 Spatial Actionness Estimation

Human Estimation. Detection of Human proposal is an important and heuristic step for action localization. We implement the human proposal detection employing the Faster R-CNN [18] pipeline with a VGG-16 model [26] pre-trained on ILSVRC dataset [27]. Faster R-CNN introduces a Region Proposal Network (RPN) that simultaneously predicts object bounding boxes and their corresponding objectness scores in near real-time speed. As the human detection task is a binary-classification problem, the output of the classification layer of Faster-RCNN network is revised to a two-way softmax classifier: one for the 'human' class and the other for the 'background' class. For action classes such as diving and swing, the appearance (especially for the shape and the pose) of the human changes significantly among the whole action duration. Therefore, the detection network fine-tuned on the standard PASCAL VOC 2007 dataset is unable to effectively detect the human under those circumstance. To handle the problem, we perform a data augmentation by merging the training data of the human class of PASCAL VOC 2007 and 2012, and rotating each training sample with seven different angles from $\frac{\pi}{4}$ to $\frac{7\pi}{4}$ with an interval of $\frac{\pi}{4}$. Let b_t^i denotes the bounding box for the i-th human proposal at t-th frame. The bounding box is represented as $[x, y, w, h]$, where w and h stand for width and height respectively, and (x, y) is the center. After training, for each bounding box b_*^* in the test video, a probability $S_h(b_*^*)$ can be estimated by the CNN network. By setting a probability threshold, human proposals with higher probability are kept for follow-up processing. A comparison of human detection results between original Faster R-CNN model and our refined one is showed in Fig. 2, from which it can be clearly observed that detection results from refined model are more precise and compact.

Fig. 2. Comparison of human detection results. The bounding boxes with red and green color are the ground truth and the detection results respectively. The 1-st and 3-nd columns are from the initial Faster R-CNN [18] (There is a missing detection in the 3-nd column); while the 2-nd and 4-th columns are from our fine-tuned model. (Color figure online)

Motion Estimation. Human cue provides important prior information for generating frame-level action proposals, however it is not sufficient to determine whether an action occurs, *e.g.*, human standing still. Thus we propose to further utilize motion cue for discarding false positive action proposals. The histograms of optical flow (HOF) [10] descriptor is used to describe the motion pattern of each human proposal. We construct two Gaussian Mixture Models (GMMs) $G_p(.)$ and $G_n(.)$ upon the HOFs, which represent the positive and negative proposal class respectively, to predict the probability of a motion pattern belonging to the actions or the background. HOFs calculated within bounding boxes of an Intersection-over-Union (IoU) overlapping with ground truth more than 0.5 are used as positive samples, while those with IoU overlapping less than 0.1 as negative samples. Given a test proposal b_t^i and its HOF h_i, we define the likelihood of b_t^i being a motion score using the predictions from two mixture of Gaussian models as:

$$S_m(b_t^i) = \sigma(G_p(h_i)/G_n(h_i)),\tag{1}$$

where $\sigma = 1/(1 + e^{-x})$ maps likelihood into the range $[0, 1]$. To reduce the influence induced by camera movement to optical flow calculation, we adopt the approach presented by [2] to estimate camera motion and subtract it to obtain robust optical flow.

Actionness Score Calculation. The actionness score of a bounding box b_t^i consists of two parts: human detection score and motion score, and is defined as follows:

$$S(b_t^i) = S_h(b_t^i) + \lambda_p * S_m(b_t^i),\tag{2}$$

where λ_p is the parameter that balances the human estimation and motion estimation score.

3.2 Temporal Action Path Extraction

Problem Formulation. Given action proposals on each frame, our goal is to find a set of action paths $\mathbf{P} = \{p_1, p_2, \ldots, p_i\}$, where $p_i = \{b_s^i, b_{s+1}^i, \ldots, b_e^i\}$ corresponds to a path that starts from s-th frame and ends at e-th frame. Yu and Yuan [14] formulate finding action path set $\mathbf{P}$ as a maximum set coverage problem (MSCP) and propose an optimization objective maximizing actionness score. Inspired by their work, we formulate it as a MSCP with an improved optimization objective, which simultaneously maximizes actionness score and similarity among members within the path set $\mathbf{P}$. Formally, our optimization objective can be presented as follows:

$$\max_{\mathbf{P} \in \Phi} \sum_{b_t \in \bigcup p_i} S(b_t) + \sum_{i,j} W(p_i, p_j)$$
$$s.t. \quad |\mathbf{P}| \leq N$$
$$\mathbf{O}(p_i, p_j) \leq \eta_p, \forall p_i, p_j \in \mathbf{P}, i \neq j, \tag{3}$$

where $W(p_i, p_j)$ represents the similarity between action path p_i and p_j, and its definition will be explained in subsection action-path-association; $S(b_t)$ is the actionness score of bounding box b_t (cf. Eq. 2); Φ is action-path-candidate set; η_p is a threshold. The first constraint in Eq. 3 sets the maximum number of paths contained in $\mathbf{P}$; while the second constraint facilitates $\mathbf{P}$ to avoid generating redundant action paths that are overlapped. The overlapping of two paths is evaluated by $\mathbf{O}(p_i, p_j)$, which is defined as follows:

$$\mathbf{O}(p_i, p_j) = \frac{1}{\max(t_e^i, t_e^j) - \min(t_s^i, t_s^j)} \bullet \sum_{\max(t_s^i, t_s^j) \leq t \leq \min(t_e^i, t_e^j)} o(b_t^i, b_t^j). \tag{4}$$

In Eq. 4, $o(b_t^i, b_t^j)$ is defined as $\frac{\cap(b_t^i, b_t^j)}{\cup(b_t^i, b_t^j)}$, representing for IoU of two bounding boxs b_t^i and b_t^j.

Action Path Generation. To solve the MSCP in Eq. 3, the action-path-candidate set Φ needs to be obtained first. We wish that Φ consists of spatio-temporal smooth path p_i whose consecutive elements b_t^i, b_{t+1}^i should satisfy the following two requirements:

$$o(b_t^i, b_{t+1}^i) \geq \eta_o$$
$$\|C(b_t^i) - C(b_{t+1}^i)\| + \lambda_a \|H(b_t^i) - H(b_{t+1}^i)\| \leq \eta_f, \tag{5}$$

where $o(b_t^i, b_{t+1}^i)$ represents IoU, as defined in Eq. 4; $C(b_t^i)$ and $H(b_t^i)$ stand for histograms of color (HOC) and histograms of gradient (HOG) of b_t^i, and λ_a is a trade-off balancing the weight of the two terms; η_o and η_f are thresholds. The first requirement in Eq. 5 ensures that consecutive bounding box b_t^i and b_{t+1}^i are spatially continuous; the second requirement ensures that b_t^i and b_{t+1}^i have similar appearance, thus the path p_i may follow the same actor.

To obtain Φ, we adopt the method proposed by [14] with minor modification to avoid generating much highly-overlap paths. The algorithm includes two stages: forward search and backward track. The aim of the former is to locate the end of a path; while that of the latter seeks to recover the whole path. The central idea is to maintain an updating pool of best Top-N path candidates, which

---

Algorithm 1.   Forward Search and Backward Track

Input: bounding box score $S(b_t^i)$
Output: action path $p_k, k = 1, 2, \ldots, N$
1: $\tau_k = 0, b^k = \emptyset, k = 1, 2, \ldots, N$
2: **for** $t = 1 \to T$ **do**
3:    **for** $i = 1 \to N_t^b$ **do**
4:      $\tau(b_t^i) = \max_{b_{t-1}^j} \tau(b_{t-1}^j) + S(b_t^i)$
5:    **end for**
6:    step1: update each candidate (τ_k, b^k) with b_t^i that connects with b^k and has the largest score $\tau(b_t^i)$
7:    step2: update (τ_N, b^N) as $(\tau(b_t^i,), b_t^i)$, if $\tau(b_t^i) > \tau_N$
8: **end for**
9: backward trace to locate $b_t^k, t = t_s, t_{s+1}, \ldots, t_e$ in p_k

---

is represented as $\Phi = (\tau_k, b^k), k = 1, 2, \ldots, N$, where τ_k is the score of path k and obtained by accumulating $S(b_t^k)$ of b_t^k it passes by; b^k is the bounding box of the end of k-th path. In the forward search, it also records an accumulated actionness score of each $b_t^i : \tau(b_t^i) = \max_{b_{t-1}^j} \tau(b_{t-1}^j) + S(b_t^i)$, where b_{t-1}^j and b_t^i satisfy the two requirements in Eq. 5. Given b_t^i at frame t, we update path candidate pool according to the following two steps: first, for each candidate $(\tau_k, b^k), k = 1, 2, \ldots, N$, if there exists any b_t^i connecting to b^k, then b^k will be replaced by b_t^i that has the largest $\tau(b_t^i)$; second, if the accumulated score of b_t^i is larger than the score of N-th proposal, i.e. $\tau(b_t^i) > \tau_N$, then (τ_N, b^N) is updated as $(\tau(b_t^i), b_t^i)$. After the forward search, a backward trace is performed to recover each b_t^k on the candidate path (τ_k, b^k). More specifically, for path p_k, we obtain $\{b_t^k : t = t_s, t_{s+1}, \ldots, t_e\}$ by solving the equation: $\tau_k = \sum_{t_s \leq t \leq t_e} S(b_t^k)$.

The pseudo-code of forward-backward search is illustrated in Algorithm 1. It takes bounding box score $S(b_t^i)$ as input data and outputs action paths $p_k, k = 1, 2, \ldots, N$. The lines 1 to 8 describe forward search and line 9 corresponds to backward trace. In line 3, N_t^b denotes the number of bounding box on frame t.

Action Path Association. Once obtaining Φ, the MSCP in Eq. 3 can be solved. According to [19], the maximum set coverage problem is NP-hard but a greedy-search algorithm can achieve an approximation ratio of $1 - 1/e$. Here, we present a greedy-search solution to address the problem. In the beginning, we search for the candidate p_i with the largest action score τ_k in Φ, then add it into path set **P**. Supposing that **P** has contained k action paths, we enumerate the rest paths in Φ and find the one that maximizes the flowing equation as the $k + 1$-path p_i:

$$\arg\max_i \sum_{b \in p_i \cup p_1 \cup \ldots \cup p_k} S(b) + \sigma(1/k \cdot \sum_{j=1,2,\ldots,k} W(p_i, p_j)). \tag{6}$$

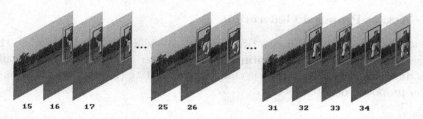

Fig. 3. Examples of tracking-by-detection results. The bounding boxes with green and red color are the groundtruth and the detected frame-level human bounding boxes respectively, and those with blue color are obtained by our tracking-by-detection strategy. All the missing human targets before frame 34 are perfectly located by the tracking approach. (Color figure online)

In Eq. 6, $W(p_i, p_j)$ represents the similarity of action path p_i and p_j, and is defined as: $W(p_i, p_j) = 1/(\|C(p_i) - C(p_j)\| + \lambda_a \|H(p_i) - H(p_j)\|)$, where $C(p_*)$ and $H(p_*)$ represent the cluster centers of HOC and HOG of bounding boxes from path p_* respectively. The larger value of $W(p_i, p_j)$, the more likely that the paths p_i and p_j follow the same actor. To reduce redundant paths in set $\mathbf{P}$, the newly added path p_i should satisfy the constraint in Eq. 4.

Action Path Completion. As human detection may miss hitting in some frames, the track obtained by connecting the paths in $\mathbf{P}$ will have temporal gaps. To get a temporal-spatial continuous track of an actor, we fill the gaps by using tracking-by-detection approaches [9]. We train a linear SVM as frame-level detector. The initial set of positives consist of bounding boxes in set $\mathbf{P}$, while negatives compose of bounding boxes excluded from set $\mathbf{P}$ and boxes that are randomly selected around positives with the IoU less than 0.3. Given the detection region b_t on frame t, we intend to find the most likely location on frame $t + 1$ where the human detection is missed. Firstly, we map b_t to b'_{t+1} with the shift of the median of optical-flow inside region b_t; secondly, construct a search region $\overline{b'_{t+1}}$ by extending the height and width of b'_{t+1} to half past one times of original length; thirdly, scan $\overline{b'_{t+1}}$ with a set of windows whose ratio between width and length varies in a range $[0.8, 1.2]$ to adapt possible size change of an actor. The best region b_{t+1} is selected as the one that maximizes the following equation:

$$b_{t+1} = \arg \max_{\gamma \in N(\overline{b'_{t+1}})} S_f(\gamma), \qquad (7)$$

where $N(\overline{b'_{t+1}})$ represents the window set produced by scanning $\overline{b'_{t+1}}$ and $S_f(\cdot)$ is the SVM detector whose input feature is chosen as the combination of HOC and HOG. After obtaining b_{t+1}, we update the SVM detector by adding b_{t+1} as a positive sample and boxes around b_{t+1} with the IoU less than 0.3 as negatives. An example of how the tracking approach supplementing missing detections is illustrated in Fig. 3.

3.3   Action Proposal Generation

The spatio-temporal continuous track can be considered as an action tube that focuses on an actor from appearing until disappearing. For each action tube, if its duration is larger than a specified threshold (*e.g.* 20), we regard it as an action proposal, denoted as $\mathscr{T}$.

4   Experiment

In this section, we describe the details of the experimental evaluation of the proposed approach, including datasets and evaluation metrics, implementation details, an analysis of the proposed approach and the overall performance comparison with state of the art methods.

4.1   Datasets and Evaluation Metric

We evaluate the performance of the proposed action proposal approach on two publicly available action-detection datasets: UCF-Sports [28] and UCF-101 [29].

UCF-Sports. UCF-Sports dataset consists of 150 short videos of sports collected from 10 action classes. It has been widely used for action localization. The videos are truncated to contain a single action and bounding box annotation is provided for each frame.

UCF-101. UCF-101 dataset has more than 13000 videos that belong to 101 classes. In a subset of 24 categories, human actions are annotated both spatially and temporally. Compared with UCF-Sports, only a part of videos (74.6%) are trimmed to fit the motion.

Evaluation Metric. To evaluate the quality of the action proposal $\mathscr{T}$, we follow the metric proposed by [11]. More specifically, the estimation is based on the mean IoU value between action proposal $\mathscr{T}$ and ground truth $\mathbf{G}$, which is defined as: $IoU(\mathbf{G}, \mathscr{T}) = \frac{1}{|\mathbf{C}|} \sum_{t \in \mathbf{C}} o(\mathbf{G}_t, \mathscr{T}_t)$, where $\mathbf{G}_t$ and $\mathscr{T}_t$ are the detection bounding box and ground truth on t-th frame respectively; $o(.,.)$ is the IoU value that is defined in Eq. 4; $|\mathbf{C}|$ is the set of frames where either the detection result or the ground truth is not null. An action proposal is considered as true-positive if $IoU(\mathbf{G}, \mathscr{T}) \geq \eta$, where η is a specified threshold. In the following passage, η is set as 0.5 if not specified.

4.2   Implementation Details

The human estimation is implemented under the Caffe platform [30] and based on the Faster R-CNN pipeline with a VGG16 model for parameter initialization as described in Sect. 3.1. We use a four-step alternating training strategy [18] to optimize two pipelines (*i.e.* RPN and Fast RCNN) of the whole network. For training the RPN pipeline, the same settings of scales and aspect ratios are used

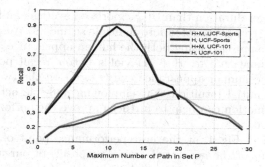

Fig. 4. Recall vs. maximum number of path in set **P** under different test settings.

as in [18]. For training the Fast RCNN pipeline, the mini-batch size is set to 128, and the ratio of positive to negative samples is set to 1:4. The network is trained with Stochastic Gradient Descent (SGD) with an initial learning rate of 0.001 and drop by 10 times at every the 5-th epochs, and the momentum and weight decay are set as 0.9 and 0.0005 respectively.

For the motion estimation of the bounding boxes, the number of components of GMMs is set to the same as the number of action categories. For constructing GMMs, we use randomly selected 1/3 of the video clips in UCF-101 for training and test on UCF-Sports dataset. While test on UCF-101 dataset, all the video clips of UCF-Sports are used for training. This setting is for a fair comparison with existing non-learning based methods which test on the whole dataset. The number of action paths in a candidate set Φ is set to 50 for UCF-Sports and 100 for UCF-101, as the latter one has a longer duration of action videos on average, and hence may contains more action-path segments. The value of N in Eq. 3 (*i.e.* the maximum number of paths in set **P**) is set as 12 for UCF-Sports and 18 for UCF-101. For each video clip, we propose at least one path set **P**, while a path set **P** is generated, the paths $\{p_i, i = 1, 2, \ldots, N\}$ in **P** are removed from the candidate set Φ and the greedy search algorithm is performed again to find a new path set $\mathbf{P}'$ until the duration of the longest path p_i' ($p_i' \in \mathbf{P}'$) is less than 10.

4.3   Analysis of the Proposed Approach

We analyze the performance of the proposed approach from different aspects, including the sensitivity of the parameter N (*i.e.*, the maximum number of path in set **P**), the influence of actionness estimation based on human appearance and motion cues, and the number of generated proposals.

Figure 4 shows the recall performance of our approach using different actionness estimation schemes by varying the value of N. From Fig. 4, it can be observed that the proposed approach achieves the best performance when the value of N is in the range [9, 14] on UCF-Sports, and the performance degrades significantly when N is far from this range. The optimal value of N for UCF-101 is larger than that for UCF-Sports. The reason is probably that the video clips of

UCF-101 have longer duration than UCF-Sports on average, and thus the action path is more likely to be separated into multiple segments. We can also observe that the proposed approach using both the human appearance and motion cues for actionness estimation (*i.e.* H + M) yields better recall performance than that using only the human appearance cue (*i.e.* H) on the two datasets. This demonstrates our initial intuition that employing multiple action-related cues for actionness estimation can help to further improve the performance of action proposal generation.

As also shown in Fig. 4, at the best performance point of recall on UCF-Sports, the number of the generated action proposals of our approach is only 13, and it is significantly less compared with state-of-the-art methods on the same recall performance level (see Table 1). The notable improvement is mainly due to the precise human estimation from our fine-tuned Faster-RCNN model, and the modified forward-backward search algorithm for generating candidate set Φ. Compared to [14], the improved optimization objective leverages appearance similarity among paths for effectively separating different actors. Figure 5 illustrates the improvement by our approach. It can be observed that the action path generated by the proposed approach is correctly associated to the same actor. More examples of the action-proposal generation results on UCF-Sports and UCF-101 are shown in Figs. 8 and 9, respectively.

Fig. 5. Examples of action-path generation results. The 1-st row shows the results obtained from [14] (The action path contains an irrelevant actor within the first few frames); the 2-nd row is our results, where the main actor is correctly tracked.

The runtime of the proposed approach includes three parts: (i) spatial actionness estimation: Faster RCNN for human estimation takes around 0.1 seconds per frame (s/f) and GMM-HOF for motion estimation takes around 1 s/f; (ii) temporal action path extraction: the average runtime of this step is 0.09 s/f; and (iii) action path completion takes 0.5 s/f. In summary, the average runtime of the approach is 1.69 s/f. We conduct the runtime analysis on the UCF-Sports dataset with an image resolution of 720 × 404, and based on hardware configurations of an Nvidia Tesla-K80 GPU, 3.4 GHz CPU and 4 GB memory.

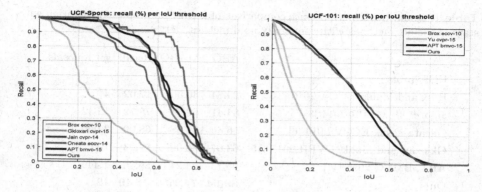

Fig. 6. Recall vs. IoUs on UCF-Sports and UCF-101 datasets.

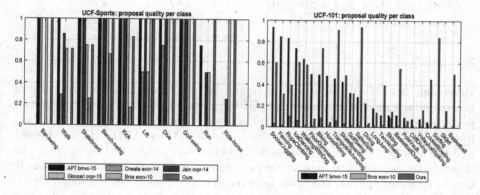

Fig. 7. Recall performance on each action category on UCF-Sports and UCF-101 datasets.

4.4   Overall Performance

We compare the performance of the action proposal generation of the proposed approach with state-of-the-art methods on UCF-Sports and UCF-101. We vary the value of IoU (η) in $[0, 1]$, and plot recall as a function of η. Figure 6 shows Recall vs. IoU curves of difference approaches. It is clear that our approach obtains a significant performance gap over the state-of-the-art methods on UCF-Sports (The recall of ours is above 0.7 when η is even at 0.7, while the others are below 0.4.), and achieves very competitive performance on UCF-101. Figure 7 also shows the recall performance for each action category. We can observe that our approach presents superior recall performance on almost all action categories except for few classes (*e.g.*, Walk and Kick) on UCF-Sports, and on UCF-101, ours greatly outperforms the comparison methods on action classes such as Biking, Surfing and TrampolingJumping.

From Fig. 7, it can be also noticed that the performance on UCF-101 is inferior to that on UCF-Sports. The reason is probably because the testing action

Table 1. Quantitative performance comparison of the action proposal generation with state-of-the-art methods with commonly used metrics.

	ABO	MABO	Recall	#Proposals
UCFSprots				
Brox and Malik, ECCV 2010 [31]	29.84	30.90	17.02	4
Jain *et al.*, CVPR 2014 [8]	63.41	62.71	78.72	1,642
Oneata *et al.*, ECCV 2014 [6]	56.49	55.58	68.09	3,000
Gkioxari and Malik, CVPR 2015 [15]	63.07	62.09	87.23	100
APT, BMVC 2015 [11]	65.73	64.21	89.36	1,449
Ours	**89.64**	**74.19**	**91.49**	12
UCF 101				
Brox and Malik, ECCV 2010 [31]	13.28	12.82	1.40	3
APT, BMVC 2015 [11]	40.77	39.97	35.45	2,299
Ours	**63.76**	**40.84**	**39.64**	18

video clips of UCF-101 have more dynamically and continuously varying size of actors and are more untrimmed than UCF-Sports. Finally, we report the overall performance on the two datasets in Table 1 using several commonly used metrics, including ABO (Average Best Overlap), MABO (Mean ABO over all classes) and the average number of proposals per video. The results further confirm the superior action proposal generation performance achieved by our approach. It should be noted that on the same level recall performance, our approach generates relatively much smaller number of action proposals for each video clip than

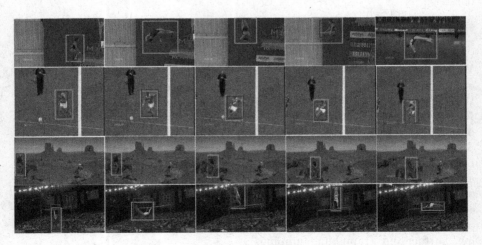

Fig. 8. Examples of action-proposal generation results on UCF-Sports. The bounding boxes with green and red color are the ground truth and the action proposal, respectively. (Color figure online)

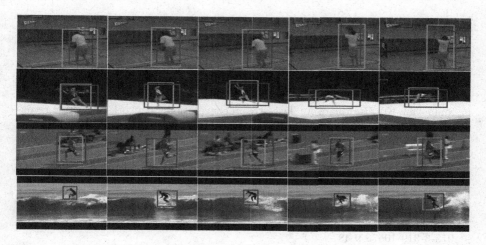

Fig. 9. Examples of action-proposal generation results on UCF-101. The bounding boxes with green and red color are the ground truth and the action proposal, respectively. (Color figure online)

the other methods, which is especially important for reducing the computational complexity for the follow-up applications such as action recognition and action interaction modeling.

5   Conclusions

A novel framework for action proposal generation in video has been presented in this paper. Given an unconstrained video clip as input, it generates spatial-temporal continuous action paths. The proposed approach is built upon action-ness estimation leveraging both human appearance and motion cues on frame-level bounding boxes, which are produced by a Faster-RCNN network trained on augmented datasets. Then we search spatial-temporal action paths via linking, associating, and tracking the bounding boxes across frames. We formulate the association of action paths belonging to the same actor as a maximum set coverage problem and propose a greedy search algorithm to solve it. Experiments on two challenging datasets demonstrate that our approach produces more accurate action proposals with remarkably less proposals compared with the state-of-the-art approaches. Based on the observation on the experimental results, the proposed approach is especially effective when the action video clip contains only one actor. In the future, we will explore using CNN networks for better actionness estimation of video frames, and consider recurrent neural networks for modeling the action paths for action recognition.

Acknowledgement. The work was partially supported by Shenzhen Peacock Plan (20130408-183003656), Science and Technology Planning Project of Guangdong Province, China (No. 2014B090910001) and China 863 project of 2015AA01 5905.

References

1. Wang, H., Kläser, A., Schmid, C., Liu, C.L.: Action recognition by dense trajectories. In: CVPR (2011)
2. Wang, H., Schmid, C.: Action recognition with improved trajectories. In: ICCV (2013)
3. Simonyan, K., Zisserman, A.: Two-stream convolutional networks for action recognition in videos. In: NIPS (2014)
4. Uijlings, J.R., van de Sande, K.E., Gevers, T., Smeulders, A.W.: Selective search for object recognition. Int. J. Comput. Vis. (IJCV) **104**, 154–171 (2013)
5. Ma, S., Zhang, J., Ikizler-Cinbis, N., Sclaroff, S.: Action recognition and localization by hierarchical space-time segments. In: ICCV, pp. 2744–2751 (2013)
6. Oneata, D., Revaud, J., Verbeek, J., Schmid, C.: Spatio-temporal object detection proposals. In: Fleet, D., Pajdla, T., Schiele, B., Tuytelaars, T. (eds.) ECCV 2014. LNCS, vol. 8691, pp. 737–752. Springer, Heidelberg (2014). doi:10.1007/978-3-319-10578-9_48
7. Bergh, M., Roig, G., Boix, X., Manen, S., Gool, L.: Online video seeds for temporal window objectness. In: ICCV (2013)
8. Jain, M., Gemert, J., Jégou, H., Bouthemy, P., Snoek, C.: Action localization with tubelets from motion. In: CVPR (2014)
9. Weinzaepfel, P., Harchaoui, Z., Schmid, C.: Learning to track for spatio-temporal action localization. In: ICCV (2015)
10. Wang, H., Kläser, A., Schmid, C., Liu, C.L.: Dense trajectories and motion boundary descriptors for action recognition. Int. J. Comput. Vis. (IJCV) **103**, 60–79 (2013)
11. Van Gemert, J.C., Jain, M., Gati, E., Snoek, C.G.: APT: action localization proposals from dense trajectories. In: BMVC (2015)
12. Prest, A., Ferrari, V., Schmid, C.: Explicit modeling of human-object interactions in realistic videos. IEEE Trans. Pattern Anal. Mach. Intell. (PAMI) **35**, 835–848 (2013)
13. Kläser, A., Marszałek, M., Schmid, C., Zisserman, A.: Human focused action localization in video. In: Kutulakos, K.N. (ed.) ECCV 2010. LNCS, vol. 6553, pp. 219–233. Springer, Heidelberg (2012). doi:10.1007/978-3-642-35749-7_17
14. Yu, G., Yuan, J.: Fast action proposals for human action detection and search. In: CVPR (2015)
15. Gkioxari, G., Malik, J.: Finding action tubes. In: CVPR (2015)
16. Tran, D., Yuan, J., Forsyth, D.: Video event detection: from subvolume localization to spatiotemporal path search. IEEE Trans. Pattern Anal. Mach. Intell. (PAMI) **36**, 404–416 (2014)
17. Zitnick, C.L., Dollár, P.: Edge boxes: locating object proposals from edges. In: Fleet, D., Pajdla, T., Schiele, B., Tuytelaars, T. (eds.) ECCV 2014. LNCS, vol. 8693, pp. 391–405. Springer, Heidelberg (2014). doi:10.1007/978-3-319-10602-1_26
18. Ren, S., He, K., Girshick, R., Sun, J.: Faster R-CNN: towards real-time object detection with region proposal networks. In: NIPS (2015)
19. Nemhauser, G.L., Wolsey, L.A., Fisher, M.L.: An analysis of approximations for maximizing submodular set functions–I. Math. Program. **14**, 265–294 (1978)
20. Siva, P., Xiang, T.: Weakly supervised action detection. In: BMVC, vol. 2, p. 6 (2011)
21. Laptev, I., Pérez, P.: Retrieving actions in movies. In: ICCV (2007)

22. Gaidon, A., Harchaoui, Z., Schmid, C.: Temporal localization of actions with actoms. IEEE Trans. Pattern Anal. Mach. Intell. (PAMI) **35**, 2782–2795 (2013)
23. Wang, H., Oneata, D., Verbeek, J., Schmid, C.: A robust and efficient video representation for action recognition. Int. J. Comput. Vis. (IJCV) **119**, 1–20 (2015)
24. Hare, S., Saffari, A., Torr, P.H.: Struck: structured output tracking with kernels. In: ICCV (2011)
25. Kalal, Z., Mikolajczyk, K., Matas, J.: Tracking-learning-detection. IEEE Trans. Pattern Anal. Mach. Intell. (PAMI) **34**, 1409–1422 (2012)
26. Simonyan, K., Zisserman, A.: Very deep convolutional networks for large-scale image recognition (2014). arXiv preprint: arXiv:1409.1556
27. Russakovsky, O., Deng, J., Su, H., Krause, J., Satheesh, S., Ma, S., Huang, Z., Karpathy, A., Khosla, A., Bernstein, M., et al.: ImageNet large scale visual recognition challenge. Int. J. Comput. Vis. (IJCV) **115**, 211–252 (2015)
28. Rodriguez, M.D., Ahmed, J., Shah, M.: Action mach a spatio-temporal maximum average correlation height filter for action recognition. In: CVPR (2008)
29. Soomro, K., Zamir, A.R., Shah, M.: UCF101: a dataset of 101 human actions classes from videos in the wild. In: CoRR (2012)
30. Jia, Y., Shelhamer, E., Donahue, J., Karayev, S., Long, J., Girshick, R., Guadarrama, S., Darrell, T.: Caffe: convolutional architecture for fast feature embedding (2014). arXiv preprint: arXiv:1408.5093
31. Brox, T., Malik, J.: Object segmentation by long term analysis of point trajectories. In: ECCV (2015)

Markov Chain Monte Carlo Cascade for Camera Network Calibration Based on Unconstrained Pedestrian Tracklets

Louis Lettry[1]([⊠]), Ralf Dragon[1], and Luc Van Gool[1,2]

[1] CVL, ETH Zurich, Zürich, Switzerland
lettryl@vision.ee.ethz.ch
[2] VISICS, KU Leuven, Leuven, Belgium

Abstract. The presented work aims at tackling the problem of externally calibrating a network of cameras by observing a dynamic scene composed of pedestrians. It relies on the single assumption that human beings walk aligned with the gravity vector. Usual techniques to solve this problem involve using more assumptions such as a planar ground or assumptions about pedestrians' motion. In this work, we drop all these assumptions and design a probabilistic layered algorithm that deals with noisy outlier-dominated hypotheses to recover the actual structure of the network. We demonstrate our process on two known public datasets and exhibit results to underline the effectiveness of our simple but adaptable approach to this general problem.

1 Introduction

External calibration of a camera network is a process preceding many other tasks in computer vision such as scene reconstruction or human gesture analysis. Manual techniques are inconvenient as they require moving a calibration pattern in the common field of view or manually annotating corresponding points in multiple images. Automatic techniques exist to circumvent this manual part by automatically detecting corresponding keypoints or regions. However, when the baseline grows or the scene becomes dynamic, precision and recall of all kinds of such correspondences drop dramatically.

The use of dynamic objects finds its use in many different fields of work. Surveillance applications usually cannot rely on background correspondences of the scene as people are moving in front, or because of the lack of stable detectable keypoints. A variety of work has already been done in this area, relying on assumptions about different aspects of human beings such as average height, motion or a planar ground.

In this work, we aim at demonstrating the possibility to solve this problem by using the smallest assumption of human walking: Human beings stand against a common gravity vector while walking. Following the paradigm that *intra-camera* correspondences (between multiple time steps) are more reliable than *inter-camera* ones (e.g. pedestrian correspondences), we fit a plane through an

© Springer International Publishing AG 2017
S.-H. Lai et al. (Eds.): ACCV 2016, Part II, LNCS 10112, pp. 400–415, 2017.
DOI: 10.1007/978-3-319-54184-6_25

Fig. 1. Views from two cameras C^a and C^b at times t_1 and t_2 (stacked with transparency). In each camera view, a walking person defines a plane (marked in red) between his head and foot points in two time steps. Planes between multiple cameras are related by a homography H, which is used during our external camera calibration approach. Green arrows show gravity vector examples. (Color figure online)

estimated gravity vector of a tracked person at two time instances (Fig. 1). Using a probabilistic sampling framework, we establish plane correspondences between multiple camera pairs at different time steps and intervals in order to estimate the relative pose between two cameras. On a coarser level, all pairwise pose estimations will be fused into a geometrically consistent network.

We believe the contributions of our work to be multifold:

- A layered Markov chain Monte Carlo algorithm for camera calibration to work with noisy outlier-dominated data.
- Usage of foot-head planes for external camera calibration.
- Extension to the Shortest Triangle Paths Algorithm to incorporate stability constraints.

We present the current state of the art in this field in the Sect. 2. Section 3 presents the underlying mathematical model used for relative pose estimation based on homography extraction and decomposition over four points on a plane. It is followed by the in-depth presentation of our algorithm in Sect. 4. Afterwards will come the results presentation in Sect. 5 before concluding in Sect. 6.

2   Related Work

Camera calibration is extensively studied as knowledge about the scene geometry simplifies many computer vision tasks, e.g. multi-view object tracking or searching correspondences along the epipolar line for depth estimation. If we assume we have correspondences between image points $\bar{x}_i^c$ in different cameras

C^j, we can use the regular structure-and-motion approach: combine all $\bar{x}_i^c$ in a measurement matrix which is factorized into the underlying 3D points $\boldsymbol{X}_i$ and the motion of the cameras.

Manual techniques have been created such as [1] which uses this technique with one light source manually moved through the scene, providing one inter-camera correspondence for many frames. In order to provide many correspondences per frame, calibration patterns could be used such as [2] for intrinsic parameters estimation or [3] in the case of a camera network. However, such manual procedures can be cumbersome or even impossible to conduct in certain situations, e.g. surveillance setups where cameras are not reachable. An automatic and reliable procedure is thus needed for such situations.

Automatic inter-camera correspondences have been the solution developed to overcome those manual procedures. These techniques rely on detecting keypoints and establishing correspondences based on appearance of a patch around the extract points [4]. Non-linear optimizations can then be used to improve the precision and quality of such estimations [5]. However, determining inter-camera correspondences automatically becomes especially hard if cameras have a small field of view overlap (or even none in a network of cameras), or if they watch in different directions (also called *wide baseline*) as analyzed for example by [6]. To focus on the calibration approach, many methods take given correspondences as input [7,8]. Very few methods do not require such knowledge, as [9].

Even though dynamic objects add another layer of complexity, they have been used and analyzed to extract information of all kinds. Pedestrians, for example, have been studied in many different fields of computer vision such as in detection task [10] or for tracking purposes [11]. They also have been extensively used as observations of inter-camera correspondences in camera calibration [7–9,12–20]. Although being intrinsically hard elements to work with due to their intra-class variety and per-instance non rigid deformations, they can be used in conjunction with various assumptions or priors.

Many different methods relying on pedestrian observations have been presented to solve the intrinsic parameters estimation problem. [17] suggested to extract vanishing points and line on a single pedestrian. Since vanishing points are sensitive to noise, other works have built upon it to robustify this approach, such as [18] which detects leg crossing events in order to extract more accurate foot-head points, or [14] which proposes a probabilistic approach in the form of a Markov-chain Monte-Carlo process to handle noise. [13] worked on a different assumption which incorporates a prior about the pedestrian motion. An important shadow model is added by [12] to extract more accurate points, with the same goal [15] used a human model.

External camera calibration has also been addressed using pedestrians as basic observations. [21] proposed to observe soccer players with PTZ cameras for calibration helped by the particular ground markings of soccer fields. [20] also calibrated PTZ camera networks but by tracking a single pedestrian walking on a plane, requiring a known inter-camera correspondence. On another side, [9] showed a work not requiring inter-camera correspondences by working directly on foreground blobs instead of pedestrians. All these works assumed a planar ground where people walked, [7] went away from this assumption and proposed camera network calibration on uneven terrains.

The underlying optimization process in camera calibration can take very different shapes to overcome noisy data. For example as a fully probabilistic approach as in [14]. Certain methods filter outliers based on robust analysis [7]. Others may want to extract only very reliable points by adding different models to compensate for noise as [12] which learns a complex shadow model to increase the precision of point extraction or [15].

In this work, we do not focus on precise point extraction or assume reliable and accurate information. This is motivated by the will of being robust in every situation, even when early processings in the pipeline, such as background subtraction or tracking, may fail. By extension, we do not use complexe non-linear refinement as it ask for accurate data in the first place.

The closest works to our approach are [9,14]. [14] proposed a Markov chain Monte Carlo approach but focused only on intrinsic camera calibration of one camera. We present a 3-layer cascade of Markov chain Monte Carlo in the different setup of external network calibration. [9] also follows the idea of analyzing inter-camera correspondences prior to network calibration but relies on a stable ground plane estimation from person heights.

3   Pose Estimation Using Tracklet Correspondences

Our pose estimation is based on the decomposition of homographies $\mathbf{H_{ab}}$ into

$$\mathbf{H_{ab}} = \mathbf{K_b}(\mathbf{R_{ab}} + \frac{1}{d} t_{ab} n^{\mathsf{T}}) \mathbf{K_a^{-1}} \tag{1}$$

where $\mathbf{R_{ab}}$, t_{ab} are rotation and translation of cameras C^b wrt. C^a, and n and d are orientation and distance of the underlying plane wrt. C_a which maps points according to $\bar{x}_b = \mathbf{H_{ab}}\bar{x}_a$. Thus, if $\mathbf{H}$ can be estimated precisely, the external camera parameters are known (up to scale which is encoded in d). Furthermore, any plane visible in both cameras can be used to estimate $\mathbf{R_{ab}}$, t_{ab}, just n and d are plane-dependent.

For our scenario, an intuitive idea would be to select the foot points as $\bar{x}$ from four pedestrians in two views and compute $\mathbf{H}$ (see Fig. 2c). However, the planarity might not be valid and, more important, all foot point correspondences have to be true which is unlikely for large baselines with low precision. Formally, this plane configuration would be four inter-camera correspondences, one time instance (intra-camera correspondence), and two cameras, or $\mathcal{C}_{41} = (4, 1, 2)$ as done in [21]. Since $\mathbf{H_{ab}}$ has eight degrees of freedom, the product of the configuration elements has to be 8. Since the number of cameras is 2, the configuration space is quite limited, so only $\mathcal{C}_{22} = (2, 2, 2)$ and $\mathcal{C}_{14} = (1, 4, 2)$ would be alternative minimal sampling sets.

Out of these, $\mathcal{C}_{14}$ (Fig. 2a) also intuitively aims at determining the ground plane, but this time from one point visible over time. Compared to $\mathcal{C}_{41}$, it has the advantage that only one true inter-camera correspondence is needed. However, the configuration is degenerate if the four intra-camera correspondences are on a line, which often occurs during human walking.

We focus on the $\mathcal{C}_{22}$ configuration, which means two points in two frames for two cameras (Figs. 1 and 2b). Compared to C_{14}, it has the disadvantage that two inter-camera correspondences are needed. However, by establishing correspondences between pedestrians instead of points, with stably-localizable head and foot points, we only need *one* inter-camera pedestrian correspondence and *two* time instances where it is seen.

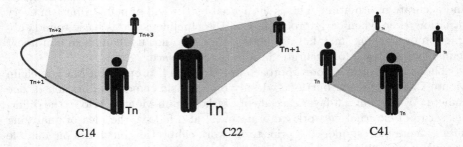

Fig. 2. Different configurations for **H** estimation (magenta plane). (a) shows $\mathcal{C}_{14} = (1, 4, 2)$ an example of a pedestrian walking (blue track) from which we select four points. (b) $\mathcal{C}_{22} = (2, 2, 2)$ is the situation used in this paper, with two points extracted at two timesteps and (c) presents $\mathcal{C}_{41} = (4, 1, 2)$ an estimation of a ground plane using four foot points at the same timestep. (Color figure online)

4 Markov Chain Monte Carlo Cascade for Likelihood Maximization

The presented algorithm takes as inputs m synchronized sequences produced by cameras C^i forming a network N. We assume the intrinsic calibration K^i known. For automatic temporal synchronization, audio signals could be cross-correlated, and for intrinsic parameters, automatic techniques, as presented in Sect. 2, could be used.

These sequences will be in first place pre-processed to extract pedestrian tracklets. These tracklets will be used to produce the so-called foot head points as explained in Sect. 4.1, similarly to [12,14,20].

As a second step, tracklets in different views will be associated in pedestrian pair hypotheses. These hypotheses will be the basic elements of the pairwise pose estimation process (Sect. 4.2) which aims at estimating the relative pose density for all camera pairs. This will be conducted as a Markov chain Monte Carlo procedure which follows the idea of [14]. Once this density is estimated for every camera pair, we will fuse the pairwise pose estimates into a consistent network by a triangular structural term (Sect. 4.3).

Our algorithm is a 3-layer cascade of Markov chain Monte Carlo sampling. Each layer approximates different posterior densities and uses their estimations to feed on the next layer. The first layer is at the pedestrian observation level, which will suggest relative poses for a second layer at a camera pair level. We formulate this global network optimization as a maximum likelihood problem which will be solved by a last layer of sampling.

4.1 Pre-processing

The first step is to detect and track pedestrians to produce pedestrian tracklets. For detection, we use a standard deformable part model as presented in [10] and used the already trained algorithm provided online and trained on the Pascal VOC dataset [22]. Having the detection bounding boxes, we create the tracklets using the flow algorithm proposed in [11] and kindly distributed online by the authors. Please note that the tracklets might be incomplete or falsely developed. Our approach is designed to simply not consider these for the calibration in a later steps. The jth tracklet in camera C^i is noted t_j^i. For computational purpose we filter out all short tracklets with less than 2 s overlap.

Based on these tracklets, we will extract the commonly called foot head points for every pedestrian at every frame they are seen. To extract such points in a frame f, we use a naive background subtraction method in the form of a grayscale per-pixel median filter. Albeit being naive and extremely simple, this background filtering method has been sufficient. We then compute the principal components of the foreground enclosed inside the detection bounding box. As humans are mainly distributed along the vertical axis, itself colinear to the gravity vector, we take the first principal axis v_{pa} and the center of mass of the foreground p_{cm} to extract two points using it. The head point is simply computed as $p_{cm} + v_{pa}$ and the corresponding foot point $p_{cm} - v_{pa}$. This naive point extraction method is motivated by the will of having a robust algorithm as accurate points or tracklets may not be available in dynamic environment where lots of occlusions can occur.

4.2 Pairwise Pose Density Estimation

We will first derive the estimation of the probability density of a relative pose $(\mathbf{R}, t)$ between a camera pair (C^a, C^b), given all correspondences in-between:

$$p_{cc}(\mathbf{R}, t \mid C^a, C^b) \qquad (2)$$

In this section, we assume that there is no side information from the network. p_{cc} is modeled as Parzen density over previous estimates of $\mathbf{R}, t$. In order to iteratively refine p_{cc}, we sample a correspondence from all hypotheses which, in turn, will be used to compute another $\mathbf{R}', t'$ sample, as explained later in Sect. 4.2. To guide multinomial correspondences sampling, we use p_{cp} as described in the following. We apply Bayes' theorem to transform (2) into:

$$p_{cc}(\mathbf{R}, t \mid C^a, C^b) = \frac{p_{cp}(C^a, C^b \mid \mathbf{R}, t) p(\mathbf{R}, t)}{p(C^a, C^b)} \qquad (3)$$

The denominator $p(C^a, C^b)$ is usually considered constant and we do the same for $p(\mathbf{R}, t)$ which would be a prior on the relative pose. We devise p_{cp} as:

$$p_{cp}(C^a, C^b \mid \mathbf{R}, t) = \prod_{(t_i^a, t_j^b) \in (C^a, C^b)} p_{pp}(t_i^a, t_j^b \mid \mathbf{R}, t), \tag{4}$$

Following is the definition of the pedestrian pair probability with the reprojection error modeled by the Blake-Zisserman distribution:

$$p_{pp}(t_i^a, t_j^b \mid \mathbf{R}, t) = \sqrt[(\delta d)^\phi]{\prod_{f=1}^{\delta d} e^{-\frac{e(t_i^a, t_j^b, f \mid \mathbf{R}, t)^2}{2\sigma^2}}} + \epsilon, \tag{5}$$

where $e(t_i^a, t_j^b, f \mid \mathbf{R}, t)$ is the reprojection error between t_i^a and t_j^b, evaluated at the foot head points of frame f. δd denotes the tracklet length. $\phi \in [0, 1]$ is an independence parameter. As we expect our pedestrian pairs to be correlated, we used $\phi = 0.5$ in our experiments. $\epsilon = 0.01$ represents the uniform noise in the data. The reprojection standard deviation σ has been set to a standard value of 5 pixels.

The reprojection error e at frame f is defined as average error of the respective head and foot points in both cameras C^a and C^b. To evaluate e for a head or foot point correspondence (x^a, x^b), we use $\mathbf{R}$ and t to triangulate (x^a, x^b) to the 3D point X. After re-projecting X into both camera views, the Euclidean distance to x^a and x^b is used for e.

Relative Pose Guided Sampling. Having sampled a particular pair of tracklets (t_i^a, t_j^b) in the previous section, it will be used to generate the next relative pose $\mathbf{R}', t'$ by sampling with respect to:

$$\begin{aligned} p_{rt}(\mathbf{R}, t \mid t_i^a, t_j^b) &= \frac{p_{pp}(t_i^a, t_j^b \mid \mathbf{R}, t) p(\mathbf{R}, t)}{p(t_i^a, t_j^b)} \\ &\propto p_{pp}(t_i^a, t_j^b \mid \mathbf{R}, t) p_{cc}(\mathbf{R}, t \mid C^a, C^b) \\ &\propto p_{pp}(t_i^a, t_j^b \mid \mathbf{R}, t) p_{cp}(C^a, C^b \mid \mathbf{R}, t) \end{aligned} \tag{6}$$

The theorem of Bayes is used here again, we assume the denominator constant again. Note that you could also use it as a prior on the pedestrian pair correspondence from other information (e.g. an appearance prior). We incorporated the prior $p(\mathbf{R}, t) = p_{cc}(\mathbf{R}, t \mid C^a, C^b) \propto p_{cp}(C^a, C^b \mid \mathbf{R}, t)$ (thanks to (3)) as an indication of the overall likelihood of a relative pose with respect to the camera pair.

Density p_{rt} will also be modeled as a Parzen density of the previously visited $\mathbf{R}, t$. Every time correspondences are selected to suggest a new relative pose, they firstly produce a completely new $\mathbf{R}, t$ as explained in Sect. 3 as an exploration step and incorporate it into the Parzen density estimate. Then $\mathbf{R}', t'$ is sampled from p_{rt}. This procedure allows us to balance exploration and exploitation with the goal to find likely solution for (3).

4.3   Network Configuration Optimization

We now have a description of (3) for every pair of cameras (C^a, C^b). We want to combine them to recover the structure of the network, in other words we want to find the most likely set of relative poses for each edge that produces a consistent network. By consistent we mean relative poses that produces triangle stable network. We define triangle stability by:

$$e_\Delta = \|I_d - P^{ab} \circ P^{bc} \circ P^{ca}\|_f \tag{7}$$

where I_d is the identity matrix, $\|\cdot\|_f$ the Frobenius norm [23] and P^{ij} are 4×4 pose matrices from C^i to C^j. The smaller e_Δ is, the more consistent the triangle is. Due to the unknown scale that exists between the relative poses, we cannot directly concatenate them. To overcome this problem we locally solve for the scale, each of the three relative poses brings an unknown scale, we fix one to 1 and use a least square solution to obtain the two left. After rescaling the translation component of the relative poses, they can be used for comparison.

One could just select the most likely relative pose for each camera pair but it would probably violate the triangle constraints ($e_\Delta \gg 0$). We formalize this problem as a maximization for the relative pose set $\mathcal{P} = \{P^{ab} \forall (C^a, C^b)\}$ which fulfills camera pairs likelihood p_{cp} and network constraints p_Δ:

$$arg\,max_\mathcal{P} p(N \mid \mathcal{P}) = \prod p_{\mathrm{cp}}(C^a, C^b \mid P^{ab})$$
$$\cdot \prod p_\Delta(P^{ab}, P^{bc}, P^{ca}) \tag{8}$$

The first product of p_{cp} is our data term coming directly from the previous step (Eq. 4) and reflecting the pedestrian observations. The second product is a structural term based on e_Δ which is modeled as a Gaussian:

$$p_\Delta(P^{ab}, P^{bc}, P^{ca}) = e^{-\frac{\|I_d - P^{ab} \circ P^{bc} \circ P^{ca}\|_f^2}{2\beta^2}} \tag{9}$$

We empirically found $\beta = 0.25$ for good results. Setting it too low blocks the exploration into local minima and too high does not guide the sampling anymore.

4.4   Gibbs Metropolis Hastings Sampling

We now show how we can maximize (8) using a random walk algorithm. Due to nonlinearity and high dimensionality, it is extremely hard to solve this maximization problem by a direct approach or exhaustive testing. We propose to use the Gibbs sampling approach tinted with Metropolis Hastings acceptance ratio to walk through the state space. Our Gibbs sampling approach creates a network state vector S^N of $n = 1/2 \cdot m(m - 1)$ (m = number of cameras in N) random variables P^{ab}, each corresponding to the relative pose of one camera pair. At every iteration, every random variable P^{ab} of the network state vector S^N is updated by sampling it from the distribution

$$p(P^{ab} \mid S^N \setminus \{P^{ab}\}) \tag{10}$$

In our work, we composed this distribution using the camera pair data term (3) and a product of the triangle stability term (9) which leads the sampling around locations that produce consistent network configurations:

$$p(P^{ab} \mid S^N \setminus \{P^{ab}\}) = p_{cc}(P^{ab} \mid C^a, C^b)$$
$$\cdot \prod_{i \notin \{a,b\}} p_\triangle(P^{ab}, P^{bi}, P^{ia}) \tag{11}$$

Sampling from this distribution gives us a new state $P^{ab\prime}$. In order to guide this sampling more strongly towards the optimal solution, we spice the standard Gibbs sampling by computing an acceptance ratio α as follows:

$$\alpha = \frac{p(N \mid S^{N\prime})}{p(N \mid S^N)} \tag{12}$$

where $S^{N\prime}$ is the network state vector where P^{ab} has been replaced by $P^{ab\prime}$. The actual state vector S^N is updated based on the value of α. If α is bigger than one, which would mean accepting this new state increases the probability of having a correct network, we accept the change. Otherwise it means the change will decrease the quality of the current estimate. In this case, we accept the change proportionally to α (small decrease in quality have more chances to be accepted than big ones). This process is summarized in Algorithm 1

> **Data:** Discrete estimation of densities $p_{cc}(P^{ab} \mid C^a, C^b)$
> **Result:** Best network configuration S^B
> $\forall (C^a, C^b) : P^{ab} \leftarrow arg\,max\, p_{cp}(P^{ab} \mid C^a, C^b)$;
> $S^N \leftarrow \{P^{ab}\}$;
> $S^B \leftarrow S^N$;
> **for** $n\_iterations$ **do**
> > **for** $\forall P^{ab} \in S^N$ **do**
> > > $P^{ab\prime} \leftarrow$ sample from $p(P^{ab} \mid S^N \setminus \{P^{ab}\})$;
> > > $\alpha \leftarrow \frac{p(N \mid S^{N\prime})}{p(N \mid S^N)}$;
> > > **if** $rand() < \alpha$ **then**
> > > > $S^N \leftarrow S^N \setminus \{P^{ab}\} \cup P^{ab\prime}$;
> > >
> > > **end**
> > > **if** $p(N \mid S^N) > p(N \mid S^B)$ **then**
> > > > $S^B \leftarrow S^N$;
> > >
> > > **end**
> >
> > **end**
>
> **end**

Algorithm 1: Outline of the global network optimization sampling algorithm.

By the random walk, we explore the density in Eq. (8). Likely solutions are sampled preferably, but to overcome local minima, unlikely solutions are also explored. As final network configuration result, we use the most-probable explored state according to (8).

4.5   Smallest Stable Triangular Spanning Tree

As not all cameras are connected with each other (no common pedestrians observations) and some estimates are very hard (unstable camera pair configuration), we add a final selection step that will select only the best relative poses. Indeed, we computed the relative pose for every camera pair, yet we only need a subset of it in order to be able to calibrate it up to one unknown scale. We used an augmented version of the shortest triangle paths algorithm presented by [24, 25]. This algorithm produces triangle paths which are triangular connected, meaning it is enough for estimating all the unknown scales down to one global scale.

For the sake of conciseness, we refer to [25], and briefly explain our extensions: we incorporate our triangle stability probabilities combined with our pairwise likelihoods as edge weights on top of their graphical model. This allows us to find paths that correctly explain pedestrian observations as well as having a structurally stable network. We finally differ from their approach by initializing the algorithm from multiple different entry nodes instead of the one with highest probability, and by selecting the most probable paths set as solution.

Fig. 3. The first images of all 7 cameras of the PETS 2009 sequence.

5   Experimental Results

As input data, we take the sequences PETS 2006 S1-T1-C and 2009 S2.L1 (Fig. 3) consisting of 3'021 frames per 4 cameras, and 795 frames for each 7 cameras respectively. The cameras record a central scene from 360 degrees. As evaluation metric, we use relative camera pose differences in percentage to the groundtruth. For rotation comparison, between estimate $\mathbf{R}$ and ground truth $\mathbf{R}'$, we based our distance measure on [26] and used the Φ_5 distance function $|\mathbf{I_d} - \mathbf{R}'\mathbf{R}^{\mathsf{T}}|_f$. As scale invariant translation measure, we compute the angular difference between the translation vectors.

5.1   Structure-for-Motion as Baseline Comparison

To demonstrate the difficulty of calibrating these sequences, we test the state-of-the-art structure-from-motion pipeline from VisualSfM [4,5] to compensate for it not being able to take as input grountruth intrinsic calibration, we guided its estimation by providing the ground truth in the EXIF information of the image. Manual verification of estimated intrinsic proved it to be correctly estimated. Firstly, a standard Structure-from-Motion is conducted. The first images of all cameras are matched and their relative pose is verified with an epipolar model.

Fig. 4. Top: Only frame pair of PETS2009 with an inlier-only correspondence set. Bottom: Typical failure containing overfitted outliers.eps

Although we tried several ways of optimizing the results, out of the 21 relative inter-camera poses, only one was estimated from inlier correspondences (Fig. 4) leading to early breakdown of the algorithm.

We then proceeded to use the same input for VisualSfM as our algorithm uses. The complete list of corresponding foot head points for every hypothesis is fed to VisualSfM. Unfortunately, due to the number of outliers dominating the inliers, only a small subset of the cameras are estimated. Multiple repetitions have been conducted for the calibration and the best, selected firstly on the number of cameras then on the actual error measurements, is shown in Table 1. In order to provide a comparison baseline, we added the pedestrian correspondence knowledge for VisualSfM and used only foot head points of true manually annotated pedestrian correspondences as input. Again only partial networks are estimated. The results are summarized in Table 2.

Table 1. Baseline results produced using VisualSfM with the same input as our algorithm.

Dataset	Pets2009 S2.L1		Pets2006 S1-T1-C	
Remarks	Only **4** cameras		Only **2** cameras	
	R	t	**R**	t
Mean	15.0	35.8	7.1	4.8
Std	9.5	28.3	0	0
Min	3.4	7.4	7.1	4.8
Max	21.2	64.2	7.1	4.8

Table 2. Baseline results produced using VisualSfM and pedestrian correspondence knowledge.

Dataset	Pets2009 S2.L1		Pets2006 S1-T1-C	
Remarks	Only **5** cameras		Only **3** cameras	
	R	*t*	**R**	*t*
Mean	8.9	5.3	5.9	2.7
Std	4.4	1.1	2.2	2.0
Min	2.6	3.2	4.6	0.6
Max	18.0	6.8	8.5	4.6

5.2   Camera Calibration Using Our Cascade of Markov-Chain Monte-Carlo

To evaluate our algorithm, we compare our estimation to the dataset groundtruths. Two different scenarios are presented, the first one named *whole* network will compare every estimated relative poses whereas *minimal* network will evaluate only the minimal relative pose set computed using the triangle path algorithm from Sect. 4.5. The pairwise pose estimation process is iterated for 5000 iterations and the global network optimization is conducted for 2500 iterations. On non-optimized Matlab code, the pairwise density estimation took in average 1 h per camera pair and the global network optimization a few hours. We believe it to be optimizable and parallelizable.

Table 3 presents the error measures for Pets2009 and Pets2006 datasets.

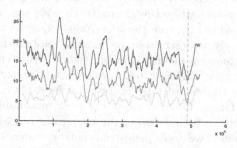

Fig. 5. Plot of energy during a global optimization. Red shows the pairwise energy, green the network energy and blue the total energy. The minimal energy is denoted by the magenta line. (Energies have been smoothed for readability purposes). (Color figure online)

We can see that the triangle path extraction allows us to increase the quality of the network on every aspect. Note that Pets2006 dataset has only four cameras resulting in six different camera pairs, when the *minimal* network solution needs five camera pairs to cover the whole network, limiting the selection of better relative poses and thus the amelioration in results. By comparing the *minimal* columns of Table 3 and Tables 3 to 2, we can see that our algorithm is able to correctly estimate relative poses with an error similar to what our baseline with groundtruth correspondences knowledge is able to achieve, however it is to be noted that our approach produces estimates for every camera. Also note that our algorithm estimates all cameras when the baseline for Pets2009 only managed to produce estimates for 5 cameras.

As the presented algorithm belongs to the random walk algorithm family, it is interesting to look at the energy variation as shown in Fig. 5. It can be seen that the global minimum may not be at the minimum of either the pairwise or network energies and sometimes accepting worse relative poses can lead to improving the overall solution by avoiding local minima.

Table 3. Rotational error relative to groundtruth as percentage to the grountruth.

Dataset	Pets2009 S2.L1				Pets2006 S1-T1-C			
Network	*whole*		*minimal*		*whole*		*minimal*	
	R	*t*	**R**	*t*	**R**	*t*	**R**	*t*
Mean	15.7	8.1	10.5	6.1	13.2	6.6	10.3	5.7
Std	10.7	6.5	5.5	4.0	6.9	5.0	7.3	5.2
Min	2.4	0.6	2.4	0.6	3.4	0.8	3.4	0.8
Max	37.5	23.9	20.3	14.4	21.6	13.6	19.0	10.7

A collateral result from network external calibration is the possibility to match pedestrians in different views. We computed the Jaccard index between inlier pedestrian set produced by the groundtruth pose and our estimated pose. We achieve in average 31.3% on the Pets2009 sequence and 25.6% on the Pets2006. Unfortunately, pedestrian correspondence are not found reliably when the relative pose estimation error is too big (when <5%, the average Jaccard index is 64.6% but when >15% it becomes 0%) but remarkably, the triangle stability term allows the use of false correspondences for calibration. Figure 7 shows some typical falsely found correspondences: reasonable results but false due to occlusion, and overfitting during the triangular reprojection.

In a last experiment, we took pedestrian correspondence estimates (pedestrian estimated with likelihood above 1% as shown in Fig. 6) and used their foot head points as input in VisualSfM to obtain a robust relative pose estimation. This is the same process as for our baseline, except we do not take the inlier pedestrians from the groundtruth but from our estimation, to see if our algorithm is able to produce such knowledge accurately. The results are shown in Table 4. In Pets2009 sequence, only a subset of cameras managed to be correctly estimated with errors in the same range as our estimates and the baseline. Pets2006 produced no correct results for our estimates due to too many wrong estimated correspondences. It can be seen that non-linear optimizations do not improve our estimates significantly.

Table 4. VisualSfM results as a post processing over our pedestrian inlier estimation for Pets2009.

Dataset	Pets2009 S2.L1	
Remarks	Only 4 cameras	
	R	t
Mean	9.6	5.8
Std	5.5	2.3
Min	3.5	0.8
Max	18.0	6.5

Fig. 6. Inlier-correspondence examples after our optimization. Magenta and yellow lines show respectively corresponding head and foot points. First row shows Pets2009 sequence and the bottom one: Pets2006. (Color figure online)

Fig. 7. Typical false correspondence examples after our optimization, labeled as in Fig. 6.

6   Conclusion and Future Work

We have presented a probabilistically justified algorithm in the form of a cascade of Markov chain Monte Carlo algorithm and applied it to the task of estimating the external calibration of a camera network by observing unconstrained pedestrians. This algorithm has many advantages, its probabilistic formulation providing it an adaptability property for whoever would like to incorporate more priors (e.g. appearance prior). A collateral result of a good estimation is the pedestrian correspondences deduction which can be useful in many surveillance situations and used as prior information for other processes. Its main quality is its robustness against noisy data due to inexact point extraction or drifting tracklets and against outlier dominated hypotheses. Lastly, we think the simplicity of this algorithm allows it to be improved on different aspects such as the relative pose estimation in itself but also by addressing wider problems such as camera synchronization or intrinsic calibration.

Acknowledgement. This research was supported by the SNF project "Tracking in the Wild" CRSII2_147693/1.

References

1. Svoboda, T., Martinec, D., Pajdla, T.: A convenient multi-camera self-calibration for virtual environments. PRESENCE: Teleoperators Virtual Environ. **14**, 407–422 (2005)
2. Zhang, Z.: A flexible new technique for camera calibration. IEEE Trans. Pattern Anal. Mach. Intell. **22**, 1330–1334 (2000)
3. Baker, P., Aloimonos, Y.: Complete calibration of a multi-camera network. In: IEEE Workshop on Omnidirectional Vision, Proceedings, pp. 134–141 (2000)
4. Wu, C.: Towards linear-time incremental structure from motion. In: Proceedings of the 2013 International Conference on 3D Vision, 3DV 2013, pp. 127–134. IEEE Computer Society, Washington, D.C. (2013)
5. Wu, C., Agarwal, S., Curless, B., Seitz, S.M.: Multicore bundle adjustment. In: IEEE Conference on Computer Vision and Pattern Recognition, CVPR 2011, pp. 3057–3064. IEEE (2011)
6. Mikolajczyk, K., Schmid, C.: A performance evaluation of local descriptors. IEEE Trans. Pattern Anal. Mach. Intell. **27**, 1615–1630 (2005)
7. Junejo, I.N.: Using pedestrians walking on uneven terrains for camera calibration. Mach. Vis. Appl. **22**, 137–144 (2011)
8. Chen, T., Bimbo, A.D., Pernici, F., Serra, G.: Accurate self-calibration of two cameras by observations of a moving person on a ground plane. In: AVSS, pp. 129–134. IEEE Computer Society (2007)
9. Liu, J., Collins, R.T., Liu, Y.: Robust autocalibration for a surveillance camera network. IEEE Winter Conference on Applications of Computer Vision, pp. 433–440 (2013)
10. Felzenszwalb, P.F., Girshick, R.B., McAllester, D., Ramanan, D.: Object detection with discriminatively trained part-based models. IEEE Trans. Pattern Anal. Mach. Intell. **32**, 1627–1645 (2010)

11. Pirsiavash, H., Ramanan, D., Fowlkes, C.: Globally-optimal greedy algorithms for tracking a variable number of objects. In: Computer Vision and Pattern Recognition CVPR (2011)
12. Rother, D., Patwardhan, K.A., Sapiro, G.: What can casual walkers tell us about a 3D scene? In: ICCV, pp. 1–8. IEEE (2007)
13. Krahnstoever, N., Mendonça, P.R.S.: Autocalibration from tracks of walking people. In: Proceedings of British Machine Vision Conference, BMVC 2006, pp. 4–7 (2006)
14. Krahnstoever, N., Mendonça, P.R.S.: Bayesian autocalibration for surveillance. In: ICCV, pp. 1858–1865. IEEE Computer Society (2005)
15. Micusik, B., Pajdla, T.: Simultaneous surveillance camera calibration and foot-head homology estimation from human detections. In: 2010 IEEE Conference on Computer Vision and Pattern Recognition (CVPR), pp. 1562–1569 (2010)
16. Kusakunniran, W., Li, H., Zhang, J.: A direct method to self-calibrate a surveillance camera by observing a walking pedestrian. In: DICTA, pp. 250–255. IEEE Computer Society (2009)
17. Lv, F., Zhao, T., Nevatia, R.: Self-calibration of a camera from video of a walking human. In: ICPR, vol. 1, pp. 562–567 (2002)
18. Lv, F., Zhao, T., Nevatia, R.: Camera calibration from video of a walking human. IEEE Trans. Pattern Anal. Mach. Intell. **28**, 1513–1518 (2006)
19. Liu, J., Collins, R.T., Liu, Y.: Surveillance camera autocalibration based on pedestrian height distributions. In: British Machine Vision Conference (BMVC) (2011)
20. Possegger, H., Rüther, M., Sternig, S., Mauthner, T., Klopschitz, M., Roth, P.M., Bischof, H.: Unsupervised calibration of camera networks and virtual PTZ cameras. In: Proceedings of Computer Vision Winter Workshop (CVWW) (2012)
21. Puwein, J., Ziegler, R., Ballan, L., Pollefeys, M.: PTZ camera network calibration from moving people in sports broadcasts. In: WACV, Breckenridge, Colorado (2012)
22. Everingham, M., Van Gool, L., Williams, C.K.I., Winn, J., Zisserman, A.: The pascal visual object classes (VOC) challenge. Int. J. Comput. Vis. **88**, 303–338 (2010)
23. Golub, G.H., Van Loan, C.F.: Matrix Computations, 3rd edn. Johns Hopkins University Press, Baltimore (1996)
24. Bajramovic, F., Denzler, J.: Global uncertainty-based selection of relative poses for multi camera calibration. In: Proceedings of the British Machine Vision Conference, pp. 74.1–74.10. BMVA Press (2008). doi:10.5244/C.22.74
25. Bajramovic, F., Brückner, M., Denzler, J.: An efficient shortest triangle paths algorithm applied to multi-camera self-calibration. J. Math. Imaging Vis. (JMIV) **43**, 1–14 (2011)
26. Huynh, D.Q.: Metrics for 3D rotations: comparison and analysis. J. Math. Imaging Vis. **35**, 155–164 (2009)

Scale-Adaptive Deconvolutional Regression Network for Pedestrian Detection

Yousong Zhu[1,2(✉)], Jinqiao Wang[1,2], Chaoyang Zhao[1,2], Haiyun Guo[1,2], and Hanqing Lu[1,2]

[1] National Laboratory of Pattern Recognition, Institute of Automation, Chinese Academy of Sciences, Beijing, China
{yousong.zhu,jqwang,chaoyang.zhao,haiyun.guo,luhq}@nlpr.ia.ac.cn
[2] University of Chinese Academy of Sciences, Beijing, China

Abstract. Although the Region-based Convolutional Neural Network (R-CNN) families have shown promising results for object detection, they still face great challenges for task-specific detection, *e.g.*, pedestrian detection, the current difficulties of which mainly lie in the large scale variations of pedestrians and insufficient discriminative power of pedestrian features. To overcome these difficulties, we propose a novel Scale-Adaptive Deconvolutional Regression (SADR) network in this paper. Specifically, the proposed network can effectively detect pedestrians of various scales by flexibly choosing which feature layer to regress object locations according to the height of pedestrians, thus improving the detection accuracy significantly. Furthermore, considering CNN can abstract different semantic-level features from different layers, we fuse features from multiple layers to provide both local characteristics and global semantic information of the object for final pedestrian classification, which improves the discriminative power of pedestrian features and boosts the detection performance further. Extensive experiments have verified the effectiveness of our proposed approach, which achieves the state-of-the-art log-average miss rate (MR) of 6.94% on the revised Caltech [1] and a competitive result on KITTI.

1 Introduction

Pedestrian detection aims to locate all pedestrian instances with various poses, scales and occlusions in an image. Over the last decade, it has become a hot topic for its wide applications, such as smart vehicles, video surveillance and robotics.

Recently, a lot of efforts have been devoted to pedestrian detection based on boosted decision forests [2–6], deformable part model [7–9], and deep learning models [10–12]. No matter what kinds of methods, the scale of pedestrian which largely affects the feature representation or even dominates the detection performance still has not been well solved. In traditional methods, sliding windows and image pyramids are often used to capture objects in different scales. However, image pyramids lead to high computational complexity. For CNN-based methods, brute-force learning (single scale) and image pyramids (multi-scale)

© Springer International Publishing AG 2017
S.-H. Lai et al. (Eds.): ACCV 2016, Part II, LNCS 10112, pp. 416–430, 2017.
DOI: 10.1007/978-3-319-54184-6_26

are the most used solutions. Brute-force learning simply forces the network to directly learn scale invariance, which is difficult to learn strong convolutional filters. While multi-scale input always takes a large amount of GPU memories, and spends more time to train and test. Since pedestrians with different scales may show quite different appearances, which lead to varied discriminative power with features extracted from the same descriptor. Especially for small pedestrians, general feature extraction often leads to weak classification. Therefore, how to design a scale adaptive detector is critical for boosting the detection performance.

Among the recent state-of-the-art generic object detectors [13–16], Fast R-CNN [15] and its variant Faster R-CNN [16] are the most prevalent pipelines. One common problem of these solutions is that the last convolutional layer is too small and coarse, which means the features pooled from this layer are lack of sufficient representation capacity for small objects. An intuitive solution is to upsample the feature maps to a proper size. Recently, Long *et al.* [17] proposed an in-network upsampling layer for pixelwise prediction. Noh *et al.* [18] designed a deconvolutional network which contains unpooling and deconvolution operations to decode the convolutional feature maps for generating accurate segmentation results. Similarly to [18], Badrinarayanan *et al.* [19] also proposed an encoder-decoder architecture to achieve pixel-wise segmentation. All these works show that the finer upsampling or decoding, the more accurate predictions. As analyzed above, a refined feature map from the upsampling layers or deconvolutional layers could help to obtain a more accurate portrayal for the small objects. Note that for small objects, the features pooled from a coarse feature map are filled with repeated values which are lack of discriminative representation ability.

Another application is image super-resolution [20], where a coarse to fine deconvolutional layer could capture more rich structural and local information for a finer reconstruction. Therefore, the deconvolutional layers are to obtain a better representation of local and structural information for small pedestrians. It makes sense to design a scale-adaptive network for pedestrian detection. In this way, the features of large and small objects are pooled from different layers for training different regressers respectively, which we called scale-adaptive deconvolutional regression (SADR).

In addition, before the arrival of R-CNN, Dollár *et al.* [4] aggregated multiple hand-crafted channels (ACF) to train the cascaded adaboost. Zhang *et al.* [6] applied many of filters in ACF channels to obtain a more rich feature representation. All of these works indicate that the more rich features, the better classification. Recently, feature fusion from different CNN layers has shown the effectiveness to enhance the discriminability [10, 21–23]. Actually, different layers in a neural network contain different levels of discriminative information. The lower layers always represent the local characteristics, whereas the deeper layers focus on the global semantic information. An intuitive idea is to fuse features of different layers to learn a strong and powerful classifier. Therefore, we investigate the effects of fusing features from different layers on pedestrian classification.

To sum up, the main contributions of this work are as follows:

(1) We propose a novel scale-adaptive deconvolutional regression (SADR) network for pedestrian detection, which could flexibly detect pedestrians with different size.
(2) By computing the classification and regression loss respectively, we integrate multi-layer outputs of CNN network to boost the detection performance.
(3) Extensive experiments on the challenging Caltech dataset well demonstrate the effectiveness of the proposed approach. We achieve the state-of-the-art result with 6.94% miss rate, which is significantly better than 9.29% by CompACT-Deep [24].

2   Related Work

In this section, we mainly review some existing object detection and pedestrian detection approaches.

Object Detection. Most recent state-of-the-art object detection methods follow the pipeline of R-CNN [13], which first generated object proposals by some unsupervised algorithms (*e.g.* Selective Search [25], Edge Boxes [26] and MCG [27]) from the input image and then classified each proposal into different categories. Since the feature extraction in R-CNN is time-consuming and the training process is implemented through a multi-stage pipeline, two subsequent models, *i.e.* Fast R-CNN [15] and Faster R-CNN [16] were proposed to improve the computational efficiency and integrate the multi-stage training process into an unified pipeline. Moreover, Faster R-CNN introduced a Region Proposal Network (RPN) to reduce the time of proposal generation.

Pedestrian Detection. In the literature of pedestrian detection, lots of top performing pedestrian detectors based on hand-crafted features are explored. The Integral Channel Features (ICF) [3] and Aggregated Channel Feature (ACF) [4] efficiently computed features such as local sums, histograms, and Haar features and their various generalizations using integral images. Zhang *et al.* [5] designed informed filters by incorporating prior information as to the appearance of the up-right human body. Cai *et al.* [24] proposed complexity-aware cascaded detectors, which combined features of very different complexities. Deformable part-based models [8] learned a mixture of local templates for each part to deal with appearance variations. Many recent works using convolutional neural networks (CNN) to improve the performance of pedestrian detection [10–12,28,29]. Ouyang and Wang [12] integrated feature extraction, part deformation handling, occlusion handling and classification into a joint deep model. Tian *et al.* [29] used semantic tasks to assist pedestrian detection. Sermanet *et al.* [10] exploited two contextual regions centered on each object for pedestrian detection. Hosang *et al.* [28] firstly applied the R-CNN framework [13] to pedestrian detection and achieved promising performance on Caltech [30] and KITTI [31] dataset.

Scale Processing. A few works intend to deal with the scale problem. In most cases, image pyramids (multi-scale inputs) are always used to solve this problem, but it is very time-consuming. More recently, Li *et al.* [32] designed two sub-networks to learn the universality and specificity of large-scale and small-scale proposals, which resulted in more training time. Yang *et al.* [33] proposed scale-dependent pooling to handle the scale variation. However, they used the earlier convolutional layers to model the small objects, which might be too weak to make a strong decision. In addition, in order to improve classification performance they also introduced the cascaded AdaBoost Classifiers, which increased the complexity. In this paper, we propose a scale-adaptive deconvolutional regression architecture which is different from [32,33], and in order to decrease the complexity we just use one classifier.

3   The Proposed Approach

3.1   Overview

Figure 1 shows the overall framework of the proposed pedestrian detection network. Based on the framework of Faster R-CNN [16], the proposed approach involves two steps: pedestrian candidates generation and pedestrian/background classification. During our implementation, we found that the Region Proposal Network (RPN) serves well as a candidate generator, which achieves 99% recall on Caltech and 96% on KITTI. Thus here we focus on improving the detection accuracy for the second stage. On the one hand, to effectively deal with pedestrians with different scales, we introduce the deconvolutional layers to adaptively upsample the feature map for small pedestrians. In this way, we can adaptively pool RoIs (Region of Intersets) from corresponding layers for regression according to the size of proposals, instead of just pooling from the last convolutional layer. Compared to Fast R-CNN, this scale-adaptive deconvolutional regression (SADR) architecture can more precisely represent pedestrians of different scales and the features used for bounding box regression are more powerful.

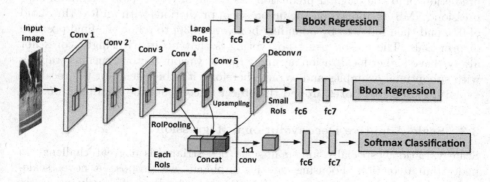

Fig. 1. The architecture of the proposed detection network.

On the other hand, to further enhance the discriminative power of classification, we concatenate features RoI-pooled from multiple layers, including deconvolutional and some convolutional layers. This kind of multi-scale feature fusion strategy can effectively capture some fine-grained details by pooling from multiple layers, which is especially important for classification. Conversely, features which are pooled directly from the last convolutional layer may not contain sufficient information for classification.

Finally, the proposed network ends up with three output layers. The first two output layers operate on different RoI feature vectors and output the predicted coordinate tuples for small and large RoIs respectively. The last output layer is the standard softmax layer which predicts the classification score of each RoI.

3.2   Network Architecture

In the proposed detection network, we use the pre-trained VGG16 model [34] to initialize the proposed network. All the convolutional layers and max pooling layers of the VGG16 network are used to encode features before the deconvolutional layers to decode features from the input image. All of the fc layers are initialized from VGG16 at the beginning. We randomly initialize the 1×1 convolution layer by drawing weights from a zero-mean Gaussian distribution with standard deviation 0.01. Given an image, we firstly apply RPN to generate a number of proposals. Then the proposed detection network takes the whole image and the proposals as input. In the forward pass, the scale-adaptive deconvolutional regression (SADR) verifies the height of each proposal and pools the region of interest from the corresponding layers according to the height, the details will be discussed in Sect. 3.3. We also concatenate the RoI-pooled features from the outputs of multi-layers to score each proposal.

After the forward pass, each proposal gets a regressed coordinate tuple and a score, which denote the original predictions. Box refinement [35] is followed to the final inference. That is, we take the original predictions as the input and forward into the network again. Thus, we obtain a new classification score and a new regressed box, which denote the new predictions. Then, we merge the new predictions and the original predictions as the final result. Non-maximum suppression (NMS) is applied on the union set of predictions with an IoU threshold of 0.3, and then followed by bounding box voting [36] to refine the final position of proposals. The reason using box voting is that boxes with high scores not always have higher localization accuracy due to various factors, such as training with suboptimal examples and so on. Therefore, it is beneficial for the recall by exploiting predicted boxes around the object to compute the final position.

3.3   Scale-Adaptive Deconvolutional Regression

Scale variation, especially for small-scale pedestrians, is a great challenge in pedestrian detection. Focusing on this problem, we propose a novel scale-adaptive deconvolutional regression (SADR) network which effectively integrates small-scale and large-scale regression into a unified framework.

As we known, the max pooling and the stride in convolution operations can reduce the spatial resolution of an input image over layers. As a result, the feature map at the conv5-3 in VGG16 turns out to be 1/16 of the original image. Actually, in Fast R-CNN [15], we can compute the minimal size of object which has no repeated RoI-pooled features using the following equation:

$$S_{min} = s_s \times s_r \tag{1}$$

where s_s and s_r are the stride of specific layer and the output size of RoI pooling operation respectively. The default value of s_r is 7×7. Here we use the height to denote the scale or size of an instance, as the height doesn't vary significantly while the width is more sensitive to pedestrian's pose [30]. Based on the assumption, the minimal height of bounding box is 112 pixels for conv5-3, 56 for the first deconvolutional layer and 28 for the second deconvolutional layer. So the proposed SADR architecture can flexibly choose the appropriate layers to do regression according to the height of pedestrians. The box regression loss is calculated as:

$$L_{loc}(F) = \begin{cases} L_{loc}(F_{c_5}), & h \in [112, \infty) \\ L_{loc}(F_{d_1}), & h \in [56, 112) \\ L_{loc}(F_{d_2}), & h \in (0, 56) \end{cases} \tag{2}$$

where F_{c_5}, F_{d_1} and F_{d_2} denote the features pooled from conv5-3, the 1st and 2nd deconvolutional layers respectively, h means the height of proposals. L_{loc} is the regression loss for each branch. In fact, the Caltech test set (*reasonable*) only evaluates the pedestrians with height greater than 50 pixels, therefore we just use the 1st deconvolutional layer to tackle the instances with height ranging from 50 to 112 pixels.

As shown in Fig. 1, each regression branch is followed by 2 4096-d *fc* layers with *ReLU* activations and *dropout* layers so as to learn a set of scale-specific parameters. In the fine-tuning process, we first divide the input proposals into three groups depending on the height and then feed them into corresponding RoI pooling layers so as to pool features from corresponding layers for regression. Since each regression branch learns from scale-specific samples, the branch will capture the rich information for this scale, resulting in a more accurate detection model.

The advantages of SADR are two-fold. On the one hand, the features extracted from deconvolutional layers could capture more rich structural information which are more powerful than the earlier convolutional layers. On the other hand, a single network is designed to adaptively select corresponding layers for regression according to the height of RoIs, which makes the training process more concise and fast. In this way, the proposed SADR effectively avoids upsampling the input images to handle specific instances.

3.4   Multi-layer Feature Fusion

As analysis previously, using reduplicative RoI-pooled features is hard to learn a strong classifier and regresser. We concatenate the RoI-pooled features from multiple layers, including deconvolutional and some convolutional layers. This kind of multi-scale feature fusion strategy can effectively capture some fine-grained details by pooling from multiple layers, which are especially important for classification. These multi-layer features are used to predict its scores to pedestrian.

The implementation of fusing multi-layer features is closely related to those used in [21]. Different from [21], we find it also works well without the complex operations of L2-normalized and scaled. That's to say, we directly concatenate each RoI-pooled feature along the channel axis and reduce the dimension to $512 \times 7 \times 7$ with a 1×1 convolution. This kind of integration for multi-layer features can effectively boost the performance of pedestrian detection.

3.5   Multi-task Loss

For training the proposed detection network, we use a multi-task loss for joint object classification and scale-adaptive bounding box regression. The final loss function is defined as:

$$L(p, k^*, t, t^*) = L_{cls}(p, k^*) + \sum_{i=1}^{n} \lambda_i [k_i^* \geq 1] L_{loc\_i}(t_i, t_i^*) \tag{3}$$

where $p = (p_0, p_1)$ is a discrete probability distribution over 2 categories (pedestrian or not). k^* is the true category label, n is the number of scales, $k_i^* \subseteq k^*$ is the true label of RoIs corresponding to i^{th} scale. t_i and t_i^* are the predicted tuple and the true tuple for bounding box regression respectively. L_{loc_i} is the standard smoothed L_1 loss. λ_i is the predefined hyper-parameter to balance the losses of different tasks, here we set $\lambda_1 = \cdots = \lambda_n = 1$ as same to Fast R-CNN.

It should be noted that the proposed network achieves regression and classification in a different manner. Only the regression distinguishes the size of objects, while the classification loss treats all objects equally by exploiting identical features regardless of the size of objects. That's because classification and regression are two different sub-tasks. So in backpropagation, derivatives for classification loss can be back-propagated to all the previous layers.

4   Experiments

4.1   Datasets and Metrics

Caltech Pedestrian Dataset. The Caltech dataset [30] is one of the most prevalent datasets for pedestrian detection. It consists of 10 hours of 640×480 30 Hz video in an urban traffic environment. The raw annotations amount to a total of 350k bounding boxes and 2300 unique pedestrians. The standard training set and test set extract one out of each 30 frames, which results in 4024 frames with 1014 pedestrians for evaluating. In most case, researchers can leverage more data for training by extracting one out of three or four frames. Recently, Zhang et al. [1] revised the original annotations and released a new high quality ground truth for training and testing. In this paper, we use the new annotations of Caltech10x for training and evaluated on the new aligned test set. In the standard Caltech evaluation, the log-average miss rate (MR) which is averaged over the FPPI range of $[10^{-2}, 10^0]$ is used to evaluate the performance of detectors.

KITTI Dataset. The KITTI object detection benchmark [31] consists of 7481 training images and 7518 test images. Due to the diversity of scale, occlusion and truncation of objects, the dataset evaluates at three levels of difficulty, *i.e.*, easy, moderate and hard, where the difficulty is differentiated by the minimal scale of object and the occlusion and truncation of the object. The benchmark follows the PASCAL protocol and use Average Precision (AP) to measure the detection performance, where 50% overlap thresholds are adopted for pedestrian.

4.2   Evaluation on Caltech Dataset

Implementation Details. We use the pre-trained VGG16 model to initialize the proposed detection framework. For the RPN stage, each location in a feature map generates 10 bounding boxes with one aspect ratio of 2.0, where 2.0 indicates the ratio between height and width of the box. In both RPN and the proposed detection network, the scale of input image is set to be 720 pixels on the shortest side and the negative examples have an IoU threshold ranges from 0 to 0.1. We use stochastic gradient descent with momentum of 0.9 and weight decay of 0.0005 to train our detection network. Each SGD mini-batch is constructed from 2 images which are randomly chosen from the whole training set. The foreground-to-background in a mini-batch is set to be 1:3, thus ensuring that 25% of training examples is foreground (*fg*) RoIs.

For training detection network, we use the same way as [16]. In stage 1, we update all the parameters except the first four convolutional layers. We fine-tune the network for about 4 epochs with initial learning rate of 0.001 which is decayed by 0.1 after 2 epochs. In stage 2, we only update the parameters of deconvolution and *fc* layers. We fine-tune the network for about 10 epoches with a fixed learning of 0.0001. The whole network is trained on a single NVIDIA GeForce GTX TITAN X GPU with 12 GB memory.

Analysis on SADR. We first evaluate the effectiveness of the proposed SADR architecture. Table 1 shows that the Faster R-CNN baseline [16] which just uses the feature map of conv5-3 to do classification and regression achieves 10.59% miss rate, which outperforms most of state-of-the-art detectors showed in Fig. 3. We observe that the main factor affecting performance is small objects, since the large objects have already achieved 1.10% miss rate. Therefore we upsample the last convolutional feature map by inserting a deconvolutional layer between conv5-3 and fc6 in VGG16. The features pooled from the 1st deconvolutional layer are fed into the classifier and regresser. As we predicted, the overall performance improves a lot. Especially for the small objects, the miss rate improves from 11.65% to 10.29%, which verifies that a bigger feature map is better for

Table 1. Miss rate(MR) of baseline and our scale-adaptive deconvolution regression (SADR) based model. S, L and A denote the height group of $[50, 112)$, $[112, \infty)$ and $[50, \infty)$. Baseline: Faster R-CNN [16]; d_1: the 1st deconvolutional layer.

Methods	SADR	S	L	A
Baseline [16]	No	11.65%	1.10%	10.59%
Baseline [16] + d_1	No	10.29%	1.55%	9.41%
Baseline [16] + d_1	Yes	**9.04%**	**0.0%**	**8.27%**

locating small objects. We also observe that the performance of large objects degrades slightly. It maybe lost some spatial information when pooling from the deconvolutional layer, since the last convolutional layer can normally pool object with minimum height of 112 pixels. With a flexible scale-adaptive strategy, the best performance is achieved by the proposed SADR method both in small and large objects. On the one hand, compared with the baseline Faster R-CNN, the performance of large objects is further improved and achieves zero miss rate, which implies that the features pooled from the deconvolutional layer are more discriminative for classification than the last convolutional layer. It indirectly indicates that features extracted from deeper structures are more powerful to differentiate the object category. On the other hand, in contrast to the second model in Table 1, the performance of small objects is further improved from 10.29% to 9.04%, which confirms that the SADR architecture can more easily learn the inter-scale variances as the small-scale and large-scale branches just focus on their own specificity respectively.

Analysis on Multi-layer Feature Fusion. We also compare different feature fusion strategies for the classification performance. In this section, we use the same SADR architecture, just to verify the effectiveness of multi-layer feature fusion for classification. Based on the analysis of Table 1, we use features pooled from conv5-3 to do bounding box regression for candidates with height greater than 112 pixels and use the first deconvolutional layer for height ranging from 50 to 112 pixels.

As shown in Table 2, the first line is the baseline Faster R-CNN, the rest of models all use the proposed SADR architecture. As the analysis previously, the model in line 2 outperforms baseline for two reasons. First, features extracted from the finer feature map are more discriminative for classification. Second, the SADR architecture is superior in learning the inter-scale variances. We observe that the performance almost remains unchanged by blending Deconv1 with Conv5, while the fusion of Deconv1 and Conv4 benefits a little more. This result is caused by two factors, the layer depth and down-sampling factor. In general, a deeper layer with lower down-sampling factor can better represent the object, there is a tradeoff between depth and down-sampling factor. This observation also stands in CCF [37]. The fusion of these three layers achieves the best performance, which confirms that the more rich features, the better

Table 2. Comparison performance for different feature fusion strategies. We only use two scale-adaptive regression branches. Line 1 is the baseline Faster R-CNN; line 2 to line 5, using Conv5 to do regression for height between $[112, \infty)$ and Deconv1 for height less than 112. Deconv1: features pooled from the first deconvolution; Conv5: features pooled from conv5-3 in VGG16; Conv4: features pooled from conv4-3 in VGG16; MR: log-averaged miss rate over the FPPI range of $[10^{-2}, 10^{0}]$; BR: box refinement, first introduced in [36] as iterative localization.

Deconv1	Conv5	Conv4	MR	+BR	Δ
	✓		10.59%	8.96%	+1.63%
✓			8.27%	8.48%	−0.21%
✓	✓		8.25%	8.48%	−0.23%
✓		✓	7.96%	7.90%	+0.06%
✓	✓	✓	**7.40%**	**6.94%**	+0.46%

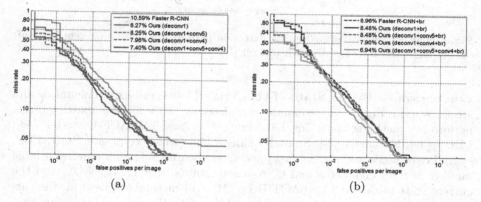

Fig. 2. Average miss rate on Caltech test set. (a) The comparison of different feature fusion strategies; (b) Adding box refinement.

classification. In fact, different layers in convolutional neural networks contain different levels of structural information, while integrating them can capture fine-grained details.

Box Refinement. As shown in Table 2, we observe that MR improves a lot and reaches 8.96% after adding box refinement for baseline. We all know that box refinement just simply revises the position of final detection boxes, it has no effects on the score of bounding box. In addition, combined with Table 1, we can conclude that the main factor influencing the performance of a detector is the regression part of the network, *i.e.*, the ability of localization for small objects. However, for model in line 2, the performance degrades slightly after adding box refinement, which means the accuracy of localization is good enough. Figure 2 shows the average miss rate for different feature fusion strategies.

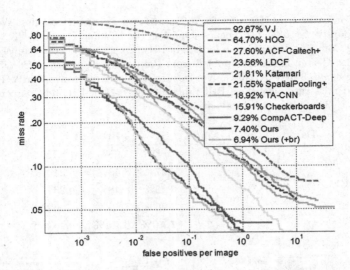

Fig. 3. The comparison of pedestrian detection performance with all recent state-of-the-art methods on revised Caltech test set [1].

Comparison with the State-of-the-Arts. The overall experimental results are shown in Fig. 3. We compare our detector with all the existing best-performing methods, including hand-crafted models, like ACF-Caltech+ [38], LDCF [38], Katamari [39], SpatialPooling+ [40] (which combines HOG, LBP, spatial covariance and optical flow) and Checkboards [6](which requires a large number of filter channels), and CNN-based models, like TA-CNN [29] and the current state-of-the-art CompACT-Deep [24]. Our method outperforms the current best detector CompACT-Deep by 1.89%, with the help of box refinement strategy, we further lower the MR to 6.94%. Since we use the new revised Caltech10x to train our model, we just evaluate on the new testing set.

4.3 Evaluation on KITTI Dataset

We also evaluate our method on the more challenging KITTI dataset [31]. Since KITTI contains more small objects with minimum height 25 pixels, we insert two deconvolutional layers in our network to ensure the features pooled from corresponding layers are more powerful. Therefore, we group the object proposals into 3 levels based on their height, $i.e.$, $[25, 56)$ for the 2nd deconvolutional layer, $[56, 112)$ for the 1st deconvolutional layer and $[112, \infty)$ for conv5-3. The scale of input image is set to be 450 pixels on the shortest side, other parameter configurations are the same as Caltech.

Table 3 and Fig. 4 show the performance on KITTI. It can be observed that our method achieves a competitive result, which outperforms the SDP + RPN [33] on Easy and Moderate subsets. In contrast with the CompACT-Deep [24], which is the state-of-the-art on Caltech, our methods improves 12.94%, 11.96%

Table 3. Comparison to state-of-the-art on KITTI Pedestrian. The evaluation metric is average precision (AP).

Methods	Easy (%)	Moderate (%)	Hard (%)
Ours	83.63	70.70	64.67
SDP + RPN [33]	80.09	70.16	64.82
3DOP [41]	81.78	67.47	64.70
Mono3D [42]	80.35	66.68	63.44
Faster R-CNN* [16]	78.86	65.90	61.18
SDP + CRC(ft) [33]	77.74	64.19	59.27
CompACT-Deep [24]	70.69	58.74	52.71
FilteredICF [6]	67.65	56.75	51.12

Note: * indicates anonymous submission and this paper is submitted to *ECCV16*.

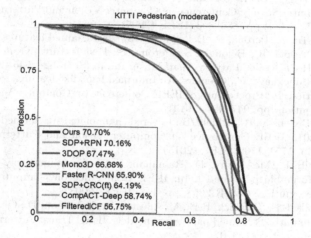

Fig. 4. Comparison to state-of-the-art on KITTI pedestrian (moderate).

and 11.96% on Easy, Moderate and Hard subsets respectively. Our detector on KITTI consists of 3 scale-specific regression branches, which means more parameters need to be learn than Caltech. However, KIITI has less training data which only contains 7481 training images covering 4487 pedestrians, our network can achieve a better result with more training images.

5 Conclusion

In this paper, we propose a novel scale-adaptive deconvolutional regression (SADR) network for pedestrian detection, which could flexibly detect pedestrians with different size. Since each regression branch learns from scale-specific examples, the proposed network has the ability to capture inter-scale differences, resulting in a more sophisticated detection model. By computing the classification and regression loss respectively, we integrate features pooled from multi-layers to further boost the detection performance. Extensive experiments on the public pedestrian dataset clearly demonstrate the superiority of the proposed method, we achieve a state-of-the-art result with MR 6.94% on Caltech dataset and a promising result on KITTI.

Acknowledgment. This work was supported by 863 Program 2014AA015104, and National Science Foundation of China 61273034, and 61332016.

References

1. Zhang, S., Benenson, R., Omran, M., Hosang, J., Schiele, B.: How far are we from solving pedestrian detection? In: CVPR (2016)
2. Dalal, N., Triggs, B.: Histograms of oriented gradients for human detection. In: IEEE Computer Society Conference on Computer Vision and Pattern Recognition, CVPR 2005, vol. 1, pp. 886–893. IEEE (2005)
3. Dollár, P., Tu, Z., Perona, P., Belongie, S.: Integral channel features (2009)
4. Dollár, P., Appel, R., Belongie, S., Perona, P.: Fast feature pyramids for object detection. IEEE Trans. Pattern Anal. Mach. Intell. **36**, 1532–1545 (2014)
5. Zhang, S., Bauckhage, C., Cremers, A.: Informed haar-like features improve pedestrian detection. In: Proceedings of IEEE Conference on Computer Vision and Pattern Recognition, pp. 947–954 (2014)
6. Zhang, S., Benenson, R., Schiele, B.: Filtered channel features for pedestrian detection. In: 2015 IEEE Conference on Computer Vision and Pattern Recognition (CVPR), pp. 1751–1760. IEEE (2015)
7. Felzenszwalb, P., McAllester, D., Ramanan, D.: A discriminatively trained, multiscale, deformable part model. In: IEEE Conference on Computer Vision and Pattern Recognition, CVPR 2008, pp. 1–8. IEEE (2008)
8. Felzenszwalb, P.F., Girshick, R.B., McAllester, D., Ramanan, D.: Object detection with discriminatively trained part-based models. IEEE Trans. Pattern Anal. Mach. Intell. **32**, 1627–1645 (2010)
9. Felzenszwalb, P.F., Girshick, R.B., McAllester, D.: Cascade object detection with deformable part models. In: 2010 IEEE Conference on Computer Vision and Pattern Recognition (CVPR), pp. 2241–2248. IEEE (2010)
10. Sermanet, P., Kavukcuoglu, K., Chintala, S., LeCun, Y.: Pedestrian detection with unsupervised multi-stage feature learning. In: Proceedings of IEEE Conference on Computer Vision and Pattern Recognition, pp. 3626–3633 (2013)
11. Ouyang, W., Wang, X.: A discriminative deep model for pedestrian detection with occlusion handling. In: 2012 IEEE Conference on Computer Vision and Pattern Recognition (CVPR), pp. 3258–3265. IEEE (2012)
12. Ouyang, W., Wang, X.: Joint deep learning for pedestrian detection. In: Proceedings of IEEE International Conference on Computer Vision, pp. 2056–2063 (2013)
13. Girshick, R., Donahue, J., Darrell, T., Malik, J.: Rich feature hierarchies for accurate object detection and semantic segmentation. In: Proceedings of IEEE Conference on Computer Vision and Pattern Recognition, pp. 580–587 (2014)
14. He, K., Zhang, X., Ren, S., Sun, J.: Spatial pyramid pooling in deep convolutional networks for visual recognition. IEEE Trans. Pattern Anal. Mach. Intell. **37**, 1904–1916 (2015)

15. Girshick, R.: Fast R-CNN. In: Proceedings of IEEE International Conference on Computer Vision, pp. 1440–1448 (2015)
16. Ren, S., He, K., Girshick, R., Sun, J.: Faster R-CNN: towards real-time object detection with region proposal networks. In: Advances in Neural Information Processing Systems, pp. 91–99 (2015)
17. Long, J., Shelhamer, E., Darrell, T.: Fully convolutional networks for semantic segmentation. In: Proceedings of IEEE Conference on Computer Vision and Pattern Recognition, pp. 3431–3440 (2015)
18. Noh, H., Hong, S., Han, B.: Learning deconvolution network for semantic segmentation. In: 2015 IEEE International Conference on Computer Vision (ICCV) (2015)
19. Badrinarayanan, V., Handa, A., Cipolla, R.: Segnet: A deep convolutional encoder-decoder architecture for robust semantic pixel-wise labelling (2015). arXiv preprint arXiv:1505.07293
20. Dong, C., Loy, C.C., He, K., Tang, X.: Image super-resolution using deep convolutional networks (2015)
21. Bell, S., Zitnick, C.L., Bala, K., Girshick, R.: Inside-outside net: detecting objects in context with skip pooling and recurrent neural networks (2015). arXiv preprint arXiv:1512.04143
22. Hariharan, B., Arbeláez, P., Girshick, R., Malik, J.: Hypercolumns for object segmentation and fine-grained localization. In: Proceedings of IEEE Conference on Computer Vision and Pattern Recognition, pp. 447–456 (2015)
23. Zagoruyko, S., Lerer, A., Lin, T.Y., Pinheiro, P.O., Gross, S., Chintala, S., Dollár, P.: A multipath network for object detection (2016). arXiv preprint arXiv:1604.02135
24. Cai, Z., Saberian, M., Vasconcelos, N.: Learning complexity-aware cascades for deep pedestrian detection. In: Proceedings of IEEE International Conference on Computer Vision, pp. 3361–3369 (2015)
25. Uijlings, J.R., van de Sande, K.E., Gevers, T., Smeulders, A.W.: Selective search for object recognition. Int. J. Comput. Vis. **104**, 154–171 (2013)
26. Zitnick, C.L., Dollár, P.: Edge boxes: locating object proposals from edges. In: Fleet, D., Pajdla, T., Schiele, B., Tuytelaars, T. (eds.) ECCV 2014. LNCS, vol. 8693, pp. 391–405. Springer, Heidelberg (2014). doi:10.1007/978-3-319-10602-1_26
27. Arbeláez, P., Pont-Tuset, J., Barron, J., Marques, F., Malik, J.: Multiscale combinatorial grouping. In: Proceedings of IEEE Conference on Computer Vision and Pattern Recognition, pp. 328–335 (2014)
28. Hosang, J., Benenson, R., Omran, M., Schiele, B.: Taking a deeper look at pedestrians. In: CVPR (2015)
29. Tian, Y., Luo, P., Wang, X., Tang, X.: Pedestrian detection aided by deep learning semantic tasks. In: Proceedings of IEEE Conference on Computer Vision and Pattern Recognition, pp. 5079–5087 (2015)
30. Dollar, P., Wojek, C., Schiele, B., Perona, P.: Pedestrian detection: an evaluation of the state of the art. IEEE Trans. Pattern Anal. Mach. Intell. **34**, 743–761 (2012)
31. Geiger, A., Lenz, P., Urtasun, R.: Are we ready for autonomous driving? The KITTI vision benchmark suite. In: 2012 IEEE Conference on Computer Vision and Pattern Recognition (CVPR), pp. 3354–3361. IEEE (2012)
32. Li, J., Liang, X., Shen, S., Xu, T., Yan, S.: Scale-aware fast R-CNN for pedestrian detection (2015). arXiv preprint arXiv:1510.08160

33. Yang, F., Choi, W., Lin, Y.: Exploit all the layers: fast and accurate CNN object detector with scale dependent pooling and cascaded rejection classifiers. In: Proceedings of IEEE International Conference on Computer Vision and Pattern Recognition (2016)
34. Simonyan, K., Zisserman, A.: Very deep convolutional networks for large-scale image recognition (2014). arXiv preprint arXiv:1409.1556
35. He, K., Zhang, X., Ren, S., Sun, J.: Deep residual learning for image recognition (2015). arXiv preprint arXiv:1512.03385
36. Gidaris, S., Komodakis, N.: Object detection via a multi-region and semantic segmentation-aware CNN model. In: Proceedings of IEEE International Conference on Computer Vision, pp. 1134–1142 (2015)
37. Yang, B., Yan, J., Lei, Z., Li, S.Z.: Convolutional channel features. In: Proceedings of IEEE International Conference on Computer Vision, pp. 82–90 (2015)
38. Nam, W., Dollár, P., Han, J.H.: Local decorrelation for improved pedestrian detection. In: Advances in Neural Information Processing Systems, pp. 424–432 (2014)
39. Benenson, R., Omran, M., Hosang, J., Schiele, B.: Ten years of pedestrian detection, what have we learned? In: Agapito, L., Bronstein, M.M., Rother, C. (eds.) ECCV 2014. LNCS, vol. 8926, pp. 613–627. Springer, Heidelberg (2015). doi:10.1007/978-3-319-16181-5_47
40. Paisitkriangkrai, S., Shen, C., Hengel, A.: Strengthening the effectiveness of pedestrian detection with spatially pooled features. In: Fleet, D., Pajdla, T., Schiele, B., Tuytelaars, T. (eds.) ECCV 2014. LNCS, vol. 8692, pp. 546–561. Springer, Heidelberg (2014). doi:10.1007/978-3-319-10593-2_36
41. Chen, X., Kundu, K., Zhu, Y., Berneshawi, A., Ma, H., Fidler, S., Urtasun, R.: 3D object proposals for accurate object class detection. In: NIPS (2015)
42. Chen, X., Kundu, K., Zhang, Z., Ma, H., Fidler, S., Urtasun, R.: Monocular 3D object detection for autonomous driving. In: CVPR (2016)

Author Index

Printed in the United States
By Bookmasters